Foundations of the Pricing of Financial Derivatives

The Frank J. Fabozzi Series

Foundations of the Pricing of Financial Derivatives

Theory and Analysis

ROBERT E. BROOKS
DON M. CHANCE

WILEY

Published by John Wiley & Sons, Inc., Hoboken, New Jersey.
Published simultaneously in Canada.

For general information on our other products and services or for technical support, please contact our Customer Care Department within the United States at (800) 762-2974, outside the United States at (317) 572-3993 or fax (317) 572-4002.

Wiley also publishes its books in a variety of electronic formats. Some content that appears in print may not be available in electronic formats. For more information about Wiley products, visit our web site at www.wiley.com.

Library of Congress Cataloging-in-Publication Data:

Names: Brooks, Robert E., 1960- author. | Chance, Don M., author.
Title: Foundations of the pricing of financial derivatives : theory and analysis / Robert E. Brooks & Don M. Chance.
Description: Hoboken, New Jersey : Wiley, [2024] | Series: Frank J. Fabozzi series | Includes bibliographical references and index.
Identifiers: LCCN 2023044922 (print) | LCCN 2023044923 (ebook) | ISBN 9781394179657 (cloth) | ISBN 9781394179671 (adobe pdf) | ISBN 9781394179664 (epub)
Subjects: LCSH: Derivative securities—Prices.
Classification: LCC HG6024.A3 B765 2024 (print) | LCC HG6024.A3 (ebook) | DDC 332.64/57—dc23/eng/20231027
LC record available at https://lccn.loc.gov/2023044922
LC ebook record available at https://lccn.loc.gov/2023044923

Cover Design: Wiley
Cover Image: © Jeremy Horner/Getty Images

SKY10062106_120823

Contents

Preface

Teaching graduate students in finance must be the second best job in the world, next to parenting. Finance is such a rich and exciting subject, and it is uncommon to teach a student who is not interested. Now, we did not say they all put in the maximum effort, but they seldom act as if they are bored. After all, everyone wants to know more about money. Finance professors are the envy of their relatives, and they typically command a great deal of attention at family reunions and receptions with everyone hoping to learn just one iota of information that might make them financially better off.

The authors of this book have taught graduate students in finance for many decades (we stopped counting while we were still young!), and we can attest that our students have taught us a great deal, too. But as longtime teachers of advanced masters and doctoral students, we have been concerned that it is possible to get one of these impressive advanced degrees and circumvent, or just never have, the opportunity to get familiar with the principles of financial derivative pricing. Corporate finance and asset pricing are the core of what we do in the academic finance field. But if a corporate finance or asset pricing specialist is not exposed to derivative pricing, they are missing out on a rich body of knowledge that can help them do their jobs. Or at least half-way understand when a derivatives person presents a paper.

This book had its origin in a set of teaching notes that one of the authors (Chance) began writing in 1996. These notes were designed to fill what he saw as a void in instructional material at the advanced level. Well, it was either that or he just wanted it done his way. These notes were posted on the Web and eventually grew in number to almost 60. The notes were short, often less than 10 pages, tightly contained treatments of various topics. He received a great deal of recognition from complete strangers around the world, and in time, the notes were morphed into this book. He thought it would be easy just to turn the notes into a book. But that led to two problems.

The first was that we had to remove the notes from the Web, leading to some disappointment from fans. Although we had given away much of this material, publishers expect that you will protect their investment by not giving it away any longer. The second problem was that a multiplicity of notation was used in the notes. They were never written as a unified whole. As such, we had a great deal of cleaning up to do. And, we suppose, a third problem was that the notes did not cover the entire field, so yes, we had a lot more writing to do.

This book in manuscript form has been class-tested three times. This course was a doctoral seminar that was open to finance masters and doctoral students and also STEM students across the university. All of these students contributed a great deal to catching errors, forcing us to rethink how we said something, and in some cases contributing end-of-chapter problems. This book would be nowhere near ready for prime time were it not for them. Because these students endured rough drafts of this book, I would like to thank them by name: Brecklyn Groce, Dennel McKenzie, Jeremy Vasseur, Paul Mahoney, Tengfei Zhang, Nha Tran, Nur Faisal, Jason Priddle, Mehdi Khorram, Mengmeng Liu,

Phuc An Vinh Nguyen, Santoshi Rimal, Cameron Roman, Gillian Sims, Pujan Shrestha, Yuanyi Zhang, Aihuan Zhang, Fouad Hasan, Junior Betanco, Ravi Joshi, and Yingying Guo. We also thank Chance's research assistant, Stephanie Hoskins, for additional comments.

A special note of thanks is extended to Chance's PhD student, Amber Schreve, who carefully read and edited the entire book, catching a number of errors and typos, questioning sentences that might not have made complete sense to the reader, and offering many suggestions for improvements. Amber's background in teaching math to undergraduates challenged us to strive for the highest level of clarity that we could.

We never set out to write this book. As noted, it sprung from Chance's teaching notes. And we also agreed not to attempt to compete with highly technical books on quantitative finance and financial engineering written by STEM scholars. What we wanted to do was create a book that was within reach of PhD and advanced masters' students in finance, many of whom do not have the technical backgrounds of STEM students. So, do not expect to find everything on the subject here. But you will find a broad, relatively technical overview of the most important knowledge you need to know to build a solid foundation for understanding the pricing of financial derivatives.

Like all authors, we think we accomplished that. We know, in all honesty, that there are inevitably failures that even another 10 years of editing and class-testing would not completely eliminate. We accept full responsibility for any such deficiencies and promise to consider them if the book goes into a future edition. If you want to communicate with us on any such matters, please send an email to dchance@lsu.edu and rbrooks@frmhelp.com.

The least likely people in the world to email us with a problem in the book are our families. And they deserve credit for just being there and not complaining that we were writing another book when it seemed like we just finished the last one. Don would like to thank his wife, Jan, and their adult and married daughters, Kim and Ashley, and their families. Robert would like to thank his wife, Ann, and their adult and married six children and similarly their constantly increasing number of grandchildren.

Robert E. Brooks
Tuscaloosa, AL

Don M. Chance
Baton Rouge, LA

CHAPTER 1

Introduction and Overview

Finance is the study of money, something commonly used as a medium of exchange. It involves the measurement and management of money: how much we had, how much we have, and how much we expect to have in the future. Much of what we study and do in finance, however, is about making money, so its focus tends to be on the future. Yet, the future is unknown, and the unknown is about risk. People take risks in order to earn money. Finance is essentially the study of the risk and return of money.

In order to do what we do in finance, we must measure money. In fact, measurement in terms of monetary units, such as US dollars, is a core activity of finance, and there is considerably more to measuring money than just counting it. The challenge in finance is in measuring the value of assets that are not cash but can be expressed in terms of cash. An *instrument* is a generic term that refers to a tool that measures something. A *financial instrument* is a type of instrument defined as a tool that measures something in dollars or other currency units. In our context, positive-valued financial instruments are called *assets* and negative-valued financial instruments are called *liabilities*. In general, assets and liabilities are both categorized as instruments or financial instruments because our focus is on finance. Thus, an instrument can reference either assets or liabilities.

Some instruments trade in a market where we can observe their prices, but does that mean that we do not question whether these prices are good prices, in the sense of fairly and accurately reflecting what something should be worth? If you need to buy a used car and you find a 10-year-old car with 150,000 miles on it selling for $50,000, does that mean you would pay that price? No, it is likely you would believe that price to be too high. Perhaps $5,000 is a better price. What we have just done is a valuation of the car. We may have gotten it from some service, observed the prices of similar cars, or simply said that $5,000 is the amount we would pay, meaning that we would willingly part with the consumption of $5,000 of other goods and services to obtain the car.

Likewise, securities that trade in markets have prices that are observable, but that does not mean that we accept those prices as fair. These securities need to be valued, meaning to assign a number to them that represents what one thinks is a fair price. If the value assigned by the investor exceeds the price at which the security is trading in the market, the security is attractively priced and would suggest that the investor should buy the security. If the value assigned by the investor is below the price at which the security is trading in the market, the security is unattractively priced and would suggest that the investor should sell the security if they own it, short sell it if they do not own it, or simply not trade it at all.

We now explain our motivation for writing *Foundations of the Pricing of Financial Derivatives*.

1.1   MOTIVATION FOR THIS BOOK

Many finance courses focus on valuing stocks and bonds. Yet, there is also another family of financial instruments known as *derivatives*, and valuing derivatives is one of the most technical subjects in finance. It requires not only setting up a model of the prices of assets that trade in the market but also establishing a means by which one can connect the derivative to the asset on which the derivative is based. There is a great deal of technical knowledge that must transfer from instructor to student. Much of that knowledge can seem cryptic and inaccessible, though that could be a bit of an overreaction from the fear of learning something new. However, those who know this subject reasonably well can easily fall into the trap of assuming that those who do not know this subject well should find the subject easy. That is the pitfall of being a scholar. A scholar thinks that material in which they have expertise in is not that difficult, when in fact, it really is quite challenging. What a scholar should do in conveying knowledge, however, is to recall how it was when they were learning it. In other words, putting oneself in the student's shoes and empathizing with the student will result in the most successful learning environment. That is indeed one of the overriding objectives of this book: to teach some seemingly complex material in a very user-friendly way.

For without a doubt, teaching and learning advanced material in finance is challenging to the instructor and to the student. Indeed, one of the greatest challenges for instructors in advanced graduate courses in finance is to cover a large body of highly technical information in a relatively short course of study. Well-prepared students make it a lot easier, but student preparation is often not at the desired level, and classes are frequently filled with students with varying degrees of preparation. In an ideal world, such students would have previously had courses in probability, calculus, linear algebra, coding, stochastic processes, econometrics, numerical analysis, non-parametric statistics, differential equations, microeconomics and macroeconomics, and last but certainly not least, finance. It is common for faculty members and students alike to complain that students are inadequately prepared for the technical rigors of advanced graduate study in finance. This book is an effort to address this problem by leveling the base of preparation.

The degree of preparation of finance students is typically a function of the program in which the student is enrolled. Graduate study in finance can generally be done in one of three types of programs. One is an MBA, which usually comprises two years of study, the first consisting of a core set of business courses, of which a broad survey of finance is typically one component. The second year of an MBA is composed of a few required courses in general business but largely permits students to tailor their programs toward their specialized interests. Many students choose to take second-year courses in finance. The MBA is usually the marquee program at top-tier universities, a large money-maker, and is designed to draw students with degrees from all undergraduate disciplines. Hence, the first-year finance course starts at the very foundations of the subject with such topics as time value of money and discounted cash flow valuation, ultimately moving on to understanding financial markets, the relationship between risk and return, market efficiency, and corporate capital structure and dividend policy. Though some MBA students have technical backgrounds, most do not. Hence, MBA students will often struggle with advanced finance courses that are particularly quantitative.

A second form of masters' level study in finance is the specialized master's degree in finance, often called an MS or master of science, and sometimes MSF for master of science in finance. Such a program provides concentrated graduate study in finance, typically over a period of one to two academic years. There may be certain core or required courses, and students are usually allowed to take electives in their preferred areas of finance. Students in this type of program will almost always have previously studied finance and will tend to have more technical backgrounds than the average MBA student.

The third type of program in finance is the doctor of philosophy or PhD. This degree, requiring a minimum of four years, is an intensive research-oriented program that requires all students to achieve a high level of understanding of theoretical models and empirical research methods.[1] It is here that the greatest problem lies in giving students a sufficient level of technical knowledge without sacrificing the time they need to devote to seminars in the various areas of finance. When students are accepted into PhD programs in finance, they are typically required to have a solid foundation in math. But, the definition of a "solid foundation in math" can vary, and merely taking some math courses and making good grades is not necessarily enough. Most finance PhD students, even those with strong math backgrounds, learn something new about math while in their PhD programs. Students who have been accepted into PhD programs are often advised to take more math courses before starting the program. They usually do, but it does not often help nearly as much as one might think.

Let us be clear. Finance is not mathematics. Mathematics is a set of tools used in finance. But just as one cannot build furniture efficiently without knowledge of how to use tools, one cannot understand finance without having the necessary tools of mathematics.

And, as noted, one of the most technical subjects in finance is derivative pricing theory. Sometimes it can be even difficult to keep terms straight because terms can mean different things in different settings. Throughout this book, we will use price and value interchangeably as is financial industry custom. Technically, the concept of price refers to the monetary amount that is exchanged when something is traded, irrespective of what one thinks the item is worth. Value refers to an instrument's non-observed monetary amount as assigned by a market participant. The individual may or may not be using a formal mathematical model. Thus, technically, formal models in finance, such as the capital asset pricing model (CAPM) and the Black-Scholes-Merton option pricing model (BSMOPM), should have been termed the capital asset valuation model (CAVM) and the Black-Scholes-Merton option valuation model (BSMOVM). Theoretical models are just a means of expressing one's view on value. As we will see later, arbitrage activity typically moves observed market prices to the arbitrageur's value. Hence, we will stick with financial industry custom even though value will always have higher levels of epistemic uncertainty when compared to the market price.

Although the majority of finance faculty and PhD students will not specialize in derivatives, there is no doubt that a solid understanding of derivative pricing theory is an important element of doctoral-level education in finance. Derivative pricing theory, in particular the Black-Scholes-Merton model, has had a tremendous impact on finance. It has provided a framework for understanding not only standard derivatives, such as options, but also it has shown us that derivatives can explain many other topics and relationships in finance, such as callable bonds, convertible bonds, credit risk, and corporate capital structure. The impact of derivative pricing theory has been so great that Nobel Prizes were awarded in 1995 to Myron Scholes and Robert Merton with special recognition to the late Fischer Black for work on this subject. Yet, with the increasing need for students to take so many courses in econometrics and statistics, there is often little room in a doctoral program for such a course.

Our goal is to introduce the vast financial derivatives markets to PhD students and others in hopes that it will stimulate your interest in research related to financial derivatives, as well as aid in your future research agenda, even if your agenda is not explicitly financial derivatives. By way of introduction, we present selected derivatives market prices in Table 1.1.[2] Derivatives market prices are unique and often convey cloaked information. On completion of this book, you will be better able to rightfully interpret what information is and is not conveyed.

Table 1.1 Panel A shows natural gas futures prices. Notice that the key descriptor is the delivery month. We denote the current year as Y1. Thus, the December Y1 last traded futures price is $2.896/MMBtu (million British thermal units). Note that futures prices generally rise for longer maturities with the exception of September.[3] The aggregate open interest is the number of either long or short positions currently outstanding. Each contract is for 10,000 MMBtu. Thus, the natural gas futures market currently represents 12,689,550,000 MMBtus.

Table 1.1 Panel B shows option prices of the SPY, which is the exchange-traded fund of the S&P 500 Index. With options, the key descriptors are more complex requiring both the maturity date expressed in days to maturity here and strike price. For example, the 114-day call price with strike price $280/share is $10.40/share and the corresponding put price is $9.76/share. Notice that with longer maturities, similar strike options have higher prices. Further, call prices decline for higher strike prices, whereas put prices rise for higher strike prices. You will learn why later in this book.

Table 1.1 Panel C shows interest rate swap rates. For example, a five-year swap was quoted at 2.022%. The mid fixed rate is the average of the bid and ask rate for the fixed leg of a fixed-for-floating interest rate swap. Notice that this fixed rate initially declines and then subsequently rises for longer maturities.[4]

Table 1.1 Panel D shows 1-month secured overnight financing rate (SOFR) futures data.[5] The data provided here are rate-based and hence show the implied interest rates rather than quoted prices. The interest rate is simply 100 minus the quoted price. For example, the November Y1 rate of 5.400 implies a quoted price of 94.600. Each SOFR futures contract is for $5,000,000 notional amount on 1-month interest rates. At this point, just notice the aggregate open interest is 704,883. *Open interest* is the number of long contracts outstanding. Thus, the total number of contracts outstanding, representing both long and short positions as this is an exchange-traded contract, is 1,409766 [= 2(704,883)].

TABLE 1.1   Panel A. Selected Derivatives Markets Prices: Selected Natural Gas Futures Prices

Description	Price	Change	Open Interest	Volume
JunY1	2.615	+.033	9,781	2,311
JulY1	2.620	+.036	353,254	22,906
AugY1	2.627	+.034	92,888	4,108
SepY1	2.618	+.031	177,123	2,905
OctY1	2.651	+.030	122,641	2,001
NovY1	2.727	+.027	81,686	1,057
DecY1	2.896	+.026	99,738	878
...	...	...	...	...

Note: Aggregate Open Interest = 1,268,955; Volume = 38,910.

TABLE 1.1 Panel B. Selected Option Prices on S&P 500 Index Exchange-Traded Fund (SPY)

Maturity (Days)	Strike Price	Call Price	Put Price
30	279	5.97	4.94
30	280	5.43	5.39
30	281	4.85	5.82
51	279	7.39	6.22
51	280	6.80	6.63
51	281	6.18	7.06
79	279	9.33	7.59
79	280	8.67	8.02
79	281	7.93	8.39
114	279	11.04	9.30
114	280	10.40	9.76
114	281	9.96	10.04
. . .	. . .	. . .	. . .

Note: Spot SPY = 280.15.

TABLE 1.1 Panel C. Interest Rate Swaps

Tenor	Mid Fixed Rate
1 Year	2.350
2 Year	2.115
3 Year	2.032
5 Year	2.022
10 Year	2.175
20 Year	2.352
30 Year	2.378

TABLE 1.1 Panel D. Secured Overnight Financing Rate Futures Contracts

Description	Implied Rate	Change	Open Interest	Volume
AugY1	5.305	UNCH	275,240	1,362
SepY1	5.325	+0.005	105,264	2,839
OctY1	5.455	+0.005	94,418	22,034
NovY1	5.400	+0.005	106,863	13,293
DecY1	5.395	UNCH	47,427	4,810
JanY2	5.480	−0.005	32,062	7,439
FebY2	5.300	−0.005	19,308	5,216
. . .	. . .	. . .	. . .	. . .

Note: Aggregate Open Interest = 704,883; Volume = 75,845.

As each contract is for $5,000,000 notional, the aggregate notional amount represented by this one market is $7,048,830,000,000 or over $7 trillion. The *notional amount* or simply *notional* is the implied principal on which interest calculations are based. Thus, without explaining all this data in great detail, clearly the derivatives industry is large and involves numerous interesting complexities worthy of investigation.

The purpose of this book is to provide detailed training in the pricing of financial derivatives in a lean and efficient manner. It endeavors to convey the mathematical foundations of derivative pricing theory in a compact way. And although there is a great deal of mathematical formality, it is far less so than there would be in a more formal mathematical course in the subject. There are many great books of that genre, but their audience is quite a bit different. Indeed, no pure mathematician or financial engineer will likely give this book much praise, and that is of no concern. It is not an attempt to turn the student into a quant. What it does attempt to do, however, is to take a finance PhD student who in all likelihood is not going to specialize in derivatives and give that person the foundational layers that will pay off in a better understanding of the role that derivatives play in finance. This book does not incorporate the latest advanced mathematical knowledge. Its goal is more modest: to lay a solid foundation in a lean and efficient manner.

But enough about the book. Let us now begin to lay that foundation. A good place to start is to define a derivative.

1.2   WHAT IS A DERIVATIVE?

This book is about derivatives, so to get started we need to know just exactly what a derivative is. There is a very basic definition of a *derivative* that goes like this:

> *A derivative is a financial instrument in which the payoff is derived from the value of some other asset.*

The payoff is the unknown future cash payment required by the derivative contract. This definition works relatively well, but there are situations in which it breaks down. It does this by encompassing instruments that most people would not consider derivatives. For example, mutual funds and exchange-traded funds derive their values from the securities they hold, and most people would not consider them derivatives.

The Financial Accounting Standards Board, which is the accounting industry association that sets standards for financial accounting, has given us a good definition. In 1998, it created a new standard, which was called *FAS* (Financial Accounting Standard) 133 and is now called *ASC* (Accounting Standards Codification) 815, which laid out a new set of procedures for derivatives accounting. It provides a reasonably strong and more detailed definition, which is as follows:

> *A derivative is a contract with one or more underlyings and one or more notional amounts. Its value changes as the value of the underlying changes. Its initial value is either zero or an amount smaller than that required by other transactions to obtain the same payoff. At expiration it settles either by delivery or an equivalent cash amount.*

As it turns out, this definition works relatively well but does leave out a bit. Note that the definition says that a derivative is a *contract*. Let us not forget that a contract is a legal document and enforceable by law. Though not mentioned in the definition, a contract has two parties, so perhaps we should add that there are two parties, and we should probably identify them. One party is the buyer of the derivative, sometimes called the *long*, whose wealth benefits when the value of the derivative increases. The long or buyer is also sometimes known as the *holder*, in particular when the derivative is an option. The opposite party is called the *short*, whose position benefits when the value of the derivative decreases. The short is the seller of the contract and is sometimes called the *writer*, particularly if the derivative is an option. These parties to whom we have been referring are also sometimes called *counterparties*. When we prepend the word *counter*, we are implying a relationship. Each party is counter to the other. For the most part, however, we can use the terms *party* or *counterparty* interchangeably.

The *ACS* definition says that the derivative has an *underlying*, which can sometimes be an asset, such as a stock, bond, currency, or commodity, or it can be something else such as an interest rate.[6] It can even have more than one underlying. The statement says that the value of the derivative changes with the value of the underlying, but note that it does not specify whether the value changes linearly or not (we shall see that there are both linear and nonlinear cases). The statement also mentions that the initial value of the derivative is either zero or a smaller amount than the value of the underlying, which implies that derivatives have a tremendous amount of leverage. One either invests no money or a smaller amount than would be required to obtain the equivalent exposure in the underlying. And as mentioned, at expiration, the derivative settles up either by delivering the underlying or by having one party, the long or short, exchange an equivalent cash amount with the other party, the corresponding short or long.

The reference to a notional amount captures the requirement that a derivative is based on a certain amount of the underlying. This amount might be shares of stock, face value of bonds, units of currency, or, say, barrels of oil. We should also add that a derivative provides either a right or an obligation to engage in a transaction at a future date. Options provide a right, and forwards, futures, and swaps provide an obligation. We shall see what these points mean in later chapters.

So, let us now try to rewrite the definition of a derivative. We shall also do a little paraphrasing to keep the accountants at bay.

> *A derivative is a legally enforceable contract between two parties, the buyer or long and the seller or short, who has at least one underlying and a notional amount for each underlying. The value of the derivative changes with the value of the underlying(s). The initial amount of money one must put down to engage in the derivative contract is either zero or a smaller amount than required to obtain equivalent exposure in the underlying(s). When it expires, the derivative either settles by delivery of the underlying(s) or the parties exchange the cash equivalent.*

Alas, it is virtually impossible to give a one-sentence definition, but this four-sentence definition works quite well. As we get into the specifics of certain types of derivatives, we will have to add some features, but for now, we are set.

Now that we have defined a derivative let us go back and contrast it with the market for the underlying. Oftentimes, one would just call this type of market the stock market, bond market, currency market, or commodity market. It is common in derivatives lingo to call the market for the underlying the *spot market*. This term refers to transactions that are done "on the spot," meaning that one pays for it and receives it immediately.[7]

This procedure contrasts with derivatives, which always refer to transactions that will be conducted in the future.

So where are these things called *derivatives*? Well, they are created and traded in markets. There are two types of these markets, though the differences are becoming blurred. First, there are *exchange-listed derivatives*, sometimes called *listed derivatives*, which are standardized instruments that trade on exchanges. An exchange is an entity, usually in the form of a corporation or nonprofit, that provides a physical or electronic facility for trading. When we think of an exchange, we tend to think of something like the New York Stock Exchange, but derivatives exchanges also exist, such as the Chicago Mercantile Exchange and the Chicago Board Options Exchange. Also, derivatives can trade on exchanges that are more known as stock exchanges, such as Euronext and the Korea Exchange.

What we mean by the notion of standardized instruments is that the exchanges have decided on most of the terms and conditions, such as which underlyings will have derivatives available for trade, when these derivatives will expire, and how many units of the underlying are covered by a single derivative contract. There are also other terms that are germane to certain types of derivatives that are specified by the exchange. What the exchange does not specify, however, is the price, which is negotiated between the parties on each trade.

An exchange also has rules as to who can trade on the exchanges, and it provides clearing services, which references a system of bookkeeping that matches the counterparties and ensures that the money passes from one party to the other as appropriate. The exchanges also ensure that the parties that make money will always be paid, with the funds coming from the parties that lose money. The exchange guarantees that if the parties losing money cannot pay the parties making money, the exchange will cover through its clearinghouse. In this way, exchanges provide a guaranty that essentially eliminates credit risk.

The other type of derivative is the *over-the-counter*, sometimes called *OTC*, or customized derivative, which trades in an informal market. This type of transaction is essentially one between any two parties that is not conducted on an exchange. As such, the parties can customize the transaction with any terms and conditions they want, as long as they do not break a law. Corporations commonly create derivatives with their banks to manage various risks they face.

Derivatives markets, whether exchanges or over-the-counter, rely on a set of firms or individuals called *dealers*. Dealers stand ready to take either side of a derivative. They do this by quoting a bid price, the price they are willing to pay, and an ask price, the price they are demanding to sell. The ask is higher than the bid, so the dealer has a profit built into the quote. Thus, the investor has a built-in loss as the investor must buy high at the ask price and sell low at the bid. When a dealer takes on a transaction, it has acquired exposure; however, it does not generally carry that exposure, which would be risking its own survival on market direction. Instead, it lays off the risk by finding an offsetting transaction elsewhere in the market. Dealers are, thus, wholesalers of risk. They can easily do virtually any transaction in any market, quickly, and with low cost.

1.3   OPTIONS VERSUS FORWARDS, FUTURES, AND SWAPS

There are two general classes of derivatives that are distinguished by the fact that one class has payoffs that are linear in the underlying, while the other has payoffs that are nonlinear in the underlying. The former are forwards, futures, and swaps. Their payoffs are linearly related to the underlying in that the relationship between what the derivative pays when it expires and the value of the underlying is the same for any value of the underlying. Note that we are not saying that the payoff is the same for any value of the underlying. We are saying that the relationship is the same. This means that we can write the payoff function of the derivative as a linear function of the value of the underlying.

Derivative instruments whose payoffs are linear—forwards, futures, and swaps—are also known as *symmetric instruments*. A symmetric instrument is one in which the payoff for a given change in the underlying is the same in absolute value as the payoff if the underlying changes by the same magnitude but in the opposite direction. For example, if the underlying moves up by 5 and the derivative payoff is 10, then a move of −5 will result in a payoff of −10. The up-front cost to enter into a forward or swap is zero. You essentially pay for the large gains by bearing the risk of large losses.

Options compose the family known as *nonlinear derivatives*. For them, the payoff of the derivative and the value of the underlying are not related in a linear manner, so we cannot write the payoff function as a linear function. As we shall see, however, they are related in what we call a piece-wise linear manner, meaning that the relationship is linear over one range of the underlying but different over another. You can think of that description as a straight line with one slope connected to another straight line with a different slope. We shall get into the details of what these explanations mean in later chapters.

Derivative instruments with payoffs that are nonlinear—options and other option-like instruments—are also known as *asymmetric instruments*, because the payoffs are not symmetric. That is, the payoff for a given move in the underlying is not equivalent in absolute value to the payoff for the same move in the opposite direction of the underlying. So, for example, if the underlying goes up by 5 and the derivative pays 10, a move of −5 will not result in a payoff of −10 on the derivative. The cost to enter an asymmetric instrument is nonzero. You essentially pay for the large gains by paying the price, also called a *premium*, up front or you bear the risk of large losses by receiving this price or premium up front.

This book will start with options. In fact, most of the book is really about options, because as it turns out, options are far more difficult to model than are forwards, futures, and swaps. Everything we know about options will apply to forwards, futures, and swaps, but we shall not need to go to such mathematical lengths to derive our models for those instruments. For the most part, their models are obtained by simple present value calculations.

1.4 SIZE AND SCOPE OF THE FINANCIAL DERIVATIVES MARKETS

The derivatives market is massive in size and global in scope. Measuring the overall size of the derivatives markets, however, is a bit difficult. Generally, derivatives markets are broken down into the over-the-counter (OTC) markets and exchange-traded markets. The OTC markets are typically measured by either notional amount or gross market value. Recall the notional amount, or notional for short, reflects the characteristic that a derivative is based on a certain amount of the underlying. This amount might be shares of stock, face value of bonds, units of currency, barrels of oil, and so forth. Gross market value is the current absolute value of one side of the derivatives position. The overall gross market value of derivatives is zero because for every long position there has to be a short position. The exchange-traded derivatives markets is generally measured with either open interest or trading volume.

Each measure has its advantages and disadvantages. The notional amount is a number that grossly overstates the amount of money that is transacted or at risk. Derivative payments are based on the notional amount, but most derivatives do not involve paying the notional itself. The notional amount, however, is a fairly accurate number, because it is written into the contracts. Market value, however, must be estimated either from market prices, from educated guesses, or from using models. Yet, market value reflects more accurately the amount of money at risk. Most of this book is about estimating market value.

Figure 1.1 Panel A presents the notional amount of OTC derivatives outstanding approximately the decade before and after the global financial crisis of 2008–2009. Clearly, the global financial crisis starting in 2008 significantly affected the growth of the OTC market. The gross market value illustrated in Figure 1.1 Panel B shows a spike in December 2008 that reflects the significant gyrations in all world markets, and the subsequent declines indicate a move out of OTC derivatives. Figure 1.1 Panel C illustrates the exchange-traded open interest. The rise in recent years reflects the push by regulatory bodies for more centralized clearing in response to the global financial crisis. In summary, the size of the derivatives market, measured in hundreds of trillions of dollars, warrants further investigation.

Table 1.2 Panel A presents annual volume by region expressed as a percentage of total volume for a recent year based on information provided by the Futures Industry Association. North America and Asia Pacific have gone back and forth in ranking between number one and number two for a while, but the Asia Pacific region now dominates. Table 1.2 Panel B shows the volume by category for the same year. Equity and equity index derivatives dominate the rankings. Foreign exchange and interest rates are generally next in the rankings. Thus, the derivatives markets span the globe as well as span numerous underlying instruments.

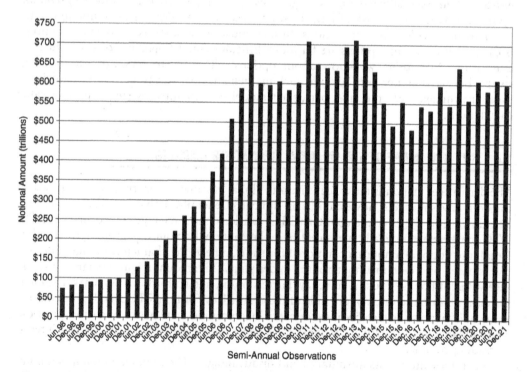

FIGURE 1.1   Panel A. Over-the-Counter Notional Amount Outstanding

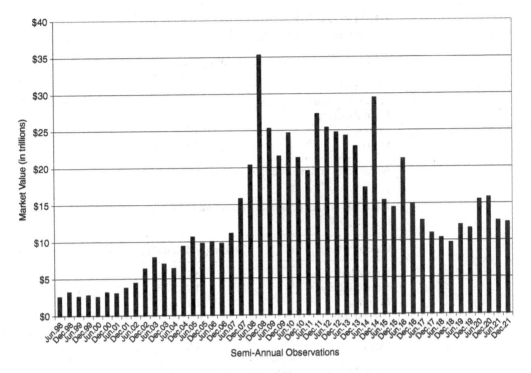

FIGURE 1.1   Panel B. Gross Market Value Outstanding

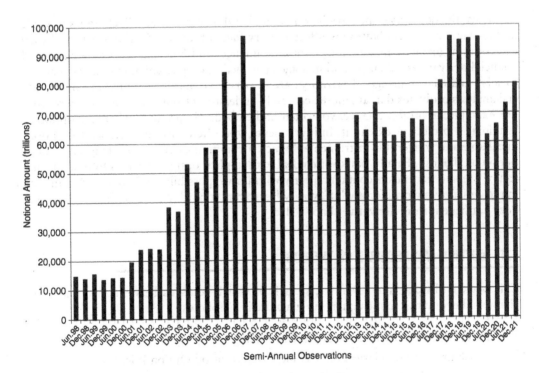

FIGURE 1.1   Panel C. Exchange-Traded Open Interest

TABLE 1.2   Panel A. Global Futures and Options Trading Volume Percentage by Region

Region	Trading Volume (%)
Asia-Pacific	60
North America	20
Latin America	10.2
Europe	5.7
Other	4.1

Note: Total volume = 83,847,697,472.

TABLE 1.2   Panel B. Global Futures and Options by Category

Category	Trading Volume (%)
Equity Index	58
Individual Equity	16
Currency	9.1
Interest Rates	6.1
Energy	2.5
Other	8.3

Note: Total volume = 83,847,697,472.

Thus, the derivatives industry is truly global and massive in size. Techniques to value and manage derivatives have now affected every facet of finance. Within the three broad categories of finance—investments, corporate finance, and financial services—mastery of financial derivatives valuation and management methodologies enhances one's capacities to solve financial challenges. Derivatives are so pervasive that it is common to find financial analysis problems that are more accurately understood from a derivatives perspective. For example, in project finance, you often find embedded options, such as the option to terminate a project or extend it. In investments, you often encounter embedded options in callable and convertible instruments. In fact, a common stock can be analyzed as a call option on the firm. Finally, banks often deal with implicit options, such as from loans that can be prepaid to deposits containing early withdrawal possibilities. Thus, if you hang out in the derivatives world for a while, you will also find it hard to name a finance issue that does not benefit from derivatives knowledge.

1.5   OUTLINE AND FEATURES OF THE BOOK

This book is divided into seven parts, each of which contains between two and seven chapters. Beyond this first chapter, the structure is as follows:

Part I: Basic Foundations for Derivative Pricing
   Chapter 2: Boundaries, Limits, and Conditions on Option Prices
   Chapter 3. Elementary Review of Mathematics for Finance

As previously suggested, the book does contain many equations. Some are numbered, and some are not. The ones that are not numbered are generally intermediate stages of a derivation, but if those equations need to be referenced later, then they are numbered. An effort is made to provide supporting actual examples in practice as well as illustrated calculations. Most of the calculations you would do from these models are fairly simple. A few end-of-chapter problems are provided so you can confirm your understanding of the basic concepts. Equations showing calculations are typically not numbered.

A technical book such as this one employs a large number of symbols. Every effort has been made to use symbols in a unique manner, but some replication is inevitable. For example, p is an excellent symbol for probability but also an excellent symbol for the price of a put. The lower case d is an excellent symbol for the down factor in the binomial model, but it is widely used as a factor in the Black-Scholes-Merton model, in the form d_1 and d_2. The Greek letter rho with the symbol ρ is commonly used for correlation but rho is also a partial derivative of the Black-Scholes-Merton model. The letter X is often a variable in mathematics, but it is commonly used as the exercise price of an option. Tolerance for symbol duplication is a necessary evil in material of this sort. Symbols must, therefore, be context specific. There should never be a situation in which a symbol takes on two meanings in the same context.

1.8   FINAL THOUGHTS AND PREVIEW

Although your relatives might disagree, the study of finance and in particular financial derivatives is fascinating. If you do not believe this, take a look at Bernstein (1992, 1996, 2007), Mehrling (2005), Derman (2004), and Taleb (2001, 2007), all of which are highly readable books directed to the general public that give interesting insights into the body of human knowledge about the subject we call finance. You will find that the study of finance reveals profound insights into the behavior of people as they endeavor to grow their wealth. Studying finance does not provide a road to riches, but it does provide useful guidelines that can tilt the odds in favor of the investor, even the small investor who may feel out of their league.

The study of finance will challenge you as much as just about any subject ever has. Perhaps you should think of it as a rollercoaster ride. There is an uphill part, and there may be several. But there is also a downhill part that will be found when subjects begin to make sense because a strong foundation has been laid. There may even be thrills and chills that will leave you exhilarated when it is all over. If you have come this far in your education, please do not fear the ride that awaits, as this book will hold your hand for the duration.

As noted, most of this book will deal with options. In Chapter 2, we shall more formally introduce the concept of an option and identify certain boundaries, limits, and conditions on option prices. These statements are not pricing models. They are simply general statements to which the pricing models must conform, or investors would exploit them until prices adjust and do conform. They tell us the ground rules within which we must operate and show us why these rules exist.

So, let us get started.

QUESTIONS AND PROBLEMS

1 What is money?

2 What is at the core activity of finance?

3 Provide a working definition of a financial derivative.

4 Contrast a symmetric derivative instrument with an asymmetric derivative instrument.

5 Contrast exchange-listed derivatives with over-the-counter derivatives.

6 "Finance is mathematics." Evaluate and explain whether this statement is true or false.

NOTES

1. There is yet another type of graduate finance program, which is the masters' or doctorate in financial mathematics, quantitative finance, or financial engineering. These programs are usually conducted in engineering or math departments and schools, and they are designed to train students to become specialists in the subject. This book deals with the same topics as do those programs, but its audience is different.
2. The data are based on information provided by Bloomberg® on May 29, 2019.
3. We will explore explanations for this pattern in Chapter 22.
4. See Chapter 26 for more details on interpreting patterns in swap rates.
5. In 2023, the London interbank offered rate (LIBOR) was phased out. In terms of trading volume, the LIBOR-related Eurodollar futures contract was the most successful futures contract ever launched. SOFR replaced LIBOR.
6. Note that the use of the word *underlying* is grammatically incorrect. *Underlying* is a participle, which is a noun or verb that is used as an adjective, and therefore must modify a noun. In the world of derivatives, however, the word *underlying* is used as a noun. Words often enter the English language through the highway of finance, and they do not always conform to the correct rules of the language, but they become good examples of the dynamic nature of the English language. Incidentally, the term *underlier* is also sometimes used.
7. Another term for the spot market is the *cash market*.

Basic Foundations for Derivative Pricing

CHAPTER 2

Boundaries, Limits, and Conditions on Option Prices

Before one can begin to examine models for option pricing, it is necessary to understand certain fundamental principles that govern the prices of options. These option principles do not give specific option prices except in a very limited sense; rather, they define the bounds within which option prices must lie. In addition, they define relationships among different options, such as those differing by exercise price and those differing by time to expiration. Some of these principles also define relationships between the prices of puts and calls that have the same exercise prices and expiration dates. Finally, some option principles define relationships between options that can be exercised before expiration versus those that can be exercised only at expiration.

An option is a derivative that grants the buyer, who is called the *long*, the right to either buy or sell the underlying at a fixed price either at or before a specific date. The buyer is also said to be the owner or holder of the option because they own and hold the right to buy or sell the underlying. This right is granted by the seller, who is called the *short* or the writer. We stated that an option provides either the right to buy the underlying or the right to sell. Be sure you understand that it is one or the other. The right to buy is referred to as a *call*, and the right to sell is referred to as a *put*. The buyer and the seller decide on whether the option will be a call or a put.

The definition of an option specifies that the holder can either buy the underlying if a call, or sell the underlying if a put, at a fixed price. This fixed price is called the *exercise price* and sometimes the *strike price*, *strike*, or *striking price*.

The option has an expiration date, beyond which the right to buy or sell the underlying no longer exists. When the holder of the option decides to pay the exercise price and acquire the underlying if a call, or deliver the underlying and receive the exercise price if a put, the buyer is said to be exercising the option. Note in the definition of an option that we specified that the right to buy or sell exists either at the expiration date or before the expiration. In other words, some options can be exercised only at the expiration, and some can be exercised any time up to the expiration. The former is called a *European* or *European-style* option, and the latter is called an *American* or *American-style* option. These terms reflected the places where these types of options traded in the early 1900s. Today, both types of options trade on both continents and around the world.

So, let us now begin our examination of the basic boundaries, limits, and conditions on option prices. These boundaries are like the ground rules, akin to the edges of a sports playing field and the regulations that govern the game. They also enable us to build toward understanding how options are priced.

2.1   SETUP, DEFINITIONS, AND ARBITRAGE

We position ourselves at time t, and let S_t be the asset price today, T be the expiration of the derivative, S_T be the asset price at expiration, X be the exercise price, r_c be the annualized, continuously compounded risk-free rate, σ be the annualized volatility, and $\tau = T - t$ be the time to expiration.[1] Let c_t be the price of a European call at time t and p_t be the price of a European put at time t. Let C_t be the price of an American call at time t and P_t be the price of an American put at time t. Other notation will be introduced as needed. The asset is assumed to make no payments such as dividends that might be paid if the asset were a stock, but we shall relax that assumption at appropriate points.[2] The results demonstrated herein are largely intuitive, but formal proofs and discussions are covered in Stoll (1969), Merton (1973a, 1973b), Smith (1976), Cox and Rubinstein (1985), and Chance and Brooks (2016).

Before starting, let us introduce the concept of *arbitrage*. Let us start with a definition.

> *Arbitrage is a market condition in which two assets or combinations of assets that produce the same results at a future date sell for different prices prior to or at that future date. When an arbitrage condition exists, it creates an opportunity for a person to buy the lower-priced asset or combination of assets and sell the higher-priced asset or combination of assets, thereby netting a positive cash flow at the initiation of the transaction, with the long and short positions offsetting and eliminating the risk, resulting in no negative potential cash flow at any future date. As a result, an arbitrage generates money at the start and has no obligation to pay any money later. It is, thus, free money.*

Virtually the entire corpus of derivative pricing theory assumes that arbitrage opportunities do not exist. If they did, we assume that such opportunities would be exploited. The buying of the lower-priced asset or combination of assets and the selling of the higher-priced asset or combination of assets will result in an increase in the cost of the former and a decrease in the cost of the latter until the prices converge.

Let us take a look at a simple example of arbitrage. Suppose someone offers you an opportunity to bet a certain amount of money on a game in which the outcome is based on movements in the S&P 500. Specifically, if the S&P 500 goes up the next day, you receive 10. If it goes down, you receive 5. Let us assume it costs 6 to play the game. Now, consider a second game that pays 20 if the S&P 500 goes up and 10 if it goes down, but this game costs 11 to play. It should appear that either the first game is overpriced or the second game is underpriced. If someone else is offering the second game to the public for 11, you too can offer it for 11. But if someone would pay you 11, why not just cut the cost and the payoffs in half, thereby offering the second game for 5.50 with payoffs of 10 for heads and 5 for tails. So, you offer the first game for 6 and buy into the second game for 5.50. If the market goes up, you collect 20 on the second game and pay out 20 on the first. If the market goes down, you collect 10 on the second game and pay out 10 on the first. The end result is completely offset, so you have a hedged position and yet you collect $6.00 - 5.50 = 0.50$ at the start. That is your money to keep.

Of course, your instincts say that this could not happen. Someone would notice that the two games are identical and yet are selling for different prices. But in the financial markets, there are millions of prices quoted at any one time. Almost none of them will present arbitrage opportunities, but on occasion a few will.

When an arbitrage opportunity presents itself, the person taking advantage of it, called an *arbitrageur* or sometimes just an *arb*, will buy the cheaper opportunity and sell the more expensive one, pushing up the price of the former and driving down the price of the latter until the two prices are equal.[3]

In the financial markets, arbitrage opportunities are rare, but they do exist very briefly. By assuming that they are quickly exploited, we invoke a very powerful statement about human behavior that requires few, if any, restrictive assumptions. We simply assume that people will take money that is offered if there is no risk involved. We do not have to know whether they would take the risk to make more money. Here there is no risk, and yet there is money for the taking. We are not required to make any strong assumptions about human behavior to know that people would act on such opportunities.

So, throughout this book, we shall assume that market prices equal the prices that are implied by models that assume that all arbitrage opportunities have been exploited.

Before exploring boundaries, limits, and other conditions, we introduce selected terminology. More detailed explanations will be given later. We now present moneyness terminology that is independent of whether you buy or sell the option. Let S_t be the price of the underlying at time t and X be the exercise price of the call option, then, when $S_t > X$, we say that the call is *in-the-money* and when $S_t < X$, we say that the call is *out-of-the-money*. When $S_t = X$, the call is said to be *at-the-money*. Now suppose X is the exercise price of the put option, then, when $S_t < X$, we say that the put is *in-the-money* and when $S_t > X$, we say that the put is *out-of-the-money*. When $S_t = X$, the put is said to be *at-the-money*.

2.2 ABSOLUTE MINIMUM AND MAXIMUM VALUES

By absolute minimum and maximum values, we wish to define bounds within which the option prices must lie. We do not rule out the possibility that the actual option prices might have a higher minimum or lower maximum value that we can establish later.

The first absolute minimum value that we can easily specify is that an option cannot have a negative value. One can never be forced to exercise an option. Hence, it can never be worth less than zero. This result is true for both calls and puts as well as for American and European options,

$$c_t \geq 0 \text{ and } C_t \geq 0$$
$$p_t \geq 0 \text{ and } P_t \geq 0.$$

(2.1)

Now, let us think about the maximum values. Both the European call and the American call cannot cost more than the value of the underlying asset. Either call option allows the holder the right to buy the asset so the holder of the option would not pay more than the cost of the asset to acquire the right to buy the asset. A weaker statement might place this upper bound at infinity, because that is the upper bound on the asset price, but there is no reason to impose such an extreme upper bound as the current value of the asset is more precise.

A put reaches its maximum value when the asset is worthless. A European put then is worth the present value of the exercise price as its holder has the right to exercise the option at expiration and claim X dollars at that time. Thus, its current worth is the present value of X.[4] For an American put, however, there is no reason to wait as it can be immediately exercised for a cash flow of X.

The maximum values of calls and puts are stated formally as

$$c_t \le S_t \quad \text{and} \quad C_t \le S_t$$
$$p_t \le Xe^{-r_c\tau} \quad \text{and} \quad P_t \le X_t.$$

(2.2)

Suppose you call your broker for a quote on a one-year, at-the-money ($S_t = X = 100$), European call and to your surprise the broker quotes you 101 per share. As this is a clear violation of the upper boundary condition, there is an arbitrage profit available. The quoted call price is too high; hence, you would sell the call and buy the stock for a net cash flow of 1 (= 101 − 100). If the option expires out-of-the-money, sell the stock and receive nonnegative proceeds ($S_t \ge 0$). If the option expires in-the-money, then when the option is exercised by the buyer, you receive the exercise price ($X = 100$) and deliver the stock you previously purchased. Thus, in all future states of the world, there is no possibility of having to pay money. This is clearly an attractive transaction that will rarely, if ever, be available.

Instrument	Now Time (t)	Expiration (time T, out-of-the-money $S_T < X$)	Expiration (time T, in-the-money $X < S_T$)
Short the call	101	0	$-S_T + X$
Long the stock	−100	S_T	S_T
Net proceeds	1	S_T	X

2.3 THE VALUE OF AN AMERICAN OPTION RELATIVE TO THE VALUE OF A EUROPEAN OPTION

Because an American option permits the holder to exercise the option at expiration as well as any time prior to expiration, its value must be at least as great as that of the corresponding European option:

$$C_t \ge c_t \quad \text{and} \quad P_t \ge p_t.$$

(2.3)

2.4 THE VALUE OF AN OPTION AT EXPIRATION

The value of an option when it is exercised is called the *exercise value*. At expiration, both a European and an American call are worth their exercise values, which is the same for both, because the American call has no time remaining and, therefore, no benefit to exercising it early,

$$c_T = C_T = \max(0, S_T - X).$$

(2.4)

To exercise the option, the holder will pay X and will receive an asset worth S_T. Thus, if $S_T > X$ and an instant before expiration the call is selling for less than $S_T - X$, it can be purchased and exercised, resulting in an immediate gain to the holder of $S_T - X$ less the price of the call. The ability to earn this risk-free profit will induce trading of this sort that will result in the call price increasing in value until it is equal to $S_T - X$. If $S_T \leq X$, the option should not be exercised and, hence, it expires with no value.

Similarly, at expiration both a European and an American put are worth the exercise value,

$$p_T = P_T = \max(0, X - S_T). \tag{2.5}$$

To exercise the put, the holder will need to acquire the underlying, which can be done an instant before expiration either by buying it in the open market at S_T or by simply short selling it. If the put holder exercises the put, the holder can purchase shares at S_T and subsequently sell them at X. Thus, if $S_T < X$ and the put is selling for less than $X - S_T$, it can be purchased and exercised resulting in an immediate gain to the holder of $X - S_T$ less the price of the put. The ability to earn this risk-free profit will induce trading of this sort that will result in the put price increasing in value until it is equal to $X - S_T$. If $S_T > X$, the option should not be exercised and, hence, it expires with no value.[5]

Alternatively, suppose the stock is trading at 90 and a 100-strike put is trading for 11 on the expiration date. In this case, the market put price is too high relative to the exercise value (10). The arbitrageur should sell the put receiving 11. The put buyer will exercise this option. As the seller, the arbitrageur will have to buy the stock at the strike price of 100 and subsequently sell it in the market at 90 for a loss of 10. The proceeds from selling the put more than offset this loss, netting 1.

2.5 THE LOWER BOUNDS OF EUROPEAN AND AMERICAN OPTIONS AND THE OPTIMALITY OF EARLY EXERCISE

We previously identified zero as the minimum values of European and American options. It is possible, however, to establish higher minima. By establishing a minimum price that is higher than zero, we are able to place more restrictive limits on the prices of options, which enables us to narrow the range over which the option price can exist.

2.5.1 The Lower Bound of a Call

For an American call, the lower bound can be initially stated as the option's exercise value. At any time $t < T$,

$$C_t \geq \max(0, S_t - X). \tag{2.6}$$

If $S_t > X$ and the call is selling for less than $S_t - X$, it can be purchased and exercised resulting in an immediate gain to the holder of $S_t - X$ less the price of the call. The ability to earn this risk-free profit will induce trading of this sort that will result in the call price increasing in value until it is worth at least $S_t - X$. In the other case, $S_t < X$, the option should not be exercised and, hence, we can say only that its minimum value is zero.

For a European call, such a statement is not possible because it cannot be exercised immediately. It is possible, however, to make a stronger statement, via simple arbitrage arguments. Suppose we construct two portfolios, A and B, with portfolio A consisting of a long position in a European call, and a zero-coupon bond with face value equal to the exercise price, and current value equal to the present value of the exercise price. Portfolio B consists of a unit of the asset. Table 2.1 shows the current values of these portfolios and their values at expiration.

TABLE 2.1 Establishing a Lower Bound for a European Call

Instrument	Current Value	Value at Expiration $S_T \leq X$	$X < S_T$
Portfolio A			
European call	c_t	0	$S_T - X$
Zero-coupon bond	$Xe^{-r_c\tau}$	X	X
Total	$c_t + Xe^{-r_c\tau}$	X	S_T
Portfolio B			
Asset	S_t	S_T	S_T

Note that Portfolio A performs as well as Portfolio B when $S_T > X$ (because the Xs cancel) and performs better when $S_T \leq X$. Thus, portfolio A is said to dominate Portfolio B. Consequently, the current value of A must be at least as great as the current value of B, which can be stated as $c_t + Xe^{-r_c\tau} \geq S_t$. Rearranging, we can write this statement as $c_t \geq S_t - Xe^{-r_c\tau}$. For the case where $c_t < S_t - Xe^{-r_c\tau}$ and $S_t < Xe^{-r_c\tau}$, however, it makes little sense to state that a call price must exceed some negative value for we already know that its absolute minimum is zero. Consequently, we can state formally that

$$c_t \geq \max(0, S_t - Xe^{-r_c\tau}). \tag{2.7}$$

For an American call, we previously noted that $C_t \geq \max(0, S_t - X)$. It is obvious, however, that $S_t - Xe^{-r_c\tau}$ is greater than $S_t - X$ except at the expiration where $\tau = 0$.[6] Combined with the fact that the American call price is at least as great as the European call price, we can then state that the lower bound for the European call must also hold for the American call,

$$C_t \geq \max(0, S_t - Xe^{-r_c\tau}). \tag{2.8}$$

To better understand how arbitrage activity influences option prices, consider the following situation: Suppose the stock price is $S_t = 102$, the strike price is $X = 100$, the continuously compounded risk-free interest rate is $r_c = 5.01\%$, the time to maturity τ is 0.608 based on its expiration in 222 days ($222/365 = 0.608$), and the quoted American call price is 4.95. How would an arbitrageur trade? The first step is to assess whether there is a boundary violation. Note in this case $S_t - Xe^{-r_c\tau} = 102 - 100e^{-0.0501(222/365)} = 5.0$. Thus, the quoted call price is 0.05 below the lower boundary. The arbitrageur seeks to pocket this 0.05 through a series of current trades. Note in this case $S_t - Xe^{-r_c\tau} - C_t = 0.05$ and the trading strategy is embedded in this expression. Notice that you can generate this result by transactions that generate these specific cash flows; namely, short sell the stock at $S_t = 102$, buy a zero-coupon bond at $Xe^{-r_c\tau} = 97$, and buy the call at $C_t = 4.95$. Table 2.2 provides a cash flow table that clearly demonstrates that these transactions result in arbitrage profits today with no subsequent chance for future negative cash flows.

The Cash Flow Today column results in the receipt of 0.05 today per share of stock. The column $S_T \leq X$ is known to be nonnegative because we assumed the terminal stock price is less than or equal to the strike price.

Note that short selling pressure will drive the stock price down, and call buying pressure will drive the call price up. In well-functioning markets, we expect the net cash flow

TABLE 2.2 Numerical Illustration Establishing a Lower Bound for a European Call

		Cash Flow at Expiration	
Instrument	Cash Flow Today	$S_T \leq X$	S_T
Short sell stock	$+S_t = 102$	$-S_T$	$-S_T$
Buy bond	$-Xe^{-r_c\tau} = -97$	$+X$	$+X$
Buy call	$-c_t = -4.95$	0	$S_T - X$
Net	0.05	$100 - S_t$	0

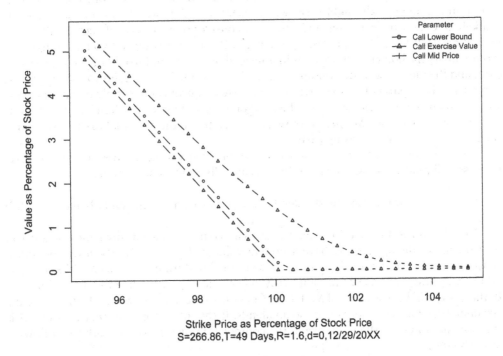

FIGURE 2.1 Illustration of Lower Bound for a Non-Dividend-Paying Call (Expressed as Percentage)

today to quickly move so as to be nonpositive. Also, remember the arbitrageur purchased the American call; hence, they do not have to be concerned with being forced to exercise early. Therefore, the early exercise feature cannot be harmful to the arbitrage trader.

Figure 2.1 illustrates what we have asserted thus far. As is common in industry practice, the strike price, call price, and lower bound are normalized by the underlying price. That is, these values are divided by the underlying price. By normalizing values, we can easily compare option information across different instruments. Thus, the call mid price is the average of the bid and ask call price for an American call option on an index ETF (266.86) on the last trading day of a recent year divided by the index ETF price. This anecdote shows with actual data that the call price remains above the lower bound. Also, note that for calls, the exercise value is below the lower bound. As such, the exercise value is not binding in real markets. Specifically, the higher lower bound is binding and nothing else.

The market risk-free interest rate is 1.6%, the time to maturity is 49 days, and there are no anticipated dividends. Please note that these graphs look the opposite of what you may have seen before: Here we normalize by the underlying price. The opposite image would appear if normalizing by the strike.

2.5.2 Early Exercise of an American Call

From this result, we can discern that the American call will never be exercised early because the minimum value of an in-the-money call is $S_t - Xe^{-r_c\tau}$, which is more than its value if exercised, $S_t - X$. In other words, an American call is worth more by simply selling this call in the market. This point may seem somewhat counterintuitive when one considers that a deep-in-the-money call might seem worth exercising. A holder of such a call might be unlikely to expect further gains, but one must consider that exercise of such a call would simply result in the holder possessing the asset. If the asset is indeed going no higher, it would satisfy the holder no more to hold it instead of the call and would be out the exercise price and the interest it could continue to earn if they waited until expiration to exercise. It should also be apparent that exercise of a call early is equivalent to simply paying someone for an asset before it is necessary and then forgoing the right to change one's mind about its purchase at a later date. We shall soon see, however, that if the asset makes cash payments, then it may be worth exercising early.

Thus, for the case where the asset makes no cash payments, the absence of early exercise will render the American and European calls equivalent in value:

$$c_t = C_t \text{ if there are no dividends or cash payments on the underlying.} \qquad (2.9)$$

Now let us assume that the asset makes cash payments, such as dividends on a stock, over the life of the option that have a present value of D_t. To derive the lower bound, it is necessary that we change the current value of the bond from $Xe^{-r_c\tau}$ to $Xe^{-r_c\tau} + D_t$. The bonds will, thus, have a current value of $X + D_t e^{r_c\tau}$ at expiration in either case. Portfolio B, the asset, will have a value of $S_T + D_t e^{r_c\tau}$ in either case at expiration, which reflects the accumulation and reinvestment of the dividends. It should be apparent that Portfolio A is still dominant but the slight change in the composition of A leaves us with the following lower bound,

$$c_t \geq Max(0, S_t - D_t - Xe^{-r_c\tau}). \qquad (2.10)$$

Unlike the case of no dividends, we now see that the lower bound of a European call can be less than $S_t - X$, the exercise value of the American call. This case would occur if $X(1 - e^{-r_c\tau}) < D_t$, meaning that the interest on the exercise price is less than the present value of the dividends. That is,

$$S_t - Xe^{-r_c\tau} - D_t < S_t - X$$

$$X - Xe^{-r_c\tau} < D_t$$

$$X(1 - e^{-r_c\tau}) < D_t.$$

In such a case, we cannot state an equivalence of the European and American call prices. In the case where the inequality is reversed, however, we have a sufficient condition for no early exercise of the American call and such a call would be priced as a European call. That is,

$$S_t - Xe^{-r_c\tau} - D_t > S_t - X$$

$$X - Xe^{-r_c\tau} > D_t$$

$$X(1 - e^{-r_c\tau}) > D_t.$$

So, if the interest on the exercise price is more than the present value of the dividends, then there is absolutely no reason to exercise the call early. This leads to the condition,

$$C_t = c_t \text{ if } X(1 - e^{-r_c\tau}) > D_t. \tag{2.11}$$

The American call clearly has a positive probability of early exercise except in this special case of sufficiently small dividends relative to the present value of the exercise price. This result establishes the fact that it may be optimal to exercise an American call early to capture a dividend. When the call is exercised early, the holder throws away the time value and claims the exercise value. To avoid throwing away any more time value than necessary, however, it is always optimal to exercise only at the last instant before the asset goes ex-dividend. Also, note that if the condition in Equation (2.11) is not met, it does not automatically imply that the option will be exercised early.

Figure 2.2a illustrates the influence of actual dividends on the call lower bound. Recall in Figure 2.1 that the exercise value is less than the lower bound for a 49-day call option. For this longer dated option, there are two future quarterly dividends before this 203-day option expires. The present value of these dividends was estimated to be 2.326, based on

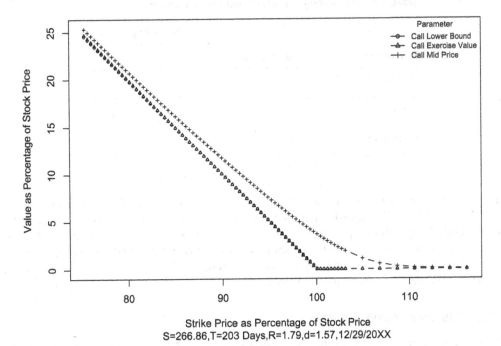

FIGURE 2.2a　Illustration of Dividend Impact on Call Lower Bound (Expressed as Percentage)

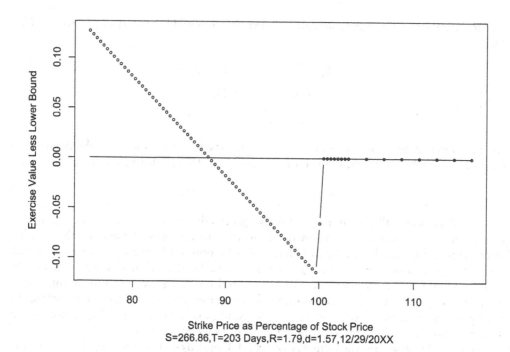

FIGURE 2.2b   Call Lower Bound Less Call Exercise Value

TABLE 2.3   Establishing a Lower Bound for a European Put

		Value at Expiration	
Instrument	Current Value	$S_T \leq X$	S_T
Portfolio E			
European put	p_t	$X - S_T$	0
Asset	S_t	S_T	S_T
Total	$p_t + S_t$	X	S_T
Portfolio F			
Zero-coupon bond	$Xe^{-r_c\tau}$	X	X

a market risk-free interest rate of 1.79%. Thus, the present value factor is 0.990094 or $e^{-r_c\tau} = e^{-0.0179(203/365)} = 0.990094$. Based on Equation (2.11), $X(1 - e^{-r_c\tau}) = D_t$ at 88% of the underlying value or $[2.326/(1 - 0.990094)]/266.86$. As shown in Figure 2.2b, the exercise value is higher than the lower bound for strike prices below 88% of the underlying price. The observed call prices appear to be converging to the call exercise value for lower strike prices, which are deep in-the-money calls.

2.5.3   The Lower Bound of a Put

Now we look at the lower bound of a European put. First, we consider the case of no dividends. We construct Portfolio E, consisting of a European put and a unit of the asset, and Portfolio F, consisting of a zero-coupon bond with face value X and current value of $Xe^{-r_c\tau}$.[7] Table 2.3 shows the outcomes.

Portfolio E clearly dominates Portfolio F, matching its outcome in one case and beating it in the other. Thus, the current value of Portfolio E must be no less than the current value of Portfolio F, giving us $p_t + S_t \geq Xe^{-r_c\tau}$. Rewriting this expression to isolate the put and noting that a negative lower bound is dominated by a lower bound of zero gives

$$p_t \geq \max(0, Xe^{-r_c\tau} - S_t). \qquad (2.12)$$

For an American put, however, the lower bound is still $\max(0, X - S_t)$ because this value exceeds the European lower bound. Consequently,

$$P_t \geq \max(0, X - S_t). \qquad (2.13)$$

Recall that if $S_t < X$ the put is in-the-money, if $S_t > X$ the put is out-of-the-money, and if $S_t = X$ the put is at-the-money.

Figure 2.3 illustrates the results for an American put option. We again normalize values by the underlying price. With this actual data, we observe that the put bid price remains above the exercise value. Remember for calls, the exercise value is below the lower bound. This is not true for puts. As such, the exercise value is binding in real markets due to arbitrage activity.

For the case of a European put on an asset that makes cash payments, such as a dividend-paying stock, we modify portfolio F so that its bond has a face value of $Xe^{-r_c\tau} + D_t$. This makes its payoff be $X + D_t e^{r_c\tau}$. Then the payoff of portfolio E will include $D_t e^{r_c\tau}$. Thus, Portfolio E is still dominant, but the resulting boundary now becomes

$$p_t \geq \max(0, Xe^{-r_c\tau} + D_t - S_t). \qquad (2.14)$$

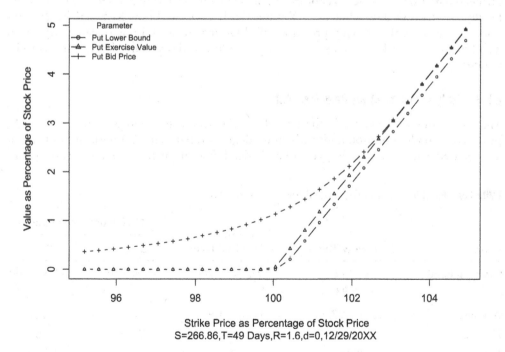

FIGURE 2.3 Illustration of Lower Bound for a Non-Dividend-Paying Put (Expressed as Percentage)

The plus sign on the dividends implies that they have a positive effect on put options, which is easy to rationalize. When a firm pays a dividend, it reduces its ability to grow, which is harmful to holders of calls, who benefit only from growth in the asset, but the dividend benefits holders of puts who gain from less growth, which keeps the stock price down.

Consider the following situation similar to the previous example: Suppose the stock price has fallen and now is $S_t = 97.93$, the strike price is $X = 100$, the risk-free interest rate is $r_c = 5.01\%$ continuously compounded, annualized, the time to maturity is 0.608 (or more precisely 222 days in a 365-day year), and the quoted European put price is 0.95. Now, however, there is one anticipated dividend in 111 days of $D_{tD} = 1.96$ where the appropriate risk-free interest rate for this dividend is 4.17%. How would an arbitrageur trade? The first step is to assess whether there is a boundary violation. Note in this case the value of the lower bound can be computed as

$$Xe^{-r_c\tau} + D_t - S_t = 100e^{-0.0501(222/365)} + 1.96e^{-0.0417(111/365)} - 97.93$$

$$= 97 + 1.93 - 97.93 = 1.0.$$

Thus, the quoted put price of 0.95 is below the lower boundary. The arbitrageur seeks to pocket this 0.05 through a series of current trades. Note in this case $Xe^{-r_c\tau} + D_t - S_t - c_t = 0.05$ and the trading strategy is embedded in this expression. Notice that you can generate this result by executing transactions that produce these specific cash flows, namely, short sell the zero-coupon bond ($Xe^{-r_c\tau} + D_t = 98.93$), buy stock ($S_t = 97.93$), and buy the put ($-p_t = -0.95$). Table 2.4 clearly demonstrates that these transactions result in arbitrage profits today with no subsequent chance for future negative cash flows.

Note in Table 2.4 that the net cash flow today is 0.05. The positive cash flow on the dividend date from the stock is captured today through short selling a bond or borrowing. Finally, the net cash flow when $S_T > X$ is positive because we know $S_T > 100$ by assumption. Further note that buying pressure will drive the stock and put price up. Again, in well-functioning markets, we expect the net cash flow today to quickly move so as to be nonpositive.

2.5.4   Early Exercise of an American Put

The lower bound dominance in the case of an American call, where $S_t - Xe^{-r_c\tau} > S_t - X$, provided a simple condition under which we demonstrated that an American call will not be exercised early except in the event of a dividend. For an American put, it is sufficient to

TABLE 2.4   Establishing a Lower Bound for a European Put

Instrument	Cash Flow Today	Cash Flow at Dividend Date	Cash Flow at Expiration $S_T \leq X$	Cash Flow at Expiration $X < S_T$
Short sell bond	$Xe^{-r_c\tau} + D_t$ $= 98.93$	$-D_{tD} = -1.96$	$-X = -100$	$-X = -100$
Buy stock	$-S_T = -97.93$	$+D_{tD} = 1.96$	$+S_T$	$+S_T$
Buy put	$-p_t = -0.95$	0	$X - S_T$	0
Net	0.05	0	0	$S_T - 100$

demonstrate that the case of an asset price falling to zero will trigger early exercise. The holder of the put will gain no more by waiting because the asset can go down no further, nor can it ever attain a positive value again. It is not true, however, that the holder must wait until the asset falls to zero. The exact point of early exercise is a complex matter to determine, and we do not cover it in this chapter. We shall see under what conditions American puts are exercised early after we have developed a full pricing model.

If the put were American, the existence of dividends paid by the stock renders it less likely to be exercised early. An American put can clearly be sold for at least its minimum European value of $Xe^{-r_c\tau} + D_t - S_t$ or exercised for $X - S_t$. If $D_t \geq X(1 - e^{-r_c\tau})$, then it cannot be worth more to exercise it. Clearly, high dividends make a put more attractive alive than exercised. If it is, however, optimal to exercise a put early, it will be done immediately after the asset goes ex-dividend. One might as well wait an instant to let the stock price drop before exercising the put.

In summary, the lower bound for an American put has three conditions as illustrated in the following expression,

$$P_t \geq \max(0, X - S_t, Xe^{-r_c\tau} + D_t - S_t).$$

Clearly, as time passes, the present value of the strike price and dividends increase until an ex-dividend date. On that date, the stock price typically falls by the dividend amount and the present value of the dividends also falls by the dollar dividend paid. Thus, a dividend payment generally has no significant impact on the third condition ($Xe^{-r_c\tau} + D_t - S_t$) but the second condition increases with the drop in stock price ($X - S_t$).

Prior to expiration, both American put and call option prices will exceed the exercise values because sellers of the options will bear the risk that the options will be worth substantially more than their current exercise values by the time expiration has arrived. The option price, thus, is said to consist of two components: the exercise value and the time value, the latter reflecting the premium that disappears as expiration nears. The full price of the option—the exercise value plus the time value—is the objective of developing an option pricing model, a topic to which much of this book is devoted.

Figure 2.4 illustrates the influence of dividends on the put lower bound. Recall in Figure 2.3 that the exercise value is higher than the lower bound for a 49-day put option. Recall from Figure 2.2 that this longer dated option has two future quarterly dividends with present value 2.326. Clearly the exercise value and lower bound are much closer together due to the influence of dividends.

2.6 DIFFERENCES IN OPTION VALUES BY EXERCISE PRICE

The results in this section will enable us to understand the effect of exercise price on the pricing of options. That is, of two otherwise identical calls or puts, differing only by exercise price, which option will be priced higher? Our intuition should tell us that the call with the lower exercise price and the put with the higher exercise price should have higher prices. For the call, the exercise price is the hurdle over which the underlying must get to justify exercise. Thus, the lower the hurdle the easier it is to get over it. Moreover, for a given level of the underlying above the exercise price, the more valuable is the call when exercised. Hence, the lower-exercise price call should be worth more before expiration. For a put, the argument is reversed. To exercise the put, the underlying must get under the

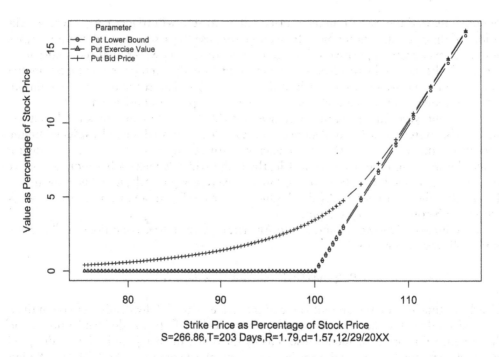

FIGURE 2.4   Illustration of Dividend Impact on Put Lower Bound (Expressed as Percentage)

exercise price. Hence, the higher is the exercise price, the easier it is to get under it, and for a given level of the underlying below the exercise price, the higher is the exercise price, the higher the payoff of the put. Hence, puts with higher exercise prices should be worth more before expiration. Let us first take a formal look at calls.

2.8.1   Calls Differing by Exercise Price

Consider two European calls differing only by exercise price. The first call has an exercise price of X_L and a price of c_{tL} and the second call has an exercise price of X_H and a price of c_{tH}. The subscripts L and H stand for "low" and "high," respectively, but you should not get in mind that c_{tL} is lower than c_{tH}. In fact, it will not be, as we show here.

So let us construct portfolio G, consisting of a long position in the call with exercise price X_L and a short position in the call with exercise price of X_H, and portfolio H consisting of zero-coupon bonds with face value of $X_H - X_L$ and current value of $(X_H - X_L)e^{-r_c\tau}$. We use these portfolios to establish an upper bound for the difference in the prices of the two calls, as shown in Table 2.5.

The first result to notice is that the payoff to Portfolio G is nonnegative, which means that the initial value of the call portfolio must be nonnegative. Therefore,

$$c_{tL} \geq c_{tH}. \tag{2.15}$$

This result means simply that the call with the lower exercise price must sell for at least as much as the call with the higher exercise price. If the calls are American, this result still holds if we can prove that the payoff of the calls is never negative. We need not worry

TABLE 2.5 Establishing an Upper Bound for the Difference in the Prices of Two European Calls Differing Only by Exercise Price

Instrument	Current Value	Value at Expiration		
		$S_T \leq X_L$	$X_L < S_T < X_H$	$S_T \geq X_H$
Portfolio G				
Long European call (X_L)	c_{tL}	0	$S_T - X_L$	$S_T - X_L$
Short European call (X_H)	$-c_{tH}$	0	0	$-(S_T - X_H)$
Total	$c_{tL} - c_{tH}$	0	$S_T - X_L$	$X_H - X_L$
Portfolio H				
Zero-coupon bond	$(X_H - X_L)e^{-r_c\tau}$	$X_H - X_L$	$X_H - X_L$	$X_H - X_L$

about the call we hold for we would never exercise it early if it were to our disadvantage. If the call we are short is exercised early, then it must be the case that $S_t > X_H$, which means that $S_t > X_L$ and we could exercise our long call early, capturing a gain of $X_H - X_L$, the maximum payoff at expiration in the case of a European call. Thus, early exercise makes no difference and we can state that[8]

$$C_{tL} \geq C_{tH}. \qquad (2.16)$$

So, the lower exercise price American call is worth at least as much as the higher exercise price American call.

The second result we notice is that Portfolio H dominates Portfolio G. Consequently, we have

$$c_{tL} - c_{tH} \leq (X_H - X_L)e^{-r_c\tau}. \qquad (2.17)$$

This statement establishes an upper bound on the spread between the call prices. It says that the spread between the prices of the two call options differing only by exercise price is no more than the present value of the difference in the exercise prices.

If the calls are American, then we are required to modify Portfolio H such that its current value is $X_H - X_L$ and its face value is $(X_H - X_L)e^{r_c\tau}$. If our short call is exercised early, we simply exercise our long call, which is even deeper in-the-money, and capture a value of $X_H - X_L$. This money is then invested at the risk-free rate. Without adjusting Portfolio H, we might have Portfolio G dominating Portfolio H due to the interest earned on the reinvestment of $X_H - X_L$ if early exercise occurs. With Portfolio H worth $X_H - X_L$ today, however, it will grow to a value that is at least as great as that of Portfolio G in the event of early exercise. Consequently, for American calls the rule becomes

$$C_{tL} - C_{tH} \leq X_H - X_L. \qquad (2.18)$$

If there is no possibility of early exercise, as is the case when the asset makes no payments, the upper bound on the American spread comes down to the upper bound on the European spread, which is a smaller number.

Consider the following situation for two American calls: Suppose the lower 100 strike (X_L) call price is $C_{tL} = 17$ and the higher 110 strike (X_H) call price is $C_{tH} = 6.9$. Based solely on this data, how would an arbitrageur exploit this data? Again, the first step is to assess whether there is a boundary condition violation. Because these options are American, the boundary can be assessed based on Equation (2.18) as $C_{tL} - C_{tH} = 17.0 - 6.9 = 10.1 > X_H - X_L = 110 - 100 = 10$. Thus, these quoted call prices suggest a 0.1 arbitrage opportunity. The arbitrageur seeks to pocket this 0.1 through a series of current trades. Note in this case $C_{tL} - C_{tH} - (X_H - X_L) = 0.1$ and the trading strategy is embedded in this expression. Notice that you can generate this result by executing transactions that generate these specific cash flows, namely, sell the low strike call ($C_{tL} = 17$), buy the high strike call ($C_{tH} = 6.9$), and lend the difference in strike prices ($X_H - X_L = 10$). If the buyer of the X_L call decides to exercise their right to buy the asset at some presently unknown point in time, t', then the arbitrageur will exercise their right to buy the asset at X_H as well as sell bonds. Table 2.6 clearly demonstrates that these transactions result in arbitrage profits today with no subsequent chance for future negative cash flows.

As interest will be earned on buying bonds, we indicate this future value as *FV*. Note in Table 2.6 that the net cash flow today is 0.1. Buying pressure will drive up the high strike call and selling pressure will drive down the low strike call. For every future possible outcome, the net cash flow is positive assuming interest rates are positive. Further note that trading pressure will drive the option prices in such a way that the positive net cash flow will no longer exist. Again, in well-functioning markets, we expect the net cash flow today to quickly move so as to be nonpositive.

2.6.2 Puts Differing by Exercise Price

Now consider two European puts differing only by exercise price. The first put has an exercise price of X_L and a price of p_{tL} and the second put has an exercise price of X_H and a price of p_{tH}. Construct Portfolio I, consisting of a short position in the put with exercise

TABLE 2.6 Numerical Illustration Establishing an Upper Bound for the Difference in the Prices of Two American Calls Differing Only by Exercise Price

Instrument	Cash Flow Today	Early Exercise (t')	Cash Flow at Expiration $S_T \leq X_L$	$X_L < S_T \leq X_H$	$S_T > X_H$
Sell X_L call	$C_{tL} = 17.0$	$-(S_{t'} - X_L)$ $= 100 - S_{t'}$	0	$-(S_T - X_L)$ $= 100 - S_T$	$-(S_T - X_L)$ $= 100 - S_T$
Buy X_H call	$-C_{tH} = -6.9$	$+(S_{t'} - X_H)$ $= S_{t'} - 110^*$	0	0	$+(S_T - X_H)$ $= S_T - 110$
Buy $X_H - X_L$ bond	$-(X_H - X_L)$ $= 10$	$FV_{t'}(X_H - X_L)$ > 10	$FV_T(X_H - X_L)$ > 10	$FV_T(X_H - X_L)$ > 10	$FV_T(X_H - X_L)$ > 10
Net	$C_{tL} - C_{tH}$ $-(X_H - X_L)$ $= 0.1$	> 0	> 10	> 0	Interest > 0

Note: The arbitrageur will exercise early the high strike call only if it is in-the-money and trading for less than its exercise value; otherwise, they will just sell it. If $S_{t'}$ is less than 110, then the net column for some early exercise date (t') will be positive, even if X_H call is worthless.

price X_L and a long position in the put with exercise price of X_H, and Portfolio J consisting of zero-coupon bonds with face value of $X_H - X_L$ and current value of $(X_H - X_L)e^{-r_c\tau}$. We use these portfolios to set an upper bound for the difference in the prices of the two puts as shown Table 2.7.

The first result we should notice is that the payoff to Portfolio I is nonnegative, which means that the initial value of the put portfolio must be nonnegative. In other words,

$$p_{tH} \geq p_{tL}. \tag{2.19}$$

This statement means that the put with the higher exercise price must sell for at least as much as the put with the lower exercise price. If the puts are American, this result still holds if we can prove that the payoff of the put portfolio is never negative. We need not worry about the put we hold for we would never exercise it early if it were to our disadvantage. If the put we are short in is exercised early, then it must be the case that $S_t < X_L$, which means that we could exercise our long put early, capturing a gain of $X_H - X_L$ earlier than expiration. This is the maximum payoff at expiration in the case of a European put. Thus, early exercise makes no difference, and we can state that

$$P_{tH} \geq P_{tL}. \tag{2.20}$$

The second result we notice is that Portfolio J dominates Portfolio I. Consequently, we have

$$p_H - p_L \leq (X_H - X_L)e^{-r_c\tau}. \tag{2.21}$$

This statement establishes an upper bound on the spread between the European put prices. The spread is no more than the present value of the difference in the exercise prices.

If the puts are American, then we are required to modify Portfolio J such that its current value is $(X_H - X_L)e^{-r_c\tau}$ and its face value is $X_H - X_L$. If our short put is exercised early, we simply exercise our long put, which is even deeper in-the-money, and capture a value of $X_H - X_L$. This amount is then invested at the risk-free rate. Without adjusting Portfolio J, we might have Portfolio I dominating Portfolio J due to the interest earned on the reinvestment of $X_H - X_L$. With Portfolio J worth $X_H - X_L$ today, however, it will

TABLE 2.7 Establishing an Upper Bound for the Difference in the Prices of Two European Puts Differing Only by Exercise Price

Instrument	Current Value	Value at Expiration		
		$S_T \leq X_L$	$X_L < S_T < X_H$	$S_T \geq X_H$
Portfolio I				
Long European put (X_H)	p_{tH}	$X_H - S_T$	$X_H - S_T$	0
Short European put (X_L)	$-p_{tL}$	$-(X_L - S_T)$	0	0
Total	$p_{tH} - p_{tL}$	$X_H - X_L$	$X_H - S_T$	0
Portfolio J				
Zero-coupon bond	$(X_H - X_L)e^{-r_c\tau}$	$X_H - X_L$	$X_H - X_L$	$X_H - X_L$

grow to a value that is at least as great as that of Portfolio I in the event of early exercise. Consequently, for American puts the rule becomes

$$P_{tH} - P_{tL} \leq X_H - X_L. \tag{2.22}$$

Thus, the spread between the American put prices is no more than the spread between the exercise prices.

Consider the following situation for two three-month European puts: Suppose the lower 100-strike (X_L) put price is $p_L = 6.05$ and the higher 110-strike (X_H) put price is $p_H = 16$. Assume the continuously compounded risk-free rate is 5%. Based solely on this data, how would an arbitrageur exploit this situation? As before, the first step is to assess whether there is a boundary condition violation. Because these options are European puts, the boundary can be assessed based on Equation (2.21) as

$$p_{tH} - p_{tL} = 16.0 - 6.05 = 9.95 > (X_H - X_L)e^{-r_c \tau} = (110 - 100)e^{-0.05(3/12)} = 9.875778.$$

Thus, these quoted call prices suggest an approximately 0.07 arbitrage opportunity. The arbitrageur seeks to pocket this 0.07 through a series of current trades. Note in this case the boundary is violated and again the trading strategy is embedded in this expression. Rearranging we have

$$p_{tH} - p_{tL} - (X_H - X_L)e^{-r_c \tau} = 9.95 - 9.88 = 0.07.$$

Notice that you can generate this result by executing transactions that generate these specific cash flows, namely, selling the high strike put ($p_H = 16$), buying the low strike put ($p_L = 6.05$), and lending the difference in the present value of the strike prices (($X_H - X_L)e^{-r_c \tau} \cong 9.88$). Table 2.8 clearly demonstrates that these transactions result in arbitrage profits today with no subsequent chance for future negative cash flows.

Because interest will be earned on buying bonds, we buy bonds valued at the present value of the difference in strike prices. Note in Table 2.8 that the net cash flow today is 0.07. Further, buying pressure will drive up the low strike put and selling pressure will

TABLE 2.8 Numerical Illustration Establishing an Upper Bound for the Difference in the Prices of Two European Puts Differing Only by Exercise Price

Instrument	Cash Flow Today	Cash Flow at Expiration		
		$S_T \leq X_L$	$X_L < S_T \leq X_H$	$S_T > X_H$
Sell X_H put	$+p_H = 16.0$	$-(X_H - S_T)$ $= S_T - 100$	$-(X_H - S_T)$ $= S_T - 100$	0
Buy X_L put	$-p_L = -6.05$	$+(X_L - S_T)$ $= 100 - S_T$	0	0
Buy PV($X_H - X_L$) bond	$-PV(X_H - X_L)$ $= 9.88$	$+(X_H - X_L)$ $= 10$	$+(X_H - X_L)$ $= 10$	$+(X_H - X_L)$ $= 10$
Net	$p_H - p_L$ $+PV(X_H - X_L)$	0	$S_T - 100 > 0$	10

drive down the high strike put. For every future possible outcome, the net cash flow is nonnegative assuming interest rates are positive. Further note that trading pressure will drive the option prices in such a way that the positive net cash flow will no longer exist. Again, in well-functioning markets, we expect the net cash flow today to quickly move so as to be nonpositive.

2.7 THE EFFECT OF DIFFERENCES IN TIME TO EXPIRATION

Consider two European calls differing only by time to expiration. One option expires at time T_1 and has a time to expiration of $T_1 - t = \tau_1$ and the other expires at T_2 and has a time to expiration of $T_2 - t = \tau_2$, and where $\tau_2 > \tau_1$. Their respective prices will be c_{t1} and c_{t2} with similar notation for American calls as well as for European and American puts.

For a European call, it is simple to demonstrate that in the absence of dividends, the longer-term call must sell for at least as much as the shorter-term call. Suppose we are at the expiration date of the shorter-term call and the asset price is S_{T_1}. Its value is $\max(0, S_{T_1} - X)$. The longer-term option, however, has time remaining of $\tau_2 - \tau_1$, so its minimum value is $\max[0, S_{T_1} - Xe^{-r_c(\tau_2-\tau_1)}]$, which is at least as great as the value of the shorter-term expiring option. Consequently,

$$c_{t2} \geq c_{t1}, \tag{2.23}$$

with the LHS being the current price of the longer term call, and the RHS being the current price of the shorter-term call. Thus, the longer-term call cannot be worth less than the shorter-term call. In the absence of dividends, there would be no early exercise, so the previous statement would apply for American options.

$$C_{t2} \geq C_{t1}. \tag{2.24}$$

If there are dividends, then the longer term European call has a minimum value of $\max[0, S_{T_1} - D - Xe^{-r_c(\tau_2-\tau_1)}]$, which might make it seem as if that option has a minimum less than the exercise value of the expiring option. If that is the case, however, the longer-term American option would sell for at least the exercise value. Consequently, Equation (2.24) is still valid.

For a European put, we obtain a somewhat counterintuitive result. First assume no dividends. Then the expiring, shorter-term option is worth $\max(0, X - S_{T_1})$. The second option, which still has time remaining, is worth at least $\max[0, Xe^{-r_c(\tau_2-\tau_1)} - S_{T_1}]$. Thus, it might be the case that the shorter-term option is worth more. This somewhat strange result that a longer-term European put can be worth less than a shorter-term European put arises because there are conflicting sources of value arising from the time to expiration in an option. A longer time to expiration generally helps an option, whether a put or a call, in that it gives it greater time for a favorable asset price move to occur. The longer time to expiration also has an effect arising from the present value of the potential payoff at expiration. For a put, the better outcome at expiration is to exercise it, which will result in a cash inflow from the sale of the asset. A longer time to expiration reduces the present value of this inflow, however, rendering the put potentially less valuable. This disadvantage of a longer expiration can be partially, perhaps wholly, offset by the advantage of the longer time for a favorable asset price move. Puts that are deep in-the-money will tend to have this disadvantage weigh more than the advantage because their potential for exercise is greater and their potential for gains from further asset price moves is limited. All other puts will tend to have the advantage greater than the disadvantage. With the LHS being the current price of the longer-term put and the RHS being the current price of the shorter-term put, the overall result is

$$p_{t2} \gtreqless p_{t1}. \tag{2.25}$$

In other words, we cannot definitely make a statement about the relationship of a longer-term European put price to the shorter-term put price.

If the put is American, however, there is no requirement to wait to receive the exercise price at expiration. It can always be claimed now. Thus,

$$P_{t2} \geq P_{t1}. \tag{2.26}$$

If there are dividends on the asset and the puts are European, the expiring option is worth $\max(0, S_{T_1} - X)$. The other option is still alive and worth at least $\max[0, Xe^{-r_c(\tau_2 - \tau_1)} - D_1 - S_{T_1}]$. Again, it may be the case that the longer-term put is worth less, which would, of course, all depend on the various factors that affect the option price. Consequently, our previous statement for European puts for the no-dividend case holds as well if there are dividends. If the puts are American, the remaining option will always sell for at least its exercise value, which makes it worth at least as much as the expiring option. Consequently, our statement for American puts without dividends holds when dividends are introduced.

2.8 THE CONVEXITY RULE

A mathematical function, $y = f(x)$, is said to be convex if for any two values of x, $x_2 > x_1$, we have that the line between these two x values lies above or on the $y = f(x)$ graph. It is possible to derive a relationship between the prices of three options differing by exercise price. Let their exercise prices be X_L, X_M, and X_H (for low, medium, and high) and the corresponding call prices be c_{tL}, c_{tM}, and c_{tH}. For convenience, we shall call these the first, second, and third calls, respectively. Let us construct Portfolio K consisting of ω units of the first call and $1 - \omega$ units of the third call. In other words, we go long ω units of the first call and long $1 - \omega$ units of the third. Portfolio L consists of one unit of the second call. The value of ω is defined as $(X_H - X_M)/(X_H - X_L)$ so that $1 - \omega$ is $(X_M - X_L)/(X_H - X_L)$. Table 2.9 shows the outcomes.

When $S_T < X_L$, both portfolios produce the same result of zero value. When $X_L \leq S_T < X_M$, Portfolio K is better than Portfolio L, because $\omega > 0$ and $S_T > X_L$. When $X_M \leq S_T < X_H$, we can prove that Portfolio K is better by substituting the definition of ω. In the last case, where $S_T > X_H$, Portfolio K is equivalent to Portfolio L, which can also be proven by substituting the definition of ω. Putting these results together tells us that Portfolio K dominates Portfolio L. Consequently, the current value of Portfolio K must be at least as great as the current value of Portfolio L,

$$\omega c_{tL} + (1 - \omega)c_{tH} \geq c_{tM}. \tag{2.27}$$

TABLE 2.9   The Relationship Between the Prices of Three European Calls Differing Only by Their Exercise Prices

Instrument	Current Value	Value at Expiration			
		$S_T < X_L$	$X_L \leq S_T < X_M$	$X_M \leq S_T < X_H$	$S_T \geq X_H$
Portfolio K					
European call	ωc_{tL}	0	$\omega(S_T - X_L)$	$\omega(S_T - X_L)$	$\omega(S_T - X_L)$
European call	$(1-\omega)c_{tH}$	0	0	0	$(1-\omega)(S_T - X_H)$
Total	$\omega c_{tL} + (1-\omega)c_{tH}$	0	$\omega(S_T - X_L)$	$\omega(S_T - X_L)$	$S_T - \omega X_L$ $-(1-\omega)X_H$
Portfolio L					
European call	c_{tM}	0	0	$S_T - X_M$	$S_T - X_M$

TABLE 2.10   The Relationship Between the Prices of Three European Puts Differing Only by Their Exercise Prices

Instrument	Current Value	Value at Expiration			
		$S_T < X_L$	$X_L \leq S_T < X_M$	$X_M \leq S_T < X_H$	$X_H \leq S_T$
Portfolio M					
European put	ωp_{tL}	$\omega(X_L - S_T)$	0	0	0
European put	$(1-\omega)p_{tH}$	$(1-\omega)(X_H - S_T)$	$(1-\omega)(X_H - S_T)$	$(1-\omega)(X_H - S_T)$	0
Total	ωp_{tL} $+ (1-\omega)p_{tH}$	$\omega X_L - S_T$ $+(1-\omega)X_H$	$(1-\omega)(X_H - S_T)$	$(1-\omega)(X_H - S_T)$	0
Portfolio N					
European put	p_M	$X_M - S_T$	$X_M - S_T$	0	0

This statement is called the *convexity rule* because it proves that the option price is convex with respect to the exercise price.[9] If there are dividends on the asset, the rule is not affected because none of the positions above will collect dividends. If the options were American, Portfolio K would still dominate Portfolio L because the payoffs above at expiration would occur early. Thus,

$$\omega C_{tL} + (1-\omega)C_{tH} \geq C_{tM}. \tag{2.28}$$

Now, let us prove the rule for puts. Consider Table 2.10, which is based on Portfolios M and N.

It is easy to see that Portfolio M dominates Portfolio N. For the case in which $S_T \geq X_H$, both portfolios are worth zero. For the case in which $X_M \leq S_T < X_H$, Portfolio N is worth zero, whereas Portfolio M has a positive value. For the case in which $X_L \leq S_T < X_M$, we can prove that Portfolio M is worth more than Portfolio N by substituting the definition of ω. And for the case in which $S_T < X_L$, it can be shown that Portfolio M dominates Portfolio N by also substituting the definition of ω. These results mean that

$$\omega p_{tL} + (1-\omega)p_{tH} \geq p_{tM}. \tag{2.29}$$

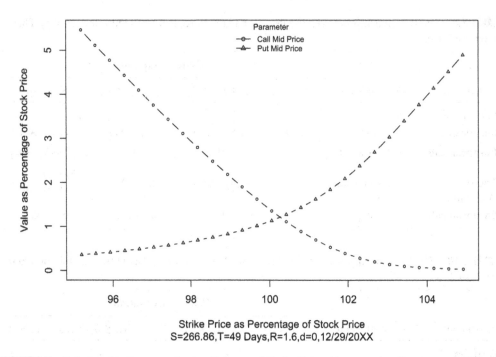

FIGURE 2.5   Relationship Between Option Prices and Strike Prices Expressed as Percentage of Stock Price

If there are dividends, the results are unaffected, as these instruments will not receive or pay dividends. If the puts are American, the outcomes above would occur early, and the same result would hold.

$$\omega P_{tL} + (1 - \omega)P_{tH} \geq P_{tM}. \tag{2.30}$$

So put options are also convex in the exercise price.

Figure 2.5 illustrates the convex relationship between call and put option mid prices and different strike price expressed as a percentage of the stock price. Note that for both the upper and lower limits, the relationship converges to linear. Hence, the convex relationships identified previously may at times be simply equal.

We now turn to addressing put-call parity.

2.9   PUT-CALL PARITY

The connection between the prices of puts and calls with the same exercise price and expiration is referred to as *put-call parity*, and it is one of the most important relationships in all of option pricing. We first begin with European options.

2.9.1   Put-Call Parity for European Options

We construct two portfolios called O and P. Portfolio O consists of one unit of the asset and one European put and Portfolio P consists of one European call and a zero-coupon bond with face value of X and current value of $Xe^{-r_c\tau}$. We assume no dividends on the underlying. Table 2.11 shows the outcomes at expiration.

Notice that Portfolio O and Portfolio P produce the same results at expiration. Thus, they must have the same value today. We, therefore, can state put-call parity as

$$S_t + p_t = c_t + Xe^{-r_c\tau}. \tag{2.31}$$

This equation indicates that a long position in the asset and the put is equivalent to a long position in the call and risk-free bonds. In other words, we note that each of the four symbols—S_t, p_t, c_t, and $Xe^{-r_c\tau}$—represents a financial instrument, respectively, the asset, a put, a call, and a zero-coupon bond. The sign, positive (implied) or negative, represents whether one is long or short. Now, let us use simple algebra to reveal some other relationships. First, let us isolate the put price,

$$p_t = c_t - S_t + Xe^{-r_c\tau},$$

which indicates that a put is equivalent to a long call, short asset, and long bonds. Likewise, a call is indicated as

$$c_t = p_t + S_t - Xe^{-r_c\tau},$$

which means that a call is equivalent to a long put, long asset, and short bonds. The asset itself can be decomposed as follows,

$$S_t = c_t - p_t + Xe^{-r_c\tau},$$

which indicates that the asset can be replicated by a long call, short put, and long bonds. Finally, the risk-free asset can be expressed as

$$Xe^{-r_c\tau} = S_t + p_t - c_t,$$

meaning that the zero-coupon bond is equivalent to the asset, a long put, and a short call. You can also take each of these equations, multiply by −1 and make similar statements about short positions in the put, call, asset, and zero-coupon bond.

TABLE 2.11 Put-Call Parity for European Options on Assets Making No Cash Payments

Instrument	Current Value	Value at Expiration	
		$S_T \leq X$	S_T
Portfolio O			
Asset	S_t	S_T	S_T
European put	p_t	$X - S_T$	0
Total	$S_t + p_t$	X	S_T
Portfolio P			
Call	c_t	0	$S_T - X$
Zero-coupon bond	$Xe^{-r_c\tau}$	X	X
Total	$c_t + Xe^{-r_c\tau}$	X	S_T

If there are dividends on the asset, put-call parity is easily established by simply modifying the zero-coupon bond in Portfolio P such that its current value is $Xe^{-r_c\tau} + D_t$. At expiration Portfolio P will, therefore, pay off $X + D_t e^{r_c\tau}$. Portfolio O's payoff in each state will be augmented by the reinvested value of the dividends, $D_t e^{r_c\tau}$. The final results will be the same in that the payoffs of the two portfolios are still equivalent, but put-call parity is now stated as

$$S_t - D_t + p_t = c_t + Xe^{-r_c\tau}. \tag{2.32}$$

In other words, the asset price is simply reduced by the present value of the dividends. This process of reducing the asset price by the present value of the dividends will arise often in derivative pricing.

Let us now examine a case where put-call parity does not hold. Suppose you observe the following data: The stock price is $S_t = 100$, the strike price is $X = 100$, a European call is at $c_t = 7.55$, a European put is at $p_t = 7.55$, and the time to expiration is 0.5 years, a 1.0 dividend will be paid in 0.25 years ($D_{t+0.25} = 1.0$ or the present value $D_t = 0.99$). The risk-free interest rate is $r_c = 5.0\%$ (continuously compounded, annualized, appropriate for both dividend and expiration). How would an arbitrageur trade? The first step is to assess whether there is a put-call parity violation. Note in this case the value of the put-call parity relation can be assessed based on the expression provided in Equation (2.32). Note that the two sides do not equal,

$$S_t - D_t + p_t = 100 - 0.99 + 6.02 = 105.03 < c_t + Xe^{-r_c\tau} = 7.55 + 97.53 = 105.08.$$

The arbitrageur seeks to pocket this 0.05 difference through a series of current trades. Rearranging the expression such that a positive number is produced, we have $c_t + Xe^{-r_c\tau} - S_t + D_t - p_t = 0.05$. To generate this cash flow, you would sell the call ($+c_t = 7.55$), short sell the bond or borrow the present value of the exercise price ($Xe^{-r_c\tau} = 100e^{-0.05(0.5)} = 97.53$), buy the stock ($-S_t = -100$), short sell the bond or borrow the present value of the dividend ($-D_t = -1e^{-0.05(0.25)} = 0.99$), and buy the put ($-p_t = -6.02$). Table 2.12 clearly demonstrates that these transactions result in arbitrage profits today with no future cash flows at all.

TABLE 2.12 Numerical Illustration Put-Call Parity for European Options on Assets Making No Cash Payments

Instrument	Cash Flow Today	Cash Flow at Dividend Date	Cash Flow at Expiration $S_T \leq X$	$X < S_T$
Sell call	$+c_t = +7.55$		0	$-(S_T - X)$ $= 100 - S_T$
Short sell bond	$Xe^{-r_c\tau} = +97.53$		$-X = -100$	$-X = -100$
Buy stock	$-S_t = -100$	$+D_{tD} = +1.0$	$+S_T$	$+S_T$
Short sell bond	$+D_t = +0.99$	$-D_{tD} = -1.0$		
Buy put	$-p_t = -6.02$		$X - S_T$ $= 100 - S_T$	0
Net	$c_t + Xe^{-r_c\tau} - S_t$ $+D_t - p_t = 0.05$	0	0	0

Note that buying pressure will drive the stock and put prices up and selling pressure will drive the call price down. Again, in well-functioning markets, we expect the net cash flow today to quickly move to nearly zero.

2.9.2 Put-Call Parity for American Options

If the options are American, put-call parity is a bit more complex, and the statements we shall be able to make a bit less precise. We first consider the case of no dividends. Portfolio P, consisting of a long call and a zero-coupon bond, is not subject to early exercise because the long call would not be exercised due to the absence of dividends on the asset. Given the potential for early exercise, it is necessary to adjust the initial value of the bonds to X instead of $Xe^{-r_c\tau}$. That is, we make the current value of the bonds equal to the exercise price, not the present value of the exercise price, which means we borrow a little more money. At expiration, the bonds will mature and pay off $Xe^{-r_c\tau}$. So, at expiration, the payoff of Portfolio P will either be $Xe^{r_c\tau}$ if $S_T \leq X$ or $S_T - X(1 - e^{r_c\tau})$ if $S_T > X$. Portfolio O, however, is subject to early exercise of the put. Let us choose an arbitrary time before expiration, j, in which the remaining time to expiration is τ_j. Suppose at that point, the asset price, S_j, is sufficiently below the exercise price to justify early exercise. In that case, the holder of the put will tender the asset and receive the exercise price. We assume that this person will reinvest the exercise price at the risk-free rate, such that it will grow to a value of $Xe^{r_c\tau_j}$ at time T. The final value at T will be

If $S_T \leq X$

0	(the call)
$Xe^{r_c\tau}$	(the bonds)
$Xe^{r_c\tau}$	(total for Portfolio P)
$Xe^{r_c\tau_j}$	(for Portfolio O with put exercised early)

If $S_T > X$

$S_T - X$	(the call)
$Xe^{r_c\tau}$	(the bonds)
$S_T - X + Xe^{r_c\tau}$	(total for Portfolio P)
$Xe^{r_c\tau_j}$	(total for Portfolio O with put exercised early)

In the first outcome, $S_T \leq X$, Portfolio P has the same value as Portfolio O. In the second outcome, $S_T > X$, P beats O because the value $S_T > X$ and $Xe^{r_c\tau_j}$ is positive. Hence, P dominates O, and we can say that $C_t + X \geq P_t + S_t$. Let us rearrange this to $C_t \geq P_t + S_t - X$ and further arrange to have the call on the left-hand side,

$$C_t \geq P_t + S_t - X. \tag{2.33}$$

This result establishes a lower limit on the call price. We can also establish an upper limit on the call price by using European put-call parity. With the conditions $p_t = c_t - S_t + Xe^{-r_c\tau}$ and $P_t > p_t$, we can say that $P_t \geq c_t - S_t + Xe^{-r_c\tau}$. With no dividends, $C_t > c_t$, we have,

$$C_t \leq P_t + S_t - Xe^{-r_c\tau}.$$

Now we can combine these results to obtain put-call parity for American options on non-dividend-paying assets.

$$P_t + S_t - Xe^{-r_c\tau} \geq C_t \geq P_t + S_t - X. \tag{2.34}$$

Unfortunately, the best we can do is place these bounds around the put price.[10]

If the asset pays dividends with present value D_t, the proof is only slightly altered. Consider the first inequality, $P_t + S_t - Xe^{-r_c\tau} \geq C_t$. If we impose dividends, it will raise the put price and lower the call price, so this relationship will still be correct. For the other inequality, $C_t \geq P_t + S_t - D_t - X$, let us pretend that it is violated. That is, suppose $C_t < P_t + S_t - D_t - X$. If we believe that the left-hand side (LHS) should be greater than the right-hand side (RHS), an arbitrage should be possible. Because the call price appears to be too low, let us buy the call. We would then sell the RHS by selling the put, shorting the stock, and buying zero-coupon bonds with face value $D_t + X$. Now, suppose the put is immediately exercised on us. We now have to pay out X to buy the shares, so let us borrow X. We then tender X to acquire the shares from the put holder, and we use the shares to cover the short sale. At expiration, we would owe $Xe^{r_c\tau}$ but our zero-coupon bond would mature and pay that same amount plus $D_t e^{r_c\tau}$. If the puts are not exercised until maturity, we will have accrued an obligation to pay all the dividends, but this amount will be covered by the value of $D_t e^{r_c\tau}$ from the maturing bonds. So, this strategy generates a cash inflow at the start and a positive or zero outflow at maturity. Therefore, it is an arbitrage opportunity and will be exploited, which will drive up the price of the call or drive down the price of the put until the correct inequality is established. Therefore,

$$P_t + S_t - Xe^{-r_c\tau} \geq C_t \geq P_t + S_t - D_t - X. \tag{2.35}$$

Note we can rearrange Equation (2.35), isolating the role of the interest rate, time to maturity, and present value of dividends, as $Xe^{-r_c\tau} \leq S_t + P_t - C_t \leq D_t + X$. For non-dividend-paying stocks with short maturity options in a low rate environment we have very tight put-call parity boundaries. Figure 2.6 illustrates the normalized Equation (2.35) or

$$1 - \frac{Xe^{-r_c\tau} - P_t}{S_t} \geq \frac{C_t}{S_t} \geq 1 - \frac{D_t + X - P_t}{S_t},$$

where the LHS is the put-call parity upper bound, and the RHS is the put-call parity lower bound. Figure 2.6a illustrates short-dated options (44 days to maturity with 5% strike range) and Figure 2.6b illustrates long-dated options (212 days to maturity with 25% strike range). Note that we use a trading day in November that contains one potential future dividend payment for the short-dated option and two potential future dividends for the long-dated option. The presence of dividends results in wider bounds.

With an understanding of option boundaries, we observe a rationality to the relationship between the underlying, calls, and puts that is not apparent to the untrained eye.

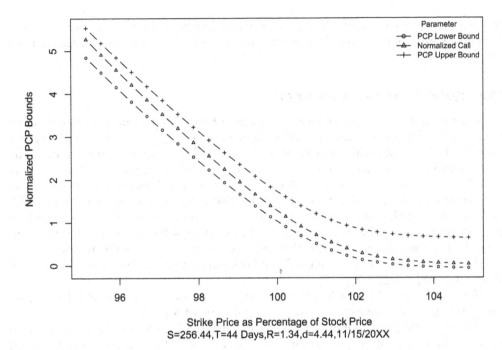

FIGURE 2.6a Illustration of Put-Call Parity Bounds for Short-Dated American Options (in Percentage)

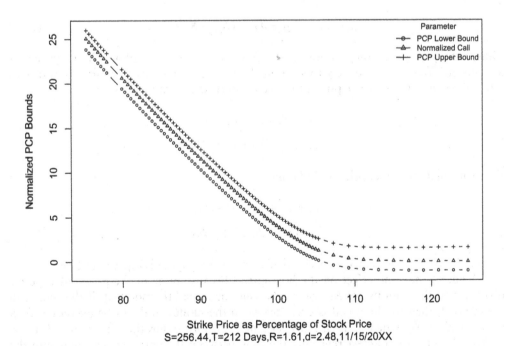

FIGURE 2.6b Illustration of Put-Call Parity Bounds for Long-Dated American Options (in Percentage)

We are just beginning to partially unravel financial derivatives mysteries. Once you finish this book, there will be plenty of remaining mysteries worthy of a lifetime of research and learning.

2.9.3 Put-Call Parity and the Box Spread

There is a special relationship between the prices of European puts and calls that differ by exercise price. Suppose one buys the call with exercise price X_L, sells the call with exercise price X_H, buys the put with exercise price X_H, and sells the put with exercise price X_L. These portfolios constitute a strategy known in the world of options trading as a *box spread*. Because the call purchased will cost more than the call that is sold, there is a net cash outflow for the calls. Likewise, the put purchased will cost more than the put sold, so there is a net cash outflow for the puts. Thus, these transactions will net out to a long position, meaning that the investor who does this strategy will pay out money at the start. Let us look at the outcomes because they produce a somewhat surprising result.

If $S_T > X_H$, both calls are exercised and both puts are out-of-the-money for a payoff of $S_T - X_L - (S_T - X_H) = X_H - X_L$.

If $X_L < S_T \leq X_H$, the long X_L call is exercised for a value of $S_T - X_L$ and the long X_H put is exercised for a value of $X_H - S_T$ for a total value of $X_H - X_L$.

If $S_T \leq X_L$, the long X_H put is exercised for a value of $X_H - S_T$ and the short X_L put is exercised for a value of $-(X_L - S_T)$ for a total of $X_H - X_L$. Thus, the box spread pays off $X_H - X_L$ in every state.

With the same payoff in each case, the value of the box spread is the present value of $X_H - X_L$,

$$c_{tL} - c_{tH} + p_{tH} - p_{tL} = (X_H - X_L)e^{-r_c\tau}. \tag{2.36}$$

Thus, it may surprise you to know that this strategy, which combines four options, is actually risk-free. It is also easy to see that the box spread is just a combination of two put-call parities. Let us write put-call parity for both sets of options as

$$p_L + S = c_L + X_L e^{-r_c\tau}$$

$$p_H + S = c_H + X_H e^{-r_c\tau}.$$

We then isolate the asset price and obtain

$$S_t = c_L + X_L e^{-r_c\tau} - p_L$$

$$S_t = c_H + X_H e^{-r_c\tau} - p_H.$$

Setting these equations equal to each other gives the box spread, Equation (2.36).

Dividends have no effect on the box spread because none of the positions produce dividends. If the options are American, the analysis is slightly modified. If the short call is exercised, then the long call is even deeper in-the-money and can be exercised for a value of $X_H - X_L$, which is then reinvested in cash until expiration. Thus, it will grow to $X_H - X_L$ plus the interest reflecting the time between the early exercise date and the expiration. In addition, the puts can have some value at expiration to add. At expiration, the puts will be worth zero at worst and $X_H - X_L$ at best, provided they are not exercised early. If the puts are exercised early, they will be worth $X_H - X_L$, which will be reinvested until expiration. In addition, the calls can have some value at expiration to add. In short, the box spread will pay off $X_H - X_L$ at expiration or possibly before expiration and there can be additional payoff from the non-exercised options. Because the options are subject to immediate early exercise, in which case they could be worth $X_H - X_L$ right now and because it is possible that at some time during the life of the options that are alive, another payoff will be received, we can say that

$$C_{tL} - C_{tH} + P_{tH} - P_{tL} \geq X_H - X_L. \tag{2.37}$$

2.10 THE EFFECT OF INTEREST RATES ON OPTION PRICES

Interest rates impart a small but positive effect on call option prices and a small but negative effect on put option prices. Consider that the holder of a European call faces a payoff at expiration of either zero or $S_T - X$. If interest rates increase, the value of the possible zero payoff is unaffected, but the present value of the X dollars paid out if the option ends up in-the-money is less. Hence, a higher risk-free rate makes a European call option more valuable as the $S_T - PV(X)$ increases.

If the options are American, the results are unaffected but do require an explanation. For American calls, there might be immediate early exercise. In that case, the extra interest from the higher rate and waiting to exercise is forgone, but this effect does not disadvantage the call. If there is early exercise at any point beyond immediacy, there is additional interest saved on the payment of the exercise price. Hence, American calls are positively related to interest rates.

For the holder of a European put option, the payoff at expiration is either zero or $X - S_T$. If interest rates increase, the value of the possible zero payoff is unaffected but the present value of the receipt of X dollars is lower. Consequently, rising interest rates decrease the value of the European put.

For an American put, exercise at any time after the present but before expiration reduces the effect of higher interest rates but does not eliminate it. Hence, American puts are negatively related to interest rates.

Note that for both American calls and puts, the effect of exercising immediately is neutral with respect to interest rates. We can easily see this by simply remembering that an option that is immediately exercised is worth $S_T - X$ for a call and $X - S_T$ for a put. There is no interest rate effect at all. In fact, an immediately exercisable option is really not an option in that it instantly turns into either the asset, if a call, or a short position in the asset, if a put, or the cash equivalent if the contract provides for settlement in cash.

There are several other explanations for the effect of interest rates on option prices. Most rely on the idea that the call is a leveraged transaction that substitutes for a stock margin trade and that the put is like an insurance policy. They lead to the same conclusion.

2.11 THE EFFECT OF VOLATILITY ON OPTION PRICES

If it is not already intuitively obvious, it is simple to demonstrate that a call option on an asset with higher volatility will be worth more, all else equal. For example, consider an option with any expiration except immediate. The options can be European or American.

Let the exercise price be X and the underlying asset price be S. Let us specify that the underlying can make only two moves, up or down. It can move up by ΔS or down by ΔS. Thus, its new price is either $S + \Delta S$ or $S - \Delta S$. Let us specify that if the up move occurs, the option expires in-the-money, is exercised, and, therefore, pays off $S + \Delta S - X$. If it moves down, the option expires out-of-the-money, and is, therefore, not exercised and worth zero.

It is easy to treat the value of ΔS as a good measure of the volatility of the underlying. The greater is ΔS, the more volatile is the asset. Suppose that we increase the volatility by making ΔS be a bit larger. Let us use the factor $\Delta^* S$ to represent a higher volatility. In that case, the payoff of the option will be $S + \Delta^* S$ in one outcome and zero in the other. We can see that the in-the-money outcome is higher, and the out-of-the-money outcome is the same as when the volatility was ΔS. We also see that these outcomes occur in the same corresponding conditions as when the volatility was lower. A similar argument applies for puts.

Hence, it should be quite clear that higher volatility leads to a higher option value. This result will hold regardless of whether there are dividends. In addition, because an American call need not be exercised early, it can be made to produce the same payoffs as a European call. Hence, the right to exercise early does not impair the positive effect of volatility.

2.12 THE BUILDING BLOCKS OF EUROPEAN OPTIONS

There are two types of options that can be seen as the building blocks of standard European options, and it is useful to introduce them in this chapter. One type is called a *binary option* or *digital option*, owing to the fact that there are only two possible payoffs, 0 or 1.

To create the building blocks of a standard European option, we first introduce an option that is quite different from a standard European option and has multiple possible payoffs. This instrument is called an *asset-or-nothing option*. If the underlying asset value at expiration exceeds the exercise price, an asset-or-nothing call option pays the asset. Otherwise, it pays nothing. If the underlying asset value at expiration is below the exercise price, an asset-or-nothing put pays the asset and nothing otherwise. This type of option is obviously not binary or digital, but it is covered with binary or digital options, because it complements them to form a standard European option.

The second type of option is called a *cash-or-nothing option*. If the underlying asset value at expiration exceeds the exercise price, a cash-or-nothing call option pays a fixed amount of money and zero otherwise. If the underlying asset value at expiration is less than the exercise price, a cash-or-nothing put option pays a fixed amount of money and zero otherwise. The standard type of cash-or-nothing option pays \$1, or one other currency unit, or it pays nothing. Hence, it is often called a binary option or a digital option.

The standard European call option can be viewed as a long position in an asset-or-nothing call and a short position in X cash-or-nothing calls where the exercise prices of the asset-or-nothing option and the cash-or-nothing option are both X:

$$S_T > X$$

Long asset-or-nothing call pays S_T.

Short X cash-or-nothing calls pay $-X$.

Total of $S_T - X$

$$S_T \leq X$$

Long asset-or-nothing call pays 0.

Short X cash-or-nothing calls pay -0.

Total of 0

As you can see, this creates a payoff identical to the European call option with strike of X. Similarly, a standard European put can be decomposed into a long position in X cash-or-nothing puts and a short position in an asset-or-nothing put, both options struck at X:

$$S_T > X$$

Short asset-or-nothing call pays 0.

Long X cash-or-nothing calls pay 0.

Total of 0

$$S_T \leq X$$

Short asset-or-nothing call pays $-S_T$.

Long X cash-or-nothing calls pay X.

Total of $X - S_T$

Asset-or-nothing and cash-or-nothing options can be thought of as the building blocks of a standard option. As such, these instruments are like atoms that make up many things with which we are most familiar. By analogy, as hydrogen and oxygen can be combined to make water, so various binary options can be combined to make other financial instruments. As you can probably imagine by now, if these fundamental options trade separately, their combined prices must equal that of a European call in the manner shown in this chapter.

2.13   RECAP AND PREVIEW

In this chapter we identified certain conditions that restrict option prices. For example, we covered upper and lower limits, the relationships between options differing by exercise price or time to expiration, and we derived the relationship between put and call option prices. We also covered the effect of interest rates and volatility on option prices. These results do not identify specific option prices, but they do give us restrictions on option prices and, thereby, enable us to narrow down prices so that we can say more than simply that prices are nonnegative.

In Chapter 3, we shall take a step away from derivatives and provide a review of some of the general mathematical results that we need to recall from previous courses. This review will be foundational material from what is hopefully previously taken math courses. You may actually just need to skim over it.

QUESTIONS AND PROBLEMS

1 In your own words, define *arbitrage*.

2 Based on the following illustration of two instruments (1 and 2) and two points in time (0 and 1), design an arbitrage strategy. Defend your answer by computing cash flows at both points in time.

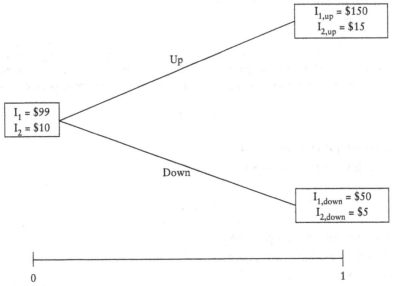

3 This problem, as well as the next one, explores American put lower boundary violations. Consider the following inputs: Suppose the stock price is $S_t = 95$, strike price is $X = 100$, risk-free interest rate is $r_c = 5.0\%$ (continuously compounded, annualized, applies to both maturities—option and dividend), time to maturity is one year, a 4.75 dividend will be paid in one month, and the quoted American put price is 4.95. Identify the arbitrage opportunity and illustrate how it would be captured.

4 Consider the same inputs as the previous problem, but now let the dividend amount increase to 4.9875 per share paid in one month and the quoted put price be 5.04. Again, the stock price is $S_t = 95$, the strike price is $X = 100$, the risk-free interest rate is $r_c = 5.0\%$, and the time to maturity is one year. Identify the arbitrage opportunity and illustrate how it would be captured.

5 This problem as well as the next one explores American put-call parity violations. Consider the following inputs: Suppose the stock price is $S_t = 100$, strike price is $X = 100$, risk-free interest rate is $r_c = 5.0\%$ (continuously compounded, annualized, applies to both maturities—option *and* dividend), time to maturity is six months, a dividend of 1 will be paid in three months, and the quoted American put price is 6.02. If the quoted American call price is 4.98, identify the arbitrage opportunity and illustrate how it would be captured.

6 Again, consider the same inputs as the previous problem except now the American call price is 8.54. Identify the arbitrage opportunity and illustrate how it would be captured.

NOTES

1. There is some debate on whether to use trading days rather than calendar days. In the US, there are approximately 252 trading days in a year.
2. We are also making a subtle assumption that the asset incurs no costs to hold. At one time, many securities had costs to hold and store, owing to the fact that physical certificates were issued, but that is no longer true as securities are now almost exclusively issued in electronic form. Most commodities such as oil and gold would incur storage costs, and these costs would factor into the pricing models, but we shall ignore these until we get to pricing forwards and futures much later in the book.
3. If there are costs involved in doing the trades, the prices will be pushed to near equality, such that transaction costs would discourage further exploitation of any small remaining price difference.
4. We assume continuous compounding and annualized rate (r_c) and express the time to expiration as a fraction of a year (τ). Thus, the present value of 1.0 is $e^{-r_c\tau}$.
5. We described what happens at expiration as involving a transaction in the underlying. Some option contracts are written to provide cash settlement, meaning that instead of actually buying or selling the underlying, the short pays the long a cash settlement, which is equal to the exercise values as given by Equations (2.4) and (2.5).
6. Throughout this book, we assume positive risk-free rates.
7. We skipped calling our portfolios C and D, so that you would not confuse C for a call and because we use D_t as the present value of the dividends paid over the life of the option.
8. Clearly, we could have also assumed no dividends, so the arguments here imply that there are dividends on the underlying.
9. A function is convex if a weighted average of two points is greater than a point in between the two points. Technically the fact that we have an inequality means that the option price function might not be convex at some point, but it will generally be the case that convexity holds. It will fail to hold only in the extreme case when all of the options are very deep-in-the-money or deep out-of-the-money.
10. Of course, we can again make various algebraic arrangements to put any of the instruments in the middle of the two inequalities to put bounds on their prices.

Elementary Review of Mathematics for Finance

The study of financial derivatives and much of financial economics at the graduate level requires a solid foundation in basic mathematical concepts, some of which may seem incredibly simple to certain well-prepared students and to others who are less well prepared not so obvious. In fact, there are entire courses devoted to mathematical economics. One of the highly acclaimed books on this subject is *Fundamental Methods of Mathematical Economics* (Chiang and Wainwright 2005), and the reader is encouraged to get a copy of that book to fill in some details of the material in this chapter.

3.1 SUMMATION NOTATION

The summation symbol is often used in finance. For example, portfolios and stock indexes are additive combinations of their constituent securities. Consider the summation expression,

$$\sum_{i=1}^{n} x_i = x_1 + x_2 + \cdots + x_n. \tag{3.1}$$

The subscript is sometimes called the index. Any variable that appears on the RHS without the index can be brought to the LHS, as in the following examples:[1]

$$\sum_{i=1}^{n} x_i y = y \sum_{i=1}^{n} x_i, \text{ and,} \tag{3.2}$$

$$\sum_{i=1}^{n} x_i a_j = a_j \sum_{i=1}^{n} x_i. \tag{3.3}$$

This step often simplifies the equation considerably. In finance, we sometimes encounter a double summation, as is the case when there are two assets. For example,

$$\sum_{i=1}^{n} \sum_{j=1}^{q} x_i a_j, \tag{3.4}$$

which is like a double loop in computer code. We start with the outer summation, letting $i = 1$. We then sum through the inner summation. Then we increment the index in the outer summation and sum through the inner summation again, continuing in that manner until both summations have been completely covered. Thus, the previous equation amounts to the following expression in expanded form,

$$x_1 a_1 + x_1 a_2 + \cdots + x_1 a_q$$

$$+ x_2 a_1 + x_2 a_2 + \cdots + x_2 a_q$$

$$+ \cdots$$

$$+ x_n a_1 + x_n a_2 + \cdots + x_n a_q.$$

Because we can move the x_i outside the second summation, we could have written this equation as

$$\sum_{i=1}^{n} x_i \sum_{j=1}^{q} a_j = (x_1 + x_2 + \cdots + x_n)(a_1 + a_2 + \cdots + a_q). \tag{3.5}$$

Also note that the following operations are appropriate and often used in finance:

$$\sum_{i=1}^{n} (x_i + y_i) = \sum_{i=1}^{n} x_i + \sum_{i=1}^{n} y_i \text{ and} \tag{3.6}$$

$$\sum_{i=1}^{n} a_j = n a_j. \tag{3.7}$$

For example, most portfolio managers will frequently need to know the values of their portfolios. The manager may own financial instruments in numerous categories, such as domestic bonds, domestic stocks, international bonds, international stocks, financial derivatives, private equity, and so forth. Within each category, the manager may own varying amounts of numerous individual instruments. Based on the principle of value additivity, the portfolio value of each category is found by summing the number of units held for each instrument ($N_{t,j,i}$) times the estimated price of the instrument ($P_{t,j,i}$), where t denotes the point in time, j denotes the category, and i denotes the specific instrument. Thus, each category's value ($C_{t,j}$) can be represented as

$$C_{t,j} = \sum_{i=1}^{N_j} N_{t,j,i} P_{t,j,i}.$$

The overall portfolio value (V_t, uppercase Greek pi) can be expressed as

$$V_t = \sum_{j=1}^{N_C} C_{t,j} = \sum_{j=1}^{N_C} \sum_{i=1}^{N_j} N_{t,j,i} P_{t,j,i}.$$

3.2 PRODUCT NOTATION

Product notation often occurs when linking rates of return. Consider the specification,

$$\prod_{i=1}^{n} x_i = x_1 x_2 \cdots x_n. \tag{3.8}$$

The following operations are appropriate,

$$\prod_{i=1}^{n} x_i y = y^n \prod_{i=1}^{n} x_i. \tag{3.9}$$

Here we simply factored out all the ys, because they are not indexed and can move to the left of the product sign. Note, however, we can write this as y raised to the power of n, inasmuch as y is multiplied n times. That is, the previous expression is $x_1 y x_2 y \ldots x_n y$ so y appears as multiplied by itself n times.

Another permissible operation is

$$\prod_{i=1}^{n} x_i a_j = a_j^n \prod_{i=1}^{n} x_i. \tag{3.10}$$

Even though a is subscripted with j, its value does not change, because there is no j in the product expression. Thus, a_j is simply a constant, just as y was a constant in Equation (3.9).

We can also do the following:

$$\prod_{i=1}^{n} x_i y_i = \prod_{i=1}^{n} x_i \prod_{i=1}^{n} y_i. \tag{3.11}$$

That is, if we write out the expression on the LHS, we have $x_1 y_1 x_2 y_2 \ldots x_n y_n$. This is clearly equal to $x_1 x_2 \ldots x_n y_1 y_2 \ldots y_n$, the expression on the RHS.

Another simplifying operation is

$$\prod_{i=1}^{n} a_j = a_j^n. \tag{3.12}$$

Here the value of a_j does not change, so the same term is simply being multiplied by itself n times.

The rate of return is an important metric in finance and can be measured in many different ways. For this illustration, we assume monthly discrete compounded rates of return based on the portfolio's estimated value at the start (V_{t-1}) and end (V_t) of the month. Thus, the monthly discrete compounded rate of return (R_t^d) is defined as

$$R_t^d = \frac{V_t - V_{t-1}}{V_{t-1}} = \frac{V_t}{V_{t-1}} - 1.$$

Note that we assume the initial value is positive. Many financial derivatives have initial values of zero presenting some challenges addressed later. The monthly total return (TR_t) is based on rearranging the monthly rate of return expression or

$$TR_t = \frac{V_t}{V_{t-1}} = 1 + R_t^d.$$

For example, consider a portfolio's return performance over the past five years measured monthly. Hence, we have 60 observations of the rate of return and the five-year total return is

$$TR_T = \frac{V_T}{V_0},$$

where the subscript in this case is 60. Based on the definition of monthly total return, we observe the following relationship based on the product notation,

$$TR_T = \frac{V_T}{V_0} = \prod_{t=1}^{60} TR_t = \prod_{t=1}^{60} \left(1 + R_t^d\right).$$

Using the properties of products above, one could easily manipulate total returns based on calendar months (January, February, and so forth) or further decompose monthly total returns into daily returns.

3.3 LOGARITHMS AND EXPONENTIALS

The logarithm of a number is an exponent to which a base raised to that power equals the number. There are two primary types of logarithms, the base e, called the *natural* or *Naperian log*, and the base 10, called the *common log*. We write the former as $\log_e$ or sometimes ln, and the latter as $\log_{10}$. In financial applications we nearly always use natural logs, which is primarily because continuously compounded interest is consistent with the natural log function.

By definition, the natural log of a value x is the power to which $e = 2.71828\ldots$ must be raised to equal x. That is, when we say that $\ln x = a$, we mean that $e^a = x$.

The notation we use for the natural log of x is sometimes written as $\ln x$ or $\ln(x)$ and occasionally $\log(x)$. Working with logarithms oftentimes greatly facilitates mathematical operations. The following are typical operations performed with logarithms:

$$\ln(xy) = \ln x + \ln y, \tag{3.13}$$

$$\ln\left(x^a\right) = a \ln x, \tag{3.14}$$

$$\ln(\frac{x}{y}) = \ln x - \ln y, \tag{3.15}$$

$$\ln e = 1, \tag{3.16}$$

$$\ln e^x = x, \text{and} \tag{3.17}$$

$$\ln 1 = 0. \tag{3.18}$$

Be aware, however, that $\ln(x + y) \neq \ln x + \ln y$.

The mathematical definition of e is

$$e = \lim_{n \to \infty} \left(1 + \frac{1}{n}\right)^n. \tag{3.19}$$

The expression e^x, which is

$$e^x = \lim_{n \to \infty} \left(1 + \frac{x}{n}\right)^n, \tag{3.20}$$

and can be approximated as

$$e^x = \lim_{n \to \infty} \left(1 + x + \frac{1}{2!}x^2 + \frac{1}{3!}x^3 + \cdots + \frac{1}{n!}x^n\right) = \sum_{i=0}^{\infty} x^i/i!. \tag{3.21}$$

The expression e^x is sometimes written $\exp(x)$. Some typical operations with the exponential function are as follows:

$$e^x e^y = e^{x+y}, \tag{3.22}$$

$$e^x e^{-y} = e^{x-y}, \text{and} \tag{3.23}$$

$$\left(e^x\right)^n = e^{nx}. \tag{3.24}$$

Rather than compute discrete rates of return, it is often much more useful to compute continuously compounded rates of return (R_t^c). Following the previous illustration, we assume monthly continuous compounding. The monthly continuous rate of return can be expressed as

$$R_t^c = \ln\left(\frac{V_t}{V_{t-1}}\right).$$

Clearly, we once again must assume an initial positive value and no possibility of nonpositive value in the future as $\ln(0)$ is undefined. Unfortunately, many instruments have a significant chance of being nonpositive in the future. The terminal value at time t can be expressed as

$$V_t = V_{t-1}e^{R_t^c}.$$

Thus, the monthly total return with continuous compounding is

$$TR_t = \frac{V_t}{V_{t-1}} = e^{R_t^c}.$$

Again, consider a portfolio's performance over the past five years measured monthly with continuous compounding. Hence, we have 60 observations of the rate of return and the five-year total return is

$$TR_T = \frac{V_T}{V_0} = \prod_{t=1}^{60} TR_t = \prod_{t=1}^{60} e^{R_t^c} = e^{\sum_{t=1}^{60} R_t^c}.$$

Using the previous product and summation properties, one could again easily manipulate total returns based on calendar months (January, February, and so forth) or further decompose monthly total returns into daily returns.

3.4   SERIES FORMULAS

The following formulas for the sums of various finite and infinite series can be useful. These expressions are not easy to memorize and do not show up often in finance, but they can be used on occasion to simplify results, and you should have an accessible list such as the following:

(assumes $|x| < 1$)

$$\sum_{i=0}^{n-1} x^i = \frac{1 - x^n}{1 - x}, \tag{3.25}$$

$$\sum_{i=m}^{n} x^i = \frac{x^m - x^{n+1}}{1 - x}, \tag{3.26}$$

$$\sum_{i=1}^{n} i = \frac{n(n + 1)}{2}, \tag{3.27}$$

$$\sum_{i=1}^{n} i^2 = \frac{n(n + 1)(2n + 1)}{6}, \tag{3.28}$$

$$\sum_{i=1}^{n} x^i = x\left(\frac{1 - x^n}{1 - x}\right), \tag{3.29}$$

$$\sum_{i=1}^{n} ix^i = x\frac{1 - (n + 1)x^n + nx^{n+1}}{(1 - x)^2}, \tag{3.30}$$

$$\sum_{i=1}^{n} i^2 x^i = x\frac{1 + x - (n + 1)^2 x^n + \left(2n^2 + 2n - 1\right)x^{n+1} - n^2 x^{n+2}}{(1 - x)^3}, \tag{3.31}$$

$$\sum_{i=1}^{\infty} ix^i = \frac{x}{(1 - x)^2} \quad \text{for } 0 < x < 1, \tag{3.32}$$

$$\sum_{k=0}^{\infty} ax^k = \frac{a}{1 - x}, \tag{3.33}$$

$$\sum_{i=0}^{\infty} x^i = \frac{1}{1 - x} \text{ assuming } |x| < 1, \tag{3.34}$$

$$\sum_{i=1}^{\infty} x^i = \frac{x}{1 - x} \text{ assuming } |x| < 1, \text{ and} \tag{3.35}$$

$$\sum_{i=1}^{\infty} \frac{x^i}{i} = \ln\left(\frac{1}{1 - x}\right) \text{ assuming } |x| < 1. \tag{3.36}$$

For example, the value of a simple annual coupon-paying bond (V_B) can be expressed as

$$V_B = \sum_{i=1}^{N} \frac{C}{(1+y)^i} + \frac{Par}{(1+y)^N},$$

where C denotes the dollar amount of coupon, Par denotes the principle or par amount of the bond, y denotes the yield to maturity expressed in decimal terms, and N denotes the number of remaining coupons (or years in this case). Rearranging, we have

$$V_B = C \sum_{i=1}^{N} \frac{1}{(1+y)^i} + \frac{Par}{(1+y)^N}.$$

Based on Equation (3.29) and $x = 1/(1+y)$, we have

$$V_B = C \sum_{i=1}^{N} x^i + Par\left(x^N\right) = Cx\left(\frac{1-x^n}{1-x}\right) + Par\left(x^N\right)$$

$$= C\frac{1}{(1+y)}\left(\frac{1 - \frac{1}{(1+y)^N}}{1 - \frac{1}{(1+y)}}\right) + \frac{Par}{(1+y)^N} = \frac{C}{y}\left[1 - \frac{1}{(1+y)^N}\right] + \frac{Par}{(1+y)^N}.$$

This result is much easier to manipulate and express in a spreadsheet or other programming platforms.

3.5 CALCULUS DERIVATIVES

The derivative of a function describes the rate at which the function changes.[2] The derivative is a function that is obtained (hence derived) from the original function. Given a function, $y = f(x)$, the first derivative is denoted as dy/dx or $f'(x)$ and is formally defined as

$$\frac{dy}{dx} = \lim_{\Delta x \to 0} \frac{\Delta y}{\Delta x} = \lim_{\Delta x \to 0} \frac{f(x + \Delta x) - f(x)}{\Delta x}. \tag{3.37}$$

An alternative notation for the first derivative is y_x, but we do not use this notation much in finance because we tend to use subscripts for other purposes, such as an indication of a point in time. Thus, an expression such as y_t, which in math might mean dy/dt, in finance often means the value of y at time t.

In words, the first derivative is the limit of the slope of the line tangent to a function at a specific point. It is the rate of change of the function at that point. The terms Δy and Δx are called *differentials* and the ratio of differentials when $\Delta x \to 0$ is the derivative.[3] For some functions, however, the derivative does not exist at a certain point, such as a derivative that contains division by x when $x = 0$.

The first derivative is a function itself. There are various rules for determining the derivative of a function. The most commonly used ones are summarized in the following subsections.

3.5.1 Derivatives of Common Algebraic Functions

We highlight several algebraic functions that are useful in various finance applications:

$$y = c\ (c \text{ a constant}),$$
$$\frac{dy}{dx} = 0\ (x \text{ is not even in the function}),$$
(3.38)

$$y = cb \text{ where } b = f(x),$$
$$\frac{dy}{dx} = c\frac{db}{dx},$$
(3.39)

$$y = b + v \text{ where } b = f(x) \text{ and } v = g(x),$$
$$\frac{dy}{dx} = \frac{db}{dx} + \frac{dv}{dx},$$
(3.40)

$$y = bv \text{ where } b = f(x) \text{ and } v = g(x),$$
$$\frac{dy}{dx} = b\frac{dv}{dx} + v\frac{db}{dx} \text{ (the product rule)},$$
(3.41)

$$y = \frac{b}{v} \text{ where } b = f(x), v = g(x), \text{ and } v \neq 0,$$
$$\frac{dy}{dx} = \frac{v\frac{db}{dx} - b\frac{dv}{dx}}{v^2} \text{ (the quotient rule)},$$
(3.42)

$$y = a^x,$$
$$\frac{dy}{dx} = a^x \ln a, \text{ and}$$
(3.43)

$$y = b^n \text{ where } b = f(x) \text{ and } n \text{ is independent of } x,$$
$$\frac{dy}{dx} = nb^{n-1}\frac{db}{dx}.$$
(3.44)

Note in the last example the simple case of $y = b^n$ where $b = x$ gives

$$dy/dx = nb^{n-1}(dx/dx) = nx^{n-1}.$$

To illustrate several of these results, recall the value of a simple bond can be expressed as

$$V_B = \frac{C}{y}\left[1 - \frac{1}{(1+y)^N}\right] + \frac{Par}{(1+y)^N},$$

where C denotes the coupon amount, y denotes the periodic yield to maturity, N denotes the number of remaining payments, and *Par* denotes the bond's par value. The first derivative of the bond value with respect to yield to maturity is

$$\frac{dV_B}{dy} = \frac{d}{dy}\left\{\frac{C}{y}\left[1 - \frac{1}{(1+y)^N}\right] + \frac{Par}{(1+y)^N}\right\}. \qquad (3.45)$$

Based on the following definitions,

$$f(y) = \frac{C}{y} = Cy^{-1},$$

$$g(y) = 1 - \frac{1}{(1+y)^N} = 1 - (1+y)^{-N}, \text{ and}$$

$$h(y) = \frac{Par}{(1+y)^N} = Par(1+y)^{-N}.$$

Thus, based on the product rule [Equation (3.41)] and Equation (3.45), we have

$$\frac{dV_B}{dy} = \frac{d}{dy}[f(y)g(y) + h(y)] = f\frac{dg}{dy} + g\frac{df}{dy} + \frac{dh}{dy}.$$

Based on Equations (3.39) and (3.45), we have

$$\frac{df}{dy} = -Cy^{-2} = -\frac{C}{y^2}.$$

Based on Equations (3.35) and (3.45), we have

$$\frac{dg}{dy} = N(1+y)^{-(N+1)}.$$

Based on Equations (3.44) and (3.45), we have

$$\frac{dh}{dy} = -NPar(1+y)^{-(N+1)}.$$

Substituting we have

$$\frac{dV_B}{dy} = f\frac{dg}{dy} + g\frac{df}{dy} + \frac{dh}{dy} = \frac{C}{y}N(1+y)^{-(N+1)} + \left[1 - \frac{1}{(1+y)^N}\right]\left(-\frac{C}{y^2}\right)$$
$$+ \left[-NPar(1+y)^{-(N+1)}\right].$$

Rearranging we have the following well-known result,

$$\frac{dV_B}{dy} = \frac{C}{y}\left\{\frac{N}{(1+y)^{N+1}} - \frac{1}{y}\left[1 - \frac{1}{(1+y)^N}\right]\right\} - \frac{NPar}{(1+y)^{N+1}}.$$

Sensitivities such as these have proven very useful in various risk management tasks, particularly related to managing bond portfolios. We leave one specific application related to bond duration and convexity to an end-of-chapter problem.

3.5.2 Derivatives of Logarithmic Functions

We review selected logarithmic functions that are useful in various finance applications:

$$y = \log_a b \text{ where } b = f(x),$$

$$\frac{dy}{dx} = \frac{\log_a e}{b}\frac{db}{dx}, \tag{3.46}$$

$$y = \ln b,$$

$$\frac{dy}{dx} = \frac{1}{b}\frac{db}{dx}, \text{ and} \tag{3.47}$$

$$y = \ln x,$$

$$\frac{dy}{dx} = \frac{1}{x}. \tag{3.48}$$

Several of these results are particularly useful when deriving various risk measures related to option valuation models.

3.5.3 Derivatives of Exponential Functions

We review selected exponential functions that are useful in various finance applications:

$$y = a^b \text{ where } b = f(x),$$

$$\frac{dy}{dx} = a^b \ln a \frac{db}{dx}, \tag{3.49}$$

$$y = e^b \text{ where } b = f(x),$$

$$\frac{dy}{dx} = e^b \frac{db}{dx}, \text{ and} \tag{3.50}$$

$$y = b^v \text{ where } b = f(x) \text{ and } v = g(x),$$

$$\frac{dy}{dx} = vb^{v-1}\frac{db}{dx} + b^v \ln b \frac{dv}{dx}. \tag{3.51}$$

Again, several of these results are particularly useful when deriving various risk measures related to option valuation models. We illustrate here an application related to coupon-bearing bonds. Based on continuously compounded discount rates, the value of a bond can be expressed as

$$V_B = C \sum_{i=1}^{N} e^{-y(i)} + Par\left(e^{-y(N)}\right).$$

Thus, the first derivative of bond value with respect to the continuously compounded yield to maturity is

$$\frac{dV_B}{dy} = C \sum_{i=1}^{N} (-i)\, e^{-y(i)} + (-N)\, Par\left(e^{-y(N)}\right) = -C \sum_{i=1}^{N} i e^{-y(i)} - (N)\, Par\left(e^{-y(N)}\right).$$

Clearly, as the yield increases, the bond value decreases.

3.5.4 Derivatives of Trigonometric Functions

Although trigonometric functions rarely make their way into finance applications, it does happen occasionally, particularly when estimating functions, such as the yield curve, or when deploying econometric models. We show here selected derivatives of trigonometric functions:

$$y = \sin b \text{ where } b = f(x),$$

$$\frac{dy}{dx} = \cos b \frac{db}{dx}, \tag{3.52}$$

$$y = \cos b \text{ where } b = f(x),$$

$$\frac{dy}{dx} = -\sin b \frac{db}{dx}, \text{ and} \tag{3.53}$$

$$y = \tan b \text{ where } b = f(x),$$

$$\frac{dy}{dx} = \sec^2 b \frac{db}{dx}. \tag{3.54}$$

One application is modeling calendar seasonality with trigonometric functions.

3.5.5 Derivatives of Inverse Functions

The derivative of an inverse functions can be expressed as

$$y = f(x) \text{ and } x = g(y) \text{ are inverse functions,}$$

$$\frac{dy}{dx} = \frac{1}{\frac{dx}{dy}}. \tag{3.55}$$

For example, based on periodic compounding, the future value (FV) of some present value (PV) can be expressed as a function of the annualized yield to maturity (y) or

$$FV = PV\left(1 + \frac{y}{m}\right)^N,$$

where m denotes the number of compounding periods per year, and N denotes time to maturity expressed in compounding periods. Alternatively, we can express the annualized yield to maturity as a function of the future value in the following form,

$$y = m\left[e^{\frac{\ln\left(\frac{FV}{PV}\right)}{N}} - 1\right].$$

Thus, $y = f(FV)$ and $FV = g(y)$. Thus, based on the previous inverse equation, we note

$$\frac{dy}{dFV} = \frac{1}{\frac{dFV}{dy}},$$

and

$$\frac{dFV}{dy} = \frac{d}{dy}\left[PV\left(1 + \frac{y}{m}\right)^N\right] = \frac{(N)\,PV}{m}\left(1 + \frac{y}{m}\right)^{N-1}.$$

Therefore,

$$\frac{dy}{dFV} = \frac{1}{\frac{dFV}{dy}} = \frac{1}{\frac{(N)PV}{m}\left(1 + \frac{y}{m}\right)^{N-1}}.$$

At times, the derivative of an inverse is easier to obtain than the derivative directly. For example, the direct approach is a bit more tedious as shown here:

$$\frac{dy}{dx} = \frac{d}{dFV}\left\{m\left[e^{\frac{\ln\left(\frac{FV}{PV}\right)}{N}} - 1\right]\right\} = m\frac{d}{dFV}\left[e^{\frac{\ln\left(\frac{FV}{PV}\right)}{N}}\right]$$

$$= \frac{m}{(N)\,FV}e^{\frac{\ln\left(\frac{FV}{PV}\right)}{N}} = \frac{m}{(N)\,FV}e^{\frac{N\ln\left(1+\frac{y}{m}\right)}{N}}$$

$$= \frac{m}{(N)\,PV\left(1 + \frac{y}{m}\right)^N}\left(1 + \frac{y}{m}\right) = \frac{m}{(N)\,PV\left(1 + \frac{y}{m}\right)^{N-1}}.$$

3.5.8 The Chain Rule

If we have a function $y = f(x)$ and x is also a function such as $x = f(g)$, then we take the derivative dy/dg in the following manner:

$$\frac{dy}{dg} = \left(\frac{dy}{dx}\right)\left(\frac{dx}{dg}\right), \tag{3.56}$$

or in other words, as the product of two derivatives. This is called the *chain rule*.

For example, manipulating the Black-Scholes-Merton option pricing model often requires use of the chain rule. For example, the cumulative standard normal distribution can be expressed as

$$y = N\left(d_1\right) = \int_{-\infty}^{d_1} \frac{e^{-\frac{x^2}{2}}}{\sqrt{2\pi}}dx.$$

The value of d_1 is

$$x = d_1 \equiv \frac{\ln\left(\frac{S}{X}\right) + \left(r_c + \frac{\sigma^2}{2}\right)\tau}{\sigma\sqrt{\tau}},$$

where S denotes the current stock price, X denotes the strike price, r_c denotes the annualized, continuously compounded, risk-free interest rate, σ denotes the volatility, and τ denotes the time to expiration. Therefore, the first derivative of $N\left(d_1\right)$, the normal probability based on d_1, with respect to the stock price can be worked out based on the chain rule or

$$\frac{dN\left(d_1\right)}{dS} = \left(\frac{dN\left(d_1\right)}{dd_1}\right)\left(\frac{dd_1}{dS}\right) = \frac{e^{-\frac{d_1^2}{2}}}{S_t\sigma\sqrt{2\pi\tau}}.$$

3.5.7 Higher-Order Derivatives

These are examples of the first derivative, the rate of change of the function at the indicated point. There is also a second derivative, which is the rate of change of the first derivative. For example, consider $y = f(x)$ and its first derivative, dy/dx. Its second derivative is written as follows:

$$\frac{d\left(\frac{dy}{dx}\right)}{dx} = \frac{d^2y}{dx^2}, \tag{3.57}$$

and sometimes as $f''(x)$. The second derivative provides further information, beyond that of the first derivative, about the characteristics of the function. For example, a positive first derivative and positive second derivative describes a function that is upward sloping with the slope increasing at an increasing rate. A positive first derivative and negative second derivative describes a function that is increasing at a decreasing rate. A negative first derivative and positive second derivative is a function that is decreasing at a decreasing rate. A negative first derivative and negative second derivative is a function that is decreasing at an increasing rate. A zero first derivative and positive second derivative is a function that is at a minimum, whereas a zero first derivative and negative second derivative is a function that is at a maximum. There are also third- and higher-order derivatives, but we rarely need them in finance.

Again, manipulating the Black-Scholes-Merton option pricing model often requires use of the higher-order derivatives. For example, the second derivative of the call option price with respect to the underlying instrument is known as *gamma* and, though rarely used, the third derivative of the call option price with respect to the underlying instrument is known as *speed*.

3.5.8 Partial Derivatives

A function containing more than one variable that can be differentiated with respect to one of the variables by treating the other variables as constants is a derivative called a *partial derivative*. Consider $y = f(x, z)$ where both x and z are variables. Then the partial derivatives are written as $\partial y/\partial x$ and $\partial y/\partial z$, and they represent the change in y for a change in either x or z, while treating the other variable as a constant.

The rules for taking partial derivatives are the same as the rules for taking ordinary derivatives. For example, consider the function $y = 2x^3 + 4z^2 + c$ where c is a constant. Then

$$\frac{\partial y}{\partial x} = 6x^2$$

$$\frac{\partial y}{\partial z} = 8z.$$

Likewise, there are second partial derivatives. Hence,

$$\frac{\partial\left(\frac{\partial y}{\partial x}\right)}{\partial x} = \frac{\partial^2 y}{\partial x^2} = 12x$$

$$\frac{\partial\left(\frac{\partial y}{\partial z}\right)}{\partial z} = \frac{\partial^2 y}{\partial z^2} = 8.$$

It is common in mathematical notation that given a function $y(x,z)$, we use an expression such as y_1 to be the partial derivative of y with respect to the first variable indicated in the expression, that is, $\partial y/\partial x$, and y_2 to be the partial derivative with respect to the second variable, that is, $\partial y/\partial z$. Likewise y_{11} is $\partial^2 y/\partial x^2$ and y_{22} is $\partial^2 y/\partial z^2$. We do not use this notation much in finance, however, owing to the use of subscripts as other indications such as a point in time or the identity of a variable.

Consider the following function: $y = 4x^3 z^2$. The partial derivatives are $\partial y/\partial x = 12x^2 z^2$ and $\partial y/\partial z = 8x^3 z$. You can also take the partial derivative of each of these derivatives with respect to the other variable. In other words,

$$\frac{\partial\left(\frac{\partial y}{\partial x}\right)}{\partial z} = \frac{\partial^2 y}{\partial x \partial z} = 24x^2 z$$

$$\frac{\partial\left(\frac{\partial y}{\partial z}\right)}{\partial x} = \frac{\partial^2 y}{\partial z \partial x} = 24x^2 z.$$

Note that they are the same. It does not matter which order you do the differentiation.

In mathematical notation these derivatives, sometimes called *cross-partial derivatives*, often are indicated with such symbols as y_{12} and y_{21}, but as noted as shown previously we do not use this notation much in finance.

3.5.9 Taylor's Theorem

Taylor's theorem appears a surprising number of times in finance. Consider a function, $f(x)$, observed at two points, x and $x + \Delta x$. In other words, the function is $f(x)$ at point x and $f(x + \Delta x)$ at point $x + \Delta x$. *Taylor's theorem* states that we can take the value of the function at $f(x)$ and add a series of terms to obtain its value at $f(x + \Delta x)$ in the following manner:

$$f(x + \Delta x) = f(x) + \frac{f'(x)}{1!}\Delta x + \frac{f''(x)}{2!}\Delta x^2 + \frac{f'''(x)}{3!}\Delta x^3 + \cdots$$

$$f(x + \Delta x) - f(x) = \frac{f'(x)}{1}\Delta x + \frac{f''(x)}{2}\Delta x^2 + \frac{f'''(x)}{6}\Delta x^3 + \cdots \qquad (3.58)$$

$$\Delta f(x) = f'(x)\Delta x + \frac{f''(x)}{2}\Delta x^2 + \frac{f'''(x)}{6}\Delta x^3 + \cdots.$$

This operation is also called a *Taylor series expansion*, which is useful when we need to study the properties of the rate of change of a function, rather than its specific value at a point. Although derivatives do the same thing, they represent the change in a function when there is an infinitesimal change in a variable. Taylor series expansions are useful when the variable change is not infinitesimal, as is often the case in finance.

If the function has more than one variable, we can apply Taylor's Theorem to a given variable at a time, holding the other variables constant. The derivatives expressed are then partial derivatives. Alternatively, we could do a Taylor series expansion for more than one variable. For example, let $y = f(x, z)$. Then a Taylor series expansion for both x and z would be written as

$$\Delta y = \frac{\partial y}{\partial x}\Delta x + \frac{\partial y}{\partial z}\Delta z$$

$$+ \frac{1}{2}\frac{\partial^2 y}{\partial x^2}\Delta x^2 + \frac{1}{2}\frac{\partial^2 y}{\partial z^2}\Delta z^2 + \frac{\partial^2 y}{\partial x \partial z}\Delta x \Delta z$$

$$+ \cdots, \tag{3.59}$$

where the additional terms would represent higher-order partial derivatives and all possible cross-partial combinations.

A Taylor series expansion is exact only in the limit, that is, where the number of terms on the RHS is infinite. In applications we often do a Taylor series expansion to the second order, that is, where terms after the second derivative are dropped, and then change the equals sign to the approximation sign, which is $\approx$.

The Taylor series approximation is a widely used tool in financial analysis. For example, various risk measures related to bonds, such as duration and convexity, are derived based on this approximation.

3.5.10 Total Differential

Consider a function $y = f(x, z)$. Suppose both x and z change. If we want to know by how much the function changes, we can obtain the answer using a function called the *total differential*, which is

$$dy = \frac{\partial y}{\partial x}dx + \frac{\partial y}{\partial z}dz. \tag{3.60}$$

This expression essentially says that the change in y is the change in x multiplied by the rate at which y changes if x changes plus the change in z multiplied by the rate at which y changes if z changes. It is exact only if the changes in x and z are very small.

Note that the total differential looks somewhat like a first-order Taylor series expansion. A first-order Taylor series expansion, however, is just an approximation. The total differential is an exact expression. The distinction lies in the fact that the Taylor series expansion applies to discrete changes in x and z (Δx and Δz) and uses partial derivatives to approximate the discrete change in y (Δy). If we let Δx and Δz be infinitesimal changes, we denote them as dx and dz. Then we restate the Taylor series expansion for the two variables that we presented previously as

$$dy = \frac{\partial y}{\partial x}dx + \frac{\partial y}{\partial z}dz$$

$$+ \frac{1}{2}\frac{\partial^2 y}{\partial x^2}dx^2 + \frac{1}{2}\frac{\partial^2 y}{\partial z^2}dz^2 + \frac{\partial^2 y}{\partial x \partial z}dxdz$$

$$+ \cdots$$

$$= \frac{\partial y}{\partial x}dx + \frac{\partial y}{\partial z}dz, \tag{3.61}$$

which is the total differential and which holds by virtue of the fact that any product such as $dxdx = dx^2$ or $dxdz$ is zero because dx is a differential and is defined as an infinitesimally small value, which when squared moves it closer to zero. In general, $dx^k \to 0$ if $k > 1$. Interestingly, this result is used a great deal in option pricing as we shall see in later chapters.

3.5.11 Total Derivative

As previously noted, a derivative is a ratio of differentials. Working with the total differential, we can divide by one of the variables and obtain the *total derivative* with respect to that variable. In other words,

$$\frac{dy}{dx} = \frac{\partial y}{\partial x} + \frac{\partial y}{\partial z}\frac{dz}{dx}$$

$$\frac{dy}{dz} = \frac{\partial y}{\partial x}\frac{dx}{dz} + \frac{\partial y}{\partial z}.$$

(3.62)

3.6 INTEGRATION

Integration is closely related to differentiation, but people usually find it to be a much more difficult concept to grasp. There are two general ways to classify an integral. One, called the *indefinite integral*, is the opposite of differentiation and is sometimes called the *antiderivative*. Given a derivative, indefinite integration attempts to find the function that when differentiated obtains the given derivative. For example, suppose we were given the expression $12x$. The indefinite integral is $6x^2 + a$ where a is an unknown constant. This specification is written as

$$\int 12x = 6x^2 + a,$$

which is true because if $y = 6x^2$, then $dy/dx = 12x$. But what if $y = 6x^2$ plus some constant c. Then again, $dy/dx = 12x$, so in finding the integral, we must allow for an indefinite constant, which is why it is called the *indefinite* integral. It is not precise.

The other interpretation of integration is as the *definite integral*, which is the area under a curve between two points on the x-axis. For example, suppose we have a function $f(x)$ and wish to know the area under the curve between the points where $x = j$ and $x = k$. We write this as

$$\int_j^k f(x)dx = F(k) - F(j),$$

(3.63)

where the F function is obtained upon integration, which we shall demonstrate next. The definite integral is defined specifically as the limit of the sum of an infinite series of rectangles drawn under the curve as follows,

$$\lim_{n\to\infty}\sum_{i=1}^n f(x_i)\,\Delta x_i = \int_j^k f(x)dx,$$

where the curve between $x = j$ and $x = k$ has been partitioned into n rectangles.[4] Note that the expression $f(x_i)\,\Delta x_i$ is the area of a rectangle with a base of length Δx_i such that

$\Delta x_i = x_i - x_{i-1}$ and height of $f(x_i)$. The expression $f(x_i)$ is simply the y value associated with a value of x of x_i. The value x_1 is $x_1 = j$ and the value x_n is $x_n = k$ and the range of x from $x_1 = j$ to $x_n = k$ has been partitioned into an infinite number of subranges. This definition is called a *Riemann integral* (pronounced Ree-mahn).

The actual process of finding the area under the curve involves determining the integral and evaluating it at the end points. We shall do that later. First let us review the major formulas and rules for integration. For indefinite integrals, the major ones are

$$\int x^n dx = \frac{1}{n+1} x^{n+1} + a, \tag{3.64}$$

$$\int e^x dx = e^x + a \quad (\text{Note}: \text{Because } de^x/dx = e^x), \tag{3.65}$$

$$\int \frac{1}{x} dx = \ln x + a, \tag{3.66}$$

$$\int [f(x) + g(x)] \, dx = \int f(x) dx + \int g(x) dx, \tag{3.67}$$

$$\int a f(x) dx = a \int f(x) dx, \tag{3.68}$$

$$\int f(b) \frac{db}{dx} dx = \int f(b) db, \text{and} \tag{3.69}$$

$$\int v \, db = bv - \int b \, dv \quad (\text{integration by parts}). \tag{3.70}$$

Tables have been constructed to provide the integrals for thousands of functions, though not all functions can be integrated.

Definite integration uses these same rules to determine the integrals but without the addition of the constant. In addition, the following rules are useful tricks in definite integration:

$$\int_j^k f(x) dx = - \int_k^j f(x) dx, \tag{3.71}$$

$$\int_j^j f(x) dx = 0, \tag{3.72}$$

$$\int_i^k f(x) dx = \int_i^j f(x) dx + \int_j^k f(x) dx, \quad i \leq j \leq k, \tag{3.73}$$

$$\int_i^j -f(x) dx = - \int_i^j f(x) dx, \tag{3.74}$$

$$\int_i^j a f(x) dx = a \int_i^j f(x) dx \quad \text{where } a \text{ is a constant, and} \tag{3.75}$$

$$\int\limits_i^j \left[f(x) + g(x) \right] dx = \int\limits_i^j f(x)dx + \int\limits_i^j g(x)dx. \tag{3.76}$$

Now let us illustrate an example of finding the area under the curve. Suppose we have the following problem:

$$\int\limits_i^j 12x\,dx.$$

The value is computed by determining the function that represents the integral of $12x$. This would be $F(x) = 6x^2$. Then we determine $F(b) - F(a)$. The process is written in the following manner:

$$\int\limits_i^j 12x\,dx = 6x^2 \Big|_a^b = 6b^2 - 6a^2.$$

Hence, given whatever choices we make of j and i (here $j = b$ and $i = a$), the area under the curve is easily computed in this manner.

3.7   DIFFERENTIAL EQUATIONS

A differential equation is an equation that contains a derivative. The objective of solving a differential equation is to determine the original function whose derivative is given by the differential equation. Differential equations that contain only ordinary derivatives are called *ordinary differential equations*. Differential equations that contain partial derivatives are called *partial differential equations*.

Taking a derivative creates a differential equation. For example, the expression

$$\frac{dy}{dx} = 3qx^4$$

is a differential equation. Oftentimes it is written as

$$dy = 3qx^4 dx.$$

The objective is to "solve" the differential equation, meaning to find the original function whose derivative is the differential equation. Solving differential equations can be very difficult. The one just shown is quite simple: We can use indefinite integration to obtain the following:

$$\int dy = \int 3qx^4 dx = 3q \int x^4 dx$$

$$y = \left(\frac{3q}{5} \right) x^5 + a.$$

Note, however, that without knowing the value of *a* we cannot be very specific about the solution, which can vary widely depending on the value of *a*. To determine a more precise solution, we often impose one or more conditions on the value of *q*, which are called *initial conditions* or *boundary conditions* and represent values we know. For example, if we know that at $x = 0$, the function value is 50, then we can substitute zero for *x* and obtain $50 = (3q/5)0^2 + a$ so $a = 50$. Or if we know that at $x = 100$, $y = 5{,}200$, then we know that $5{,}200 = (3q/5)100^2 + a$, giving us a value for *a* in terms of *q*.

Differential equations that contain partial derivatives are naturally called *partial differential equations* or PDEs. They are usually much harder to solve than are ordinary differential equations. Much of the process of solving differential equations involves classifying the equation into a given category of differential equations and then following known rules, hints, and suggestions for solutions of equations in that category. We will also sometimes exploit the fact that we have an idea of what the solution might look like. One approach to pricing options relies on solving a PDE.

3.8 RECAP AND PREVIEW

In this chapter we reviewed the rules of summations, products, logarithms, exponentials, and series formulas. In addition, we covered calculus derivatives and integrations, as well as differential equations. This material is strictly for the purposes of refreshing our knowledge from math courses that would have been previously taken.

In Chapter 4, we move on to another review, which covers concepts in probability.

QUESTIONS AND PROBLEMS

1 The following table provides a summary of a particular portfolio at year end. Demonstrate that the value of the overall portfolio can be found two ways: (1) finding the sector's value and then aggregating to the overall portfolio and (2) aggregating each position in the overall portfolio.

Instrument ID	Sector	Number of Shares	Price per Share
ABC147MZO	Technology	100	56.00
GIM256AEF	Technology	300	27.00
BRC829GFP	Technology	200	33.00
FMC061DEC	Technology	500	17.00
MRF122NAJ	Financial services	1,000	24.00
BRO010CHN	Financial services	600	54.00
NOD332BER	Financial services	300	61.00
CNC522OKS	Consumer durables	1,400	77.00
ZAF501ZTY	Consumer durables	800	27.00

2 Because the dividend discount model assumes the value of a common stock (V_S) is simply the present value at a common discretely compounded discount rate (k_{dc}) and expected annual dividend payments (D_i), we have an infinite series or mathematically,

$$V_S = \frac{D_1}{(1+k_{dc})} + \frac{D_2}{(1+k_{dc})^2} + \frac{D_3}{(1+k_{dc})^3} + \cdots = \sum_{i=1}^{\infty} \frac{D_i}{(1+k_{dc})^i}.$$

Let us assume a single growth rate (g), annual dividend payments, and apply the growth rate solely on a dividend payment date. Assuming annual discrete compounding, we can express the stock value, based on this model, as

$$V_S = \sum_{i=1}^{\infty} \frac{D_0(1+g_{dc})^i}{(1+k_{dc})^i}.$$

Assuming a growth rate less than the discount rate and based on the infinite series results presented in this chapter, express the stock value in closed form. Specifically, prove that the stock value can also be expressed as

$$V_S = \frac{D_0(1+g_{dc})}{(k_{dc}-g_{dc})}.$$

3 Derive the first derivative of the stock value with respect to the discount rate as well as the derivative of the stock value with respect to the growth rate when the stock value can be expressed as

$$V_S = \frac{D_0(1+g_{dc})}{(k_{dc}-g_{dc})}.$$

Assuming $D_0 = 1.0$, $g_{dc} = 5.0\%$, and $k_{dc} = 10\%$, compute the stock value as well as both derivatives. Justify your derivative solutions by recalculating the derivative assuming a one basis point increase in the growth rate and then a one basis point increase in the discount rate. (Note that a basis point is 0.0001, or 1/100%. For example, if a rate goes from 6% to 6.03%, it has gone up three basis points.)

4 US Treasury bonds yields (y) are quoted based on a semiannual compounding basis where the fraction (f) of the coupon period that has elapsed (measured in portion of the period) is incorporated. That is,

$$V_B = \sum_{i=1}^{N} \frac{\left(\frac{Coupon}{m}\right)Par}{\left(1+\frac{y}{m}\right)^{i-f}} + \frac{Par}{\left(1+\frac{y}{m}\right)^{N-f}},$$

where *Coupon* denotes the stated annual coupon percentage rate, m denotes the payment frequency (for US treasury bonds, $m = 2$), *Par* denotes the bond dollar par value, and N denotes the number of remaining coupon payments. Derive the closed-form expression for the bond value.

5 Modified duration and standard convexity are common bond-related risk measures. Mathematically, they are defined as follows (using the notation in the previous question):

$$ModDur_B \equiv -\frac{dV_B/V_B}{dy} = -\frac{dV_B}{dy}\frac{1}{V_B}, \text{and}$$

$$Convexity_B \equiv \frac{1}{V_B}\frac{d^2V_B}{dy^2}.$$

Based on the properties of the bond valuation equation in the previous problem, derive the modified duration and standard convexity expressions. Illustrate the solution in both series format and closed form.

6 One approach to establishing the theoretical price of a futures contract is based on the carry model. The carry model asserts the futures price that expires at time T (F_T) is simply the spot price at time t (S_t) grossed up by the risk-free interest rate (r_c) and other carry costs (cc) over the time to expiration $\tau = T - t$ or $F_T = S_t e^{(r_c+cc)\tau}$. For many risk management applications, we are interested in the first derivative, known as *delta*, and second derivative, known as *gamma*, with respect to the spot price. Solve for these values as well as interpret them.

NOTES

1. Recall LHS and RHS denotes the left-hand side and right-hand side of equations.
2. These derivatives are referred to as *calculus derivatives*, which is not typical terminology in math classes. In finance, however, options, forwards, futures, and swaps are referred to as *derivative instruments* inasmuch as they derive their values from an underlying asset or rate. Hence, we are compelled to make a distinction.
3. Note that $\Delta x \rightarrow 0$ is read, "Δx tends to zero."
4. The integral symbol, the elongated S, literally stands for sum. Definite integration is the sum over an infinite number of rectangles under the curve between the endpoints of interest.

CHAPTER 4

Elementary Review of Probability for Finance

People with an interest in finance are frequently required to apply concepts from probability and statistics. Corporate cash flows, exchange rates, commodity prices, stock prices, and interest rates are all random variables that can be usefully modeled by using probability and statistics. Unfortunately, you never have the luxury of observing the actual, objective probability distribution in the social science field of finance. The closest alternative is your perceived, subjective probability distribution. Historical frequencies of an event are never an exact representation of future outcome. Thus, we must be cognizant of the well-known disclaimer, "Please remember that past performance may not be indicative of future results." Market participants learn from the past and make behavioral adjustments that influence future observations.

This chapter provides a brief introductory review of the key foundational concepts in probability that have proven useful in financial analysis. More depth is found in almost any good book on probability.[1]

4.1   MARGINAL, CONDITIONAL, AND JOINT PROBABILITIES

Probability is a measure, meaning a numerical description or quantification, of the relative frequency of an event. The probability of observing heads in one toss of a fair coin is obviously 1/2. There are two possibilities and one fits the definition of the desired event. Technically this concept is called the *marginal probability*, which distinguishes it from the *conditional probability*, which is the probability that something will happen, given that something else has already happened. Finally, we have the concept of *joint probability*, which is the probability that two or more events will happen.

To illustrate these concepts, consider the following information. A sample of 100 people, 55 female and 45 male, is collected and examined for the frequency of brown hair. Of the 55 females, 31 have brown hair and 24 have some other color. Of the 45 males, 28 have brown hair and 17 have some other color. If this is a reliable sample, what is the probability that a person selected at random will have brown hair? This is the unconditional probability, sometimes called the marginal probability, and it is 0.59, given that 59 out of 100 have brown hair. Note that we did not condition on whether the person is male or female.

Now suppose you know that you selected a male. Then what would be the probability that the subject has brown hair? This concept is the conditional probability, specifically the probability that the subject would have brown hair, given that it is a male. Then the answer would be 28/45 = 0.62 because 62% of the males have brown hair.

Now suppose you selected someone at random and you wanted to know the probability that it would be a female with a hair color other than brown. Out of 100 subjects, 24 are females with a hair color other than brown. So, on that basis alone, we know that the joint probability would be 0.24. The joint probability is equal to the conditional probability times the marginal probability of the condition. In our example this means that we want the probability that the subject will be a female with a hair color other than brown and this will be the probability that the subject will be female, given that you know the subject does not have brown hair, times the probability that the subject does not have brown hair. The probability that the subject will be female, given that we know that the subject does not have brown hair is 24/41 because there are 41 people with a hair color other than brown and 24 are female. This is the conditional probability. The probability that the subject does not have brown hair is 41/100 because you know that out of 100 people, 41 do not have brown hair. This is the marginal probability. Multiplying the conditional probability times the marginal probability gives us (24/41)(41/100) = 24/100.

Alternatively, we could have found the conditional probability that the subject will not have brown hair given that we know it is a female, which will be 24/55, times the probability that the subject is female, 55/100, giving us the correct answer of 24/100.

Now, let us consider another simple event, such as the probability of a head on one toss of a coin being 1/2. Now suppose we toss two coins. What is the probability of a head? Now we have to define more precisely what we mean. Do we mean one head, two heads, or at least one head? We also have to be assured that the coins are independent, for if not, it does affect our answer. Let us focus on the probability of exactly one head. Assuming the two coins are independent, consider the probability of one head as restricted to the probability that we toss a head on coin one and a tail on coin two, which is (1/2)(1/2) = 1/4. This is what we mean by the joint probability with independent events. There are two ways to toss one head, however: a head on the first and a tail on the second or a tail on the first and a head on the second. So, the probability of exactly one head is really 1/4 + 1/4 = 1/2. If we want the probability of two heads, then we must toss a head on both coins. Because the events are independent, the joint probability is the product of the marginal probabilities. Thus, the probability of two heads is (1/2)(1/2) = 1/4. If we want the probability of at least one head, we require the probability of exactly one head plus the probability of two heads, which is 1/2 + 1/4 = 3/4. Interestingly, there is another way of solving the "at least one" case. Its probability is one minus the probability of no heads. The probability of no heads is (1/2)(1/2) = 1/4. So, the probability of at least one head is 1 − 1/4 = 3/4.

We can easily verify these results by summarizing all possible outcomes:

- Head on first coin, head on second coin
- Head on first coin, tail on second coin
- Tail on first coin, head on second coin
- Tail on first coin, tail on second coin

These are equally likely events so each has a probability of 1/4. The outcome of one head is represented by the second and third events so the probability is 1/4 + 1/4 = 1/2. For two heads, there is just one way to do it, the first event. So that probability is 1/4. If we want the probability of at least one head, we include the first three events, so that probability is 1/4 + 1/4 + 1/4 = 3/4. Alternatively, the probability of at least one head can be restated as 1 minus the probability of no heads. There is only one event here with no heads, the fourth event, so the answer is 1 − 1/4 = 3/4 or alternatively, 1 − the probability of a tail on both tosses, which is 1 − 1/4 = 3/4.

The probability of two events occurring, as noted, is called the *joint probability*. Defining the events as *A* and *B* and denoting probability as Pr, we state the joint probability for independent events in the following manner:

$$Pr(A \text{ and } B) = Pr(A)Pr(B), \text{ given independence of } A \text{ and } B.$$

For events that are dependent, we must make some modifications. First, let us ask what we mean by independence and dependence. The coin tosses are clearly independent. The outcome of one coin is unrelated to the outcome of the other. We can, however, map the coin tosses into values that are not independent. Consider the following experiment. Toss a coin one time. If heads occur, record the value +1; if tails occurs, record the value −1. Then toss the coin again. Once again record +1 or −1 and add it to the amount recorded after the first coin toss. We are interested in the probabilities of the numerical outcomes. Here are the equally likely possibilities:

- Head on first coin, head on second coin: sum = +2
- Head on first coin, tail on second coin: sum = 0
- Tail on first coin, head on second coin: sum = 0
- Tail on first coin, tail on second coin: sum = −2

Thus, the probability distribution is as follows:

Outcome	Probability
+2	1/4
0	1/2
−2	1/4

These are the marginal probabilities.[2] We might also want to know the conditional probabilities. For example, we might wish to know the probabilities of +2, 0, and −2 given that a head occurred on the first toss. These probabilities are as follows:

Outcome	Probability[3]
+2	1/2
0	1/2
−2	0

If a tail occurs on the first toss, the probabilities are as follows:[4]

Outcome	Probability
+2	0
0	1/2
−2	1/2

The probabilities of +2 and −2 outcomes are quite different if a head or tail has occurred on the first toss. This is what we mean by the *conditional probability*. The events, defined specifically as the sum of the +1 or −1 assigned to the head or tail tossed, are not independent. The occurrence of a head or a tail, however, is independent from one toss to the other. It is interesting to note how the same sample experiment can give rise to a quite different set of probabilities for different events.

In general, we have the following result, where *&* denotes *and*:

$$\Pr(A \& B) = \Pr(B|A)\Pr(A) = \Pr(A|B)\Pr(B).$$

Note that Pr(.) stands for the probability of the event(s) in the parentheses. When a vertical line appears in parentheses it means conditional on the event following the vertical line. To cast our event in this context, think of the two tosses as two coins, coin1 and coin2. The outcome of a +2 can occur only in the joint condition that both coins come up heads. The probability of this occurring is obtained from the conditional probabilities as follows: the probability of a head on the second coin conditional on a head on the first coin (½) times the probability of a head on the first coin (½). Alternatively, it is the probability of a head on the first coin conditional on a head on the second coin (½) times the probability of a head on the second coin (½).[5]

A probability distribution is a mathematical specification of the probabilities associated with events. For the examples shown, we easily laid out the probability distribution. Many types of random outcomes can be characterized with an exact mathematical formulation that gives either the probability of an event occurring or the probability of a range of events occurring. Coin tosses are outcomes from what is called a *binomial probability distribution*. This distribution is one of a family of distributions that are called *discrete*, which means that only a finite number of outcomes can occur. Another family of distributions is referred to as *continuous*, meaning that the number of possible outcomes is infinite. Anything that can be measured with fractional precision is continuous. For example, the return on an asset is continuous. An asset bought at 99.75 and sold at 102.5 has a return of 2.757 . . . %. There are an infinite number of possible returns, provided the return is measured with decimal precision and not rounded off to a certain number of decimal places. One major example of a continuous distribution is the familiar normal or bell-shaped distribution.

Consider a random variable, x. An expression such as $\Pr(x = a)$, representing the probability that an outcome of x is a value of a, is one example of information revealed by a discrete probability distribution. In some cases, we wish to know the cumulative probability, $\Pr(x \leq a)$. For example, consider the previous problem in which we wish to know the probability of achieving a specific total after two coin tosses where a head counts as +1 and a tail counts as −1. Suppose we wish to know the probability that the total is non-negative. Then we have $\Pr(x \geq 0)$, which equals $\Pr(x = 0) + \Pr(x = 2) = 1/2 + 1/4 = 3/4$. Alternatively, we could calculate using the complement, $\Pr(x \geq 0) = 1 - \Pr(x < 0) = 1 - \Pr(x = -2) = 1 - 1/4 = 3/4$. Note that it is not possible to obtain a total of +1 or −1. The outcomes are either +2, 0, or −2.

If a random variable is continuous, we cannot specify a probability in terms of a specific value. For example, in the standard normal distribution we cannot specify $\Pr(x = 0)$.[6] Because there are an infinite number of outcomes, the probability of any one outcome occurring is zero. We can, however, specify the probability of a range of outcomes occurring. Statements like $\Pr(x > 0)$ are quite acceptable.[7] The answer is 0.5, owing to the symmetry of the distribution and the fact that its expected value is zero. Likewise, we can use statements like $\Pr(b < x < a)$, which is easily found as $\Pr(x < a) - \Pr(x < b)$.

Because the probability of a specific value of the random variable occurring is zero in the continuous case, there is no mathematical specification that gives such results as $\Pr(x = a)$. There is, however, a mathematical specification that can lead to such statements as $\Pr(x < a)$. We start with a mathematical function referred to as the *probability density function*. If we plot such a function, specified as $f(x)$, we observe a graph of values of a variable $f(x)$ in terms of the random variable X. The specific value of $f(x)$ has no particular interpretation in a continuous distribution, but the area under the curve generated by $f(x)$ is the probability we seek. Thus, the area under the curve and to the left of a particular value $f(a)$ is $\Pr(x < a)$. Integrating the function, meaning to accumulate its values over a range of values of x, gives us the desired probability. These concepts are discussed in more detail later in this chapter. Such a function is called the *probability distribution function,* or just the *distribution function,* and sometimes the *cumulative density function.*[8]

Probability distributions are often characterized not only with a density or distribution function but also with a *moment-generating function*. This is a mathematical specification that incorporates the density or distribution function and yields what are called the moments of the distribution. The kth raw moment of a distribution is defined as $E(x^k)$. As we shall see in the next section, the first moment ($k = 1$) is the expected value and the second moment ($k = 2$) is closely related to the variance. The third moment, $E(x^3)$, is closely related to the concept of *skewness*, which measures the symmetry of a probability distribution. The fourth moment, $E(x^4)$, is closely related to the concept of *kurtosis*, which measures the extent to which a distribution is peaked or flat.

Not all probability distributions have a moment-generating function, but all have a *characteristic function*, which, although requiring use of complex numbers, can be used to yield many useful results. Characteristic functions and to some extent moment-generating functions are occasionally used in finance.

These paragraphs provide only a brief treatment but should be sufficient to refresh our memory of previous encounters with this material. In some cases, certain concepts are being encountered for the first time. The reader will in all likelihood be required to refer to more specific material to fill in gaps and extend knowledge. We now turn to the primary operations used with random variables in finance concepts, which are the determination of expectations, variances, and covariances.

4.2 EXPECTATIONS, VARIANCES, AND COVARIANCES OF DISCRETE RANDOM VARIABLES

A random variable is a variable that can take on many possible values representing uncertain outcomes whose frequencies of occurrence are governed by a probability distribution. A discrete random variable can take on only a finite number of values. For example, the number of people who respond yes to a survey or the number of laboratory mice who died following an experiment are examples of a discrete random variable. By contrast, a continuous random variable can take on an infinite number of values. For example, the height of a person selected randomly or the amount of time following an event can always be expressed with decimal places. For however many decimal places chosen, one can always add one more. Let us look at the expectations, variances, and covariances of both of these types of variables. We address discrete random variables first.

For a discrete random variable there are a finite number of outcomes that we often call *states* or *states of the world*. For example, if the event is the selection of a person and we are interested in whether that person is male or female, we would have two states. Let $x = x_m$ if that person is male and $x = x_f$ if that person is female. In general, we specify n states and n possible values of $x: x_1, x_2, \ldots, x_n$. We often need to characterize the properties of this random variable x. We let p_i be the probability that state or outcome i occurs. Note that by definition $\sum_{i=1}^{n} p_i = 1$. It may well be the case that p_i is given by some mathematical function that might more appropriately be written as $f(x)$, but here we shall just leave the probability specification in the form p_i.

4.2.1 The Expectation of a Discrete Random Variable

The expected value, sometimes called the *mean*, is the probability-weighted average value of x.[9] The expected value for a discrete distribution is the following specification:

$$E(x) = \sum_{i=1}^{n} x_i p_i. \tag{4.1}$$

Occasionally we shall have to work with a constant such as a in the following operations:

$$E(a) = \sum_{i=1}^{n} a p_i = a \sum_{i=1}^{n} p_i = a. \tag{4.2}$$

The expected value of a constant is simply the constant. Next, we multiply the constant times the random variable and are able to pull the constant out of the expectation sign,

$$E(ax) = a \sum_{i=1}^{n} x_i p_i = aE(x). \tag{4.3}$$

The Greek letter μ is often used for the expected value, but other Greek letters and many other symbols are also sometimes used as the expected value.

For example, Table 4.1 provides an analyst's forecast of potential rates of return from holding a particular tech company stock over the next year.

TABLE 4.1 Potential Rates of Return

Probability (%)	Rate of Return (%)
10	−25
20	−5
40	10
20	20
10	55

Based on Equation (4.1), we have

$$E(R) = \sum_{i=1}^{n} R_i\, p_i = -25(0.1) - 5(0.2) + 10(0.4) + 20(0.2) + 55(0.1) = 10,$$

or an expected rate of return of 10%.

4.2.2 The Variance of a Discrete Random Variable

The variance is a measure of the dispersion of the distribution. It is defined as the probability-weighted squared deviation of each possible value from the expected value. Using that construction, we see that the variance can be converted to another useful specification:

$$\mathrm{var}(x) = \sum_{i=1}^{n} \left[x_i - E(x)\right]^2 p_i = \sum_{i=1}^{n} \left[x_i - E(x)\right]\left[x_i - E(x)\right] p_i$$

$$= \sum_{i=1}^{n} \left\{x_i^2 - 2E(x)x_i + [E(x)]^2\right\} p_i = \sum_{i=1}^{n} x_i^2 p_i - 2E(x)\sum_{i=1}^{n} x_i p_i + [E(x)]^2 \sum_{i=1}^{n} p_i$$

$$= E(x^2) - 2[E(x)]^2 + [E(x)]^2 = E(x^2) - [E(x)]^2.$$

(4.4)

In other words, the variance is also the expected value of the squared value of x minus the square of the expected value of x.

When we work with constants, the variance is affected in the following manner:

$$\mathrm{var}(ax) = \sum_{i=1}^{n} \left[ax_i - E(ax)\right]^2 p_i = E\left[(ax)^2\right] - [E(ax)]^2 = E\left(a^2 x^2\right) - a^2[E(x)]^2$$

(4.5)

$$= a^2 E(x^2) - a^2[E(x)]^2 = a^2 \left\{E(x^2) - [E(x)]^2\right\} = a^2\, Var(x).$$

In other words, the variance of a constant times a random variable is the constant squared times the variance of the random variable. The variance of a constant is zero as shown by the following:

$$\mathrm{var}(a) = \sum_{i=1}^{n} \left[a - E(a)\right]^2 p_i = \sum_{i=1}^{n} (a - a)^2 p_i = 0.$$

(4.6)

Finally, we also show that the variance of a constant plus a random variable is the variance of the random variable. In other words, adding a constant does nothing to change the variance of the random variable.

$$\text{var}(a + x) = \sum_{i=1}^{n} \left[a + x_i - E(a + x)\right]^2 p_i$$

$$= \sum_{i=1}^{n} \left[a - E(a) + x_i - E(x)\right]^2 p_i = \sum_{i=1}^{n} \left[a - a + x_i - E(x)\right]^2 p_i$$

$$= \sum_{i=1}^{n} \left[x_i - E(x)\right]^2 p_i$$

$$= Var(x). \tag{4.7}$$

The square root of the variance is the standard deviation. Using σ as its symbol, we have $\sigma(ax) = a(\sigma(x))$.

Based on the same data in the last example, Table 4.2 provides supplemental information to assist in the calculation of the standard deviation.

Based on Equation (4.4), we have

$$\text{var}(R) = E(R^2) - [E(R)]^2 = 490 - 10^2 = 390,$$

or a standard deviation of 19.748% $\left(= \sqrt{390}\right)$.

4.2.3   The Covariance of Discrete Random Variables

An important concept somewhat similar to the variance is the *covariance*. It measures the extent to which two random variables move together. Now we need another random variable, which we shall refer to as y. The covariance between x and y is given as

$$\text{cov}(x, y) = \sum_{i=1}^{n} \left[x_i - E(x)\right]\left[y_i - E(y)\right] p_i = \sum_{i=1}^{n} \left[x_i y_i - E(x)y_i - x_i E(y) + E(x)E(y)\right] p_i$$

$$= \sum_{i=1}^{n} x_i y_i p_i - E(x) \sum_{i=1}^{n} y_i p_i - E(y) \sum_{i=1}^{n} x_i p_i + E(x)E(y) \sum_{i=1}^{n} p_i$$

$$= E(xy) - E(x)E(y) - E(y)E(x) + E(x)E(y)$$

$$= E(xy) - E(x)E(y). \tag{4.8}$$

TABLE 4.2   Illustration of Supplemental Information to Compute Standard Deviation

Probability (%)	Rate of Return (%)	Rate of Return^2
10	−25	625
20	−5	25
40	10	100
20	20	400
10	55	3,025
Sum = 100%	$E(R) = 10\%$	$E(R^2) = 490$

The numerical value of the covariance by itself is difficult to interpret. For example, we do not know what constitutes a large covariance, and what is a small covariance. We do know that a positive (negative) covariance means that the two variables tend to move together (opposite) in a linear fashion. A zero covariance implies that there is no linear relationship between the two variables, but it does not rule out a nonlinear relationship.[10] A more useful measure of association is the *correlation*, defined as

$$\rho(x, y) = \frac{\text{cov}(x, y)}{\sigma(x)\sigma(y)}, \tag{4.9}$$

where $\sigma(x)$ and $\sigma(y)$ are the standard deviations of x and y, respectively. The correlation ranges between -1 and $+1$. With bounds at -1 and $+1$, it is a good bit easier to interpret the magnitude of a correlation.

Based on the data in the last example along with a second instrument, Table 4.3 provides supplemental information to assist in the calculation of covariance.

Based on Equation (4.8), we have the covariance as

$$\text{cov}(R_1, R_2) = E(R_1 R_2) - E(R_1)E(R_2) = 535 - 10(10) = 435.$$

Note that interpreting 435 is difficult. Is it large or small? Also, note,

$$\text{var}(R_1) = 490 - 10^2 = 390 \text{ and}$$

$$\text{var}(R_2) = 1{,}090 - 10^2 = 990.$$

The correlation coefficient, however, is more intuitive as

$$\rho(R_1, R_2) = \frac{\text{cov}(R_1, R_2)}{\sigma(R_1)\sigma(R_2)} = \frac{435}{\sqrt{390}\sqrt{990}} = \frac{435}{19.748(31.464)} = 0.7.$$

A correlation of 0.7 is a strong positive correlation.

An important result that is sometimes seen in finance is that *the covariance of a variable with itself is the variance.* We can easily see this by letting y also be x and thereby obtaining the covariance of x with itself:

$$\text{cov}(x, y | y = x) = \text{cov}(x, x) = \sum_{i=1}^{n} \left[x_i - E(x)\right]\left[x_i - E(x)\right] p_i = Var(x). \tag{4.10}$$

TABLE 4.3 Illustration of Supplemental Information to Compute Covariance

Probability (%)	R(1) (%)	R(1)^2	R(2) (%)	R(2)^2	R(1)R(2)
10%	−25	625	20	400	−500
20%	−5	25	−15	225	75
40%	10	100	0	0	0
20%	20	400	5	25	100
10%	55	3,025	100	10,000	5,500
Sum = 100%	$E(R) = 10\%$	$E(R^2) = 490$	$E(R) = 10\%$	$E(R^2) = 1{,}090$	$E[R(1)R(2)] = 535$

We also need to know that the covariance of a random variable and a constant is zero:

$$\text{cov}(x, a) = \sum_{i=1}^{n} [x_i - E(x)][a - E(a)] \, p_i = \sum_{i=1}^{n} [x_i - E(x)](a - a) p_i = 0. \qquad (4.11)$$

The covariance concept facilitates the understanding of the variance of a combination of more than one random variable. Consider a weighted sum of variables x and y, obtained by multiplying x by a and y by b. Suppose we wish to find the variance of $ax + by$:

$$\text{var}(ax + by) = \sum_{i=1}^{n} (ax_i + by_i)^2 \, p_i - \left[\sum_{i=1}^{n} (ax_i p_i + by_i \, p_i) \right]^2$$

$$= \sum_{i=1}^{n} (ax_i + by_i)(ax_i + by_i) \, p_i - \left(a \sum_{i=1}^{n} x_i \, p_i + b \sum_{i=1}^{n} y_i p_i \right)^2$$

$$= \sum_{i=1}^{n} (a^2 x_i^2 + 2ab x_i y_i + b^2 y_i^2) \, p_i - [aE(x) + bE(y)]^2$$

$$= a^2 \sum_{i=1}^{n} x_i^2 \, p_i + 2ab \sum_{i=1}^{n} x_i y_i p_i + b^2 \sum_{i=1}^{n} y_i^2 p_i$$

$$\quad - \left\{ a^2 [E(x)]^2 + 2ab E(x)E(y) + b^2 [E(y)]^2 \right\}$$

$$= a^2 E(x^2) + 2ab E(xy) + b^2 E(y^2) - a^2 [E(x)]^2 - 2ab E(x)E(y) - b^2 [E(y)]^2$$

$$= a^2 E(x^2) - a^2 [E(x)]^2 + b^2 E(y^2) - b^2 [E(y)]^2 + 2ab E(xy) - 2ab E(x)E(y)$$

$$= a^2 \left[E(x)^2 - [E(x)]^2 \right] + b^2 \left[E(y)^2 - [E(y)]^2 \right] + 2ab[E(xy) - E(x)E(y)]$$

$$= a^2 \, Var(x) + b^2 \, Var(y) + 2ab \, Cov(x, y). \qquad (4.12)$$

Thus, we see that the variance of a weighted sum of random variables is a weighted average of the variances of the individual random variables plus twice the weighted covariance. This result is commonly used in portfolio analysis because it captures the collective volatility of correlated variables.

Now let us find the covariance between the weighted variables, that is, $\text{cov}(ax, by)$:

$$\text{cov}(ax, by) = \sum_{i=1}^{n} [ax_i - E(ax)] [by_i - E(by)] \, p_i$$

$$= \sum_{i=1}^{n} [ax_i - aE(x)] [by_i - bE(y)] \, p_i$$

$$= \sum_{i=1}^{n} [ax_i by_i - ax_i bE(y) - aE(x) by_i + ab E(x)E(y)] \, p_i$$

$$= ab \sum_{i=1}^{n} x_i y_i p_i - abE(y) \sum_{i=1}^{n} x_i p_i - abE(x) \sum_{i=1}^{n} y_i p_i + abE(x)E(y) \sum_{i=1}^{n} p_i$$

$$= abE(xy) - abE(y)E(x) - abE(x)E(y) + abE(x)E(y)$$

$$= ab(E(xy) - E(x)E(y))$$

$$= ab\mathrm{cov}(x, y). \tag{4.13}$$

We may also wish to know what happens to the covariance if we add constants to x and y. Let a and b be constants:

$$\mathrm{cov}(a + x, b + y) = \sum_{i=1}^{n} \left[a + x_i - E(a + x)\right]\left[b + y_i - E(b + y)\right] p_i$$

$$= \sum_{i=1}^{n} \left[a - E(a) + x_i - E(x)\right]\left[b - E(b) + y_i - E(y)\right] p_i$$

$$= \sum_{i=1}^{n} \left[a - a + x_i - E(x)\right]\left[b - b + y_i - E(y)\right] p_i$$

$$= \sum_{i=1}^{n} \left[x_i - E(x)\right]\left[y_i - E(y)\right] p_i = \mathrm{cov}(x, y). \tag{4.14}$$

Thus, quite logically, adding constants does not change the covariance between two variables.

Finally, we might be interested in the covariance between the sum of two random variables and a third random variable. Letting the third random variable be z, we want to know $\mathrm{cov}(x + y, z)$:

$$\mathrm{cov}(x + y, z) = \sum_{i=1}^{n} \left[x_i + y_i - E(x + y)\right]\left[z_i - E(z)\right] p_i$$

$$= \sum_{i=1}^{n} \left\{\left[x_i - E(x)\right] + \left[y_i - E(y)\right]\right\}\left[z_i - E(z)\right] p_i$$

$$= \sum_{i=1}^{n} \left[x_i - E(x)\right]\left[z_i - E(z)\right] p_i + \sum_{i=1}^{n} \left[y_i - E(y)\right]\left[z_i - E(z)\right] p_i$$

$$= \mathrm{cov}(x, z) + \mathrm{cov}(y, z). \tag{4.15}$$

Based on the data in the last example along with a portfolio composed of 50% of instrument 1 and 50% of instrument 2, Table 4.4 provides supplemental information to assist in the calculation of the covariance of sums. Here we let P denote the return on a portfolio of 50% instrument 1 and 50% instrument 2.

Based on Equation (4.4), we have the variance of the portfolio is

$$\mathrm{var}(P) = E\left(P^2\right) - [E(P)]^2 = 662.5 - 10^2 = 562.5.$$

TABLE 4.4   Illustration of Supplemental Information to Compute Covariance of Sums

Probability (%)	P (%)	P^2	R(1)*P (%)	R(2)*P
10	−2.5	6.25	62.5	−50
20	−10	100	50	150
40	5	25	50	0
20	12.5	156.25	250	62.5
10	77.5	6,006.25	4,262.5	7,750
Sum = 100%	E(P) = 10%	$E(P^2) = 662.5$	E[R(1) * P] = 512.5	E[R(2) * P] = 812.5

Based on the properties of portfolio returns, we know

$$P_i = 0.5(R_{1i}) + 0.5(R_{2i}).$$

Thus, based on Equation (4.15), we have the covariance as

$$cov(0.5R_1 + 0.5R_2, P) = cov(0.5R_1, P) + cov(0.5R_2, P).$$

From Equation (4.13), we know

$$cov(0.5R_1 + 0.5R_2, P) = 0.5cov(R_1, P) + 0.5cov(R_2, P),$$

$$cov(R_1, P) = E(R_1P) - E(R_1)E(P) = 512.5 - 10(10) = 412.5, \text{ and}$$

$$cov(R_2, P) = E(R_2P) - E(R_2)E(P) = 812.5 - 10(10) = 712.5.$$

Again, interpreting variances and covariances is difficult. From the definition of covariance, we observe the identity

$$var(P) = cov(0.5R_1 + 0.5R_2, P) = 0.5cov(R_1, P) + 0.5cov(R_2, P).$$

If we divide both sides by the variance of the portfolio, we observe the following percentage marginal contribution to risk (%MCTR) of each instrument in the portfolio:

$$\%MCTR_1 = 0.5\frac{cov(R_1, P)}{var(P)} = 0.5\frac{412.5}{562.5} = 36.67\% \text{ and}$$

$$\%MCTR_2 = 0.5\frac{cov(R_2, P)}{var(P)} = 0.5\frac{712.5}{562.5} = 63.33\%.$$

4.3   CONTINUOUS RANDOM VARIABLES

For continuous random variables, the operations are somewhat different, but the results are conceptually identical to the discrete case. Consider a random variable x with lower limit x_L and upper limit x_H. Its density function is $f(x)$. Over the entire range of outcomes, the density integrates to 1:

$$\int_{x_L}^{x_H} f(x)dx = 1. \tag{4.16}$$

Remember that the value of the density function at a point is not particularly useful. That is, for a given value of x, say x_i, the value $f(x_i)$ is merely the height of the curve specified by the density, $f(x)$. The integration of the density over a range gives us the probability of the event over that range. For example, if we wish to know if x is less than or equal to a, the integration of the density is

$$\Pr(x \le a) = \int_{x_L}^{a} f(x)dx. \tag{4.17}$$

This value is the cumulative probability that x will be lower than a. It is often called the distribution function and sometimes written as $F(a)$. For a range of a to b, the probability that the x falls in that range is found as follows:

$$\Pr(a \le x \le b) = \int_{a}^{b} f(x)dx. \tag{4.18}$$

Note that this result can also be found as

$$\Pr(a \le x \le b) = \int_{x_L}^{b} f(x)dx - \int_{x_L}^{a} f(x)dx. \tag{4.19}$$

Here we integrate from the lower limit up to b and from the lower limit to a. We then subtract the latter from the former.

In the following material, we will set the lower limit to $-\infty$ and the upper limit to $+\infty$.

4.3.1 The Expectation of a Continuous Random Variable

To obtain the expectation, we simply multiply each of the infinite values of the random variable by its density function and integrate,

$$E(x) = \int_{-\infty}^{\infty} xf(x)dx. \tag{4.20}$$

If x is actually a constant a, then we have the following result,

$$E(a) = \int_{-\infty}^{\infty} af(x)dx = a \int_{-\infty}^{\infty} f(x)dx = a. \tag{4.21}$$

Thus, the expected value of a constant is simply the constant. Now, suppose we multiply the constant times the random variable,

$$E(ax) = \int_{-\infty}^{\infty} axf(x)dx = a \int_{-\infty}^{\infty} xf(x)dx = aE(x). \tag{4.22}$$

And we simply pulled the constant out of the integration.

4.3.2   The Variance of a Continuous Random Variable

The variance of a continuous random variable X is given as

$$\text{var}(x) = \int_{-\infty}^{\infty} [x - E(x)]^2 f(x) dx. \tag{4.23}$$

The key results for the variance are given as follows. First note that the variance itself can be restated as

$$\text{var}(x) = \int_{-\infty}^{\infty} (x - E(x))^2 f(x) dx = \int_{-\infty}^{\infty} \left\{ x^2 - 2xE(x) - [E(x)]^2 \right\} f(x) dx$$

$$= \int_{-\infty}^{\infty} x^2 f(x) dx - 2E(x) \int_{-\infty}^{\infty} x f(x) dx + [E(x)]^2 \int_{-\infty}^{\infty} f(x) dx$$

$$= E(x^2) - 2[E(x)]^2 + [E(x)]^2 = E(x^2) - [E(x)]^2. \tag{4.24}$$

If x is a constant, we have the following:

$$\text{var}(a) = \int_{-\infty}^{\infty} [a - E(a)]^2 f(x) dx = \int_{-\infty}^{\infty} (a - a)^2 f(x) dx = 0. \tag{4.25}$$

Obviously, the variance of a constant is zero. Now suppose we multiply a constant times the random variable.

$$\text{var}(ax) = \int_{-\infty}^{\infty} [ax - E(ax)][ax - E(ax)] f(x) dx$$

$$= \int_{-\infty}^{\infty} a[x - E(x)]a[x - E(x)] f(x) dx = a^2 \int_{-\infty}^{\infty} [x - E(x)]^2 f(x) dx = a^2 \text{var}(x).$$

$$\tag{4.26}$$

So, the constant comes out of the variance as a square. Suppose we add a constant to the random variable.

$$\text{var}(a + x) = \int_{-\infty}^{\infty} [a + x - E(a + x)]^2 f(x) dx$$

$$= \int_{-\infty}^{\infty} [a - E(a) + x - E(x)]^2 f(x) dx = \int_{-\infty}^{\infty} [a - a + x - E(x)]^2 f(x) dx$$

$$= \int_{-\infty}^{\infty} [x - E(x)]^2 f(x) dx = \text{var}(x). \tag{4.27}$$

Adding a constant clearly has no effect on the variance.

4.3.3 The Covariance of Continuous Random Variables

The covariance of x and y is

$$\text{cov}(x,y) = \int\limits_{-\infty}^{\infty} [x - E(x)][y - E(y)] f(xy) \, dxy. \qquad (4.28)$$

In the previous equation, $f(xy)$ is the joint density of x and y, which gives the probability of outcomes of both x and y occurring. The single integral in the covariance equation means to integrate over all joint outcomes. At this point it will be useful to introduce a substitution for the integral and the density function,

$$\int\limits_{-\infty}^{\infty} f(xy) \, dxy = \int\limits_{-\infty}^{\infty}\int\limits_{-\infty}^{\infty} f(y|x) f(x) \, dxdy = \int\limits_{-\infty}^{\infty}\int\limits_{-\infty}^{\infty} f(x|y) f(y) \, dydx. \qquad (4.29)$$

Here we substitute the product of the conditional and marginal densities for the joint density. When we do so, we must have an integral and a differential for each density. Also note that any time we see any of these expressions with no other terms, it equals 1.0. In other words,

$$\int\limits_{-\infty}^{\infty} f(y|x) \, dy = \int\limits_{-\infty}^{\infty} f(x|y) \, dx = 1. \qquad (4.30)$$

Now our covariance can be written as

$$\text{cov}(x,y) = \int\limits_{-\infty}^{\infty} [x - E(x)][y - E(y)] f(xy) \, dxy$$

$$= \int\limits_{-\infty}^{\infty} [xy - xE(y) - yE(x) + E(x)E(y)] f(xy) \, dxy$$

$$= \int\limits_{-\infty}^{\infty} xyf(xy) \, dxy - E(y)\int\limits_{-\infty}^{\infty} xf(xy) \, dxy - E(x)\int\limits_{-\infty}^{\infty} yf(xy) \, dxy$$

$$\quad + E(x)E(y)\int\limits_{-\infty}^{\infty} f(xy) \, dxy$$

$$= E(xy) - E(y)\int\limits_{-\infty}^{\infty}\int\limits_{-\infty}^{\infty} xf(y|x) f(x) \, dxdy - E(x)\int\limits_{-\infty}^{\infty}\int\limits_{-\infty}^{\infty} yf(x|y) f(y) \, dxdy + E(x)E(y)$$

$$= E(xy) - E(y)\int\limits_{-\infty}^{\infty} f(y|x) \, dy \int\limits_{-\infty}^{\infty} xf(x) \, dx - E(x)\int\limits_{-\infty}^{\infty} f(x|y) \, dx \int\limits_{-\infty}^{\infty} yf(y) \, dy + E(x)E(y)$$

$$= E(xy) - E(y)E(x) - E(x)E(y) + E(x)E(y). \qquad (4.31)$$

So we have

$$\text{cov}(x, y) = E(xy) - E(x)E(y). \tag{4.32}$$

Covariances with constants are zero as indicated by the following:

$$\text{cov}(a, x) = \int_{-\infty}^{\infty} [a - E(a)][x - E(x)] f(xy) dxy = \int_{-\infty}^{\infty} (a - a)[x - E(x)] f(xy) dxy = 0. \tag{4.33}$$

Next we show the important result that the variance of a weighted combination of random variables is a weighted combination of their variances and all possible pairwise covariances.

$$\text{var}(ax + by) = \int_{-\infty}^{\infty} [ax + by - E(ax + by)]^2 f(xy) dxy$$

$$= \int_{-\infty}^{\infty} [ax - E(ax) + by - E(by)]^2 f(xy) dxy$$

$$= \int_{-\infty}^{\infty} \{a[x - E(x)] + b[y - E(y)]\}^2 f(xy) dxy$$

$$= \int_{-\infty}^{\infty} \left\{ a^2[x - E(x)]^2 + b^2[y - E(y)]^2 + 2ab[x - E(x)][y - E(y)] \right\} f(xy) dxy$$

$$= a^2 \int_{-\infty}^{\infty} [x - E(x)]^2 f(xy) dxy$$

$$+ b^2 \int_{-\infty}^{\infty} [y - E(y)]^2 f(xy) dxy + 2ab \int_{-\infty}^{\infty} [x - E(x)][y - E(y)] f(xy) dxy. \tag{4.34}$$

Recall that previously we substituted the product of the conditional and marginal densities for the joint density and split the single integral into the product of two integrals. Using these tricks, we obtain $a^2 \text{var}(x)$ for the first term, $b^2 \text{var}(y)$ for the second and $2ab \text{cov}(x, y)$. Thus,

$$\text{var}(ax + by) = a^2 \text{var}(x) + b^2 \text{var}(y) + 2ab \text{cov}(x, y). \tag{4.35}$$

The covariance between two weighted random variables is simply their covariance times the product of their weights:

$$\text{cov}(ax, by) = \int_{-\infty}^{\infty} [ax - E(ax)][by - E(by)] f(xy) dxy$$

$$= \int_{-\infty}^{\infty} \{a[x - E(x)]b[y - E(y)]\} f(xy) \, dxy$$

$$= ab \int_{-\infty}^{\infty} [x - E(x)][y - E(y)] f(xy) \, dxy = ab \, \text{cov}(x, y). \qquad (4.36)$$

This result is widely used in portfolio analysis, a point we made in the discrete case.

The covariance between the sum of a constant and random variables times another sum of a constant and a random variable is simply the covariance of the random variables:

$$\text{cov}(a + x, b + y) = \int_{-\infty}^{\infty} [a + x - E(a + x)][b + y - E(b + y)] f(xy) \, dxy$$

$$= \int_{-\infty}^{\infty} [a - E(a) + x - E(x)][b - E(b) + y - E(y)] f(xy) \, dxy$$

$$= \int_{-\infty}^{\infty} [a - a + x - E(x)][b - b + y - E(y)] f(xy) \, dxy$$

$$= \int_{-\infty}^{\infty} [x - E(x)][y - E(y)] f(xy) \, dxy = \text{cov}(x, y). \qquad (4.37)$$

One of the more complex results is the covariance between the sum of two random variables and a third random variable:

$$\text{cov}(x + y, z) = \int_{-\infty}^{\infty} [x + y - E(x + y)][z - E(z)] f(xyz) \, dxyz$$

$$= \int_{-\infty}^{\infty} [x - E(x) + y - E(y)][z - E(z)] f(xyz) \, dxyz$$

$$= \int_{-\infty}^{\infty} [xz - xE(z) - E(x)z + E(x)E(z) + yz - yE(z) - E(y)z + E(y)E(z)]$$

$$f(xyz) \, dxyz$$

$$= \int_{-\infty}^{\infty} xzf(xyz) \, dxyz - E(z) \int_{-\infty}^{\infty} xf(xyz) \, dxyz - E(x) \int_{-\infty}^{\infty} zf(xyz) \, dxyz$$

$$+ E(x)E(z) \int_{-\infty}^{\infty} f(xyz) \, dxyz$$

$$+ \int_{-\infty}^{\infty} yzf(xyz)\,dxyz - E(z) \int_{-\infty}^{\infty} yf(xyz)\,dxyz$$

$$- E(y) \int_{-\infty}^{\infty} zf(xyz)\,dxyz + E(y)E(z) \int_{-\infty}^{\infty} f(xyz)\,dxyz. \tag{4.38}$$

As we did previously when working with only two random variables, we need to convert the joint density into the product of the marginal and conditional densities. Now, however, we have three variables. We use the following relationships:

$$f(xyz) = f(y|xz)\,f(xz) = f(x|yz)\,f(yz) = f(z|xy)\,f(xy)$$

$$= f(yz|x)\,f(x) = f(xz|y)\,f(y) = f(xy|z)\,f(z). \tag{4.39}$$

As we have previously noted, the integral of any density function over its entire domain is 1.0. Thus,

$$\int_{-\infty}^{\infty} f(yz|x)dyz = \int_{-\infty}^{\infty} f(xz|y)dxz = \int_{-\infty}^{\infty} f(xy|z)dxy = 1.0. \tag{4.40}$$

Of the eight expressions for the covariance, the first is, therefore,

$$\int_{-\infty}^{\infty} xzf(xyz)\,dxyz = \int_{-\infty}^{\infty}\int_{-\infty}^{\infty} xzf(y|xz)\,f(xz)\,dydxz$$

$$= \int_{-\infty}^{\infty} f(y|xz)dy \int_{-\infty}^{\infty} xzf(xz)dxz = E(xz). \tag{4.41}$$

The second is

$$-E(z) \int_{-\infty}^{\infty} xf(xyz)\,dxyz = -E(z) \int_{-\infty}^{\infty} f(yz|x)dyz \int_{-\infty}^{\infty} xf(x)dx$$

$$= -E(z)E(x). \tag{4.42}$$

The third is

$$-E(x) \int_{-\infty}^{\infty} zf(xyz)\,dxyz = -E(x) \int_{-\infty}^{\infty} f(xy|z)dxy \int_{-\infty}^{\infty} zf(z)dz$$

$$= -E(x)E(z). \tag{4.43}$$

The fourth expression is clearly $E(x)E(z)$. The fifth expression is

$$+ \int_{-\infty}^{\infty} yzf(xyz)\,dxyz = \int_{-\infty}^{\infty}\int_{-\infty}^{\infty} yzf(z|yz)\,f(xyz)\,dxdyz$$

$$= \int_{-\infty}^{\infty} f(x|yz)dx \int_{-\infty}^{\infty} yzf(yz)dyz = E(yz). \qquad (4.44)$$

The sixth expression is

$$-E(z)\int_{-\infty}^{\infty} yf(xyz)\,dxyz = -E(z)\int_{-\infty}^{\infty} f(xz|y)dxz \int_{-\infty}^{\infty} yf(y)dy$$

$$= -E(z)E(y). \qquad (4.45)$$

The seventh expression is

$$-E(y)\int_{-\infty}^{\infty} zf(xyz)\,dxyz = -E(y)\int_{-\infty}^{\infty} f(xy|z)dxy \int_{-\infty}^{\infty} zf(z)dz$$

$$= -E(y)E(z). \qquad (4.46)$$

The eighth expression is clearly $E(y)E(z)$. Thus, overall our covariance is

$$E(xz) - E(z)E(x) - E(x)E(z) + E(x)E(z) + E(yz) - E(z)E(y) - E(y)E(z) + E(y)E(z)$$

$$= E(xz) - E(x)E(y) + E(yz) - E(y)E(z) = \text{Cov}(x,z) + \text{Cov}(y,z). \qquad (4.47)$$

4.4 SOME GENERAL RESULTS IN PROBABILITY THEORY

In the following subsections, we look at some of the general results from probability theory that are helpful in option theory. By the term *general*, we mean that these results are not dependent on any particular probability distribution. When a specific probability distribution is known, usually a stronger statement can be made.

4.4.1 The Central Limit Theorem

The *central limit theorem* is a powerful statement that tells us that sum or average of independent samples drawn from any given probability distribution becomes normally distributed in the limit, that is, as the sample size approaches infinity. Thus, provided the sample size is large enough, the central limit theorem enables us to use the rules associated with normal probability theory when drawing inferences about sample means. How large the sample size must be to rely on the central limit theorem is not known, but a common rule of thumb has always been at least 30.

4.4.2 Chebyshev's Inequality

Chebyshev's inequality, sometimes called *Chebyshev's theorem*, enables us to make a sometimes useful statement about the probability of a sample value deviating from the population mean. More precisely, let x be a random variable with mean μ and variance σ^2, which can come from any probability distribution. For any real number $z > 0$,

$$\Pr(|x - \mu| \geq z\sigma) \leq \frac{1}{z^2}, \text{assuming } z \geq 1. \tag{4.48}$$

Thus, Chebyshev's theorem gives an upper bound on how much the observed value deviates from the mean in terms of an arbitrary value z. For example, let x be the average height in inches of a randomly drawn male university student. Let $\mu = 70$ and $\sigma = 2$. For z equals various values, we have the following results:

z	$z\sigma$	Maximum $\Pr(|x - \mu| \geq z\sigma)$
5	10	0.04
4	8	0.0625
3	6	0.1111

In other words, the probability that a sample value will deviate from the mean by more than 10 inches is less than 0.04. Thus, the probability that a student will be more than 80 inches (70 plus 10 inches) is less than 4%. The probability that a student will be more than 78 inches (70 plus eight) is less than 6.25%. The probability that a student will be more than 76 inches (70 plus six) is less than 11.11%. These rules hold for any distribution, though knowing the exact distribution usually enables one to make more precise statements.[11]

4.4.3 The Law of Large Numbers

The *law of large numbers* provides information about the accuracy with which a sample mean approximates a population mean. Basically, it says that the probability that the difference between the sample mean and the population mean is greater than an arbitrarily chosen small value is zero because the sample size goes to infinity. This law holds as long as the sample consists of independently selected random variables from the same distribution. The law of large numbers more or less says that if we take sufficiently large samples, our sample mean estimate converges to the population mean.

4.4.4 The Law of Iterated Expectations (the Tower Law)

The *law of iterated expectations*, sometimes called the *Tower law*, is used to specify the expected value assessed at a given time t in terms of an expected value at another time $t + i$, which is taken as an expected value at a later time $t + j$. In other words, we are taking the expectation of an expectation. For example, suppose we are at time $t + i$ and are calculating the expected value at time $t + j$. We might denote this as $E_{t+i}(x_{t+j})$, which simply means that using the information at time $t + i$, we assess the expected value of x

to occur at a later time $t + j$. Now step back to time t and try to assess the expected value at time $t + i$. In other words, we want $E_t[E_{t+i}(x_{t+j})]$. The law of iterated expectations simply says

$$E_t \left[E_{t+i} \left(x_{t+j} \right) \right] = E_t \left(x_{t+j} \right). \tag{4.49}$$

In other words, the expectation is iterated from the later time to the earlier time.

4.4.5 The Law of Total Probability

The *law of total probability* is a simple statement about the conditional probabilities and marginal probabilities. Specifically let y and x be random variables where x is bifurcated into values greater than b and less than or equal to b. Then the law of total probability states that

$$\Pr(y) = \Pr(y|x > b)\Pr(x > b) + \Pr(y|x \le b)\Pr(x \le b). \tag{4.50}$$

The statement $\Pr(y)$ is simply any specification of y, such as $\Pr(y > k)$ or $\Pr(a \le y \le k)$. So, the probability of any event associated with y can be broken down into the probability of that event for y, conditional on an event occurring for x, and the probability of that event for y, conditional on that event not occurring for x.

4.5 TECHNICAL INTRODUCTION TO COMMON PROBABILITY DISTRIBUTIONS USED IN FINANCE

Numerous financial challenges are addressed with the use of common probability distributions. As noted previously, probability distributions are either discrete or continuous.

Discrete distributions have a countable set of possible outcomes, such as flipping a coin, where each outcome has a nonnegative chance of occurring and the sum of the probability of all outcomes is equal to one. More formally, for a discrete probability function, $p_X(a)$, the following properties hold for a specific value, x:

$\Pr(x = a) = p_X(a)$. [Read: The probability that X is a specific value, $x = a$, is $p_X(a)$.]

$p_X(x) \ge 0 \, \forall x$. [Read: The probability is nonnegative for all possible outcomes x.]

$\sum_{j=1}^{N} p_X(x_j) = 1$. [Read: The sum of the probabilities over all possible values of x is 1.]

Discrete probability distributions covered here include the binomial and Poisson distributions.

Continuous distributions have an infinitesimal probability of achieving any particular outcome, whereas the likelihood of achieving a value within a defined range of outcomes is measurable. Finally, the integral over the range of all outcomes is equal to one. More formally, for a continuous probability function, $p_c(x)$, the following properties hold:

$\Pr(a \le x \le b) = \int_{x=a}^{x=b} p_c(x)dx$. [Read: The probability that x takes a value in a range of a through b.]

$\Pr(a \leq x \leq b) \geq 0 \,\forall\, a \leq b$. [Read: The probability is nonnegative for all possible ranges of a through b.]

$\Pr(-\infty \leq x \leq +\infty) = \int_{x=-\infty}^{x=+\infty} p_c(x)dx = 1$. [Read: The integral over all possible values of x is 1.]

Continuous probability distributions covered here include the normal and lognormal distributions.

4.5.1 Binomial Distribution

The binomial distribution is used to model situations involving two random outcomes, such as a coin toss. Thus, this discrete probability distribution is binary, derived from the Latin root *bi* meaning "two." The -*nomial* extension is derived from Latin, meaning "name." Thus, the binomial distribution is based on two names, such as true and false, heads or tails, 1 or 0, up or down, and so forth. We often select one of the names to be *success* and the other name to be *failure*. The binomial distribution is a two-parameter distribution, with parameters n and p. The parameter n denotes the number of independent events. In finance, n is often the number of time steps, such as 252 for number of trading days in one particular year. The parameter p denotes the probability of observing a success, such as true, heads, 1, up, and so on. In finance, p is often the probability that a financial instrument will go up over the next time step.

The possible outcomes from a binomial distribution range from 0 to n, such as the number of observed up events over n days, where each event has probability p of success. The binomial distribution is therefore denoted $B(n,p)$. If x is said to have a binomial distribution, then the probability mass function can be expressed as[12]

$$b(i,n,p) = \Pr(i;n,p) = \Pr(x=i) = \binom{n}{i} p^i(1-p)^{n-i}, \qquad (4.51)$$

for all $i = 0,1,2,\ldots,n$, where the combination of n events with i successes is defined as

$$\binom{n}{i} = \frac{n!}{i!(n-i)!},$$

and $n!$ denotes factorial or $n(n-1)(n-2)\ldots(2)1$. The combination of n events with i successes identifies the number of different ways of observing i successes over n events.

The cumulative distribution function (CDF) can be expressed as

$$B(j;n,p) = \Pr(x \leq j) = \sum_{i=0}^{j} \binom{n}{i} p^i(1-p)^{n-i}. \qquad (4.52)$$

The mean is $n(p)$ and the variance is $n(p)(1-p)$. We will explore this distribution in greater detail in Chapter 7.

4.5.2 Poisson Distribution

The Poisson distribution is used to model countable outcomes, such as the number of companies registered in a particular year. This distribution is named after Siméon Denis Poisson (1781–1840), although Abraham de Moivre (1711) appears to be the first to develop it. The Poisson is also a discrete probability distribution. In finance, the Poisson distribution is used when one is interested in the number of times a particular event happens during a specified time span. This distribution assumes that the occurrence of one event does not affect the likelihood of observing another event as well as assumes the average number of events is constant.

The possible outcomes from a Poisson distribution are the integers from 0 to ∞, such as the number of bankruptcies observed in a given year. The Poisson distribution requires only one parameter, the average number of observed outcomes during an interval of time, denoted λ. Thus, the Poisson distribution is denoted $PD(\lambda)$. If x is said to have a Poisson distribution, then the probability mass function can be expressed as

$$pm(i, \lambda) = \Pr(i; \lambda) = \Pr(x = i) = \frac{\lambda^i e^{-\lambda}}{i!}, \tag{4.53}$$

for all $i = 0, 1, 2, \ldots, \infty$.

The cumulative distribution function (CDF) can be expressed as

$$PM(j; \lambda) = \Pr(x \leq j) = e^{-\lambda} \sum_{i=0}^{j} \frac{\lambda^i}{i!}. \tag{4.54}$$

The mean and variance are both λ.

4.5.3 Normal Distribution

The normal distribution is one of the most well-known of all distributions and is widely used in finance; hence, we devote the entire Chapter 5 to it. We provide a brief technical introduction here. The range of a variable, x, that follows a normal distribution is $-\infty < x < +\infty$. It is a symmetric two-parameter distribution typically identified with the mean, μ, and standard deviation, σ. The range for the mean is therefore $-\infty < \mu < +\infty$ and the range for the standard deviation is $\sigma > 0$. If x is said to have a normal distribution, then the probability density function (PDF) can be expressed as

$$n(x) = \frac{1}{\sigma\sqrt{2\pi}} e^{-\frac{(x-\mu)^2}{2\sigma^2}}. \tag{4.55}$$

The PDF is the derivative of the cumulative distribution function. The CDF, assuming $-\infty < a < +\infty$, can be expressed as

$$N(a) = \int_{x=-\infty}^{a} n(x)dx = \int_{x=-\infty}^{a} \frac{1}{\sigma\sqrt{2\pi}} e^{-\frac{(x-\mu)^2}{2\sigma^2}} dx. \tag{4.56}$$

The CDF range is $0 \leq N(a) \leq 1$ and provides the probability that an outcome is less than or equal to a. The PDF is simply the first derivative of the CDF.

We briefly identify several location statistics related to the normal distribution. The median is defined as the $\hat{x}$ such that $\int_{-\infty}^{\hat{x}} f(x)dx = \int_{\hat{x}}^{+\infty} f(x)dx = \frac{1}{2}$ for any PDF $f(x)$. For the normal distribution, the median is equal to μ(a constant). The mode is defined as the $\hat{x}$ such that $\frac{df(x)}{dx} = 0$ and $\frac{d^2 f(x)}{dx^2} < 0$ for any PDF. For the normal distribution, the mode is also equal to μ.

With statistical distributions, the nth noncentral moment is defined as $\mu_n'(x) = E(x^n)$ and the nth central moment is defined as $\mu_n(x) = E[(x - \mu)^n]$. The first noncentral moment is the mean or

$$Mean = \mu_1'(x) = \mu. \tag{4.57}$$

The mean is a constant and the first central moment is zero or $\mu_1(x) = 0$. The second central moment is the variance or

$$Variance = \text{var}(x) = \mu_2(x) = \sigma^2. \tag{4.58}$$

The variance is a constant and the second noncentral or raw moment is $\mu_2'(x) = \mu^2 + \sigma^2$. The third central moment is the skewness or

$$Skewness = \mu_3(x) = 0. \tag{4.59}$$

Skewness of zero implies a symmetric distribution. The third noncentral moment is $\mu_3'(x) = \mu^3 + 3\mu\sigma^2$. The fourth central moment is the kurtosis or

$$Kurtosis = \mu_4(x) = 3\sigma^4. \tag{4.60}$$

Note that the fourth moment is positive and a function of variance. The fourth noncentral or raw moment is $\mu_4'(x) = \mu^4 + 6\mu^2\sigma^2 + 3\sigma^4$. Kurtosis is often normalized by the variance squared. Thus, the kurtosis of the normal distribution is 3 and is known as mesokurtic.

Figures 4.1a–h illustrate both the probability density function as well as the cumulative distribution function of the normal distribution applied to some financial instrument price. The following series of figures illustrates the effect of increases in the standard deviation because different financial instruments have different levels of standard deviation. One common challenge is misinterpreting high standard deviation instruments, particularly when applying the lognormal distribution, which we cover next.

We see from these graphs that with the normal distribution, there is a nonzero probability that the underlying instrument value can fall below zero. Financial derivative instruments can often themselves take on negative values. Negative value is very common for symmetric derivatives such as forwards, futures, and swaps as well as short positions in asymmetric derivatives such as options, swaptions, caps, and floors. For example, the short position for any option cannot be positive as it can only result in a future liability. A call option price may be 14, but to the writer it is a liability of 14 and hence has a negative value. Because common stock has limited liability, we expect the stock price to remain nonnegative. Limited liability can be viewed as a company obtaining a long put option contract with a zero strike price. Clearly, company assets can obtain a negative value, such as is the case with environmental damage or company products found to be carcinogenic. Again, more extensive details on the normal distribution are given in Chapter 5. To address some of these challenges, many financial models are built on the lognormal distribution.

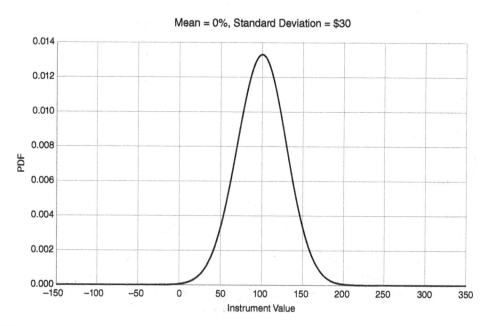

FIGURE 4.1a Normal Probability Density Function ($\mu = 0\%$, $\sigma = 30\%$)

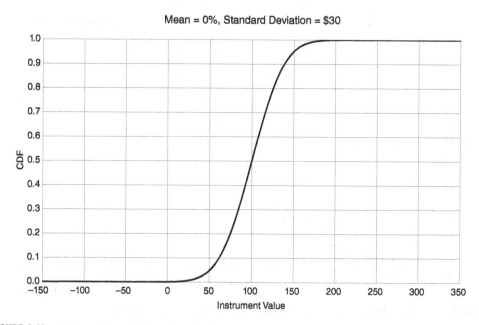

FIGURE 4.1b Normal Cumulative Distribution Function ($\mu = 0\%$, $\sigma = 30\%$)

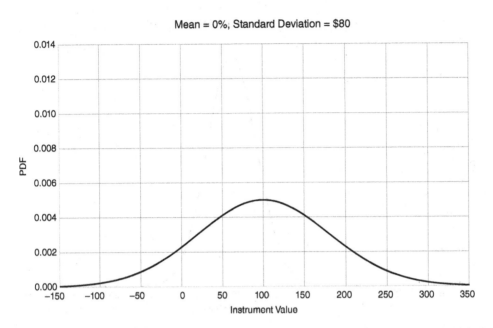

FIGURE 4.1c   Normal Probability Density Function ($\mu = 0\%$, $\sigma = 80\%$)

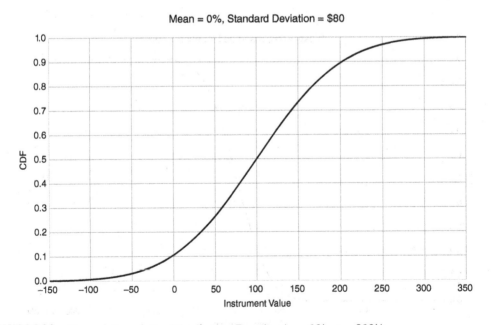

FIGURE 4.1d   Normal Cumulative Distribution Function ($\mu = 0\%$, $\sigma = 80\%$)

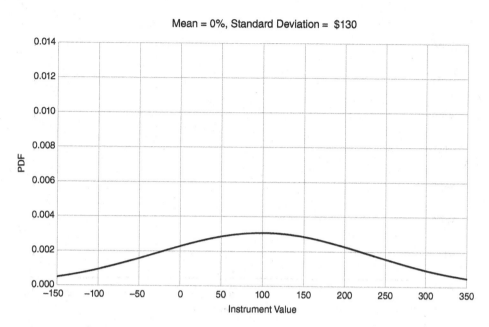

Mean = 0%, Standard Deviation = $130

FIGURE 4.1e Normal Probability Density Function ($\mu = 0\%$, $\sigma = 130\%$)

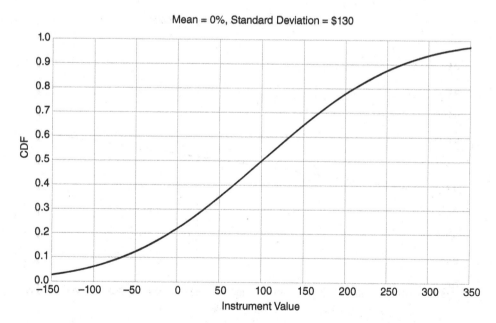

Mean = 0%, Standard Deviation = $130

FIGURE 4.1f Normal Cumulative Distribution Function ($\mu = 0\%$, $\sigma = 130\%$)

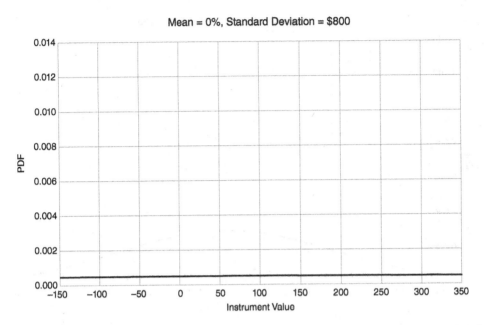

FIGURE 4.1g   Normal Probability Density Function ($\mu = 0\%$, $\sigma = 800\%$)

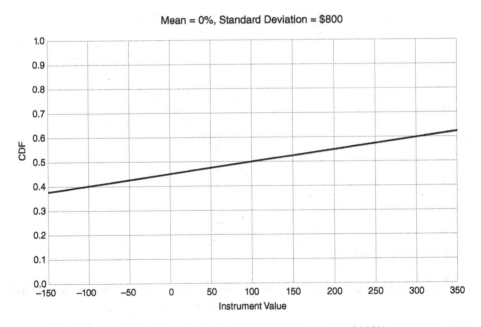

FIGURE 4.1h   Normal Cumulative Distribution Function ($\mu = 0\%$, $\sigma = 800\%$)

4.5.4 Lognormal Distribution

The lognormal distribution is related to the normal distribution in the following way: If $y = \ln(x)$ is normally distributed, x is said to be lognormally distributed. It is an asymmetric two-parameter distribution again identified with the implied normal distribution's mean, μ, and standard deviation, σ. The range for the mean is $-\infty < \mu < +\infty$ and the range for the standard deviation is $\sigma > 0$. The range of a variable, x, that follows a lognormal distribution is $0 < x < +\infty$. Note that zero is not included. If x is said to have a lognormal distribution, then the PDF can be expressed as (where the Greek lambda, λ, here is not to be confused with the Poisson distribution parameter introduced previously):

$$\lambda(x) = \frac{1}{x\sigma\sqrt{2\pi}}e^{-\frac{[\ln(x)-\mu]^2}{2\sigma^2}}. \tag{4.61}$$

The CDF, assuming $0 < a < +\infty$, can be expressed as

$$\Lambda(a) = \int_{x=-\infty}^{a} \lambda(x)dx = \int_{x=-\infty}^{a} \frac{1}{x\sigma\sqrt{2\pi}}e^{-\frac{[\ln(x)-\mu]^2}{2\sigma^2}}dx. \tag{4.62}$$

Similar to the normal distribution, the CDF range is $0 \le \Lambda(a) \le 1$ and provides the probability that an outcome is less than or equal to a.

We briefly identify several location statistics related to the lognormal distribution. The median for the lognormal distribution is

$$Median = e^{\mu}. \tag{4.63}$$

The mode is

$$Mode = e^{\mu-\sigma^2}. \tag{4.64}$$

Recall the first noncentral moment is the mean and can be expressed as

$$Mean = E(x) = \mu_1'(x) = e^{\mu+\frac{\sigma^2}{2}}\sqrt{b^2 - 4ac}. \tag{4.65}$$

The mean is a constant and the first central moment is zero or $\mu_1(x) = 0$. The second central moment is the variance or

$$var(x) = \mu_2(x) = e^{2\mu+\sigma^2}(e^{\sigma^2} - 1). \tag{4.66}$$

The variance is a constant and the second noncentral moment is $\mu_2'(x) = e^{2(\mu+\sigma^2)}$. The third central moment is the skewness or

$$Skew(x) = \mu_3(x) = e^{3\mu+\frac{3\sigma^2}{2}}(e^{\sigma^2} - 1)^2(e^{\sigma^2} + 2). \tag{4.67}$$

The skewness of the lognormal distribution is positive, which implies an asymmetric distribution, specifically positive or right skewed. The third noncentral moment is $\mu_3'(x) = e^{3\left(\mu + \frac{3}{2}\sigma^2\right)}$. Similar to variance and covariance, skewness is difficult to interpret. Thus, skewness is often normalized in the following manner:

$$NSkew \equiv \frac{\mu_3'}{Var^{2/3}} = (e^{\sigma^2} - 1)^2(e^{\sigma^2} + 2). \tag{4.68}$$

The normalized skewness is an exponentially increasing function of the normal distribution variance. Symmetrical distributions, similar to the normal distribution, will have $NSkew = 0$. If $NSkew > 0$ as is the case for the lognormal distribution, then the mean is greater than the median and the median is greater than the mode. If $NSkew < 0$ as is common with empirical finance distributions, then we have mean < median < mode. There are, however, some rare exceptions to this pattern; see Stuart and Ord (1987: 107).

The fourth central moment is the kurtosis or

$$Kurtosis(x) = \mu_4(x) = e^{3\mu + \frac{3\sigma^2}{2}}(e^{\sigma^2} - 1)^2(e^{4\sigma^2} + 2e^{3\sigma^2} + 3e^{2\sigma^2} - 3). \tag{4.69}$$

Note that the fourth moment is positive and a function of the variance. The fourth noncentral moment is $\mu_4'(x) = e^{4(\mu + 2\sigma^2)}$. Kurtosis is often normalized by the variance squared. Thus, the normalized kurtosis of the lognormal distribution is

$$NKurt = \frac{\mu_4}{Var^2} = (e^{\sigma^2} - 1)(e^{3\sigma^2} + 3e^{2\sigma^2} + 6e^{\sigma^2} + 6). \tag{4.70}$$

The kurtosis is an exponentially increasing function of the normal distribution variance. Recall that the normal distribution is mesokurtic, that is, $NKurt = 3$. Because variance is always positive, for the lognormal distribution, $NKurt > 3$ and is known as *leptokurtic*. As an aside, $NKurt < 3$ is called *platykurtic*.

Before illustrating graphically the lognormal distribution, we establish the link between the normal distribution parameters, μ and σ, and the expected value and standard deviation of the lognormally distributed variable. We make this link by illustrating with asset prices. Recall if variable x is distributed normal, denoted $x \sim N(\mu, \sigma)$, then variable y defined as $y = \exp(x)$ is distributed lognormal, denoted $x \sim \Lambda(\mu, \sigma)$. In the context of rates of return, suppose $S_T = S_t e^{R(T-t)}$. If $R \sim N(\mu, \sigma)$, then we know

$$S_T \sim \Lambda\left[\ln(S_t) + \mu(T - t), \sigma\sqrt{T - t}\right].$$

Thus, $\quad E(S_T) = S_0 e^{\left(\mu + \frac{\sigma^2}{2}\right)(T-t)}, \quad var(S_T) = S_0^2\left[e^{2(\mu+\sigma^2)(T-t)} - e^{(2\mu+\sigma^2)(T-t)}\right], \quad$ and

$SD(S_T) = \sqrt{S_0^2\left[e^{2(\mu+\sigma^2)(T-t)} - e^{(2\mu+\sigma^2)(T-t)}\right]}$. It is important to note that in finance the normal distribution parameters are typically expressed in percentage (μ and σ here), whereas the lognormal mean and standard deviation ($E(S_T)$ and $var(S_T)$ here) are typically expressed in currency units. Remember both distributions are two-parameter distributions. Often with both the lognormal and normal distributions, these two parameters are assumed to be the *normal* distribution's mean (μ) and standard deviation (σ). One could just as easily represent the lognormal distribution parameters with the lognormal distribution's mean, $E(S_T)$, and standard deviation, $SD(S_T)$. For example, suppose the normal distribution's mean is $\mu = 9.5\%$ and the standard deviation is $\sigma = 38\%$. If an asset price is trading at 100 and suppose we have a one-year horizon, we can compute $E(S_T) = 118.20$ and $SD(S_T) = 46.59$ based on the previous equations. Thus, the mean and standard deviation of the lognormally distributed asset price is expressed in dollars or other currency units.

Alternatively, the normal distribution parameters can be expressed as a function of the lognormal distribution parameters or

$$\mu = \frac{\ln\left[\frac{E(S_T)}{S_0}\right]}{T-t} - \frac{\ln\left[1 + \frac{Var(S_T)}{E(S_T)^2}\right]}{2(T-t)} \text{ and}$$

$$\sigma^2 = \frac{\ln\left[1 + \frac{Var(S_T)}{E(S_T)^2}\right]}{T-t}.$$

Figures 4.2a–h illustrate both the probability density function as well as the cumulative distribution function of the lognormal distribution applied to some financial instrument price. The following series of figures illustrates the effect of increases in the standard deviation because different financial instruments have different levels of standard deviation. With the lognormal distribution, the input parameters in finance are typically the mean and standard deviation of the continuously compounded rates of return. Thus, the parameters are expressed in percentage and not currency units.

We see from these graphs that with the lognormal distribution, there is no possibility that the underlying instrument value will fall to zero or below. Because common stock has limited liability, we expect the stock price to remain nonnegative, but we also expect that some stock prices will go to zero—an outcome not possible with the lognormal distribution. Based on the lognormal distribution, limited liability has no value. Thus, both the normal and lognormal distribution have strengths and weaknesses.

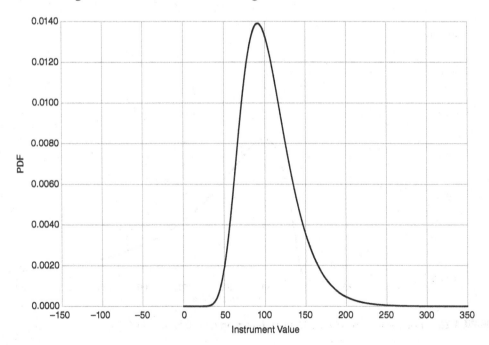

FIGURE 4.2a   Lognormal Probability Density Function ($\mu = 0\%$, $\sigma = 30\%$)

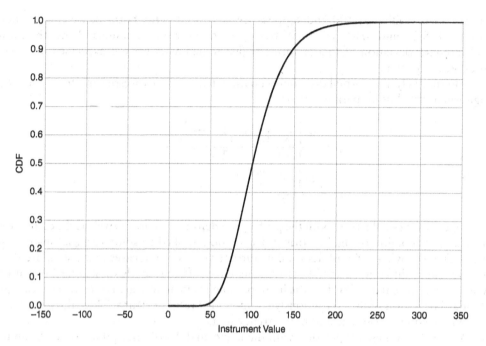

FIGURE 4.2b   Lognormal Cumulative Distribution Function ($\mu = 0\%$, $\sigma = 30\%$)

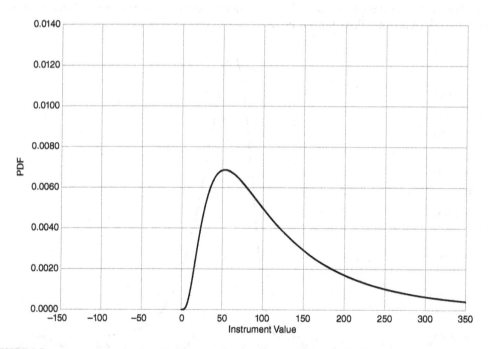

FIGURE 4.2c   Lognormal Probability Density Function ($\mu = 0\%$, $\sigma = 80\%$)

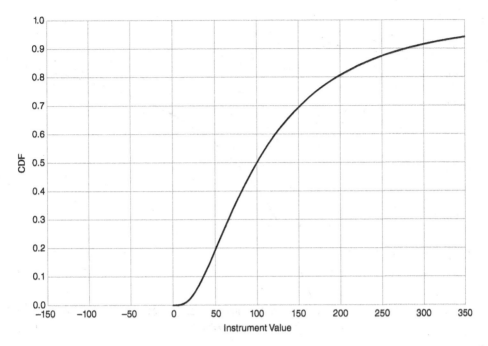

FIGURE 4.2d   Lognormal Cumulative Distribution Function ($\mu = 0\%$, $\sigma = 80\%$)

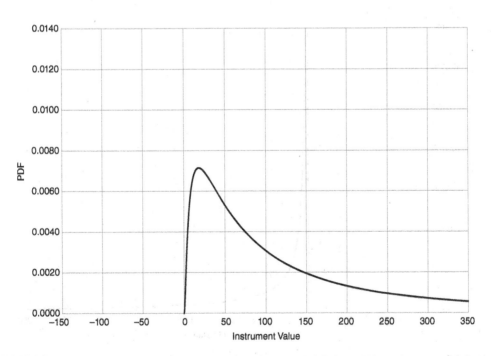

FIGURE 4.2e   Lognormal Probability Density Function ($\mu = 0\%$, $\sigma = 130\%$)

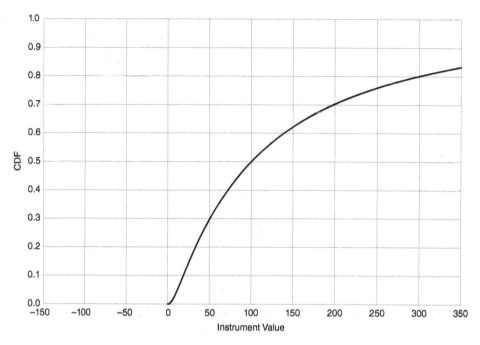

FIGURE 4.2f   Lognormal Cumulative Distribution Function ($\mu = 0\%$, $\sigma = 130\%$)

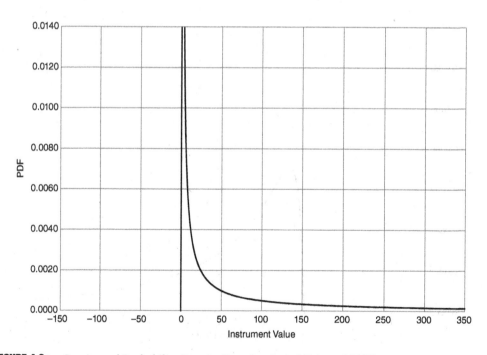

FIGURE 4.2g   Lognormal Probability Density Function ($\mu = 0\%$, $\sigma = 800\%$)

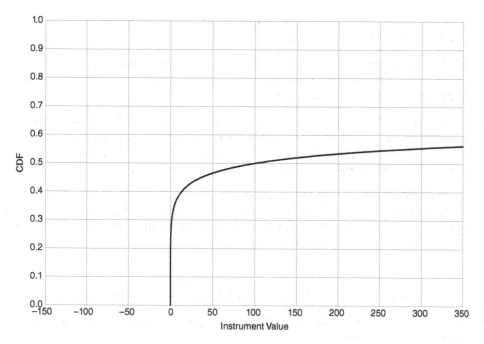

FIGURE 4.2h Lognormal Cumulative Distribution Function ($\mu = 0\%$, $\sigma = 800\%$)

4.8 RECAP AND PREVIEW

In this chapter we have provided a brief review of the foundations of probability theory, a subject widely used in understanding option valuation. We reviewed the concepts of expectation, variance, and covariance. We studied the rules for specifying these values in discrete and continuous time. We looked at the central limit theorem, the law of iterated expectations, and the law of total probability. We also examined the basic concepts behind the binomial, Poisson, normal, and lognormal distributions.

In Chapter 5, we will look more closely at the more useful probability distributions that have financial applications.

QUESTIONS AND PROBLEMS

1 Based on in-depth research, you have estimated the return distribution of two stocks for five potential scenarios that are reflected in the following table.

Probability (%)	Stock 1 (%)	Stock 2 (%)
20	7	5
20	1	6
20	4	−3
20	4	12
20	−1	5

Based on this information, compute the means and standard deviations for each stock as well as the correlation coefficient.

2 Based on the information given in the previous problem, assume you are interested in a portfolio of 50% in stock 1 and 50% in stock 2. Compute the percentage marginal contribution to the risk of each stock.

3 Suppose you are modeling an asset price with the binomial distribution where $p = 55\%$ (probability of the asset going up over the next year). You are interested in the likelihood that the asset will be up for two out of the three years. Identify the number of ways that this outcome could occur as well as its likelihood.

4 The normal distribution and lognormal distribution behave very differently as the standard deviation rises. Financial instruments with high standard deviations are very common. Complete the following table based on the mean and standard deviation of the underlying normal distribution.

Input Normal		Normal Distribution			Lognormal Distribution		
Mean (%)	Std. Dev. (%)	Mean	Median	Mode	Mean	Median	Mode
10	50						
10	100						
10	200						
10	300						

5 Based on fundamental analysis, an analyst believes that a stock's annualized, continuously compounded mean rate of return is 10% and standard deviation is 50%. Further, the analyst is comfortable with assuming a normal distribution for these rates of return. If the current stock price is 50, identify the expected stock price and standard deviation of the stock price in 10 years. Derive the implied mean and standard deviation of the normally distributed rates of return and verify your answers are correct by using your results from computing the expected stock price and standard deviation of the stock price.

NOTES

1. Those who are already well-versed in probability and statistics may benefit from an alternative perspective of finance and probability as a competitive game. See Shafer and Vovk (2001).
2. It is interesting to note that these are marginal probabilities with respect to the outcomes, +2, 0, −2, but they are joint probabilities of the coin tosses themselves. For example, the marginal probability of an outcome of +2 is ¼ but this value is the joint probability of two heads, (½)(½) = ½.
3. If we obtained a head on the first toss, we currently have a total of +1. Thus, there is a probability of ½ that we get another head, giving us a total of +2, and a probability of ½ that we get a tail, giving us a total of 0. There is zero probability that we end up with a total of 0, because we already have +1.

4. If we obtained a tail on the first toss, we currently have a total of −1. Thus, there is a probability of ½ that we get a head, giving us a total of 0, and a probability of ½ that we get another tail, giving us a total of −2. There is zero probability that we end up with a total of +2, because we already have −1 and have only one more toss.

5. In finance sometimes conditional expectations are written in the form $E(x_{t+j}|I_t)$, meaning that the expectation of x at time $t + j$ is based on the information set, I, available at time t.

6. The standard normal distribution assumes a zero mean and standard deviation of 1. More extensive details will be given later.

7. Note that because $\Pr(x = 0) = 0$, then $\Pr(x \geq 0) = \Pr(x > 0)$.

8. Although $f(x)$ is commonly used for the density function, $F(x)$ is often used for the distribution function.

9. The term *mean* is used somewhat more often to describe data drawn from samples as opposed to the concept of expected value, which is based on ex ante information.

10. For example, a sine wave is a nonlinear relationship that tends to show zero covariance between say X and $\sin(X)$. Nonetheless, there is clearly a relationship. It is simply nonlinear.

11. For example, if the distribution is normal, then we can say that the probability of being more than one standard deviation away from the mean, in each direction, is about 0.16, or 0.32 for both directions.

12. The probability mass function is the function that yields the likelihood that a discrete variable exactly equals a particular value.

CHAPTER 5

Financial Applications of Probability Distributions

In Chapter 4, we took a look at the fundamentals of probability theory and introduced some distributions. Option theory and indeed much of finance theory is based on the normal distribution as well as the related lognormal distribution; hence, we devote a full chapter to these distributions. In the limit, discrete distributions such as the binomial can be structured to converge to either the normal or lognormal distributions. The reasons for this extensive use of the normal and lognormal distributions are twofold. First, these distributions have only two parameters, because they are completely characterized by the expected value and variance. Second, financial utility is often based on two characteristics: non-satiation and risk aversion. Non-satiation is the notion that people always prefer more money to less. As such, they prefer a higher expected return, so they have a preference for the first moment of the distribution. Risk aversion is the notion that people do not like risk. Hence, they dislike the second moment of the distribution. If people care about only the first two moments, then the normal distribution is appropriate.

Of course, people may indeed care about higher-order moments. For example, positive skewness is generally a desirable feature of investments. Some financial models are based on skewness preference and skewed distributions. In the first part of the chapter, we will introduce the normal and lognormal distributions. The second part of the chapter refers to a related distribution called the *bivariate normal* and *lognormal*. As such, it might be more appropriate to call the single variable versions the univariate normal and lognormal distribution, but this is typically not done. Usually when one uses the phrase *normal distribution*, one means the univariate normal. This case is the one involving only a single variable. In the second case, the bivariate normal or lognormal, there are two normally distributed variables and they may be correlated.

5.1 THE UNIVARIATE NORMAL PROBABILITY DISTRIBUTION

The normal probability distribution, known commonly to the layperson as the *bell-shaped curve* or bell curve, was first identified in the 18th century by Abraham de Moivre (1667–1754). Its mathematical structure was developed by Carl Frederich Gauss (1777–1855) and the curve is often referred to as the *Gaussian distribution*. Recall the mathematical function that plots the normal curve, called the *probability density function* or density function for short, is

$$n(x) = \frac{1}{\sigma\sqrt{2\pi}}e^{-\left(\frac{x-\mu}{\sigma}\right)^2/2}, \tag{5.1}$$

where x is the value of the random variable, μ is its expected value, and σ is its standard deviation.[1] We use the lowercase $n()$ for the density function to remind us that it is the normal distribution. The variable has values that range from $-\infty$ to $+\infty$.

Any normally distributed random variable can be expressed as a standard normal random variable by subtracting its expected value and dividing by its standard deviation. This standardized normal variable is often referred to with the letter z and its density is written as

$$n(z) = \frac{1}{\sqrt{2\pi}}e^{-z^2/2}. \tag{5.2}$$

A normal distribution can apply to any random variable, but this is a special case that we often use that is called the *standard normal*, which has an expected value of zero and a variance of 1. For example, a normally distributed random variable with an expected value of 6 and a standard deviation of 3 can be modeled with a standard normal random variable with an expected value of 0 and a standard deviation of 1. If one observed a value of the random variable of 8, we would convert it to a standard normal by subtracting the expected value of 6 and dividing by the standard deviation of 3 to obtain $(8 - 6)/3 = 2/3$. Alternatively, if we generate a standard normal random variable with value 2/3, we can convert it to the random variable we seek to model by multiplying by 3 and adding 6 to obtain $(2/3)^*3 + 6 = 8$.

The density function gives only the height of the curve. The actual probability of a random variable lying within a particular range is provided by the distribution function, which is the cumulative density. The mathematical specification for the cumulative density function for a standard normal is

$$N(z) = \int_{-\infty}^{z} \frac{1}{\sqrt{2\pi}}e^{-x^2/2}dx. \tag{5.3}$$

This equation is interpreted as the probability that a standard normal variable will be less than z. It is sometimes written as $\Pr(x \leq z)$ or simply $N(z)$. The probability is the area under the curve. The total area under the curve, that is, the integral from $-\infty$ to $+\infty$, is 1.0. Thus $N(\infty) = 1$ and $N(-\infty) = 0$. The probability that a standard normal variable will lie between a and b $(a < b)$ is given as $N(b) - N(a)$. With any continuous variable, it is impossible to speak of the probability of obtaining a specific value such as the probability that $z = a$. The probability of observing any one value in a continuous distribution is zero. Only the probability over a range can be determined. Thus, the expressions $\Pr(z \leq a)$ and $\Pr(z < a)$ are equivalent.

Calculation of the probability for a particular range involves the evaluation of the previous integral. It is well known that the distribution function for the normal probability cannot be integrated by standard mathematical means. Instead, estimation techniques must be used. Tables for the function are widely available and found in nearly every statistics book. Fortunately, there are several other excellent and simple means of computing the normal probability.

It is a general rule that any well-behaved mathematical function with derivatives that exist up to a given order can be approximated by a polynomial function of that order. There are many such approximations of the normal probability, most of which are provided in Abramowitz and Stegun (1972). Probably the most widely used is the following:

$$N(z) = 1 - n(z) \begin{pmatrix} 0.31938153k - 0.356563782k^2 + 1.781477937k^3 \\ -1.821255978k^4 + 1.330274429k^5 \end{pmatrix}, \tag{5.4}$$

where $k = 1/(1 + 0.2316419z)$ and $n(z) = \left(1/\sqrt{2\pi}\right) \exp\left(-z^2/2\right)$. If $z < 0$, the fact that the curve is symmetric enables us to obtain $N(z)$ as $1 - N(z)$. This function is known to be accurate to at least four digits.

In addition, the Excel function "=normsdist(x)" where x is a value or cell reference can be used with reasonable accuracy. Table 5.1 provides a detailed table created with this Excel function and is useful when manually estimating values from a standard normal cumulative distribution function.

To use this table, suppose we wish to find the probability of observing a value less than 1.3520. This value is found by looking up 1.3 in the left column and moving over to the 0.05 column. We see that $N(1.35) = 0.911492$. This table provides values to six decimal places. We simply round at the 0.0025 and 0.0075. Thus, in this case, 1.352 rounds down to 1.35. Now suppose we wish to find $N(-0.5230)$. Because this table provides negative values, we simply find $N(-0.525)$ as we round up in absolute value to the next 0.005 increment. Thus, $N(-0.525) = 0.299792$. Table 5.1 is useful for classroom work as the chance for serious rounding errors is minimal. Note that $N(0.0) = 0.5$, meaning that half of the area under the curve is to the left of zero. In this table, however, there are two cells with 0.5 [$N(0.0)$ and $N(-0.0)$]. In the row with -0.0, the values are moving more negative, whereas in the row with 0.0, the values are moving more positive.

In finance, one of our primary interests is in the distribution of financial market returns. When a variable is normally distributed, statistical analysis and testing is much easier. The distribution of daily returns on the S&P 500 from a recent 66-year period is depicted in Figure 5.1. There are 16,858 returns, so the breakdowns are relatively dense.

The figure does give a modest appearance of a normal distribution, but there are some troubling factors. There are two returns of more than 10% and one of less than −20%. Let us try to determine the likelihood that such returns would occur on a given day in a normal distribution.

The average daily return is 0.000338, and the standard deviation is 0.009666. For a return of −20%, the z-statistic is $(-0.2 - 0.000338)/0.009666 = -21.2099$.[2] Thus, a return of −20% in a day is more than 21 standard deviations to the left. The probability of this occurring is 3.8724e−100, which is about 1.03295e+97 years of trading.[3] On the upside, the highest return is more than 11 standard deviations above the mean, a probability also extremely small. These returns are so extreme that they cast doubt about whether the normal distribution is appropriate.

Normal probability theory also says that only about 0.3% of all returns should be more than +/− three standard deviations of the mean. With 16,858 S&P 500 returns, we should expect that 0.003(16,858) = 50.57, or about 51 returns above and below three standard deviations. In fact, there are 112 returns above and 123 below, for a total of 235.

TABLE 5.1 Standard Normal Cumulative Distribution Function Table

Values for $N(d)$, given d: Rows give first decimal of d and columns give second and third decimals

d	0.000	0.005	0.010	0.015	0.020	0.025	0.030	0.035	0.040	0.045	0.050	0.055	0.060	0.065	0.070	0.075	0.080	0.085	0.090	0.095
-2.9	0.001866	0.001836	0.001807	0.001778	0.001750	0.001722	0.001695	0.001668	0.001641	0.001615	0.001589	0.001563	0.001538	0.001513	0.001489	0.001465	0.001441	0.001418	0.001395	0.001372
-2.8	0.002555	0.002516	0.002477	0.002439	0.002401	0.002364	0.002327	0.002291	0.002256	0.002221	0.002186	0.002152	0.002118	0.002085	0.002052	0.002020	0.001988	0.001957	0.001926	0.001896
-2.7	0.003467	0.003415	0.003364	0.003314	0.003264	0.003215	0.003167	0.003119	0.003072	0.003026	0.002980	0.002935	0.002890	0.002846	0.002803	0.002760	0.002718	0.002676	0.002635	0.002595
-2.6	0.004661	0.004594	0.004527	0.004461	0.004396	0.004332	0.004269	0.004207	0.004145	0.004085	0.004025	0.003965	0.003907	0.003849	0.003793	0.003736	0.003681	0.003626	0.003573	0.003519
-2.5	0.006210	0.006123	0.006037	0.005952	0.005868	0.005785	0.005703	0.005622	0.005543	0.005464	0.005386	0.005309	0.005234	0.005159	0.005085	0.005012	0.004940	0.004869	0.004799	0.004730
-2.4	0.008198	0.008086	0.007976	0.007868	0.007760	0.007654	0.007549	0.007446	0.007344	0.007243	0.007143	0.007044	0.006947	0.006851	0.006756	0.006662	0.006569	0.006478	0.006387	0.006298
-2.3	0.010724	0.010583	0.010444	0.010306	0.010170	0.010036	0.009903	0.009772	0.009642	0.009514	0.009387	0.009261	0.009137	0.009015	0.008894	0.008774	0.008656	0.008540	0.008424	0.008310
-2.2	0.013903	0.013727	0.013553	0.013380	0.013209	0.013041	0.012874	0.012709	0.012545	0.012384	0.012224	0.012067	0.011911	0.011756	0.011604	0.011453	0.011304	0.011156	0.011011	0.010867
-2.1	0.017864	0.017646	0.017429	0.017215	0.017003	0.016793	0.016586	0.016381	0.016177	0.015976	0.015778	0.015581	0.015386	0.015194	0.015003	0.014815	0.014629	0.014444	0.014262	0.014082
-2.0	0.022750	0.022482	0.022216	0.021952	0.021692	0.021434	0.021178	0.020925	0.020675	0.020427	0.020182	0.019940	0.019699	0.019462	0.019226	0.018993	0.018763	0.018535	0.018309	0.018085
-1.9	0.028717	0.028390	0.028067	0.027746	0.027429	0.027115	0.026803	0.026495	0.026190	0.025887	0.025588	0.025292	0.024998	0.024707	0.024419	0.024134	0.023852	0.023572	0.023295	0.023021
-1.8	0.035930	0.035537	0.035148	0.034762	0.034380	0.034001	0.033625	0.033253	0.032884	0.032519	0.032157	0.031798	0.031443	0.031091	0.030742	0.030396	0.030054	0.029715	0.029379	0.029046
-1.7	0.044565	0.044097	0.043633	0.043173	0.042716	0.042264	0.041815	0.041370	0.040930	0.040492	0.040059	0.039630	0.039204	0.038782	0.038364	0.037949	0.037538	0.037131	0.036727	0.036327
-1.6	0.054799	0.054247	0.053699	0.053155	0.052616	0.052081	0.051551	0.051025	0.050503	0.049985	0.049471	0.048962	0.048457	0.047956	0.047460	0.046967	0.046479	0.045994	0.045514	0.045038
-1.5	0.066807	0.066162	0.065522	0.064886	0.064255	0.063630	0.063008	0.062392	0.061780	0.061173	0.060571	0.059973	0.059380	0.058791	0.058208	0.057628	0.057053	0.056483	0.055917	0.055356
-1.4	0.080757	0.080011	0.079270	0.078534	0.077804	0.077079	0.076359	0.075644	0.074934	0.074229	0.073529	0.072835	0.072145	0.071460	0.070781	0.070106	0.069437	0.068772	0.068112	0.067457
-1.3	0.096800	0.095946	0.095098	0.094255	0.093418	0.092586	0.091759	0.090938	0.090123	0.089313	0.088508	0.087709	0.086915	0.086127	0.085343	0.084566	0.083793	0.083026	0.082264	0.081508
-1.2	0.115070	0.114102	0.113139	0.112183	0.111232	0.110288	0.109349	0.108415	0.107488	0.106566	0.105650	0.104739	0.103835	0.102936	0.102042	0.101155	0.100273	0.099396	0.098525	0.097660
-1.1	0.135666	0.134580	0.133500	0.132425	0.131357	0.130295	0.129238	0.128188	0.127143	0.126105	0.125072	0.124045	0.123024	0.122009	0.121000	0.119997	0.119000	0.118009	0.117023	0.116044
-1.0	0.158655	0.157448	0.156248	0.155053	0.153864	0.152682	0.151505	0.150334	0.149170	0.148011	0.146859	0.145713	0.144572	0.143438	0.142310	0.141187	0.140071	0.138961	0.137857	0.136758
-0.9	0.184060	0.182733	0.181411	0.180096	0.178786	0.177483	0.176186	0.174894	0.173609	0.172329	0.171056	0.169789	0.168528	0.167272	0.166023	0.164780	0.163543	0.162312	0.161087	0.159868
-0.8	0.211855	0.210410	0.208970	0.207536	0.206108	0.204686	0.203269	0.201859	0.200454	0.199055	0.197663	0.196276	0.194895	0.193519	0.192150	0.190787	0.189430	0.188078	0.186733	0.185394
-0.7	0.241964	0.240405	0.238852	0.237305	0.235762	0.234226	0.232695	0.231170	0.229650	0.228136	0.226627	0.225124	0.223627	0.222136	0.220650	0.219170	0.217695	0.216227	0.214764	0.213307
-0.6	0.274253	0.272589	0.270931	0.269277	0.267629	0.265986	0.264347	0.262714	0.261086	0.259464	0.257846	0.256234	0.254627	0.253025	0.251429	0.249838	0.248252	0.246672	0.245097	0.243528
-0.5	0.308538	0.306779	0.305026	0.303277	0.301532	0.299792	0.298056	0.296325	0.294599	0.292877	0.291160	0.289447	0.287740	0.286037	0.284339	0.282646	0.280957	0.279274	0.277595	0.275922
-0.4	0.344578	0.342739	0.340903	0.339071	0.337243	0.335418	0.333598	0.331781	0.329969	0.328160	0.326355	0.324555	0.322758	0.320966	0.319178	0.317393	0.315614	0.313838	0.312067	0.310300
-0.3	0.382089	0.380183	0.378280	0.376381	0.374484	0.372591	0.370700	0.368813	0.366929	0.365047	0.363169	0.361295	0.359424	0.357556	0.355691	0.353830	0.351973	0.350119	0.348268	0.346421
-0.2	0.420740	0.418786	0.416834	0.414884	0.412936	0.410990	0.409046	0.407104	0.405165	0.403228	0.401294	0.399362	0.397432	0.395505	0.393580	0.391658	0.389739	0.387822	0.385908	0.383997
-0.1	0.460172	0.458188	0.456205	0.454223	0.452242	0.450262	0.448283	0.446306	0.444330	0.442355	0.440382	0.438411	0.436441	0.434472	0.432505	0.430540	0.428576	0.426615	0.424655	0.422696
-0.0	0.500000	0.498005	0.496011	0.494016	0.492022	0.490027	0.488034	0.486040	0.484047	0.482054	0.480061	0.478069	0.476078	0.474087	0.472097	0.470107	0.468119	0.466131	0.464144	0.462157

0.0	0.500000	0.501995	0.503989	0.505984	0.507978	0.509973	0.511966	0.513960	0.515953	0.517946	0.519939	0.521931	0.523922	0.525913	0.527903	0.529893	0.531881	0.533869	0.535856	0.537843
0.1	0.539828	0.541812	0.543795	0.545777	0.547758	0.549738	0.551717	0.553694	0.555670	0.557645	0.559618	0.561589	0.563559	0.565528	0.567495	0.569460	0.571424	0.573385	0.575345	0.577304
0.2	0.579260	0.581214	0.583166	0.585116	0.587064	0.589010	0.590954	0.592896	0.594835	0.596772	0.598706	0.600638	0.602568	0.604495	0.606420	0.608342	0.610261	0.612178	0.614092	0.616003
0.3	0.617911	0.619817	0.621720	0.623619	0.625516	0.627409	0.629300	0.631187	0.633072	0.634953	0.636831	0.638705	0.640576	0.642444	0.644309	0.646170	0.648027	0.649881	0.651732	0.653579
0.4	0.655422	0.657261	0.659097	0.660929	0.662757	0.664582	0.666402	0.668219	0.670031	0.671840	0.673645	0.675445	0.677242	0.679034	0.680822	0.682607	0.684386	0.686162	0.687933	0.689700
0.5	0.691462	0.693221	0.694974	0.696723	0.698468	0.700208	0.701944	0.703675	0.705401	0.707123	0.708840	0.710553	0.712260	0.713963	0.715661	0.717354	0.719043	0.720726	0.722405	0.724078
0.6	0.725747	0.727411	0.729069	0.730723	0.732371	0.734014	0.735653	0.737286	0.738914	0.740536	0.742154	0.743766	0.745373	0.746975	0.748571	0.750162	0.751748	0.753328	0.754903	0.756472
0.7	0.758036	0.759595	0.761148	0.762695	0.764238	0.765774	0.767305	0.768830	0.770350	0.771864	0.773373	0.774876	0.776373	0.777864	0.779350	0.780830	0.782305	0.783773	0.785236	0.786693
0.8	0.788145	0.789590	0.791030	0.792464	0.793892	0.795314	0.796731	0.798141	0.799546	0.800945	0.802337	0.803724	0.805105	0.806481	0.807850	0.809213	0.810570	0.811922	0.813267	0.814606
0.9	0.815940	0.817267	0.818589	0.819904	0.821214	0.822517	0.823814	0.825106	0.826391	0.827671	0.828944	0.830211	0.831472	0.832728	0.833977	0.835220	0.836457	0.837688	0.838913	0.840132
1.0	0.841345	0.842552	0.843752	0.844947	0.846136	0.847318	0.848495	0.849666	0.850830	0.851989	0.853141	0.854287	0.855428	0.856562	0.857690	0.858813	0.859929	0.861039	0.862143	0.863242
1.1	0.864334	0.865420	0.866500	0.867575	0.868643	0.869705	0.870762	0.871812	0.872857	0.873895	0.874928	0.875955	0.876976	0.877991	0.879000	0.880003	0.881000	0.881991	0.882977	0.883956
1.2	0.884930	0.885898	0.886861	0.887817	0.888768	0.889712	0.890651	0.891585	0.892512	0.893434	0.894350	0.895261	0.896165	0.897064	0.897958	0.898845	0.899727	0.900604	0.901475	0.902340
1.3	0.903200	0.904054	0.904902	0.905745	0.906582	0.907414	0.908241	0.909062	0.909877	0.910687	0.911492	0.912291	0.913085	0.913873	0.914657	0.915434	0.916207	0.916974	0.917736	0.918492
1.4	0.919243	0.919989	0.920730	0.921466	0.922196	0.922921	0.923641	0.924356	0.925066	0.925771	0.926471	0.927165	0.927855	0.928540	0.929219	0.929894	0.930563	0.931228	0.931888	0.932543
1.5	0.933193	0.933838	0.934478	0.935114	0.935745	0.936370	0.936992	0.937608	0.938220	0.938827	0.939429	0.940027	0.940620	0.941209	0.941792	0.942372	0.942947	0.943517	0.944083	0.944644
1.6	0.945201	0.945753	0.946301	0.946845	0.947384	0.947919	0.948449	0.948975	0.949497	0.950015	0.950529	0.951038	0.951543	0.952044	0.952540	0.953033	0.953521	0.954006	0.954486	0.954962
1.7	0.955435	0.955903	0.956367	0.956827	0.957284	0.957736	0.958185	0.958630	0.959070	0.959508	0.959941	0.960370	0.960796	0.961218	0.961636	0.962051	0.962462	0.962869	0.963273	0.963673
1.8	0.964070	0.964463	0.964852	0.965238	0.965620	0.965999	0.966375	0.966747	0.967116	0.967481	0.967843	0.968202	0.968557	0.968909	0.969258	0.969604	0.969946	0.970285	0.970621	0.970954
1.9	0.971283	0.971610	0.971933	0.972254	0.972571	0.972885	0.973197	0.973505	0.973810	0.974113	0.974412	0.974708	0.975002	0.975293	0.975581	0.975866	0.976148	0.976428	0.976705	0.976979
2.0	0.977250	0.977518	0.977784	0.978048	0.978308	0.978566	0.978822	0.979075	0.979325	0.979573	0.979818	0.980060	0.980301	0.980538	0.980774	0.981007	0.981237	0.981465	0.981691	0.981915
2.1	0.982136	0.982354	0.982571	0.982785	0.982997	0.983207	0.983414	0.983619	0.983823	0.984024	0.984222	0.984419	0.984614	0.984806	0.984997	0.985185	0.985371	0.985556	0.985738	0.985918
2.2	0.986097	0.986273	0.986447	0.986620	0.986791	0.986959	0.987126	0.987291	0.987455	0.987616	0.987776	0.987933	0.988089	0.988244	0.988396	0.988547	0.988696	0.988844	0.988989	0.989133
2.3	0.989276	0.989417	0.989556	0.989694	0.989830	0.989964	0.990097	0.990228	0.990358	0.990486	0.990613	0.990739	0.990863	0.990985	0.991106	0.991226	0.991344	0.991460	0.991576	0.991690
2.4	0.991802	0.991914	0.992024	0.992132	0.992240	0.992346	0.992451	0.992554	0.992656	0.992757	0.992857	0.992956	0.993053	0.993149	0.993244	0.993338	0.993431	0.993522	0.993613	0.993702
2.5	0.993790	0.993877	0.993963	0.994048	0.994132	0.994215	0.994297	0.994378	0.994457	0.994536	0.994614	0.994691	0.994766	0.994841	0.994915	0.994988	0.995060	0.995131	0.995201	0.995270
2.6	0.995339	0.995406	0.995473	0.995539	0.995604	0.995668	0.995731	0.995793	0.995855	0.995915	0.995975	0.996035	0.996093	0.996151	0.996207	0.996264	0.996319	0.996374	0.996427	0.996481
2.7	0.996533	0.996585	0.996636	0.996686	0.996736	0.996785	0.996833	0.996881	0.996928	0.996974	0.997020	0.997065	0.997110	0.997154	0.997197	0.997240	0.997282	0.997324	0.997365	0.997405
2.8	0.997445	0.997484	0.997523	0.997561	0.997599	0.997636	0.997673	0.997709	0.997744	0.997779	0.997814	0.997848	0.997882	0.997915	0.997948	0.997980	0.998012	0.998043	0.998074	0.998104
2.9	0.998134	0.998164	0.998193	0.998222	0.998250	0.998278	0.998305	0.998332	0.998359	0.998385	0.998411	0.998437	0.998462	0.998487	0.998511	0.998535	0.998559	0.998582	0.998605	0.998628

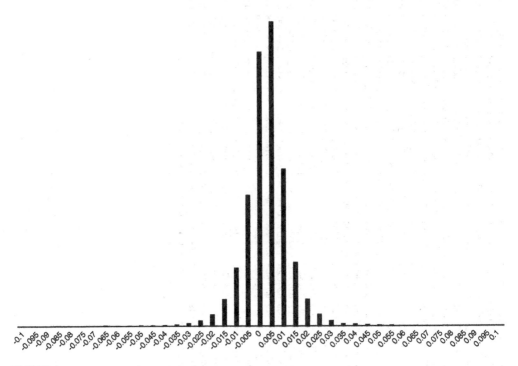

FIGURE 5.1   S&P 500 Daily Returns

So indeed, the normal distribution is not a perfect fit. We should not have expected as much, but that does not mean that it cannot serve as a useful approximation. Now, however, let us consider a closely related alternative.

Recall the discretely compounded rate of return, R_t^d, expresses the relationship between asset prices at $t-1$ and t is

$$S_{t-1}\left(1 + R_t^d\right) = S_t. \tag{5.5}$$

Hence, the return is calculated as

$$R_t^d = \left(\frac{S_t}{S_{t-1}}\right) - 1. \tag{5.6}$$

Some researchers have argued that the use of log or continuously compounded returns can reduce the nonnormality. The log return, R_t^c, expresses the relationship between a variable at two points in time as it grows continuously:

$$S_{t-1}e^{R_t^c} = S_t. \tag{5.7}$$

Hence, it is calculated as

$$R_t^c = \ln\left(\frac{S_t}{S_{t-1}}\right). \tag{5.8}$$

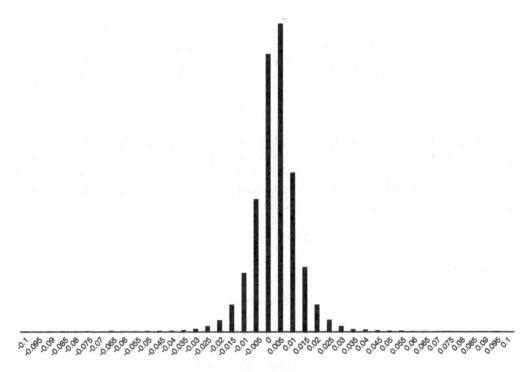

FIGURE 5.2 S&P 500 Log Returns

Some researchers believe that log returns are closer to being normally distributed. This point is partially true but only by a small amount. Log returns are slightly smaller than regular returns, thereby pulling in the largest positive returns but by reducing returns, they lower the largest negative returns. For example, the largest positive return of 11.58% (October 13, 2008) has a log equivalent of 10.96%. The largest negative return, −20.47% (October 19, 1987), has a log equivalent of −22.90%. Both returns are lower and not enough to eliminate the simple fact that there are far too many extreme returns for a normal distribution. Figure 5.2 illustrates the same data using log returns. As you can see, there is not much difference.

5.2 CONTRASTING THE NORMAL WITH THE LOGNORMAL PROBABILITY DISTRIBUTION

Based on properties of the lognormal distribution, if $S_t = S_{t-1}e^{R_t^c}$ and R_t^c is normally distributed, then we know S_t is lognormally distributed. Recall if x has a lognormal distribution, then the PDF is

$$\lambda(x) = \frac{1}{x\sigma\sqrt{2\pi}}e^{-\left(\frac{[\ln(x)-\mu]^2}{2\sigma^2}\right)}. \tag{5.9}$$

Again, the range of a lognormally distributed variable is $0 < x < +\infty$. Importantly, zero is not included.

Recall from Chapter 4, the nth noncentral moment is defined as $\mu_n'(x) = E(x^n)$ and the nth central moment is defined as $\mu_n(x) = E[(x - \mu)^n]$. Thus, subtracting a constant term from a random variable affects only the mean, not the higher central moments, such as variance, skewness, and kurtosis. We denote the *first difference* or dollar profit and loss as

$$\Delta S_t = S_t - S_{t-1}. \tag{5.10}$$

There are many finance applications where we are interested in dollar gains and losses. For example, company earnings are simply revenues less expenses. Because earnings can be zero or negative, we do not want model earnings with a distribution that does not admit zero or negative numbers.[4] Because the uncertainty at time $t - 1$ is solely the instrument price or value at time t, the expected value is

$$E(\Delta S_t) = E(S_t) - S_{t-1}. \tag{5.11}$$

The higher central moments are simply

$$var(\Delta S_t) = var(S_t), \tag{5.12}$$

$$Skew(\Delta S_t) = Skew(S_t), \text{ and} \tag{5.13}$$

$$Kurtosis(\Delta S_t) = Kurtosis(S_t). \tag{5.14}$$

Several important insights can be gained when we assume a lognormal distribution. First, the continuously compounded rate of return follows a normal distribution. Second, the normalized skewness of the rate of return should be zero, but the normalized skewness of the first difference should be positive. Third, the normalized kurtosis of the rate of return should be 3, but the normalized kurtosis of the first difference should be greater than 3.[5]

When building quality valuation models, it is important to understand empirical properties of financial data. Here we briefly review empirical data related to two exchange-traded funds (ETFs), the S&P 500 ETF (SPY) and the Technology Sector ETF (XLK), as well as Apple stock (AAPL). We explore five years of daily data. Figure 5.3 provides plots of the time series rates of return and first differences. To facilitate comparison, the first differences are based on an assumed $1 investment at the beginning of the period. For ease of comparison, we set the y-axis to be the same for all the return plots. Likewise, we set the y-axis to be the same for all the first difference plots. Intuitively, we would expect that an individual stock would be more risky than a portfolio of stocks within its sector. Further, we would expect the technology sector ETF would be more risky than a broad-based index such as SPY. The pattern of standard deviations is consistent with our intuition. Outside of basic distributional properties, it is unclear how skewness and kurtosis would behave. Interestingly, all skewness measures are negative rather than zero or positive as the distributional assumptions would imply. Further, the skewness measures are similar between returns and first differences. We would have expected the first difference skewness to be positive or at least less negative than returns. All kurtosis measures are greater than three with daily first differences being slightly lower.

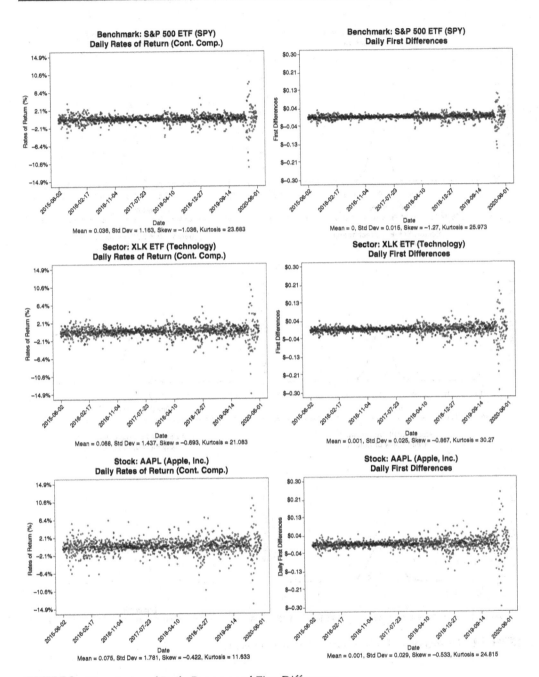

FIGURE 5.3 Time Series of Daily Returns and First Differences

Figure 5.4 provides histograms of this same data along with the corresponding normal distribution probability density function. The histograms are more spread out as we move from SPY to XLK to AAPL. The patterns are very similar when comparing returns and first differences as well as consistent with the previous univariate analysis.

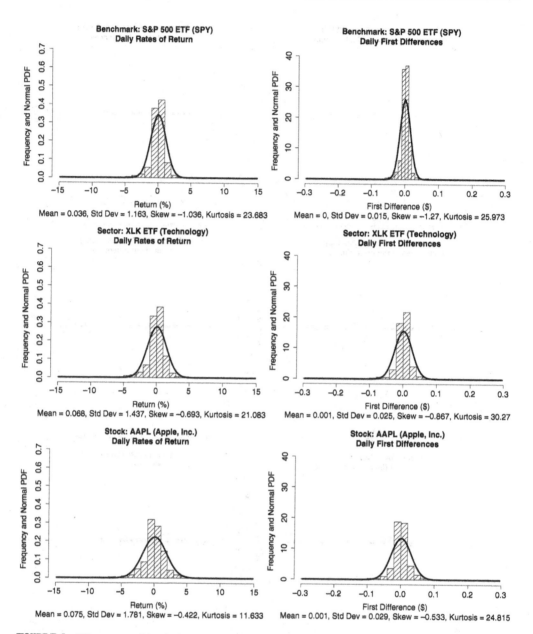

FIGURE 5.4 Histogram of Daily Returns and First Differences

5.3　BIVARIATE NORMAL PROBABILITY DISTRIBUTION

Suppose now that we have two normally distributed random variables, x and y. The expected values are μ_x and μ_y and the standard deviations are σ_x and σ_y. For bivariate normally distributed random variables, the conditional expected values of y and x are linearly related, as indicated by the following:

$$E(y|x_a) = \mu_y + \rho \left(\frac{\sigma_y}{\sigma_x} \right) (x_a - \mu_x), \tag{5.15}$$

where ρ is the correlation between y and x. This statement says that if the value of x is known, the expected value of y is given by the right-hand-side expression. This expectation of y is called the conditional expected value of y, given x_a. The terms μ_y and μ_x are the unconditional expected values. They are our best estimates of the expected values of y or x, given that we know nothing about the value of the other. If x and y are linearly related, then the correlation between x and y can be used to make a better prediction of y, given that we know the current value of x, which is x_a. If y and x are related in this manner, then the joint distribution of y and x is bivariate normal. The conditional variance of y is related to its unconditional variance by the formula

$$\sigma_{y|x}^2 = \sigma_y^2 (1 - \rho^2). \tag{5.16}$$

The probability density function for the bivariate normal is

$$f(x, y, \rho) = \frac{1}{2\pi\sigma_x\sigma_y\sqrt{1-\rho^2}}$$
$$\exp\left[-\frac{1}{2} \left(\frac{\left((x-\mu_x)/\sigma_x \right)^2 - 2\rho\left((x-\mu_x)/\sigma_x \right)\left((y-\mu_y)/\sigma_y \right) + \left((y-\mu_y)/\sigma_y \right)^2}{1-\rho^2} \right) \right]. \tag{5.17}$$

The distribution function or cumulative bivariate normal probability is

$$\Pr(x \le x_a, y \le y_b|\rho) = \frac{1}{\sigma_x\sigma_y} \int_{-\infty}^{\frac{x_a-\mu_x}{\sigma_x}} \int_{-\infty}^{\frac{y_b-\mu_y}{\sigma_y}} \frac{1}{2\pi\sqrt{1-\rho^2}}$$
$$\exp\left[-\frac{1}{2} \left(\frac{k^2 - 2\rho kj + j^2}{1-\rho^2} \right) \right] dk\,dj. \tag{5.18}$$

Because each variable x and y is individually normally distributed, each can be transformed or normalized into a standard normal random variable, which we shall call z_1 and z_2, by the relationships,

$$z_1 = \frac{x - \mu_x}{\sigma_x}, \quad z_2 = \frac{y - \mu_y}{\sigma_y}. \tag{5.19}$$

The standard normal bivariate density is then

$$f(z_1, z_2, \rho) = \frac{1}{2\pi\sqrt{1-\rho^2}} \exp\left[-\frac{1}{2}\left(\frac{z_1^2 - 2\rho z_1 z_2 + z_2^2}{1-\rho^2}\right)\right]. \tag{5.20}$$

Figure 5.5 illustrates the standard normal bivariate density in three-dimensional space.

The following relationships are useful when dealing with the bivariate normal probability distribution. Let $N(x)$ be the (univariate) normal probability for a variable x and $N_2(x, y; \rho)$ be the bivariate normal probability for the variables x and y, which can be normalized or not, with correlation ρ. Then, the following are a handful of useful rules.

$$N_2(x, y; \rho) = N_2(y, x; \rho)$$

$$N_2(-x, y; \rho) + N_2(x, y; -\rho) = N(y) \tag{5.21}$$

$$N_2(x, y; \rho) - N_2(-x, -y; \rho) = N(x) + N(y) - 1.0.$$

Computation of the bivariate normal probability is quite challenging, but an analytic approximation developed by Drezner (1978) is often used and gives a high degree of accuracy.[6] Using that approximation, let us work a problem involving bivariate standard normal random variables. An Excel routine based on this technique is in Appendix 5A. Let $x = 0.74$, $y = -1.13$, and $\rho = 0.32$. We wish to know $\Pr(x \leq 0.74,$

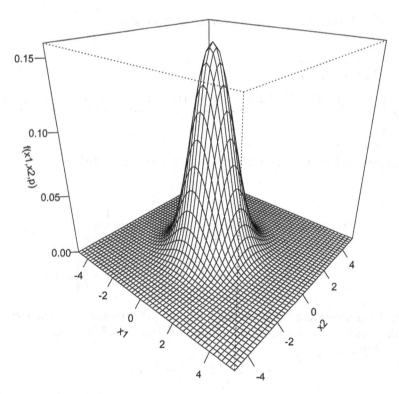

FIGURE 5.5 Bivariate Normal Density

$y \leq -1.13|0.32)$. The univariate probabilities as obtained from Excel's =normsdist() function are $N(0.74) = 0.7704$ and $N(-1.13) = 0.1292$. The bivariate normal probability is $N_2(0.74, -1.13; 0.32) = 0.1171$. Let us check out the above relationships:

$$N_2(0.74, -1.13, 0.32) = 0.1171 = N(-1.13, 0.74, 0.32) = 0.1171$$

$$N_2(-0.74, -1.13, 0.32) = 0.0529, N_2(0.74, -1.13, -0.32) = 0.0763$$

$$0.0529 + 0.0763 = 0.1292 = N(-1.13)$$

$$N_2(0.74, -1.13, 0.32) = 0.1171, N_2(-0.74, 1.13, 0.32) = 0.2175$$

$$0.1171 - 0.2175 = -0.1004$$

$$N(0.74) = 0.7704, N(-1.13) = 0.1292$$

$$0.7704 + 0.1292 - 1 = -0.1004.$$

Some special cases are worth noting. If either of the values x or y is infinite, then the bivariate probability reverts to the univariate probability. For example, $\Pr(x \leq a, y \leq \infty|\rho) = \Pr(x \leq a)$. This is because of the condition that $y \leq \infty$ has no effect because its probability is 1.0. This fact, of course, also holds if the variables are reversed. If $\rho = 0$, then the bivariate probability is the product of the two univariate probabilities of x and y, that is, $\Pr(x \leq a, y \leq b|\rho = 0) = N(a)N(b)$. This result reflects the fact that the joint probability of two independent random variables is the product of the marginal probabilities. A few other special relationships hold when $\rho = 1$, but these conditions are rarely observed. You can look these up in Abramowitz and Stegun (1972).

The bivariate normal probability generalizes into a multivariate normal probability. In finance, one occasionally sees the trivariate normal probability and there are techniques for estimating it, which involve simplification of the relationships among univariate, bivariate, and trivariate densities. For the most part, however, computation of these high-order integrals is extremely time-consuming. Monte Carlo simulation is a good way of getting these results.

5.4   THE BIVARIATE LOGNORMAL PROBABILITY DISTRIBUTION

Again if $S_t = S_{t-1}e^{R_t^c}$ and R_t^c is normally distributed, then we know S_t is lognormally distributed. In finance, we are often interested in a portfolio of instruments such as stock holdings. Unfortunately, the sum of lognormally distributed random variables does not follow any known distribution. One important property of the normal distribution is that the sum of normally distributed random variables is itself normally distributed. Thus, in practice, it is dramatically easier to assume the underlying instrument prices are normally distributed, an issue we address in Chapter 12.

Figure 5.6 illustrates a simulation of 1,000 draws from a bivariate lognormal distribution, with normal mean 0, variance 1, and correlation 0. Note that the bivariate lognormal lower bound is zero for both x and y, but it is non-inclusive. Specifically, an outcome of zero did not occur because it is not possible with the lognormal distribution. Therefore, one advantage of the lognormal distribution is that negative values are not possible. Closely related, one disadvantage of the lognormal distribution is that zero values are also not possible.

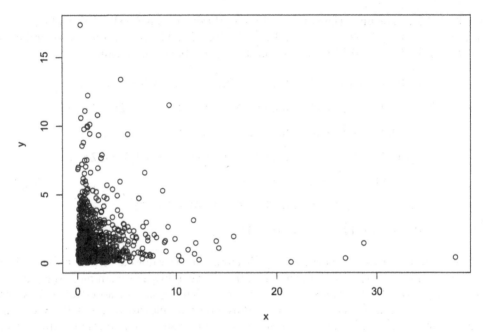

FIGURE 5.8   Bivariate Lognormal Simulation

5.5   RECAP AND PREVIEW

In this chapter, we reviewed the univariate normal probability distribution and took a look at the historical returns on the S&P 500 to see how well the distribution fits the data. We also examined the bivariate normal probability distribution, which can apply when there are two random variables.

In Chapter 6, we take a brief review of the basic concepts in valuing risky assets and derivatives.

APPENDIX 5A

An Excel Routine for the Bivariate Normal Probability

The two variables are identified as a and b and the correlation is rho. Enter them into three cells, which for the sake of illustration will be $A1$, $B1$, and $C1$. Then in a separate cell, type the function $= \text{bivar}(A1, B1, C1)$. The code, which is presented next, should be entered using the Visual Basic for Applications feature of Excel. Choose the Developer tab. Then choose Visual Basic. Then Insert, then Module. You will then have a blank space on which to enter the following code.

If you wish to receive an electronic copy of this code, please email your request to Don Chance, dchance@lsu.edu.

VB Code:

```
Sub biv1(az, bz, rhoz, bprob)
'Subroutine used to compute bivariate normal probability
Dim bp1, bp2, bp3, bp4, prob
bp1 = phiz(az, bz, rhoz)
bp2 = Application.NormSDist(az) - phiz(az, -bz, -rhoz)
bp3 = Application.NormSDist(bz) - phiz(-az, bz, -rhoz)
bp4 = Application.NormSDist(az) + Application
.NormSDist(bz) - 1 + phiz(-az, -bz, rhoz)
bprob = 0!
11 If rhoz = 0 Then bprob = Application.NormSDist(az) * Application
.NormSDist(bz) Else GoTo 12
GoTo 55
12 If az <= 0! And bz <= 0! And rhoz <= 0 Then bprob = bp1 Else GoTo 13
GoTo 55
13 If az <= 0! And bz >= 0 And rhoz >= 0 Then bprob = bp2 Else GoTo 14
GoTo 55
14 If az >= 0 And bz <= 0 And rhoz >= 0 Then bprob = bp3 Else GoTo 15
GoTo 55
15 If az >= 0 And bz >= 0 And rhoz <= 0 Then bprob = bp4 Else GoTo 16
16 GoTo 55
55 End Sub

Sub biv2(az, bz, rhoz, bprob)
'Subroutine used to compute bivariate normal probability
Dim signa, signb, rhoab, rhoba, Delta, probab, probba
If az >= 0 Then signa = 1 Else signa = -1
If bz >= 0 Then signb = 1 Else signb = -1
rhoab = (rhoz * az - bz) * signa / Sqr(az ^ 2 - 2 * rhoz * az * bz + bz ^ 2)
rhoba = (rhoz * bz - az) * signb / Sqr(az ^ 2 - 2 * rhoz * az * bz + bz ^ 2)
Delta = (1 - signa * signb) / 4
Call biv1(az, 0, rhoab, probab)
Call biv1(bz, 0, rhoba, probba)
bprob = probab + probba - Delta
End Sub

Function bivar(az, bz, rhoz)
'Function to compute bivariate normal probability
Dim bprob
If az * bz * rhoz > 0 Then GoTo 20
Call biv1(az, bz, rhoz, bprob)
GoTo 56
20 Call biv2(az, bz, rhoz, bprob)
56 bivar = bprob
End Function

Function phiz(aa, bb, rhoo)
'sub-function to compute bivariate normal probability
Dim a1, b1, fsum, i, j, f, phizz
Static w(5), x(5)
w(1) = 0.24840615: x(1) = 0.10024215
w(2) = 0.39233107: x(2) = 0.48281397
w(3) = 0.21141819: x(3) = 1.0609498
w(4) = 0.03324666: x(4) = 1.7797294
w(5) = 0.00082485334: x(5) = 2.6697604
a1 = aa / Sqr(2 * (1 - rhoo ^ 2))
b1 = bb / Sqr(2 * (1 - rhoo ^ 2))
```

```
fsum = 0
For i = 1 To 5
  For j = 1 To 5
    f = Exp(a1 * (2 * x(i) - a1) + b1 * (2 * x(j) - b1) + 2 * rhoo *
(x(i) - a1) * (x(j) - b1))
    fsum = fsum + w(i) * w(j) * f
  Next j
Next i
phizz = 0.31830989 * Sqr(1 - rhoo ^ 2) * fsum
phiz = phizz
End Function
```

QUESTIONS AND PROBLEMS

1 Based on Table 5.1, identify the following standard normal cumulative distribution values: 0.0, 0.58, −0.58, 1.65, and −1.65. Explain your results.

2 Based on Figure 5.2 and subsequent analysis, identify the key insight that is revealed.

3 Suppose a stock trading at 75 is estimated to have an annualized, continuously compounded mean rate of return of 12% and a corresponding standard deviation of 45%. Assume these returns are normally distributed. Estimate the expected change in the stock price in one year as well as the stock price standard deviation, normalized skewness, and normalized kurtosis.

4 Based solely on visual inspection of Figure 5.3, what conclusion can be drawn from comparing the time series of daily rates of return with daily first differences for the broad index ETF (SPY), the technology sector ETF (XLK), and Apple (AAPL)?

5 Based solely on visual inspection of Figure 5.4, what conclusion can be drawn from comparing the frequency distribution of daily rates of return with daily first differences for the broad index ETF (SPY), the technology sector ETF (XLK), and Apple (AAPL)?

NOTES

1. Prior to the introduction of the euro in 1999, the German currency was called the *deutschemark*. A 10-deutschemark note contained a picture of Gauss and his famous curve as well as the formula. It is likely the only piece of currency in history to contain an equation.
2. This value was calculated on a spreadsheet and is more precise than if you did it by hand.
3. By comparison, astronomers estimate the age of the universe at about 4.5 billion years, or much shorter than the expected frequency of a one-day return of less than −20% under a normal distribution of asset returns.
4. There are many other examples, such as dollar differences between purchases and sales, spreads between commodities, as well as spreads across different contract expirations.
5. Recall from Chapter 4 that the normalization process converts the skewness and kurtosis to unitless numbers by dividing by functions of variance or $var^{2/3}$ and var^2, respectively. The normal distribution has normalized skewness of 0 and normalized kurtosis of 3.
6. See Chance and Ağca (2003) for an examination of the speed and accuracy of bivariate normal approximation routines for option pricing.

CHAPTER 6

Basic Concepts in Valuing Risky Assets and Derivatives

To understand how to determine the values of financial derivatives, we need to examine some basic principles that comprise the foundation for valuing assets, after which we will extend those principles to valuing financial derivatives, especially options. A more rigorous foundation for these results is found in most classic works on the theory of finance, such as Fama (1976), Fama and Miller (1972), Huang and Litzenberger (1988), Ingersoll (1987), and Jarrow (1988).

6.1 VALUING RISKY ASSETS

Suppose there is an asset that will have a value of either 100 if a good economy occurs or 50 if a bad economy occurs one period later. The risk-free interest rate, r, for the period is 5%. Further, suppose we somehow know the probability of the good economy is 60%, and the probability of the bad economy is 40%. What should be the current price of the asset?

First, we have to start by finding the expected value of the asset, $E(S_1)$, one period later:

$$E(S_1) = \$100(0.6) + \$50(0.4) = \$80.$$

An investor who is indifferent to risk, referred to as being *risk neutral*, would value the asset (S_0) at

$$S_0 = \frac{E(S_1)}{1 + r} = \frac{\$80}{1.05} = \$76.19.$$

Let us denote this risk-neutral investor simply RNI. If RNI made this investment many times over, they would receive an average payoff of 80. If they paid 76.19 each time, they would effectively break even, being paid only a 5% return to be compensated for the time value of money. The distinguishing characteristic of RNI investors is that they do not demand compensation for the risk. They are perfectly happy to make a certain amount of money one time and lose that same amount of money another time. Over the long run, their offsetting gains and losses leave them as content as they were before they made the investments. They earn the risk-free rate in spite of assuming risk.

A risk-averse investor, identified as RAI, however, would not be satisfied to make an investment many times and earn an average return that compensates only for the time value of money. If RAI made an investment that earned $x\%$ one time and lost $x\%$ another

time, RAI would wish the investment had not been made at all. The dissatisfaction from
the loss is more painful than the satisfaction of a gain of equivalent magnitude. As such,
an RAI would pay less than 76.19. In that case, in the long run the gains would offset the
losses by more than the risk-free rate. The important question is, how much less would
an RAI pay? The answer depends on how the RAI addresses risk. The more risk averse
they are, the less they would pay. Thus, the more risk averse they are, the more they will
discount the expected value to obtain the price they will pay. People vary in their degrees of
risk aversion. Some are highly risk averse and some only moderately risk averse. A given
person may also vary in their risk aversion, based on their stage of life or accumulated
wealth.

Let us assume that the investor believes a price of 70 would be appropriate. What does
this imply about their risk aversion? At a price of 70, they have an expected return of

$$\frac{80}{70} - 1 = 0.1429.$$

Thus, their expected return is 14.29%. If they make this investment many times, they
will earn on average 14.29%. Given a 5% risk-free rate, they will earn a risk premium
of 9.29% (= 14.29% − 5%). Thus, they are pricing the investment such that they expect a
9.29% risk premium. Of course, they may make the investment only one time, in which
case they will either earn 100 − 70 = 30, which is 42.9% (= 30/70), or lose 70 − 50 = 20,
which is −28.6% (= −20/70). Note that the probability-weighted average of 42.9% and
−28.6% is 14.29%:[1]

$$(30/70)(0.6) + (-20/70)(0.4) = 0.1429.$$

Or they might make the investment only a few times and make or lose an average of some
value other than the expected return of 14.29%. But their long-run expectation is to earn
a return of 14.29%, and this rate will compensate them for taking the risk, based on their
pricing the investment at 70.

6.2   RISK-NEUTRAL PRICING IN DISCRETE TIME

One of the most important results derived from option pricing theory is that all assets
can be valued via a risk neutrality argument. Unfortunately, risk-neutral pricing has been
a source of much confusion. Those who do not understand the process are led to believe
that some mysterious sleight of hand has been done and that the resulting price is somehow
tainted by ignoring the risk. In fact, risk-neutral pricing makes no such assumption of risk
neutrality. It simply uses the assumption that no arbitrage opportunities exist rather than
the assumption that investors are risk averse to motivate the pricing process.

The point is most easily made in a simple model in which a risky asset has two future
outcomes. The outcomes themselves are known, but we do not know which will occur. We
can assign probabilities, though as we will show later, these probabilities are not required
for risk-neutral pricing. This type of framework, in which an asset has two possible future
outcomes, is sometimes called a *two-state model* but more often called a *binomial model*.
On the surface it seems that this model oversimplifies things. Naturally in the real world,
there are more than two future outcomes. That problem can be easily accommodated by

allowing multiple time periods such that each realized outcome is followed by two more possible outcomes, and so forth. But more fundamentally, a binomial framework permits a simplified but powerful accommodation of uncertainty. As long as we acknowledge that we do not know which outcome will occur, we have built uncertainty into the model.

Consider an asset currently priced at S whose price one period later can be Su with probability q or Sd with probability $1 - q$. The factors u and d represent one plus the percentage change in the asset price. Let us first consider how that asset would be priced in a world of risk-averse investors. Under risk aversion, any risky asset is priced as the discounted value of its expected future price. The expected future price of the asset is $qSu + (1 - q)Sd$. Consequently, the current price would be given as

$$S = \frac{qSu + (1 - q)Sd}{1 + k_R},$$ (6.1)

where k_R is the discount rate that reflects an adjustment for risk. For risk-averse investors, the discount rate will consist of the risk-free rate, r, plus a risk premium. The rate k_R is sometimes considered as having arisen out of an asset's covariance with the market, such as in the capital asset pricing model (recall denoted CAPM), or some alternative model of market equilibrium, but it is not necessary to assume any such type of equilibrium. It can simply be a subjective estimate of the premium a risk-averse investor expects to earn. For a risk-neutral investor, the discount rate will be the risk-free rate.

Given knowledge of S, u, and d, let us restate the price of the asset in the following manner: First, we ask if there are probability values that can be substituted for q and $1 - q$ that permit us to change k_R to the risk-free rate, r, while keeping the price exactly as it was. In the absence of arbitrage, the answer is yes, as we demonstrate next.

Consider a world of two assets: the stock and a risk-free bond. Now let us assume that $d > 1 + r$. In other words, the worst return on the stock is better than the risk-free interest rate. It would not take a genius to figure out that one could borrow money to pay the price of the stock at the risk-free rate, buy the stock, and then merely wait on the outcome. If the stock goes to Su, that is enough money to pay back the loan $S(1 + r)$ and would leave $Su - S(1 + r) = S[u - (1 + r)]$, which is positive. If $d > 1 + r$, then, if the worst-case outcome occurs, the investor will have $Sd - S(1 + r) = S[d - (1 + r)]$. Again, we assumed $d > 1 + r$, so this outcome is positive. In simple terms, if the worst outcome of the stock is better than the risk-free rate, an arbitrage profit is available. Clearly investors would take advantage of this situation, and their trading activity would likely adjust the stock price. With this new adjusted stock price u and d would change until there is no arbitrage profit available. And naturally, we could not have the risk-free rate exceed the up return, $u - 1$, on the stock, for in that case no one would buy the stock, so its value would be zero. So ultimately it must be that $u > 1 + r > d.$[2]

Given that $u > 1 + r > d$, we will now demonstrate that it is possible to find weights that we will identify as ϕ (lowercase Greek letter, pronounced "phi") and $1 - \phi$ that are consistent with the statement $\phi u + (1 - \phi)d = 1 + r$ or the equivalent statement, $\phi = (1 + r - d)/(u - d)$. If we multiply by S and do some rearranging, we obtain the following:

$$S = \frac{\phi Su + (1 - \phi)Sd}{1 + r}.$$ (6.2)

This expression looks identical to a valuation equation. If ϕ were the probability of the price Su occurring and $1 - \phi$ were the probability of the price Sd occurring, then the numerator would be the expected price. To value an asset, we discount the expected price by an appropriate rate. If the risk-free rate r was the appropriate discount rate, then S would be the price. It is a fact that S is the asset price and also a fact that Su and Sd are the possible future asset prices, but ϕ is *not* the probability of an upward move. Nonetheless, ϕ looks a lot like a probability and is used like a probability. In addition, r is not the appropriate discount rate when applying the real probability, but it looks a lot like it and is used in that manner. The numerator of Equation (6.2) is sometimes referred to as a certainty equivalent, which is an amount of cash that an investor is willing to receive rather than take a stated risk.

What we have done here is change the probabilities from their true values, q and $1 - q$, to alternative values, ϕ and $1 - \phi$, such that we can use the risk-free rate as the discount rate. These probabilities are sometimes referred to as *pseudo-probabilities* because they are not the real probabilities, but they behave like them. They are more appropriately referred to as *risk-neutral probabilities* and sometimes *martingale* or *equivalent martingale probabilities*. We shall explain the martingale concept in a few paragraphs.

The adjustment that enabled us to obtain Equation (6.2) is clearly correct and acceptable, inasmuch as it produces the correct stock price, S. What enables us to do this is the simple no-arbitrage condition that $u > 1 + r > d$. The only information we need to know is the up and down factors and the risk-free rate. The up and down factors characterize the volatility of the asset. We have made no assumptions about how investors address risk. Indeed, what we have done is perfectly compatible with risk aversion. If investors are risk averse, which is what Equation (6.1) says, then Equation (6.2) also gives the value of the asset.

What we have just illustrated is called the *arbitrage theorem*. It simply states that if there are no opportunities for arbitrage, then it is possible to state the price of the asset in terms of risk-neutral probabilities where the discounting is done at the risk-free rate. The theorem goes two ways. If it is possible to state the price of the asset in this manner, then there are no arbitrage possibilities.

Calling ϕ and $1 - \phi$ risk-neutral probabilities can be, however, a source of much confusion, inasmuch as it may imply that we are assuming investors are risk neutral, which we certainly are not. Alternatively, these probabilities are sometimes called *martingale* or *equivalent martingale probabilities*. A martingale is a stochastic process in which the expected return of the asset is the current value of the asset. Under the equivalent martingale approach, however, the future asset price is restated so as to be already discounted at the risk-free rate. That is, the right-hand side of Equation (6.2) is $\phi \widehat{S}u + (1 - \phi)\widehat{S}d$, where $\widehat{S} = S/(1 + r)$ is an adjusted asset price. In other words, the future S value is the future price already discounted.

There is a definite relationship between the binomial probability, ϕ, and the actual probability, q. Define the market price of risk of the stock as $\lambda = [E(R) - r]/\sigma$ where $E(R)$ is the expected return on the stock, which is $E(R) = qu + (1 - q)d - 1$. The stock's variance is defined as $\sigma^2 = q(1 - q)(u - d)^2$. These statements hold by definition. They are simply the expected value and volatility of a one-period binomially distributed variable, the return on the asset, that can go up to Su or down to Sd. Substituting these values into the risk premium and noting that $\phi = (1 + r - d)/(u - d)$, we obtain the result,

$$\phi = q - \lambda\sqrt{q(1 - q)}. \tag{6.3}$$

Thus, the binomial probability is the actual probability minus the risk premium times the square root term, which is related to the volatility. In short, knowing the binomial probability, $\phi = (1 + r - d)/(u - d)$, tells us everything we need to know to avoid having to know the market price of risk, λ, and the actual probability, q.

Note that we have not introduced options. The point is easily established without the use of options. When introducing options, however, it is easily shown that a call option can be replicated by positions in the asset and the risk-free bond. That being the case, the call is a redundant asset. Consequently, introducing a redundant asset does nothing to change the nature of the market, provided, of course, that the redundant asset is properly priced. But that will be true by the requirement of no arbitrage.

Though we have introduced this point in the simple binomial framework, it can also be established, albeit with more mathematical complexity in the continuous time models. This topic is covered in Chapter 12.

6.3 IDENTICAL ASSETS AND THE LAW OF ONE PRICE

Suppose there is another asset trading in a different market. If a good economy occurs, the asset will be worth 200, and if a bad economy occurs, it will be worth 100. Remember that the first asset pays 100 and 50 in the two outcomes. Notice that this second asset pays exactly two times the payoff of the good asset. Suppose that the first asset is trading for the price of 70, and the second asset is trading for 148. This situation creates an unstable condition that will bring out arbitrageurs who trade to earn risk-free profits without committing any of their own money.[3]

These arbitrageurs will execute a very simple strategy. They will sell short the second asset, thereby collecting 148 while promising to buy that asset back at 200 if the good economy occurs or 100 if the bad one occurs.[4] Now, they have 148 in cash up front. They take 140 and buy two shares of the first asset. They keep the remaining $148 - 140 = 8$.[5] Let us see what happens for both possible outcomes.

> **A good economy occurs:** They are holding two shares of the first asset, which are collectively worth 200. They take the 200 and buy one share of the second asset, which they then deliver to cover their short position.
>
> **A bad economy occurs:** They are holding two shares of the first asset, which are collectively worth 100. They take the 100 and buy one share of the second asset, which they then deliver to cover their short position.

In both outcomes, the shares they own will be precisely enough to repurchase the shares they have sold short. Thus, they are immune to the risk. They do not care which outcome occurs, because in either case they simply end up with 8 and the interest on it. They have, in effect, manufactured money without taking any risk, which is what arbitrage is.

Naturally many investors will rush to do this transaction, which will put downward pressure on the price of the second asset or upward pressure on the price of the first asset, or a combination of both. Very quickly the two asset prices must align themselves so that the second asset price is twice that of the first. Under the principle of arbitrage, prices are in equilibrium relative to each other. We do not know or care if the first asset is priced correctly or the second asset is priced correctly. We simply care whether the two prices are properly aligned, in this case such that the second asset is worth twice the price of the first, so that an opportunity to make an arbitrage profit no longer exists.

The principle of arbitrage also brings to light a result called the law of one price: Identical assets cannot sell for different prices. These assets may not appear to be identical assets, but their returns are perfectly synchronized. Regardless of which outcome occurs, the value of the second asset has to be twice that of the first. For an excellent treatment of arbitrage at an introductory level, see Billingsley (2006), and for more advanced coverage that will parallel most of the topics in the rest of this book, see Björk (1998).

6.4   DERIVATIVE CONTRACTS

Our ultimate interest is in one type of derivative, the option. Understanding option pricing is key to understanding the pricing of numerous other financial instruments. Before we get to options, however, let us take a look at a simpler type of derivative: the forward contract. Suppose we create a side bet on the outcome of the first asset. Let Trader A offer to pay Trader B 26.50 if the good economy occurs, provided that Trader B will pay Trader A 23.50 if the bad economy occurs. Suppose that Trader A borrows 70 at the risk-free rate of 5% and uses the funds to purchase the asset. They then enter into this side bet to pay Trader B 26.50 if the good economy occurs with the condition that they receive 23.50 from Trader B if the bad economy occurs. The 70 Trader B borrows will have to be paid back, thereby requiring a payment of 70(1.05) = 73.50 one period later. Trader B's overall payoff will be as follows:

> **A good economy occurs:** He holds an asset worth 100, owes 26.50, and pays back the loan value of 73.50 for a net cash flow of zero.
>
> **A bad economy occurs:** He holds an asset worth 50, receives 23.50, and pays back the loan value of 73.50 for a net cash flow of zero.

Of course, there is no particular reason this investor should do this. They committed none of their own funds at the start and end up with no money regardless of the outcome. But this example illustrates that the side bet should pay 26.50 to one party if a good economy occurs and 23.50 to the other party if the bad economy occurs. These are fair equilibrium terms for the side bet.

What if the side bet pays 26.50 if the good economy occurs and 25 if the bad one occurs? Then it should be apparent that if the bad economy occurs, Trader A will end up with $50 + 25 - 73.50 = 1.50$. If the good economy occurs, Trader A will end up with zero. So, without investing any of their own money, Trader A takes a gamble in which they either end up with no money or 1.50. Trader A will certainly do it, as will other arbitrageurs. Their collective actions force the terms of the side bet to change until the bet is a fair one for both parties. That naturally occurs if the payoff in the bad economy is 23.50.

The two payoffs to Trader A, 26.50 in the good economy and –23.50 in the bad economy, can be derived as follows:

> If the good economy occurs: $100 - x = 26.50$
>
> If the bad economy occurs: $50 - x = -23.50$

Clearly $x = 73.50$. This number is a special value and is the key to finding the equilibrium payoffs of the side bet. We could, however, have obtained that value in another and much easier way. We simply take the price of the asset, 70, and compound it at the risk-free rate, 5%,

$$70(1.05) = 73.50.$$

This side bet illustrates a type of derivative transaction known as a *forward contract*. Embedded in a forward contract is the notion of a *forward price*. We can think of the forward price, 73.50 in this example, as the price that one party agrees to pay for the asset when the contract expires. Thus, here a party going long the forward contract agrees to pay 73.50 one period later. If the good economy occurs, the party pays 73.50 and receives the asset worth 100 for a net gain of 26.50. If the bad economy occurs, the party pays 73.50 and receives the asset worth only 50 for a net loss of 23.50. Thus, a forward contract is an agreement for one party to pay the forward price and receive the asset at a later date. The other party agrees to deliver the asset and receive the forward price.

In this example, we assumed that the asset price was 70, which was the price appropriate for the highly risk-averse investor. Suppose we have a less risk-averse investor who considers the asset worth 80. In that case, the forward price would be 80(1.05)84. The forward contract would need to pay 16 to one party if the good economy occurs and 34 to the other party if the bad economy occurs. In this way, if someone did the transaction just described, that is, borrowed 80 at 5% and offered the forward contract as a seller, they would end up with $100 - 16 - 84 = 0$ if the good economy occurs and $50 - 34 - 84 = 0$ if the bad economy occurs. This result is what it should be. The person committed no personal funds and took no risk. Therefore, the gain should be zero for certain.

The point is that the forward price, whether 73.50 or 84, depends on the spot price of the underlying asset and the risk-free rate. It will also depend on how long the contract is and for some assets and some derivatives, it will depend on other factors.[6] The most important point is that we take the asset price as given. We do not care whether the investors who priced the asset are highly risk averse, slightly risk averse, or risk neutral. We assume that how investors feel about risk is fully incorporated into the price of the asset. Once the asset price is determined, we do not care what factors investors consider in obtaining that price.

With that in mind, we can treat our process of valuing derivatives as though investors are risk neutral. Consider the case where investors are highly risk averse and price the asset at 70. We noted that at a risk-free rate of 5%, these investors are pricing the asset with a risk premium of 9.29%, because the expected value of the asset is 80, which is 14.29% higher than 70, but a portion of the 14.29% is the risk-free return (5.0%). For our forward contract, however, the forward price was found to be 73.50, which is only 5% higher than the spot price of 70. Thus, the forward contract can be viewed as being priced by risk-neutral investors. That does not mean that investors are actually risk neutral. In fact, we just said they are highly risk averse in this example. But the forward price is found by taking the spot price of 70 and factoring it up to reflect only a risk-free return of 5%, thereby obtaining 73.50 as the forward price. The principle that derivative instruments are priced to prohibit the opportunity to earn an arbitrage profit turns out to be the same as the notion that we can treat participants in derivative markets as though they are risk neutral. This simply means that derivatives are priced as though investors are risk neutral. In fact, they are not risk neutral but are risk averse. Their risk aversion is embedded into the price of the underlying asset. Given the price of the underlying asset, it is easy to price the derivative, using, of course, the condition of risk neutrality.

Yet another way to look at it is that if we did not treat investors as though they were risk neutral when pricing derivatives, we would be double counting the risk. The risk is incorporated into the price of the asset. The price of the asset is then incorporated into the derivative price, which takes care of handling any adjustment for risk. We cannot treat the derivatives investor as though they require a premium as compensation for the risk. That premium is already there—in the price of the asset. We will return to this point in the last section of this chapter, where we introduce a world of risk-averse and risk-neutral investors.

6.5   A FIRST LOOK AT VALUING OPTIONS

Let us now create a different kind of derivative, a call option. We use the same example where the asset is priced at 70 and can go up to 100 with 60% probability or down to 50 with 40% probability. Suppose someone tells you that if you will give them some money today, you will acquire the right to buy the asset at 75 one period later but are not obliged to do so. Let us determine how much money you would need to pay to acquire that right.

If the asset goes up to 100 in one period, you would pay your 75 and acquire the 100 asset, netting a gain of 25. If the asset goes down to 60, it would not be worth paying 75 to acquire the asset. You would simply choose not to acquire the asset, thereby ending up with zero. This asset, which is a call option, is therefore an instrument that pays either 25 or zero one period later. We wish to know how much you would have to pay for this asset. Clearly it is somewhere between 25 and zero.

Suppose we buy two of these options and invest 47.62 in the risk-free asset, which will pay 5% interest, so that one year later, it will be worth 47.62(1.05) = 50. This 50 is the lower of the two outcomes for the asset. Now, consider the outcomes:

If the good economy occurs: You exercise the option, paying 75 and acquiring an asset worth 100 for a gain of 25. Because two options have been purchased, you have 50. You collect your 50 from the risk-free asset for a total payoff of 100.

If the bad economy occurs: The option will not be worth exercising. You will, however, collect the 50 on the risk-free asset, so your total payoff will be 50.

So, in a good economy, your total wealth is 100 and in a bad economy, your total wealth is 50. These outcomes are the same as if you had bought the asset. Therefore, the amount you should have to pay to achieve this outcome should be the same as the cost of the asset, which is 70. Because you would pay 47.62 for the risk-free asset and you bought two options, the price of each option should be

$$\frac{70 - 47.62}{2} = 11.19.$$

If the price of the option were lower than 11.19, one could buy the options and the risk-free asset and produce a payoff equal to that of the underlying asset. That means one could sell short the asset and receive net of 11.19 at the start, but one would pay less than 11.19. One period later, the options and the risk-free asset would exactly offset, thereby eliminating any uncertainty. One would then pocket the difference between 11.19 and what the option costs. If the option costs more than 11.19, the reverse arbitrage would be done: Sell the option and issue the risk-free bond, while buying the asset. In both cases, one earns an arbitrage profit. This arbitrage activity would continue until the price of the option adjusts to eliminate the opportunity. Of course, that occurs with the option price at 11.19.

6.6 A WORLD OF RISK-AVERSE AND RISK-NEUTRAL INVESTORS

It is widely accepted that we live in a world of risk-averse investors. There are certainly some exceptions. There are some irrational people who take gambles in which the odds are tilted heavily against them. Almost any form of organized betting is a good example. In addition, almost anyone who devotes the better years of their youth to pursuing a career in entertainment or sports is going up against enormous odds that almost surely have a negative expected return. Even entrepreneurship is a tremendously stacked gamble. Nonetheless, we can explain these seemingly irrational decisions because the participants are receiving a nonmonetary return. They enjoy participating in these endeavors, and that enjoyment can offset the expected monetary loss. And in some cases, they believe their odds are better than they really are.

Most financial models assume that investors are risk averse. Yet option theory makes repeated references to risk neutrality. As we explained, pricing forwards and options proceeds as if investors are risk neutral, but we stressed that they are not risk neutral. We simply said that we can treat them as if they are risk neutral, even when they are not. This point is oftentimes not very clear.

There is one way to help bring some focus to this confusing matter. Let us assume a market in which there are two sets of investors. One set is risk averse and the other is risk neutral. Let us use the example in which the asset can go up to 100 or down to 50 with respective probabilities of 60% and 40%. We stated previously that the expected payoff is 80. A risk-neutral investor would, therefore, discount 80 at 5% and arrive at a price of 76.19, which we obtained at the beginning of this chapter. Recall that we had our risk-averse investor willing to pay only 70.

Clearly these two investors can agree to trade with each other. If the risk-neutral investor is happy to pay 76.19, then they would clearly be happy paying a lesser amount. So, the risk-neutral investor would clearly pay the 70 price of the risk-averse investor. But the risk-averse investor would be happy to sell the asset at a higher price, say 76.19. So, the risk-averse investor would sell the asset to the risk-neutral investor for 76.19. Everyone would be happy, but it would be clear that the risk-averse investor would be earning far more than the expected return they demand in the face of the risk they assume. The risk-averse investor is obviously able to exploit the risk-neutral investor. The demand that risk-averse investors would exert on risk-neutral investors would eventually tell the latter that they are doing something wrong. In short, there cannot exist two prices.

Now suppose the two parties engaged in an option transaction. Assuming the risk-averse investor's price of the asset is 70, we obtain the 11.19 derived in the previous section. So, one investor thinks the option value is 11.19. The other would value the stock at 76.19. If we went through the valuation exercise, we would obtain an option price of 14.28. We saw in the previous section that the correct arbitrage-free value of the option is 11.19, and that if someone wants to trade at a higher price, an investor can engage in an arbitrage transaction to expropriate wealth from the investor who trades at any other price. As such, risk-averse investors would drive the risk-neutral investors out of the market. Or the latter would wake up and realize they are valuing assets the wrong way, because their preferences toward risk do not accurately characterize the way they really feel about risk.

6.7 PRICING OPTIONS UNDER RISK AVERSION

If risk-averse investors insisted on pricing options by incorporating their risk aversion, they should still arrive at the correct price. They would use the true probabilities, q and $1 - q$, to discount the expected option payoff. Let k_o be the discount rate of the option. For our example in which the arbitrage-free price of the option is 11.19, the option payoffs are 25 and zero, and the true probabilities are 60% and 40%, the option discount rate would be found as the solution to the following:

$$\frac{\$25(0.6) + \$0(0.4)}{1 + k_o} = \$11.19$$

$$k_o = 0.3405.$$

In other words, the discount rate on the option would be 34.05%.

The beauty of risk-neutral pricing is that we do not have to come up with the 34.05% rate, nor do we have to come up with the true probabilities. All we have to know is the asset price, S, the values of u and d that describe the asset's volatility, and the risk-free rate, r. These requirements are far less onerous.

It is important to note that risk-neutral pricing never asks how the asset price was obtained. It takes that price as being given in a market in which the buyer of the asset trades with the seller of the asset. A trade produces an asset price. Using that asset price, the price of the option can be obtained. What we do not know from option pricing is how that asset price was obtained. That is the subject of general equilibrium theory, which is a branch of economics and financial economics that studies how prices are determined in markets with buyers and sellers whose risk aversion influences how they price assets. In finance, models such as the CAPM are general equilibrium models. Models such as the arbitrage pricing theory and virtually all option pricing models are not general equilibrium models because they take market equilibrium as given without any regard to how that equilibrium occurred.[7] These types of models are called *partial equilibrium models* because they explain how a part of the market, meaning a sector, reaches equilibrium. What they explain, however, is confined to the price and a few other characteristics that we will cover in Chapter 13. They do not explain why an investor would buy or sell an option and how many options an investor would trade.

6.8 RECAP AND PREVIEW

In this chapter, we reviewed the process of valuation, which is how one assigns a value to an asset or derivative. The general rule is to discount future cash flows at a rate that reflects risk. We noted, however, that risk-neutral valuation is also perfectly legitimate, though it values an asset only in relation to another asset whose value is known. We also took an introductory look at how to value options, which is the primary theme of this book, the details of which are fleshed out later.

We have now completed Part I, which is largely review material. Part II begins our study of the formal process of valuing options. Chapter 7 introduces the binomial model, which we lightly touched on in the current chapter.

QUESTIONS AND PROBLEMS

1 "Risk-neutral pricing is flawed because it fails to incorporate risk." Evaluate and explain whether this statement is true or false.

2 Assume a given stock is trading at 100 and the total return if the up state occurs is $u = 1.1$ and the total return if the down state occurs is $d = 1/1.1$. If the investor believes there is a 55% chance of the up state occurring, calculate the risk premium if the risk-free rate is 1%.

3 Again, as in Question 2, assume a given stock is trading at 100 and the total return if the up state occurs is $u = 1.1$ and the total return if the down state occurs is $d = 1/1.1$. The investor believes there is a 55% chance of the up state occurring and the risk-free rate is 1%. Calculate the risk-neutral probability of an up move. Explain what happens to the risk-neutral probability of an up move as interest rates fall.

4 Within the single-period two-state world with u being the up factor, d being the down factor, r being the risk-free rate, and q being the probability of an up move, prove that the variance of the return, σ^2, is equal to $q(1 - q)(u - d)^2$.

5 Explain the relationship between the binomial probability, ϕ, and the actual probability, q, within a single-period binomial model.

NOTES

1. When working through your understanding of financial derivatives with numerical examples, rounding is a constant source of potential confusion. Thus, we use the exact ratios rather than rounded values.
2. The reason we see a 1 in front of the r and not in front of the u and d is that u and d are already specified as 1 plus the up and down returns. In some books and articles, r is already specified as 1 plus the risk-free rate.
3. Some arbitrageurs take little if any risk, but there are also investors who ordinarily take risks who will notice the opportunity and engage in an arbitrage transaction.
4. Short selling involves borrowing the asset from someone else, selling it to another party, and committing to repurchase the asset at a later date to return it to the original owner.
5. That money would be invested to earn the risk-free rate of interest.
6. For forward contracts on some assets, it will depend on the costs of carrying or storing those assets and any payments, such as dividends, that those assets make. For options, it depends on all of these factors as well as the volatility of the underlying asset.
7. As its name suggests, arbitrage pricing theory is typically a linear multifactor model that seeks to identify mispriced instruments. Because our focus is on financial derivatives, we do not further explore this approach.

Discrete Time Derivatives Pricing Theory

CHAPTER 7

The Binomial Model

To this point we have used a simple model in which the asset price can either move up to one level or down to another. We called this model the *two-state* or *binomial model*. In this chapter, we derive the model more formally. This work is attributed to Cox, Ross, and Rubinstein (1979) and Rendleman and Bartter (1979). An excellent book on binomial models in finance is by van der Hoek and Elliott (2006).

7.1 THE ONE-PERIOD BINOMIAL MODEL FOR CALLS

Recalling from the previous chapter, we specified an asset priced at S that can go up to Su or down to Sd. Given the historical decision to assume a lognormal distribution, early authors pursued a multiplicative binomial approach.[1] We shall refer to these two states as the up state and the down state, respectively. The basic idea is communicated in Figure 7.1. At the initial point in time, there is only one node, whereas at the next point in time, there are only two nodes. Also, at the initial point in time, there are two arcs emanating from the initial node.

This type of figure is sometimes called a *binomial tree*, although it will look a little more like a tree when we expand it later.

The risk-free rate is r. Consider a call option with exercise price X that expires in one period. The two possible values of the option at expiration are

$$c_u = \max(0, Su - X)$$
$$c_d = \max(0, Sd - X). \qquad (7.1)$$

Of course, our objective is to determine the value of the option today, c. The basic layout with the corresponding option prices inserted at each node is in Figure 7.2.

Appealing to the arbitrage principle, we shall create a portfolio consisting of the option and the underlying asset. We shall set this portfolio up so that it is hedged, meaning that its future value is known for certain. A hedged portfolio must, therefore, earn the risk-free rate. We can then solve for the price of the option that is consistent with a risk-free return on this portfolio.

Let us buy h_c units of the asset and sell one call. The value of this portfolio today is

$$V = h_c S - c. \qquad (7.2)$$

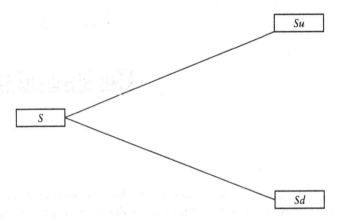

FIGURE 7.1   Binomial Process for Underlying Asset

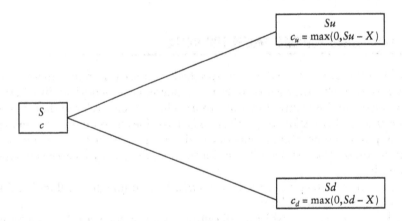

FIGURE 7.2   Binomial Process for Underlying Asset and Call Option

The payoffs of this portfolio in the two future states will be

$$V_u = h_c Su - \max(0, Su - X) = h_c Su - c_u$$
$$V_d = h_c Sd - \max(0, Sd - X) = h_c Sd - c_d. \tag{7.3}$$

Figure 7.3 illustrates the setup. The goal is to understand the linkages between the underlying asset, the call option, and this unique hedge portfolio.
If this portfolio replicates a risk-free bond, it must produce a risk-free return, meaning that
these two outcomes are the same, as specified by the condition,

$$V_u = V_d. \tag{7.4}$$

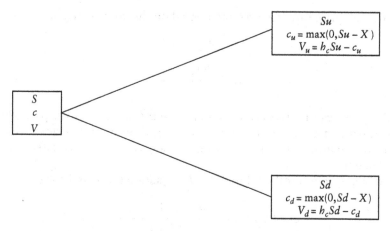

Su
$c_u = \max(0, Su - X)$
$V_u = h_c Su - c_u$

S
c
V

Sd
$c_d = \max(0, Sd - X)$
$V_d = h_c Sd - c_d$

FIGURE 7.3 Binomial Process for Underlying Asset, Call Option, and Hedge Portfolio

If we set these two values equal to each other, we have one simple equation with one unknown, $h_c Su - c_u = h_c Sd - c_d$. The solution is known as the hedge ratio and is expressed as follows:

$$h_c = \frac{c_u - c_d}{Su - Sd}. \tag{7.5}$$

Note that the sign of h_c will be positive as $c_u > c_d$ and $Su > Sd$. The logic of a positive sign for h_c is based on the fact that we prespecified that we are short one call option, as indicated by the minus sign in front of c in Equation (7.2). In effect, the call position is -1 unit. The asset position is $h_c > 0$ units. We are long the asset and short the call option. Because call values are positively related to the values of their underlying assets, it is intuitive that a short call would be offset by long units of the asset.

Thus, if the number of units of the asset that we hold is set to h_c, the two future values of the portfolio will be identical. Hence, the portfolio is risk free. To avoid arbitrage, the portfolio must be priced to earn the risk-free rate. Thus, the following condition must hold:

$$V = \frac{V_u}{1 + r}$$

or

$$V = \frac{V_d}{1 + r}. \tag{7.6}$$

Consequently, we can apply the following condition, using either V_u or V_d. We will choose V_u.

$$\frac{V_u}{1 + r} = V$$

$$\frac{h_c Su - c_u}{1 + r} = h_c S - c. \tag{7.7}$$

Therefore, the call price can be represented based on the *no-arbitrage model* as

$$c = h_c S - B_c$$

$$B_c = \frac{h_c Su - c_u}{1 + r}. \tag{7.8}$$

Thus, a call option can be replicated by purchasing h units of stock partially financed through borrowing of B_c. A call is simply a leveraged position in a stock, that is, a call is equivalent to buying stock with borrowed money. As we will see later, the degree of leverage is dynamic.

The next step is to insert the solution for h, Equation (7.8), and solve for c:

$$c = PV[E(c_T)] = \frac{\phi c_u + (1 - \phi)c_d}{1 + r}$$

$$\phi = \frac{1 + r - d}{u - d}. \tag{7.9}$$

There are quite a few steps and algebraic substitutions that lead from Equation (7.8) to Equation (7.9). Details are provided in Appendix 7A of this chapter. Thus, a call price is simply the present value of the expected future call payoffs discounted at the risk-free rate. The probabilities used in forming the expectations, however, are not the real probabilities. They are the risk-neutral probabilities and this model is known as the *risk-neutral model*. The risk-neutral model is based on the expectation taken using risk-neutral probabilities and discounting at the risk-free interest rate.

For example, suppose the underlying is a stock and its current price is 99, the exercise price is 100, the annual, discretely compounded, risk-free rate is 2%, the time to expiration is one year, $u = 1.25$, and $d = 0.8$. We can compute the call price in two ways. First, note that $c_u = \max(0, 123.75 - 100) = 23.75$ and $c_u = \max(0, 79.2 - 100) = 0$. For the no-arbitrage model, we first find the hedge ratio $h_c = (c_u - c_d)/(Su - Sd) = (23.75 - 0)/(123.75 - 79.2) = 0.5331$. Therefore, based on Equation (7.8), we have

$$c = h_c S - \frac{h_c Su - c_u}{1 + r} = 0.5331(99) - \frac{0.5331(99)1.25 - 23.75}{1 + 0.02}$$

$$= 52.7769 - 41.3933 = 11.38.$$

Alternatively, we can apply the risk-neutral model. The binomial probability of an up move is $\phi = (1 + 0.02 - 0.8)/(1.25 - 0.8) = 48.8889\%$. Therefore, based on Equation (7.9), we find the same results, or

$$c = \frac{0.4889(23.75) + (1 - 0.4889)0}{1 + 0.02} = 11.38.$$

7.2 THE ONE-PERIOD BINOMIAL MODEL FOR PUTS

Here we follow a similar pattern with puts as we did with calls in the last section. We specified an asset priced at S that can go up to Su or down to Sd (the up state and down state, respectively). This basic idea was communicated in Figure 7.1. Now consider a put option with exercise price X that expires in one period. The two possible values of the option at expiration are

$$p_u = \max(0, X - Su)$$
$$p_d = \max(0, X - Sd). \tag{7.10}$$

Recall our objective is to determine the value of the put option today. As with the call, the basic layout with the corresponding option prices inserted at each node is in Figure 7.4.

Again, appealing to the arbitrage principle, we shall create a portfolio consisting of the put option and the underlying asset. We shall set this portfolio up so that it is hedged, meaning that its future value is known for certain. In the put case, put prices move in the opposite direction to the underlying asset. Thus, if we buy a put, to hedge we will need to buy the underlying asset. A hedged portfolio must, therefore, earn the risk-free rate (r). We can then solve for the price of the option that is consistent with a risk-free return on this portfolio.

Let us buy h_p units of the asset and buy one put. The value of this portfolio today is

$$V = h_p S + p. \tag{7.11}$$

The payoffs of this portfolio in the two future states will be

$$V_u = h_p Su + \max(0, X - Su) = h_p Su + p_u$$
$$V_d = h_p Sd + \max(0, X - Sd) = h_p Sd + p_d. \tag{7.12}$$

Figure 7.5 illustrates the setup for puts. Again, the goal is to establish the linkages between the underlying asset, the put option, and this unique hedge portfolio.

Again, to form a hedge portfolio, we need the number of underlying assets to have the same sign as the number of puts. That is, we need to either buy the underlying and buy the put or short the underlying and sell the put.

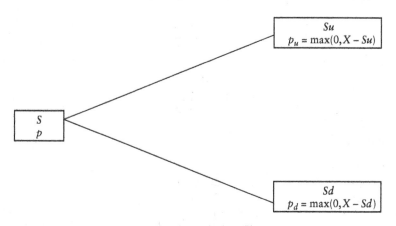

FIGURE 7.4 Binomial Process for Underlying Asset and Put Option

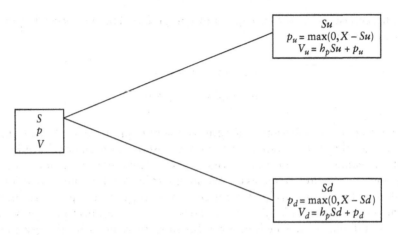

FIGURE 7.5 Binomial Process for Underlying Asset, Put Option, and Hedge Portfolio

Recall this portfolio replicates a risk-free bond and thus the two portfolio values at the next time step have to be equal or

$$V_u = V_d. \tag{7.13}$$

If we set these two values equal to each other, we have one simple equation with one unknown, $h_p Su + p_u = h_p Sd + p_d$. Thus,

$$h_p = \frac{p_d - p_u}{Su - Sd}. \tag{7.14}$$

Because the option is a put, $p_d > p_u$, and Equation (7.14) will be positive. This result should make sense. You are long the put and puts move opposite to the underlying, so you should be long the underlying. To avoid arbitrage, the portfolio must be priced to earn the risk-free rate. Again, the following condition must hold:

$$V = \frac{V_u}{1 + r}$$

or

$$V = \frac{V_d}{1 + r}. \tag{7.15}$$

We again choose V_u.

$$\frac{V_u}{1 + r} = V$$

$$\frac{h_p Su + p_u}{1 + r} = h_p S + p. \tag{7.16}$$

Therefore, the put price can be represented based on the *no-arbitrage model* as

$$p = B_p - h_p S$$

$$B_p = \frac{h_p S u + p_u}{1 + r}. \tag{7.17}$$

The remainder of the derivation proceeds exactly as with the derivation of the call. The end result or the risk-neutral model is functionally identical:

$$p = PV[E(p_T)] = \frac{\phi p_u + (1 - \phi)p_d}{1 + r}$$

$$\phi = \frac{1 + r - d}{u - d}. \tag{7.18}$$

Again, suppose the current stock price is 99, the exercise price is 100, the annual, discretely compounded, risk-free rate is 2%, the time to expiration is one year, $u = 1.25$, and $d = 0.8$. Similar to the call, we can compute the put price in two ways. First, note that $p_u = \max(0, 100 - 123.75) = 0$ and $p_d = \max(0, 100 - 79.2) = 20.8$. For the no-arbitrage model, we find the hedge ratio $h_p = (p_d - p_u)/(uS - dS) = (20.8 - 0)/(123.75 - 79.2) = 0.4669$. Therefore, based on Equation (7.16), we have

$$p = \frac{h_p S u + p_u}{1 + r} - h_p S = \frac{0.4669(99)1.25 + 0}{1 + 0.02} - 0.4669(99) = 56.6460 - 46.2231 = 10.42.$$

Alternatively, we can use the risk-neutral model. Again, we have $\phi = (1 + 0.02 - 0.8)/(1.25 - 0.8) = 48.8889\%$. Therefore, based on Equation (7.18), we find the same results or

$$p = \frac{0.4889(0) + (1 - 0.4889)20.8}{1 + 0.02} = 10.42.$$

7.3 ARBITRAGING PRICE DISCREPANCIES

If the actual market price of the option differs from the model price, an arbitrage is possible. Consider the call option case. If the call can be sold for more than the formula value, Equation (7.9), the call is overpriced. Overpriced instruments should be sold. Simply selling the call, however, hardly qualifies as an arbitrage. If the call expires in-the-money, one could incur a significant loss, even though the call was underpriced. Instead, the arbitrage should be completed, and the risk eliminated by holding an offsetting number of units of the asset.

The arbitrageur would, thus, buy h_c units of the asset for each call sold. It should be easy to see that the investment required would be less than what is given in Equation (7.2). Equation (7.2) is based on the market price equaling the model price, Equation (7.9). Hence, the amount invested is reduced by the overpricing. Convergence of the option value to its exercise value is ensured one period later because the option is expiring and can clearly be worth only its exercise value. With less money invested and the same payoff as before, the rate of return clearly exceeds the risk-free rate. If the option trades at below the formula price, it would be purchased and h_c units of the asset would be sold, creating

TABLE 7.1   Arbitrage Cash Flows Within One-Period Binomial Model for Calls

Strategy	Today	Down Event at Expiration	Up Event at Expiration
Sell call	$+c_Q = +11.43$	$-\max(0, dS - X) = 0$	$-\max(0, uS - X) = -23.75$
Buy h_c shares	$-h_c S = -52.78$	$+h_c Sd = +42.22$	$+h_c Su = +65.97$
Borrow	$+B_c = +41.39$	$-B_c(1 + r) = -42.22$	$-B_c(1 + r) = -42.22$
Net Cash flow	$+0.04^*$	0	0

*Note the quoted price is 11.43 and the model price is 11.38, a difference of 0.05. The table reports an arbitrage profit of 0.04. The 0.01 discrepancy is simply a rounding error.

TABLE 7.2   Arbitrage Cash Flows Within One-Period Binomial Model for Puts

Strategy	Today	Down Event at Expiration	Up Event at Expiration
Buy put	$-p_Q = -10.37$	$+\max(0, X - dS) = +20.80$	$+\max(0, X - uS) = 0$
Buy h_p shares	$-h_p S = -46.22$	$+h_p Sd = +36.98$	$+h_p Su = +57.78$
Borrow	$+B_p = +56.65$	$-B_p(1 + r) = -57.78$	$-B_p(1 + r) = -57.78$
Net cash flow	$+0.06^*$	0	0

*Note the quoted price is 10.37 and the model price is 10.42, a difference of 0.05. The table reports an arbitrage profit of 0.06 The 0.01 discrepancy is simply a rounding error.

a net short position. The proceeds would be invested in risk-free bonds to earn the rate r. With the option purchased at a lower than fair price, the asset and option would finance the purchase of the risk-free asset at a lower cost than it should if correctly priced, so the investor would earn an arbitrage profit.

Based on the information given in the past two examples, suppose we have the following market quotes, $c_Q = 11.43$ and $p_Q = 10.37$. Recall $S = 99$, $X = 100$, $r = 0.02$, $\tau = 1$, $u = 1.25$, and $d = 0.8$. In equilibrium, we found $c = 11.38$ and $p = 10.42$, thus, the call price is too high and the put price is too low. Arbitrageurs typically prefer to receive positive cash flow today with no chance of any future liability. Alternative arbitrage transactions result in no cash flow today, but some chance of positive cash flow in the future.

Because the quoted call price is too high, the arbitrageur would sell it and buy the synthetic call option. Buying the synthetic call entails buying the stock with borrowed money. Table 7.1 illustrates cash flows capturing the arbitrage profit available with the call option.

Thus, the arbitrageur receives 0.04 today with no chance of a future liability. Within this simple one-period binomial world, trading pressure will drive down the quoted call price and drive up the quoted stock price until the net cash flow is zero.

Turning to puts, if the quoted put price is too low, the arbitrageur would buy it and sell the synthetic put. We previously formed a hedge portfolio by buying the stock and the put. Hence, we can turn that around and see that a synthetic put would involve selling the stock and lending at the risk-free rate. Selling the synthetic put would, therefore, entail buying the stock and borrowing. Table 7.2 illustrates the cash flows that capture the arbitrage profit available with the underpriced put option.

Thus, the arbitrageur receives 0.06 today with no chance of a future liability. Within this simple one-period binomial world, trading pressure could simply drive up the quoted put price. Alternatively, buying shares can drive up the quoted stock price with some influence on the put price. Ultimately, the initial net cash flow must be zero.[2]

Regardless of the direction of the mispricing, the ability to earn an arbitrage profit would force a price alignment until the option price conforms to the model price.

7.4   THE MULTIPERIOD MODEL

The case we have covered until now spans only one period. We can easily extend the model to multiple periods and thereby accommodate options with longer lives.[3] The layout is illustrated in Figure 7.6. We let the asset move from Su to Su^2 or Sud, and from Sd to Sdu or Sd^2. Note that $Sud = Sdu$, so over two periods, there are three possible outcomes. The asset can go up twice to Su^2, up and then down or down and then up to Sud, or down twice to Sd^2. The call and put option payoffs in those states are[4]

$$c_{u^2} = \max(0, Su^2 - X) \qquad p_{u^2} = \max(0, X - Su^2)$$

$$c_{ud} = \max(0, Sud - X) \text{ and } p_{ud} = \max(0, X - Sud).$$

$$c_{d^2} = \max(0, Sd^2 - X) \qquad p_{d^2} = \max(0, X - Sd^2) \qquad (7.19)$$

The illustration is looking more like a branching tree or lattice. Two key features of the binomial model are the recombining nature of the tree and that the growth of the stock price is multiplicative. The tree is recombining because the stock price is assumed to grow multiplicatively such that $Sud = Sdu$. Clearly, the order of multiplication does not matter. The multiplicative approach presented here facilitates the convergence of the stock price to the lognormal distribution—a point we will explore in Chapter 9.[5]

Let us position ourselves in the time 1 up state, where the asset price is Su. At this point, we are now back in a one-period world. There are two possible outcomes in the

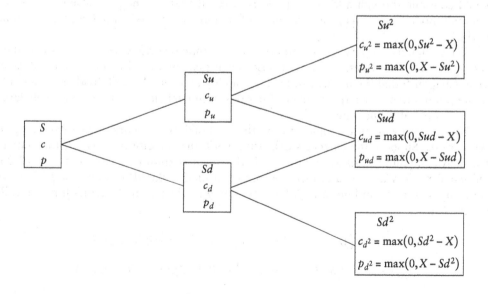

FIGURE 7.6   Two-Period Binomial Model

next period, which is the expiration. It should be easy to see that the value of the call and put at this point would be

$$c_u = \frac{\phi c_{u^2} + (1-\phi)c_{ud}}{1+r} \text{ and } p_u = \frac{\phi p_{u^2} + (1-\phi)p_{ud}}{1+r}. \tag{7.20}$$

Likewise, in the time 1 down state, the option value would be

$$c_d = \frac{\phi c_{ud} + (1-\phi)c_{d^2}}{1+r} \text{ and } p_d = \frac{\phi p_{ud} + (1-\phi)p_{d^2}}{1+r}. \tag{7.21}$$

Stepping back to time 0, the values of the call and put options are again found with Equation (7.9), where the values of c_u and p_u are given in Equation (7.20) and c_d and p_d are given in Equation (7.21). Thus, to price options in the binomial framework in a multiperiod model, we start at the end—the exercise date—and work backwards to the present.

The special case for two-period options does lend itself to a simple formula that relates the initial option value to the value two periods later, essentially skipping over the first period:

$$c = \frac{\phi^2 c_{u^2} + 2\phi(1-\phi)c_{ud} + (1-\phi)^2 c_{d^2}}{(1+r)^2} \text{ and } p = \frac{\phi^2 p_{u^2} + 2\phi(1-\phi)p_{ud} + (1-\phi)^2 p_{d^2}}{(1+r)^2}. \tag{7.22}$$

Note that the three option payoffs two periods later are each weighted by the risk-neutral probabilities, ϕ^2, $2\phi(1-\phi)$, and $(1-\phi)^2$. These are the binomial probabilities for two trials, and they add up to 1. The coefficients on the probabilities are 1, 2, and 1. These numbers are the coefficients of the two-period binomial expansion commonly expressed as $(a+b)^2$, which equals $a^2 + 2ab + b^2$. Note the *binomial coefficients* 1, 2, 1. This pattern would continue through additional periods, a result that will help us when we use the binomial model with a very large number of periods to approximate a continuous time world.[6]

The seemingly simplifying assumptions of a binomial world, whereby an asset can go to only one of two values, can send a false signal that the model is unrealistic. We shall see in Chapter 9 that the model can be made highly realistic. We will divide the finite life of an option into a very large number of periods, ultimately leading to an extremely large number of possible outcomes.

Now we provide an example of how the binomial model works when extending to two periods. Suppose the current stock price is 40, the exercise price is 40, the annual, discretely compounded, risk-free rate is 2%, the time to expiration is two years, $u = 1.25$, and $d = 0.8$. Now assume a two-period binomial model. Based on Equation (7.22), we can compute the call and put prices. First, we compute the terminal payoffs for both calls and puts as

$$c_{u^2} = \max\left(0, Su^2 - X\right) = \max\left[0, 40(1.25)^2 - 40\right] = 22.5$$

$$c_{ud} = \max\left(0, Sud - X\right) = \max\left[0, 40(1.25)0.8 - 40\right] = 0 \text{ and}$$

$$c_{d^2} = \max\left(0, Sd^2 - X\right) = \max\left[0, 40(0.8)^2 - 40\right] = 0$$

$$\bar{p}_{u^2} = \max\left(0, X - Su^2\right) = \max\left[0, 40 - 40(1.25)^2\right] = 0$$

$$\bar{p}_{ud} = \max\left(0, X - Sud\right) = \max\left[0, 40 - 40(1.25)0.8\right] = 0$$

$$\bar{p}_{d^2} = \max\left(0, X - Sd^2\right) = \max\left[0, 40 - 40(0.8)^2\right] = 14.4.$$

The binomial probability of an up move over a single period is $\phi = (1 + 0.02 - 0.8)/(1.25 - 0.8) = 48.8889\%$. Therefore, based on Equation (7.22), we find

$$c = \frac{\phi^2 c_{u^2} + 2\phi(1 - \phi)c_{ud} + (1 - \phi)^2 c_{d^2}}{(1 + r)^2}$$

$$= \frac{(0.4889)^2 22.5 + 2(0.4889)(1 - 0.4889)0 + (1 - 0.4889)^2 0}{(1 + 0.02)^2} = 5.17$$

and

$$p = \frac{\phi^2 p_{u^2} + 2\phi(1 - \phi)p_{ud} + (1 - \phi)^2 p_{d^2}}{(1 + r)^2}$$

$$= \frac{(0.4889)^2 0 + 2(0.4889)(1 - 0.4889)0 + (1 - 0.4889)^2 14.4}{(1 + 0.02)^2} = 3.62.$$

Alternatively, the two-period binomial model can be viewed as three one-period binomial models that apply the no-arbitrage condition. The call results are illustrated in Figure 7.7 and the put results are illustrated in Figure 7.8.

We turn now to address American options where early exercise can enhance the worth of an option.

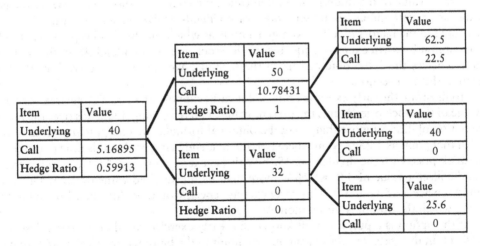

FIGURE 7.7 Two-Period Binomial Model for Calls Illustrated

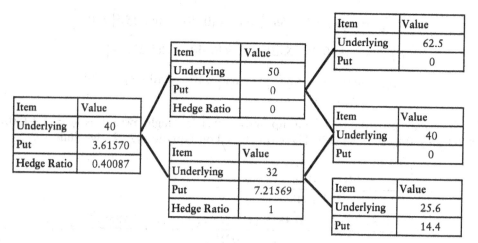

FIGURE 7.8   Two-Period Binomial Model for Puts Illustrated

7.5   AMERICAN OPTIONS AND EARLY EXERCISE IN THE BINOMIAL FRAMEWORK

If the options are American style, they can be exercised early. It is well known that American call options will not be exercised early unless there is a dividend or some other cash amount paid by the asset, in which case early exercise could be justified immediately before the cash is paid.[7] To accommodate the possibility of early exercise, there are a variety of methods that can be used in the binomial model. One extremely useful, but somewhat over-simplified method, is to assume the dividend is a constant rate of the value of the asset.[8] This approach, however, would imply a dividend at every time step. There are some limited situations in which that might be appropriate, such as for an option on a stock index where dividends are paid by the constituent stocks at different times. An alternative approach, however, is to subtract the present value of all dividends over the life of the option from the current value of the underlying—often called the escrow method. Then we let the net of the stock price minus the present value of dividends evolve through the binomial tree according to the factors u and d. At a given node at which the dividend is paid, we decide if the option is worth exercising just before the stock goes ex-dividend. If so, the exercise value replaces the value obtained using the formula. We explore the escrow method in detail in the next section.

To illustrate the early exercise decision for calls, suppose at a point in the tree, we have a value of the stock price minus the present value of all remaining dividends over the life of the option of 42.[9] Suppose that using the binomial formula, we compute the value of the call at that point as 2.25. Assume there is a 3.0 dividend being paid at this time point; then, the stock price with the dividend is 45.[10] With the exercise price at 42, we could exercise it and collect a value of 3.0, which is more than its unexercised value of 2.25. Thus, we would replace 2.25 with 3.0. This early exercise check would be done at all points in the tree in which the option is in-the-money.

If the option is a put, it is well known that early exercise could occur regardless of a dividend. In that case, at every in-the-money point in the binomial tree, we would need to check to see if the put is worth more exercised or not. If it is worth more exercised, we

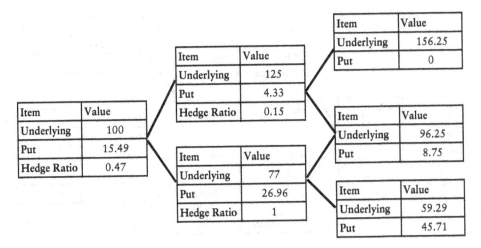

FIGURE 7.9 Two-Period Binomial Model for European Put Option Illustrated

insert the exercise value at that point into the tree as the option value. If not, we use the computed value obtained by the formula. If there is a dividend, it will reduce the frequency of early exercise because dividends drive the stock price down, which makes puts worth more. Exercising early throws away this benefit. If early exercise is justified when there are dividends, it would occur just after a dividend. That is, if we are going to exercise the put, we might as well wait until the stock falls.

Now to illustrate this early exercise feature, we consider a two-period binomial model when there are no dividends with the following parameters, $S = 100$, $X = 105$, $r = 1.0\%$, $u = 1.25$, $d = 0.77$, and two years to expiration. Figure 7.9 illustrates the underlying stock price, put price, and hedge ratio for a European put option.[11] Note that at time step 1, the hedge ratio is 1 because both subsequent up and down moves at time 2 are in-the-money.

If the put option is American, then we need to check at each time step whether it is more valuable to exercise the option early. At time 1, when a down move has occurred, the exercise value is 28 [= max(0, $X - Sd$) = max(0, 105 − 77)]. Thus, the European put price of 26.96 is replaced with 28 as the put buyer would prefer to receive 28 from early exercise rather than 26.96 from just holding the put option. This 1.04 increase in value results in the initial put price rising 0.52 to 16.01 as well as the hedge ratio rising to 0.49. The early exercise feature is illustrated in Figure 7.10.

We now further explore the role of dividends on both European and American options.

7.8 DIVIDENDS AND RECOMBINATION

Dividends pose a significant problem with the multiplicative binomial model presented here. Figure 7.11 illustrates the loss of the recombining property in the presence of a cash dividend at time 1. Recall that the stock price falls by the dollar dividend amount on the ex-dividend date. As discussed in Chapter 2, optimal early exercise may occur either right before the ex-dividend date for calls or right after the ex-dividend date for puts. Note that mathematically,

$$(Su - D_1)\, d = (Sd - D_1)\, u \text{ only if } D_1 = 0.$$

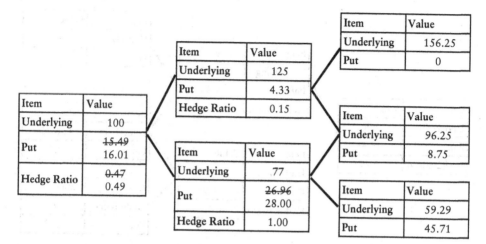

FIGURE 7.10 Two-Period Binomial Model for American Put Option Illustrated

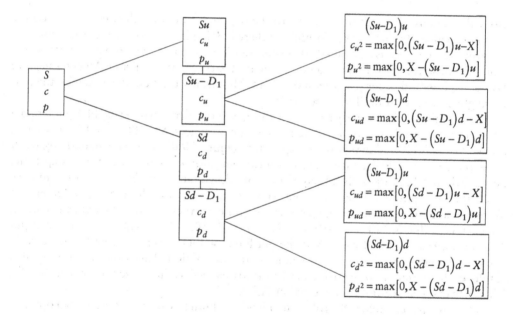

FIGURE 7.11 Two-Period Binomial Model with Discrete Dividends

Thus, the presence of discrete dividends poses a significant computational problem. The binomial tree fails to recombine and is known as a *bushy tree*. When a binomial tree does recombine, the number of nodes after N time steps is $N + 1$. Unfortunately, when a binomial tree fails to recombine the node count explodes with more time steps.[12] That is, the number of nodes after N time steps is 2^N. There are numerous potential ways to address this challenge.

One efficient method is what is known as the *escrow method*. Suppose the firm places in escrow the present value of the dividend payments over the life of the option. That is, the present value of the dividends is set aside, and the remaining stock value is modeled within the binomial framework. We define a new variable commonly known as the stock price without dividends denoted S'. Therefore, the actual stock, S, can be decomposed into two components, the present value of the dividend payments, $PV(D)$, and the rest of the stock price, which excludes the dividends over the life of the option, S'.[13] Figure 7.12 illustrates the escrow method.

We illustrate the incorporation of dividends with the escrow method in a two-period binomial example. Suppose the current stock price is 105, the exercise price is 100, the annual, discretely compounded, risk-free rate is 1%, the time to expiration is two years, $u = 1.25$, and $d = 0.77$. Assume a dividend after one year of 2.0. Thus, the present value of the dividend is

$$PV(D) = \frac{D_1}{(1+r)^{\Delta t}} = \frac{2}{(1+0.01)^1} = 1.98.$$

Therefore, the stock price without dividends is

$$S' = S - PV(D) = 105 - 1.98 = 103.02.$$

Based on the escrow method and corresponding single-period binomial model, we produce the results for European options provided in Figure 7.13.

The values for American options are different. At each node, we appraise whether the exercise value exceeds the binomial model value. At time 1, we have value adjustments as the call should be exercised right before going ex-dividend and the put should be exercised right after going ex-dividend. Based on this approach, Figure 7.14 illustrates the American option prices.

Thus, at time 1 after an up move has occurred, the call buyer will exercise the option right *before* the ex-dividend event, buy the stock at 100, and sell the stock at 130.77,

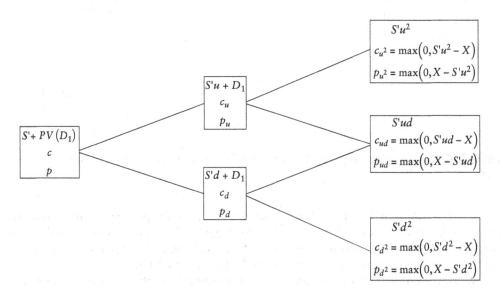

FIGURE 7.12 Two-Period Binomial Model with Dividends Based on Escrow Method

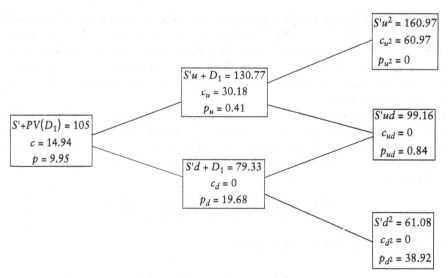

FIGURE 7.13 Numerical Example of Two-Period Binomial Model with Dividends for European Options

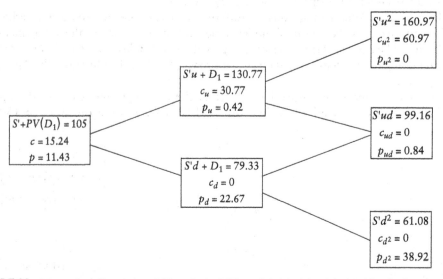

FIGURE 7.14 Numerical Example of Two-Period Binomial Model with Dividends for American Options

netting 30.77. At time 1, when a down move has occurred, the put buyer will exercise the option right *after* the ex-dividend event, buy the stock at 77.33 (= 79.33 − 2), and sell the stock via the put at 100, netting 22.67. Therefore, both the call and put prices are higher for the American options when compared to the European options.

We now examine cases where the binomial framework is difficult at best to apply.

7.7   PATH INDEPENDENCE AND PATH DEPENDENCE

European options are one family of a class of options that are described by a characteristic we call *path independence*. Path independence is when it does not matter which path was taken to get to a point. For example, in a two-period model, one could end up in the middle point by going up and then down or down and then up. For a European option and indeed all path-independent options, it does not matter which path was taken. You arrive at the same final asset price, the option payoff is determined by that asset price, and the path the asset took to get there is irrelevant. Hence, we can ignore the intermediate steps and, as such, use formulas such as Equation (7.22).

Path-dependent options are an entirely different story. American options are path-dependent, provided there is a nonzero probability of early exercise. For example, if we had an American put, it might be exercised when the asset goes down as previously illustrated. Hence, for the asset to go down and then up does not produce the same result as when it goes up and then down. As noted, there are a number of other types of path-dependent options that are the subjects of advanced topics in options.

For example, with a typical home mortgage, homeowners have the option to refinance whenever they wish. They are often motivated to do so when interest rates decline. Thus, the home mortgage value is based, in part, on this early exercise option. When interest rates fall, the homeowner may choose to refinance, whereas if interest rates rise, the homeowner is likely not to refinance.

7.8   RECAP AND PREVIEW

In this chapter, we covered one of the simplest but most important methods of valuing options: the binomial model. We showed how the model clearly illustrates the process by which a dynamically adjusted portfolio enables one to assign a value to an option that must hold to prevent arbitrage. We showed how this process works in one- and two-period models. We illustrated how the early exercise of American options is accommodated within the binomial model.

In Chapter 8, we continue our coverage of the binomial model by looking at the Greeks, which are the sensitivities of the options to the various factors that affect the value of an option.

APPENDIX 7A

Derivation of Equation (7.9)

We start with Equation (7.7), restated here as

$$\frac{h_c S u - c_u}{1 + r} = h_c S - c.$$

Now substitute for h, using Equation (7.5),

$$\frac{\left(\frac{c_u - c_d}{Su - Sd}\right) Su - c_u}{1 + r} = \left(\frac{c_u - c_d}{Su - Sd}\right) S - c.$$

Rearranging, we have

$$\left(\frac{c_u - c_d}{Su - Sd}\right)S - \frac{\left(\frac{c_u - c_d}{Su - Sd}\right)Su - c_u}{1 + r} = c.$$

Then we multiply through by $1 + r$,

$$\left(\frac{c_u - c_d}{Su - Sd}\right)S(1 + r) - \left(\frac{c_u - c_d}{Su - Sd}\right)Su + c_u = c(1 + r).$$

Then we cancel S,

$$\left(\frac{c_u - c_d}{u - d}\right)(1 + r) - \left(\frac{c_u - c_d}{u - d}\right)u + c_u = c(1 + r).$$

Using the common denominator $u - d$, we obtain

$$\frac{c_u(1 + r) - c_d(1 + r) - c_u u - c_d u + c_u u - c_u d}{u - d} = c(1 + r)$$

$$\frac{c_u(1 + r - d) + c_d[u - (1 + r)]}{u - d} = c(1 + r).$$

Now, let us define ϕ as in Equation (7.9),

$$\phi = \frac{1 + r - d}{u - d}.$$

Then $1 - \phi$ is

$$1 - \phi = \frac{u - (1 + r)}{u - d}.$$

So the solution is

$$c = \frac{\phi c_u + (1 - \phi)c_c}{1 + r}.$$

which is Equation (7.9).

APPENDIX 7B

Pascal's Triangle and the Binomial Model

Pascal's triangle has an infinite number of rows. An illustration of Pascal's triangle for the first five rows is provided in Figure 7B.1.

The elements of each row can be obtained from the row above it. For example, look at the 2 in the middle of the third row. It is obtained by summing the 1 above it to the left and the 1 above it to the right. Notice either 4 in the fifth row; each is obtained by summing the 1 above it to the left (or right) and the 3 above it to the right (or left). With the exception of the 1 in the first row, every number in the table is the sum of whatever appears above it to the left and above it to the right.[14]

Pascal's triangle contains the coefficients of the binomial expansion of $(a + b)^n$. Here we illustrate the expansion for $n = 0, \ldots, 4$ below, which encompasses the portion of the triangle illustrated above. To link Pascal's triangle to option pricing, let us change the variables a and b to the up and down factors, u and d as illustrated in Table 7B.1.

So how does Pascal's triangle come into play in option pricing? It tells us how many ways there are to reach a point after n periods. For example, in a one-period binomial model, there is only one way to go up and one way to go down. Note the coefficients 1 and 1 for the $n = 1$ case. In the two-period model, the outcomes are up up, up down, down up, and down down. There is only one way to go up twice and one way to go down twice. But if you go up and down, you end up the same place as if you had gone down and then up. That is, $Sud = Sdu$. This would put you at the middle point of the binomial tree, as in Figure 7.4. There are two ways to get to that middle point. Thus, we get a coefficient of 2 on the ud term in row 2.

For a three-period binomial tree, you can go up three times, up twice and down once, down twice and up once, and down three times. There is only one way to go up three times:

FIGURE 7B.1 Pascal's Triangle (First Five Rows)

TABLE 7B.1 Coefficients of the Binomial Expansion

n	$(u + d)^n$	Coefficients
0	1	1
1	$u + d$	1, 1
2	$u^2 + 2ud + d^2$	1, 2, 1
3	$u^3 + 3u^2d + 3ud^2 + d^3$	1, 3, 3, 1
4	$u^4 + 4u^3d + 6u^2d^2 + 4ud^3 + d^4$	1, 4, 6, 4, 1

up up up. There are three ways to go up twice and down once: up up down or up down up or down up up. Thus, we get a coefficient of 3 on the u^2d term in the table for $n = 3$. There are three ways to go down twice and then up: down down up, down up down, and up down down. Hence, we get a coefficient of 3 applied to ud^2. Finally, there is only one way to go down three times, down down down; hence, the coefficient is 1 on d^3.

These coefficients enable us to fill out the details of the binomial model. We could even use them to skip some computational steps. For example, we previously explained how to derive the binomial value by starting at the expiration and working backward to the present. We showed the special case of a two-period model in which we skip the intermediate steps between the expiration and the present, Equation (7.22). Now, consider the four-period case. Using the $n = 4$ row in the binomial expansion, we can derive the time 0 value from the possible time 4 values as follows:

$$c = \frac{\phi^4 c_{u^4} + 4\phi^3(1 - \phi)c_{u^3d} + 6\phi^2(1 - \phi)^2 c_{u^2d^2} + 4\phi(1 - \phi)^3 c_{ud^3} + (1 - \phi)^4 c_{d^4}}{(1 + r)^4}. \quad (7.23)$$

This calculation works for the path-independence case, such as for European options. For path-dependent cases, such as for American options and some complex options, one would have to start at the expiration and roll backwards to the present. We cannot skip or consolidate any paths.

For example, suppose you have the following parameters for a four-period binomial model for European options, $S = 40$, $X = 40$, $r = 2.0\%$ per annum, $u = 1.25$, $d = 0.8$ and $\tau = 4$ years. Thus, we once again have $\phi = (1 + r - d)/(u - d) = (1 + 0.02 - 0.8)/(1.25 - 0.8) = 0.4889$. The terminal payoffs for a call option are given as follows:

$$c_{u^4} = \max\left(0, Su^4 - X\right) = \max\left[0, 40\left(1.25^4\right) - 40\right] = 23.616$$

$$c_{u^3d} = \max\left(0, Su^3d - X\right) = \max\left[0, 40\left(1.25^3 0.8\right) - 40\right] = 14.4$$

$$c_{u^2d^2} = \max\left(0, Su^2d^2 - X\right) = \max\left[0, 40\left(1.25^2 0.8^2\right) - 40\right] = 0$$

$$c_{ud^3} = \max\left(0, Sud^3 - X\right) = \max\left\{0, 40\left[(1.25)0.8^3\right] - 40\right\} = 0$$

$$c_{d^4} = \max\left(0, Sd^4 - X\right) = \max\left[0, 40\left(0.8^4\right) - 40\right] = 0.$$

The call value is

$$c = \frac{\phi^4 c_{u^4} + 4\phi^3(1 - \phi)c_{u^3d} + 6\phi^2(1 - \phi)^2 c_{u^2d^2} + 4\phi(1 - \phi)^3 c_{ud^3} + (1 - \phi)^4 c_{d^4}}{(1 + r)^4}$$

$$= \frac{\left[\begin{array}{c}0.4889^4(57.656) + 4(0.4889^3)(1 - 0.4889)(22.5) + 6(0.4889^2)(1 - 0.4889)^2(0) \\ +4(0.4889)(1 - 0.4889)^3(0) + (1 - 0.4889)^4(0)\end{array}\right]}{(1 + 0.2)^4}$$

$$= 8.01.$$

We leave as a problem to find both the European and American put prices.

QUESTIONS AND PROBLEMS

1 Suppose you have the following parameters related to a single-period binomial model, $S = 32$, $X = 30$, $r = 4.0\%$ (annual, discrete compounding), $\tau = 0.5$ years, $u = 1.25$, and $d = 0.75$. Suppose the call price in the market is 5.39 and the put price is 2.80. Calculate the option prices based on the binomial model and explain how any arbitrage would be captured.

2 Assume a single-period binomial model for a put option can be expressed as

$$p = \frac{\phi p_u + (1 - \phi)p_d}{1 + r} \text{ where}$$

$$\phi = \frac{1 + r - d}{u - d}.$$

Prove that the call option value can also be represented as

$$p = \frac{h_p Su + p_u}{1 + r} - h_p S, \text{ where}$$

$$h_p = \frac{p_d - p_u}{Su - Sd}.$$

3 Based on the information given in the chapter, we have the following market quotes, $c_Q = 11.43$ and $p_Q = 10.37$. Recall $S = 99$, $X = 100$, $r = 0.02$, $\delta = 1$, $u = 1.25$, and $d = 0.8$. Further, we found $c = 11.38$ and $p = 10.42$, thus the call price is too high and the put price is too low. Recall, arbitrageurs typically prefer to receive positive cash flow today with no chance of any future liability. Based on put-call parity, demonstrate the appropriate arbitrage transaction.

4 Suppose you have the following parameters for a four-period binomial model for European options, $S = 40$, $X = 40$, $r = 2.0\%$ per annum, $u = 1.25$, $d = 0.8$, and $\tau = 4.0$ years. Calculate the European put value.

5 Again, suppose you have the following parameters for a four-period binomial model for European options, $S = 40$, $X = 40$, $r = 2.0\%$ per annum, $u = 1.25$, $d = 0.8$, and $\tau = 4.0$ years. Calculate the American put value. Further, compare the behavior of the hedge ratio for European and American puts.

6 [Contributed by Paul Maloney] A portfolio manager takes a fairly simplified approach to stock valuation. They narrow it down to just two scenarios, a good year and a bad year. The stock is currently priced at 50. They believe there is a 34% chance the stock will fall to 35. What is their estimate of the upside if the risk-free rate is 3.0%? (Note: They are using risk-neutral valuation.)

NOTES

1. An alternative approach would specify an asset priced at S can go up to $S + u$ or down to $S + d$, an additive binomial approach.
2. There is another arbitrage opportunity based on put-call parity, but we leave that one as an end-of-chapter problem.

3. As we shall see in Chapter 9, we can also divide an option's life into a larger number of smaller periods. Thus, we can accommodate a fixed life, subdivided into smaller and smaller periods. Here, however, we simply extend the life of the option.

4. Henceforth, we shall always use c_{ud} to represent both c_{ud} and c_{du}.

5. It is important to note that recombining trees can also be produced through an additive process, such as $S + u + d = S + d + u$ with $d < 0$. The additive binomial lattice facilitates the convergence of the underlying variable to the normal distribution. Recall the multiplicative binomial lattice facilitates the convergence of the underlying variable to the lognormal distribution.

6. The general binomial expansion is $(a + b)^n$. The coefficients are known as the elements of Pascal's triangle. This point is illustrated in Appendix 7B.

7. If the asset is a stock and the cash payment is a dividend, then exercise would occur immediately before the stock goes ex-dividend. Of course, early exercise is not necessarily optimal.

8. Furthermore, the relationship can be expressed as either a continuous yield or a discrete yield.

9. The only time points that we have to check are those in which the option is in-the-money.

10. This number would be even larger if there were additional dividends over the life of the option. You would add the present value of these dividends.

11. We have rounded the final answers to ease exposition, but all calculations are based on a spreadsheet with more precise values.

12. Specifically, the node growth rate is $2^N - 2^{N-1} = 2^{N-1}$. For example, if $N = 0$, then the growth rate is $2^{10} - 2^9 = 1,024 - 512 = 512$.

13. Note the remainder stock price S' is simply the present value of dividends after the expiration of the option.

14. The ones on all rows except the first also follow this rule. If nothing appears above left or above right, simply treat it as zero.

CHAPTER 8

Calculating the Greeks in the Binomial Model

Option pricing models are primarily useful for identifying the fair value of an option, but that is not their only use. They also provide measures that facilitate the management of option portfolios. In particular, dealers that make markets in options need to be able to hedge the risk.[1] Hence, they must know how sensitive their options are to changes in market conditions. Knowing these sensitivities enables them to balance that risk using other instruments. These sensitivities are often known as the *Greeks*. This term arose because they are primarily identified with Greek letters, such as delta and gamma. End users, such as corporate executives, portfolio managers, and regulators, have all found these measures and other Greeks to be very useful. For example, in Chapter 18 we will see an application of option theory to gain insights on corporate default risk. Within this framework, the Greeks provide useful information regarding sensitivities of default risk to changes in market conditions. Numerous financial instruments contain option-like features where understanding the related Greeks will aid in financial decision-making.

8.1 STANDARD APPROACH

Let S be the current price of the underlying asset and u and d be the up and down factors. The discrete risk-free rate per period is r, and X is the exercise price of the option. We illustrate both calls and puts in the binomial model for two periods in Figure 8.1.

We are interested in the sensitivities of the option value to the underlying, to time to expiration, to the volatility, and to the risk-free interest rate. These four parameters are subject to change over time; hence, risk managers need to understand how their option values change along with other assets for which they are responsible.

Recall that put-call parity (PCP) can be expressed generically as

$$c = S - PV(X) + p, \tag{8.1}$$

where $PV(X)$ denotes the present value of X. In this and future chapters the interest rate quotation convention, such as different ways of discrete compounding and even continuous compounding, can vary. Hence, for convenience, we will simply rely on a generic present value function. PCP is based on static arbitrage transactions. Therefore, put-call parity should hold with any model for option pricing. Thus, we will rely on the linkage between puts and calls whenever convenient.

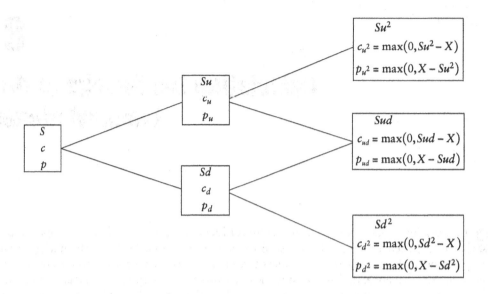

FIGURE 8.1   The Binomial Model (Two Periods)

8.1.1 Delta

The sensitivity of call and put options to the value of the underlying is called the *delta*. Of course, it is intuitive that a put moves opposite to the underlying as it benefits from decreases in the price of the underlying. Thus, its delta must be negative. Here, we take the negative of the traditional measure of the put delta to represent it as a positive number, simplifying various binomial model interpretations.[2] The call and put deltas are formally defined as

$$\Delta_c = \frac{\partial c}{\partial S} \text{ and } \Delta_p = -\frac{\partial p}{\partial S}. \qquad (8.2)$$

We have specified the deltas in terms of partial derivatives. At this point, we do not have a continuous and differentiable model of the option value, which will come later with the Black-Scholes-Merton model. We can observe the following relationship between call and put deltas based on put-call parity, Equation (8.1), and the fact that the partial derivative of anything with itself is one. Thus, we know (recall our put delta is signed)

$$\Delta_c = 1 - \Delta_p. \qquad (8.3)$$

In discrete time, we approximate the deltas with the following:

$$\Delta_c \simeq \frac{\Delta c}{\Delta S} \text{ and } \Delta_p \simeq -\frac{\Delta p}{\Delta S}, \qquad (8.4)$$

where the delta symbols in the numerator and denominator reflect the change in the prices of the options and asset from their current prices. The problem in measuring delta in a binomial tree is that technically we must assume that the asset price changes without an increment of time. Time is a variable and if we advance further into the tree, we have not held time constant. With delta, we are attempting to measure the sensitivity of the option to a single effect, the change in the value of the underlying. From the binomial tree diagram, we see that the asset price can change only with an increment of time, not at a given instant in time. One way to deal with this problem is to have enough binomial periods so that the length of each time step is very short. There is also another way to fit the tree to deal with the problem, which we explore in the second half of this chapter. For now, let us measure delta the best we can, given the form of the tree we have used to now.

As we shall see later when covering the Black-Scholes-Merton model, in a continuous time world, the option price is nonlinear with respect to the asset price, meaning that the option price change is not the same upward as it is downward for a given change in the underlying. This result is due to the convexity of the relationship between the option price and the asset price. That convexity is evident in the binomial model as well. Although it is not apparent in Figure 8.1, the option price change is not the same upward as it is downward. In continuous time, we deal with this problem by using the first derivative, which is the limit of the rate of change in the option price for an infinitesimal change in the asset price. The first derivative linearizes the change in the option price in relation to the change in the asset price. In the binomial model, the usual approach to measuring delta is to take the average of the two possible call price changes and divide by the average of the two possible asset price changes:[3]

$$\Delta_c = \frac{(1/2)(c_u - c) + (1/2)(c - c_d)}{(1/2)(Su - S) + (1/2)(S - Sd)} = \frac{c_u - c_d}{Su - Sd}$$

$$\Delta_p = 1 - \Delta_c = -\frac{(1/2)(p_u - p) + (1/2)(p - p_d)}{(1/2)(Su - S) + (1/2)(S - Sd)} = \frac{p_d - p_u}{Su - Sd}. \tag{8.5}$$

Recall that in Chapter 7, our derivation of the binomial model revealed that this is the correct measure of delta because it is the number of shares that should be held to hedge in relation to a short position in one call option. This idea is the same notion of delta in the continuous time world of the Black-Scholes-Merton model. Because the delta is $\Delta c / \Delta S$, if we hold $\Delta c / \Delta S$ units of the asset and the asset changes by ΔS, the position in the asset changes by $(\Delta c / \Delta S)\Delta S = \Delta c$. If we hold one short call, then it changes by $-\Delta c$ and the net result is an offset.

8.1.2 Gamma

A short option position hedged with delta units of a long asset position will be a perfect hedge provided that the effect of the asset price change is accurately captured by the delta. In the real, non-binomial world, this result will occur only if the asset price change is infinitesimal. Hence, delta-hedged positions are at least partially vulnerable to virtually any change in the asset price, however large that change may be. In other words, there are essentially no changes that are truly infinitesimal in the real world. In addition, the delta will change with any change in time. If the delta changes, the hedged position must be revised to reflect a new number of shares that will balance the asset price risk. If the delta is not revised quickly enough, the hedged position will lose its effectiveness. The *gamma* reflects the change in the delta given a change in the asset price and captures this effect. Technically gamma is somewhat meaningless in a binomial world unless we impose some limitations on our ability to trade to adjust the hedge position whenever the asset price changes. We often use the binomial model to approximate a continuous time world, however, so we need to know how to estimate the gamma in the binomial model.[4]

The gamma is formally defined as

$$\Gamma = \frac{\partial \left(\frac{\partial c}{\partial S} \right)}{\partial S} = \frac{\partial^2 c}{\partial S^2}. \tag{8.6}$$

Taking the second derivative of Equation (8.1), we note that the gamma of the call and put are equal or

$$\Gamma_c = \Gamma_p. \tag{8.7}$$

In the binomial model, we measure the call gamma by first noting that the call delta can change from the above measure, Δ_c, either to the new delta if the asset goes up, which we call Δ_{cu}, or to the new delta if the asset goes down, which we call Δ_{cd}. Using the formula for the call delta, these deltas at time 1 are defined as follows:

$$\Delta_{cu} = \frac{c_{u2} - c_{ud}}{Su^2 - Sud}$$

$$\Delta_{cd} = \frac{c_{ud} - c_{d2}}{Sud - Sd^2}. \tag{8.8}$$

In the binomial model, the gamma is estimated by averaging the change in the delta, $(1/2)(\Delta_u - \Delta) + (1/2)(\Delta - \Delta_d) = (1/2)(\Delta_u - \Delta_d)$, and dividing by the average change in the asset price. For the numerator,

$$(1/2)(\Delta_u - \Delta_d) = (1/2) \left(\frac{c_{u2} - c_{ud}}{Su^2 - Sud} \right) - (1/2) \left(\frac{c_{ud} - c_{d2}}{Sud - Sd^2} \right). \tag{8.9}$$

For the denominator,

$$(1/4)(Su^2 - Su) + (1/4)(Su - Sud) + (1/4)(Sud - Sd) + (1/4)(Sd - Sd^2)$$

$$= (1/4)(Su^2 - Sd^2). \tag{8.10}$$

So now we have the following estimate of the call gamma:

$$\Gamma_c = \frac{\left(\frac{c_{u2} - c_{ud}}{Su^2 - Sud} \right) - \left(\frac{c_{ud} - c_{d2}}{Sud - Sd^2} \right)}{(1/2)(Su^2 - Sd^2)}. \tag{8.11}$$

Following this logic, the put gamma can be estimated in a similar fashion:

$$\Gamma_p = \frac{\left(\frac{p_{d2} - p_{ud}}{Sud - Sd^2} \right) - \left(\frac{p_{ud} - p_{u2}}{Su^2 - Sud} \right)}{(1/2)(Su^2 - Sd^2)}. \tag{8.12}$$

Note that in order to calculate the call or put gamma, you need the asset and option prices two time periods ahead. Thus, one important limitation of this approach is that it is impossible to calculate the gamma one time step before expiration. This point in time, which would often be very close to expiration, would tend to have a high gamma unless the option is deep in- or out-of-the-money. Therefore, there can be a critical need to estimate the gamma at that very point at which it cannot be estimated. One solution is to use a large number of time steps of very short length.

8.1.3 Theta

The *theta* is the effect of the change in time on the option price. It captures the time value decay, meaning that it reflects the rate at which the option price changes when nothing else changes. We know that option prices move toward their exercise values, and at expiration, they equal their exercise value. The formal specification of theta for a continuous time world is

$$\Theta_c = \frac{\partial c}{\partial t} \text{ and } \Theta_p = \frac{\partial p}{\partial t}, \tag{8.13}$$

where t is the current point in time. If the remaining time is, say, $\tau = T - t$, then $\partial t = T - \partial \tau$. Sometimes theta is evaluated in terms of $\partial \tau$, but we need not do so here as the transformation is simple.

Observing the binomial tree diagram, it should be apparent that an estimate of theta is obtained as

$$\Theta_c = \frac{c_{ud} - c}{2\Delta t} \text{ and} \tag{8.14}$$

$$\Theta_p = \frac{p_{ud} - p}{2\Delta t}, \tag{8.15}$$

where Δt is the length of each time step. Note that we are measuring the change in option price without the effect of an asset price change. Two time steps are required for the asset to return to where it started.[5] In other words, to measure the time sensitivity at time 0, we take the difference in the price two periods later at the same asset price, c_{ud}, and subtract the current option price, dividing that difference by $2\Delta t$, because two time steps have elapsed.

8.1.4 Vega and Rho

Vega and *rho* are the effects of the volatility and interest rate, respectively, on the option price.[6] Unfortunately, vega and rho cannot be measured as easily as delta, gamma, and theta. To understand why, it is necessary to understand the role of volatility and the interest rate in a typical option pricing model.

The volatility and interest rate are assumed constant over the life of the option. When working in a continuous time world, we can certainly take the derivative of the Black-Scholes-Merton model with respect to the volatility or interest rate, but their interpretations are simply that the option price will be different if we use a different volatility or interest rate. This point should be very clear in the binomial model. The binomial tree changes only as a result of time and the asset price change. The volatility and interest rate affect the up and down factors and the risk-neutral probability, but these parameters are constant throughout the tree. It is possible, however, to allow the up and down factors and the risk-free rate to change, but this is the subject of more advanced models.

If we want to measure the sensitivity of the binomial option price to a different volatility or interest rate, we would simply recalculate the binomial option value with a different volatility or interest rate, keeping in mind that the sensitivity is different depending on whether we use a higher or lower volatility or interest rate. We would try both an increase and a decrease in the volatility and interest rate. We would then average the option price difference and divide by the average difference of the volatility or interest rate to get a reasonable estimate of the vega or rho.[7]

8.2   AN ENHANCED METHOD FOR ESTIMATING DELTA AND GAMMA

Because delta and gamma should technically be measured without a change in time, a modified method of fitting the binomial tree can be used to achieve this property. Suppose we wish to fit a binomial tree with N time steps. With this modification we fit the tree with $N + 2$ time steps and position the current asset price in the middle of the second time step, in the following manner. In our example in the previous section, we assess the Greeks assuming we are at point in time 2 with one up move. Figure 8.2 illustrates the binomial tree for a call where the shade indicates actual potential paths for the current stock price (Sud).

Now let us position ourselves right in the center where the current asset price is Sud, with the corresponding option price being c_{ud}. We shall see that this binomial tree offers several advantages over the standard approach. As we shall learn in the next sections, it will enable us to calculate both delta and gamma without imposing a time change.[8]

8.2.1   Delta

Positioned at the current asset price of S_{ud}, we can now measure the effect of a different asset price on the option price without moving forward in time. Notice that at this time point, there are two other asset prices, Su^2 and Sd^2. We can observe the effect of the asset

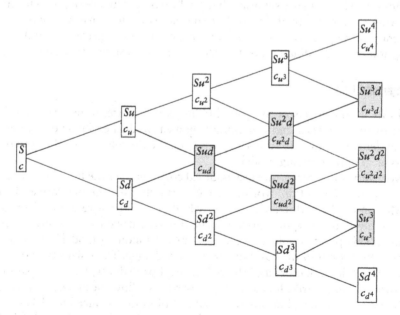

FIGURE 8.2   Two-Period Expanded Binomial Model

price being Su^2 or Sd^2 at the same time point and measure delta as the average option price change divided by the average asset price change.

$$\Delta_c = \frac{(1/2)\left(c_{u^2} - c_{ud}\right) + (1/2)\left(c_{ud} - c_{d^2}\right)}{(1/2)\left(Su^2 - Sud\right) + (1/2)\left(Sud - Sd^2\right)} = \frac{c_{u^2} - c_{d^2}}{Su^2 - Sd^2} \text{ and} \tag{8.16}$$

$$\Delta_p = \frac{(1/2)\left(p_{ud} - p_{u^2}\right) + (1/2)\left(p_{d^2} - p_{ud}\right)}{(1/2)\left(Su^2 - Sud\right) + (1/2)\left(Sud - Sd^2\right)} = \frac{p_{d^2} - p_{u^2}}{Su^2 - Sd^2}. \tag{8.17}$$

So, now our estimate of delta does not permit any passage of time.

8.2.2 Gamma

Remember that gamma measures the change in delta, given a change in the asset price. From the point at which the asset is at Sud, we can think of the gamma as being the difference in the change in option price over the change in asset price for the asset price in the node above and for the asset price in the node below. That is,

$$\Gamma_c = \frac{\left(\frac{c_{u^2} - c_{ud}}{Su^2 - Sud}\right) - \left(\frac{c_{ud} - c_{d^2}}{Sud - Sd^2}\right)}{(1/2)\left(Su^2 - Sd^2\right)} \text{ and} \tag{8.18}$$

$$\Gamma_p = \frac{\left(\frac{p_{d^2} - p_{ud}}{Sud - Sd^2}\right) - \left(\frac{p_{ud} - p_{u^2}}{Su^2 - Sud}\right)}{(1/2)\left(Su^2 - Sd^2\right)}. \tag{8.19}$$

The numerator measures the average difference in the delta and the denominator measures the average asset price change. Note that we can now measure gamma at any time step, including the one before expiration.

8.2.3 Theta

As in the standard case, theta will require two time steps ahead of the current price. The theta would then be measured the same as in the standard case:

$$\Theta_c = \frac{c_{u^2d^2} - c_{ud}}{2\Delta t} \text{ and} \tag{8.20}$$

$$\Theta_p = \frac{p_{u^2d^2} - p_{ud}}{2\Delta t}. \tag{8.21}$$

8.2.4 Vega and Rho

The same issues concerning the estimation of vega and rho that were discussed previously would apply here. The tree does not technically permit the volatility or the interest rate to change. The tree could be recomputed with small changes in volatilities and interest rates to obtain estimates of the vega and rho, but these measures would technically mean the sensitivity if vega and rho were different but fixed, as opposed to variable.

8.3 NUMERICAL EXAMPLES

Now let us take a look at some numerical examples. Let us first look at a simple problem with two binomial time steps. Later we extend the problem to more time steps. We use the following inputs: $S = 100$, $X = 100$, $\tau = 2$ (time to expiration), $\sigma = 0.4$, $r = 0.07$ (discrete), and $n = 2$. With the inputs expressed in these terms, we have to convert them to their binomial analogs. Thus, we now introduce a reasonable and widely used formula to convert the volatility and risk-free rate of the asset return to the binomial parameters,

$$u = e^{\sigma \sqrt{\tau/n}}$$

$$d = 1/u$$

$$r(per\,period) = (1 + r)^{\tau/n} - 1.$$

Note here that d is the reciprocal of u, but that does not have to be the case. Research on the binomial model has produced many formulas for converting volatility of the underlying to the u and d factors. In the limit, all of them produce the desired result. This is just one of the more widely used formulas. The calculations are

$$u = e^{0.4\sqrt{2/2}} = 1.4918$$

$$d = 1/1.4918 = 0.6703$$

$$r(per\,period) = (1 + r)^{T/n} = (1.07)^{2/2} - 1 = 0.07.$$

We obtain $u = 1.4918$, $d = 0.6703$, and the per-period risk-free rate is 0.07. The length of a time step is 2 (years)/2 = 1. That is, each time step is one year. The standard binomial tree is as shown in Figure 8.3.[9]
The standard Greeks are computed as follows:

$$\Delta_c = \frac{c_u - c_d}{Su - Sd} = \frac{55.72 - 0}{149.18 - 67.03} = 0.6783 \text{ and } \Delta_p$$

$$= \frac{p_d - p_u}{Su - Sd} = \frac{26.43 - 0}{149.18 - 67.03} = 0.3217,$$

$$\Gamma_c = \frac{\left(\frac{c_{u2} - c_{ud}}{Su^2 - Sud}\right) - \left(\frac{c_{ud} - c_{d2}}{Sud - Sd^2}\right)}{(1/2)(Su^2 - Sd^2)} = \frac{\left(\frac{122.55 - 0}{222.55 - 100}\right) - \left(\frac{0 - 0}{100 - 44.93}\right)}{(1/2)(222.55 - 44.93)} = 0.0113 \text{ and}$$

$$\Gamma_p = \frac{\left(\frac{p_{d2} - p_{ud}}{Sud - Sd^2}\right) - \left(\frac{p_{ud} - p_{u2}}{Su^2 - Sud}\right)}{(1/2)(Su^2 - Sd^2)} = \frac{\left(\frac{55.07 - 0}{100 - 44.93}\right) - \left(\frac{0 - 0}{222.55 - 100}\right)}{(1/2)(222.55 - 44.93)} = 0.0113,$$

$$\Theta_c = \frac{c_{ud} - c}{2\Delta t} = \frac{0 - 25.34}{2(1)} = -12.67 \text{ and } \Theta_p = \frac{p_{ud} - p}{2\Delta t} = \frac{0 - 12.68}{2(1)} = -6.34.$$

Note that from Equations (8.3) and (8.7), the put-call parity relationship for delta and gamma hold. The tree for the more precise method is as shown in Figure 8.4:
The positioning of this tree in the context of time merits an explanation. The current state is the middle position of time 2 where the asset is at 100 and the call is at 25.34.[10] In general

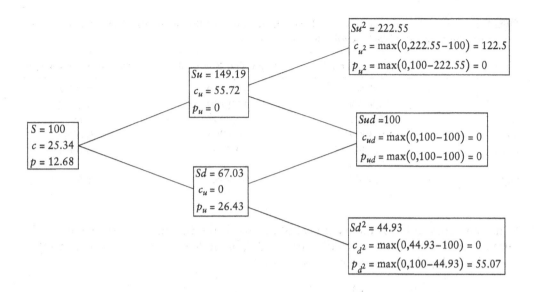

FIGURE 8.3 Standard Binomial Greeks Example

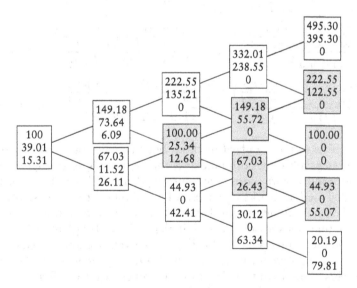

FIGURE 8.4 Expanded Binomial Tree Example

terms, the current state is where the asset price is S_{ud}. To obtain the states above and below, where the asset price is S_{u^2} and S_{d^2}, we have to back up and generate states S_u and S_d and then back up to generate S. Once we have this information, we can generate the prices S_{u^2} and S_{d^2}. The option still expires in two years, and with two periods, each time step is one year. It is important to understand that the time steps before the current step, time 2, are artificial. They do not imply a past history. And in general, S_{ud} does not have to equal

TABLE 8.1   Prices, Deltas, Gammas, and Thetas from Standard Binomial (500 Time Steps), Expanded Binomial, and Black-Scholes-Merton Models

Price/Greek	Standard Binomial	Expanded Binomial	Black-Scholes-Merton
c	27.7750	27.7750	27.7546
p	15.1189	15.1189	15.1102
Δ_c	0.6989	0.6995	0.6992
Δ_p	0.3009	0.3005	0.3008
$\Gamma_c = \Gamma_p$	0.0062	0.0061	0.0062
Θ_c	−7.7853	−7.7853	−7.7751
Θ_p	−1.8668	−1.8668	−1.8655

S and only does so if d is the reciprocal of u. That is the case in this example, given the formula we chose to convert volatility to u and d, but it does not have to be in general.

Delta and gamma are computed as follows:

$$\Delta_c = \frac{c_{u^2} - c_{d^2}}{Su^2 - Sd^2} = \frac{135.21 - 0}{222.55 - 44.93} = 0.7612,$$

$$\Delta_p = \frac{p_{d^2} - p_{u^2}}{Su^2 - Sd^2} = \frac{42.41 - 0}{222.55 - 44.93} = 0.2388,$$

$$\Gamma_c = \frac{\left(\frac{c_{u^2} - c_{ud}}{Su^2 - Sud}\right) - \left(\frac{c_{ud} - c_{d^2}}{Sud - Sd^2}\right)}{(1/2)(Su^2 - Sd^2)} = \frac{\left(\frac{135.21 - 25.34}{222.55 - 100}\right) - \left(\frac{25.34 - 0}{100 - 44.93}\right)}{(1/2)(222.55 - 44.93)} = 0.0049, \text{ and}$$

$$\Gamma_p = \frac{\left(\frac{p_{d^2} - p_{ud}}{Sud - Sd^2}\right) - \left(\frac{p_{ud} - p_{u^2}}{Su^2 - Sud}\right)}{(1/2)(Su^2 - Sd^2)} = \frac{\left(\frac{42.41 - 12.68}{100 - 44.93}\right) - \left(\frac{12.68 - 0}{222.55 - 100}\right)}{(1/2)(222.55 - 44.93)} = 0.0049.$$

Theta is the same as in the standard method, but this will not always be the case. The period two time steps ahead in both models is the expiration, so the option value will be the same, but for models with more time steps, the thetas will not always be the same.

The expanded method works better for a limited number of time steps, but with the incredible speed of the computers of today, most binomial computations can be done with a large number of time steps reasonably fast. In that case, the length of each time step would be very short, and the standard method will give relatively accurate measures of delta and gamma. Let us work a problem with a large number of time steps. We will see in Chapter 9 that these answers should converge to those obtained from the Black-Scholes-Merton model. Working the same problem with 500 time steps, we get the following results in Table 8.1, which we show alongside the Black-Scholes-Merton values.[11]

The differences are, as expected, quite small.

8.4   DIVIDENDS

As explained in Chapter 7, dividends pose a significant problem with the multiplicative binomial model. The binomial Greeks presented here would require the binomial lattice to be adjusted based on the escrow method. Once the dividend-adjusted binomial lattice was constructed, the computation of binomial Greeks would proceed as described in this chapter.

8.5 RECAP AND PREVIEW

In option valuation, the Greeks are the changes in the value of an option for changes in the inputs that determine the value of an option. They are sensitivity measures and are important in the valuation and trading of options. We examined the various Greeks known as the delta, gamma, theta, vega, and rho, and we showed how they can be calculated in the binomial model. We also presented numerical examples.

In Chapter 9, we show how the binomial model can be extended to a continuous-time world in which the time step is extremely small and how the value ultimately converges to the Black-Scholes-Merton option value.

QUESTIONS AND PROBLEMS

1 Suppose you are given the following information: $S = 48.08$, $X = 50$, $r_a = 4.0\%$ (discrete, annual compounding), one year to expiration, two time steps, $u = 1.2808$, and $d = 0.7808$. Complete the following binomial tree.

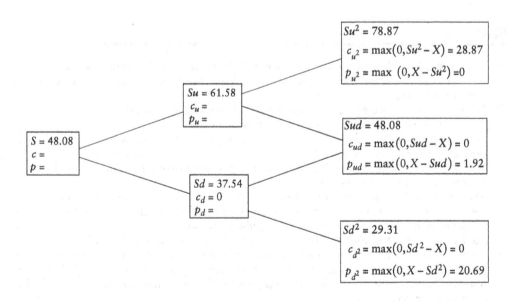

2 Based on the results from the previous problem, compute the delta, gamma, and theta based on the standard method.

3 Based on the data from the previous two problems and the following binomial tree, compute the delta, gamma, and theta using the extended method.

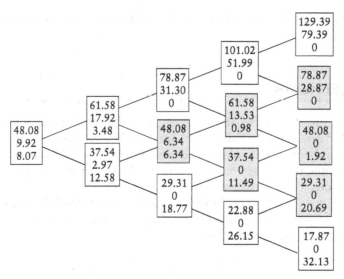

4 Based on put-call parity, derive the relationship between call and put deltas, gammas, and thetas. Assume interest rates are quoted on a continuous compounding basis.

5 Verify that your results in problem 3 are consistent with the put-call parity results in problem 4.

NOTES

1. A dealer stands ready to buy or sell options at the behest of a client. The dealer takes the opposite position of the client and lays off the risk by hedging. In essence, dealers are wholesalers or intermediaries of risk.
2. In Chapter 14, we revert to the traditional put delta with the negative sign. Thus, in this introductory material, we report put deltas as positive numbers.
3. Note that the price changes are expressed as $c_u - c$ and $c - c_d$. We are taking the average absolute price changes. Thus, we need to have $c - c_d$ rather than $c_d - c$ because the latter is negative. We take the same approach for S. Further, we take the negative of the traditional put delta so it is represented as a positive number.
4. These points will become clearer when we cover the Greeks in a continuous time world.
5. Here we note that if the down factor is not the inverse of the up factor, meaning a non-recombining tree, then $Sud \neq S$, and this method will not be as precise. Nonetheless, with enough time steps, Sud will often be close enough to S to ignore any such problems.
6. Vega is referred to as an option Greek as the Greek letter nu, ν, looks similar to the letter v for volatility. Vega is not a Greek letter, as say delta, gamma, theta, and so on. Nonetheless, in the history of options someone decided that vega sounded like a good term for the effect of volatility on an option price.
7. We should also make sure that the change in volatility and interest rate is very small, given that we are attempting to approximate a derivative.
8. Also, with this type of tree, if we need to measure gamma one period prior to expiration, we can do so. Otherwise, we could not.
9. We do not illustrate the calculations for how the option prices are obtained because this is covered in Chapter 7.
10. Note that this price is the same current price we used in the standard method.
11. We have not covered the Black-Scholes-Merton model yet but showing the computed values from that model gives us a preview of how the binomial model will converge to it.

CHAPTER 9

Convergence of the Binomial Model to the Black-Scholes-Merton Model

It is well known that the binomial model converges to the Black-Scholes-Merton model when the number of time periods increases to infinity and the length of each time period is infinitesimally short. This proof was provided in Cox, Ross, and Rubinstein (1979). Their proof, however, is unnecessarily long and relies on a specific case of the central limit theorem. This complication is evident on pp. 250 and 252 of their article, which refers to the fact that skewness becomes zero in the limit. Also, their results are derived only for the special case where the up and down factors are given by specific formulas they obtain that allow the distribution of the asset return to have the same parameters as the desired lognormal distribution in the limit. Effectively, their distribution converges to the lognormal in the limit. They go on to show that the binomial model then converges to the Black-Scholes-Merton model under their assumptions.[1]

The Cox-Ross-Rubinstein proof is elegant but far too specific. A more general proof of the convergence of the binomial to the Black-Scholes-Merton model is provided by Hsia (1983). His paper, published in *The Journal of Financial Research*, received virtually no attention, but it is clearly the most general overall proof. It imposes no restrictions on the choice of up and down parameters. Moreover, the proof is much shorter, easier to follow, and requires fewer cases of taking limits.[2]

9.1 SETTING UP THE PROBLEM

We start with our ultimate objective, the Black-Scholes-Merton model for calls and puts,

$$c_t = S_t N(d_1) - Xe^{-r_c\tau}N(d_2), \tag{9.1}$$

$$p = c - S + Xe^{-r_c\tau} = Xe^{-r_c\tau}N(-d_2) - SN(-d_1), \tag{9.2}$$

$$d_1 = \frac{\ln(S_t/X) + (r_c + \sigma^2/2)\tau}{\sigma\sqrt{\tau}}, \text{ and} \tag{9.3}$$

$$d_2 = \frac{\ln(S_t/X) + (r_c - \sigma^2/2)\tau}{\sigma\sqrt{\tau}}, \tag{9.4}$$

where S_t is the current asset price at time t, X is the exercise price, r_c is the continuously compounded risk-free rate, τ is the time to expiration, and σ^2 is the variance of the continuously compounded return on the asset. $N(d_i)$ is the cumulative normal probability for $i = 1$ and 2 as defined.[3]

For now, we focus solely on calls as we can use put-call parity to arrive at the result for puts. The binomial model for the call where the option's life is divided into n time periods is

$$c = \frac{\sum_{j=0}^{n} \binom{n}{j} \phi^n (1 - \phi)^{n-j} \max(0, u^j d^{n-j} S - X)}{(1 + r)^n}, \tag{9.5}$$

where $\binom{n}{j}$ is $n!/[j!(n-j)!]$ is known as the *binomial coefficient*. The binomial coefficient represents the number of paths the asset can take to reach a certain point in a binomial tree where j denotes the number of up moves, ϕ is the risk-neutral probability of an up move and equals $(1 + r - d)/(u - d)$, u and d are one plus the per-period return on the asset if it goes up and down, respectively, and r is the discrete risk-free interest rate per period. The numerator is the expected payout of the option at expiration under the risk-neutral binomial probability, and the denominator discounts the risk-neutral expected payoff to the present.

This expression can be simplified. For some outcomes, $\max(0, u^j d^{n-j} S - X)$ is zero. Let a represent the minimum number of upward moves required for the call to finish in the money. That is, a is the smallest integer $(a \in n)$ such that $u^a d^{n-a} S > X$.[4] Then for all $j < a$, $\max(0, u^j d^{n-j} S - X) = 0$, and for $j > a$, $\max(0, u^j d^{n-j} S - X) = u^j d^{n-j} S - X$. Now we need count only the binomial paths from $j = a$ to n so we can write the model as

$$c = \frac{\sum_{j=a}^{n} \binom{n}{j} \phi^j (1 - \phi)^{n-j} (u^j d^{n-j} S - X)}{(1 + r)^n}. \tag{9.6}$$

Now let us break this up into two terms based on S and X:

$$c = S \left[\frac{\sum_{j=a}^{n} \binom{n}{j} \phi^j (1 - \phi)^{n-j} u^j d^{n-j}}{(1 + r)^n} \right] - X(1 + r)^{-n} \left[\sum_{j=a}^{n} \binom{n}{j} \phi^j (1 - \phi)^{n-j} \right]. \tag{9.7}$$

Let us call the two terms in the large brackets B_1 and B_2. Specifically,

$$B_1 = \frac{\sum_{j=a}^{n} \binom{n}{j} \phi^j (1 - \phi)^{n-j} u^j d^{n-j}}{(1 + r)^n} \quad \text{and} \tag{9.8}$$

$$B_2 = \sum_{j=a}^{n} \binom{n}{j} \phi^j (1 - \phi)^{n-j}. \tag{9.9}$$

Equation (9.9) is the formula for the probability of a or more successes in n trials if the probability of success on any one trial is ϕ. B_1 is similar but cannot be expressed quite as easily without redefining the probability. Note the following:

$$\frac{\phi^j(1-\phi)^{n-j}u^j d^{n-j}}{(1+r)^n} = \left[\left(\frac{u}{1+r}\right)\phi\right]^j \left[\left(\frac{d}{1+r}\right)(1-\phi)\right]^{n-j}. \tag{9.10}$$

Now, it will be useful to define a modified version of the binomial probability, call it $\phi^* = [u/(1+r)]\phi$ and $1 - \phi^* = [d/(1+r)](1-\phi)$. Thus, we can write Equation (9.10) as $\phi^{*j}(1-\phi^*)^{n-j}$. Thus, B_1 is a binomial probability as stated but with the probability of each trial being ϕ^*. Now, we can write the binomial model in compact form as

$$c = SB_1 - X(1+r)^{-n}B_2, \tag{9.11}$$

where

$$B_1 = \sum_{j=a}^{n} \binom{n}{j} \phi^{*j}(1-\phi^*)^{n-j} \text{ and as before} \tag{9.12}$$

$$B_2 = \sum_{j=a}^{n} \binom{n}{j} \phi^j(1-\phi)^{n-j}. \tag{9.13}$$

Based on put-call parity, we note that the put price is

$$p = c - S + X(1+r)^{-n} = SB_1 - X(1+r)^{-n}B_2 - S + X(1+r)^{-n}$$
$$= X(1+r)^{-n}(1 - B_2) - S(1 - B_1). \tag{9.14}$$

Thus, we can easily move from the results for calls to arrive at the results for puts. Our objective is to get this formula to converge to the Black-Scholes-Merton formula. Obviously, we shall have to get B_1 and B_2 to converge to $N(d_1)$ and $N(d_2)$, respectively.

Let us do the simple part first, the risk-free rate. Recall that $(1 + r)^{-n}$ is the present value factor for n periods where the per-period rate is r. The per-period rate can be related to an annual rate applied for τ years by the relationship $(1 + r) = (1 + r_a)^{1/n_y}$ where n_y is the number of periods per year. Then,

$$(1 + r) = (1 + r_a)^{1/n_y}$$
$$(1 + r)^n = (1 + r_a)^{(1/n_y)n}$$
$$(1 + r)^n = (1 + r_a)^{\tau} \text{ because } \tau = n/n_y. \tag{9.15}$$

Thus, the present value factor for τ years is $(1 + r_a)^{-\tau}$. Now we can convert discrete discounting to continuous discounting as follows:

$$\ln(1 + r_a)^{-\tau} = -\tau \ln(1 + r_a)$$
$$e^{\ln PV} = e^{-\tau \ln(1+r_a)}$$
$$PV = e^{-r_c \tau} \text{ where } r_c = \ln(1 + r_a). \tag{9.16}$$

Thus, our present value factor is equivalent to $\exp(-r_c\tau)$ with the interest rate continuously compounded. So the binomial formula is equivalent to

$$c = SB_1 - Xe^{-r_c\tau}B_2. \tag{9.17}$$

Now let us proceed to turn this binomial formula into the Black-Scholes-Merton formula.

Because we require $Su^a d^{n-a} > X$ for the option to exercise, it follows that

$$\ln S + a\,\ln u + (n-a)\ln d > \ln X$$
$$\ln S + a\,\ln u + n\ln d - a\ln d > \ln X$$
$$a(\ln u - \ln d) > \ln X - \ln S - n\ln d$$
$$a > \frac{\ln(X/Sd^n)}{\ln(u/d)} = \frac{\ln(X/S) - n\ln d}{\ln(u/d)}. \tag{9.18}$$

By the nature of its definition as the minimum number of steps for the option to expire in-the-money, a has to be an integer, but Equation (9.18) will not likely produce an integer. For example, with $S = 100$, $X = 100$, $u = 1.10$, $d = 0.95$, and $n = 10$, we have $a > 3.4988$. This means that it would take at least four upward moves for the option to finish in-the-money. We shall handle this complication by introducing a filler variable, ι, which is a Greek lowercase iota, such that

$$a = \frac{\ln(X/S) - n\ln d}{\ln(u/d)} + \iota, \tag{9.19}$$

where ι is a number added to the result from Equation (9.18) to make a an integer. In the limit, this variable will not matter as it will converge to zero with an infinite number of time steps.

To illustrate the results thus far, suppose $S = 25$, $X = 40$, $u = 1.10$, $d = 0.90$, $n = 10$, and $r = 3.0\%$. Based on this example, Table 9.1 illustrates the following potential outcomes along with probabilities.

With the binomial tree, the total number of outcomes is 2^n or $2^{10} = 1{,}024$ here. The total probability whether modified or not is 100%. Finally, based on the risk-neutral approach, the expected terminal value of the asset is $S(1 + r)^n$ or $25(1 + 0.031)^{10} = 33.93$ here. Further, we have

$$a > \frac{\ln(X/S) - n\ln d}{\ln(u/d)} = \frac{\ln(40/25) - 10\ln 0.9}{\ln(1.1/0.9)} = 7.5926 = a^*,$$
$$\iota = a - a^* > 8 - 7.5926 = 0.4074, \text{ and}$$
$$a = \frac{\ln(X/S) - n\ln d}{\ln(u/d)} + \iota = a^* + \iota = 7.5926 + 0.4074 = 8.$$

TABLE 9.1 Illustration of 10 Period Binomial Model
Outcomes and Probabilities

Counter	n choose j	$B_1(j)$	$B_2(j)$	S
0	1	0.00001	0.0000	8.72
1	10	0.0001	0.0005	10.65
2	45	0.0015	0.0039	13.02
3	120	0.0092	0.0196	15.92
4	210	0.0374	0.0652	19.45
5	252	0.1041	0.1485	23.77
6	210	0.2012	0.2349	29.06
7	120	0.2668	0.2549	35.52
8	45	0.2322	0.1815	43.41
9	10	0.1197	0.0766	53.05
10	1	0.0278	0.0145	64.84
Sum	1024	1.0000	1.0000	33.93*

*Sum of the product of B_2S or the expected terminal asset price.

Thus, $S > X$ only for outcomes where up has occurred eight or more times. Thus, we can compute the binomial probabilities from Equations (9.8) and (9.9) as

$$\phi = \frac{1+r-d}{u-d} = \frac{1+0.031-0.9}{1.1-0.9} = 0.6550,$$

$$B_1 = \frac{\sum_{j=a}^{n} \binom{n}{j} \phi^j (1-\phi)^{n-j} u^j d^{n-j}}{(1+r)^n} = \frac{\sum_{j=8}^{10} \binom{10}{j} 0.6550^j (1-0.6550)^{10-j} 1.1^j 0.9^{10-j}}{(1+0.031)^{10}}, \text{ and}$$

$$= 0.2322 + 0.1197 + 0.0278 = 0.3797$$

$$B_2 = \sum_{j=a}^{n} \binom{n}{j} \phi^j (1-\phi)^{n-j} = \sum_{j=8}^{10} \binom{10}{j} 0.6550^j (1-0.6550)^{10-j}$$

$$= 0.1815 + 0.0766 + 0.0145 = 0.2726.$$

Thus, the option prices are

$$c = SB_1 - Xe^{-r_c\tau} B_2 = 25(0.3797) - 40(0.7369)(0.2726) = 1.46 \text{ and}$$

$$p = c - S + Xe^{-r_c\tau} = 1.46 - 25 + 40(0.7369) = 5.93.$$

9.2   THE HSIA PROOF

Now we proceed to work through Hsia's elegant and straightforward proof. We appeal to the famous DeMoivre-LaPlace limit theorem, which says that a binomial distribution converges to the normal distribution if $np \to \infty$ as $n \to \infty$. For B_1 we need

$$B_1 \to \int_a^\infty f(j)dj, \tag{9.20}$$

where $f(j)$ is the density for a normal distribution. Because j is not a standard normal distribution, however, let us convert j to a standard normal variable by defining $z = (j - E(j))/\sigma_j$. Then we would have

$$\int_a^\infty f(j)dj = \int_d^\infty f(z)dz, \text{ where } z = \frac{j-E(j)}{\sigma_j}. \tag{9.21}$$

Following Hsia, however, we define

$$d = -\frac{a - E(j)}{\sigma_j}, \tag{9.22}$$

which allows us to write the above, due to the symmetry of the normal distribution, as

$$\int_{-\infty}^d f(z)dz. \tag{9.23}$$

Thus,

$$B_1 \rightarrow \int_a^\infty f(j)d_j = \int_{-\infty}^d f(z)dz = N(d), \tag{9.24}$$

where j has been converted to z, a standard normal. What is happening here is that the $N(d_1)$ term in the Black-Scholes-Merton model is a cumulative normal probability. Therefore, it reflects the probability of the variable ranging from $-\infty$ to d_1, yet it must capture a probability related to the option being exercised, which is a probability of the asset price being *greater*, not less, than the exercise price. This transformation solves the problem. Furthermore, an identical procedure would be applied to B_2.

Now let S_T be the asset price at expiration. Of course, this value is not known but we can describe its stochastic properties. After n periods and j up moves, $S_T/S = u^j d^{n-j}$. Thus, the log (continuously compounded) return on the asset over the life of the option is

$$\ln\left(\frac{S_T}{S}\right) = j\ln u + (n-j)\ln d = j\ln\left(\frac{u}{d}\right) + n\ln d. \tag{9.25}$$

Now take the expectation of (9.25),

$$E\left[\ln\left(\frac{S_T}{S}\right)\right] = E(j)\ln\left(\frac{u}{d}\right) + n\ln d. \tag{9.26}$$

Rearranging to isolate j, we have

$$E(j) = \frac{E\left[\ln\left(\frac{S_T}{S}\right)\right] - n\ln d}{\ln\left(\frac{u}{d}\right)}. \tag{9.27}$$

The variance of the log return on the asset over the life of the option is

$$\text{var}\left[\ln\left(\frac{S_T}{S}\right)\right] = \text{var}(j)\left[\ln\left(\frac{u}{d}\right)\right]^2. \tag{9.28}$$

Thus,

$$\text{var}(j) = \frac{\text{var}\left[\ln\left(\frac{S_T}{S}\right)\right]}{\left[\ln\left(\frac{u}{d}\right)\right]^2}. \tag{9.29}$$

Recall from Equation (9.19) that a is defined as the lowest integer j such that the call option is in-the-money where ι is an adjustment factor forcing a to be the next integer or $a = \frac{\ln(X/S) - n\ln d}{\ln(u/d)} + \iota$. Therefore, based on the mean and variance results above, the upper bound defined in Equation (9.22) can be expressed as

$$d = -\frac{a - E(j)}{\sigma_j} = -\frac{\dfrac{\ln\left(\frac{X}{S}\right) - n\ln d}{\ln\left(\frac{u}{d}\right)} + \iota - \dfrac{E\left[\ln\left(\frac{S_T}{S}\right)\right] - n\ln d}{\ln\left(\frac{u}{d}\right)}}{\left\{\dfrac{\text{var}\left[\ln\left(\frac{S_T}{S}\right)\right]}{\left[\ln\left(\frac{u}{d}\right)\right]^2}\right\}^{1/2}} = \frac{\dfrac{\ln\left(\frac{S}{X}\right) + E\left[\ln\left(\frac{S_T}{S}\right)\right]}{\ln\left(\frac{u}{d}\right)} - \iota}{\dfrac{\sqrt{\text{var}\left[\ln\left(\frac{S_T}{S}\right)\right]}}{\ln\left(\frac{u}{d}\right)}}. \tag{9.30}$$

From the properties of the binomial distribution, it is known that $\text{var}(j) = n\phi(1 - \phi)$ where ϕ is the probability per outcome. So, after some rearranging and substituting this variance result only in the second term below, we have

$$d = \frac{\ln\left(\frac{S}{X}\right) + E\left[\ln\left(\frac{S_T}{S}\right)\right]}{\sqrt{\text{var}\left[\ln\left(\frac{S_T}{S}\right)\right]}} - \frac{\ln\left(\frac{u}{d}\right)}{\sqrt{\phi(1 - \phi)}} \frac{\iota}{\sqrt{n}}. \tag{9.31}$$

As n goes to infinity, the second term disappears. Our discrete binomial process is then starting to converge to a continuous lognormal process, for which it is known that $\text{var}[\ln(S_T/S)] = \sigma^2\tau$. Thus, we have

$$d = \frac{\ln\left(\frac{S}{X}\right) + E\left[\ln\left(\frac{S_T}{S}\right)\right]}{\sigma\sqrt{\tau}}. \tag{9.32}$$

We need this expression to equal d_1 and d_2 as defined in the Black-Scholes-Merton formula when the probabilities are ϕ^* and ϕ, respectively, as defined previously. This means that we need

$$E\left[\ln\left(\frac{S_T}{S}\right)\right] = \left(r_c + \frac{\sigma^2}{2}\right)\tau \text{ if the probability is } \phi^*$$

$$E\left[\ln\left(\frac{S_T}{S}\right)\right] = \left(r_c - \frac{\sigma^2}{2}\right)\tau \text{ if the probability is } \phi. \tag{9.33}$$

Now let us work on the first case. First recall that $\phi^* = [u/(1+r)][(1+r-d)/(u-d)]$. After some careful rearranging, we solve for $1+r$,

$$1+r = \frac{1}{\phi^*\left(\frac{1}{u}\right) + (1-\phi^*)\left(\frac{1}{d}\right)}. \tag{9.34}$$

Recall that $(1+r)^n = (1+r_a)^\tau$ so[5]

$$(1+r)^n = (1+r_a)^\tau = \frac{1}{\left[\phi^*\left(\frac{1}{u}\right) + (1-\phi^*)\left(\frac{1}{d}\right)\right]^n}. \tag{9.35}$$

Now note that S/S_T can be expressed as follows:[6]

$$\frac{S}{S_T} = \left(\frac{S_0}{S_1}\right)\left(\frac{S_1}{S_2}\right) \cdots \left(\frac{S_{n-1}}{S_n}\right) = \prod_{i=1}^{n}\left(\frac{S_{i-1}}{S_i}\right). \tag{9.36}$$

where $S = S_0$ and $S_T = S_n$. The expectation of this expression would be

$$E\left(\frac{S}{S_T}\right) = E\left[\prod_{i=1}^{n}\left(\frac{S_{i-1}}{S_i}\right)\right] = \prod_{i=1}^{n} E\left(\frac{S_{i-1}}{S_i}\right). \tag{9.37}$$

Note that our ability to express the expected value of a product as the product of the expected values comes from the independence of the asset prices. That is, we know that $cov(xy) = E(xy) - E(x)E(y)$. If x and y are independent, however, then $cov(xy) = 0$ and $E(xy) = E(x)E(y)$.

Now recall that our probability for B_1 is ϕ^*. Because $S_i = S_{i-1}u$ with probability ϕ^* and $S_i = S_{i-1}d$ with probability $1 - \phi^*$, then

$$E\left(\frac{S_{i-1}}{S_i}\right) = \phi^*\left(\frac{1}{u}\right) + (1-\phi^*)\left(\frac{1}{d}\right). \tag{9.38}$$

Thus,

$$E\left(\frac{S}{S_T}\right) = \prod_{i=1}^{n}\left[\phi^*\left(\frac{1}{u}\right) + (1-\phi^*)\left(\frac{1}{d}\right)\right] = \left[\phi^*\left(\frac{1}{u}\right) + (1-\phi^*)\left(\frac{1}{d}\right)\right]^n. \tag{9.39}$$

Inverting Equation (9.39) gives

$$\left[E\left(\frac{S}{S_T}\right)\right]^{-1} = \left[\phi^*\left(\frac{1}{u}\right) + (1-\phi^*)\left(\frac{1}{d}\right)\right]^{-n}. \tag{9.40}$$

Because $(1+r)^\tau = [\phi^*(1/u) + (1-\phi^*)(1/d)]^{-n}$, then $(1+r)^\tau = E(S/S_T)^{-1}$ or $(1+r)^{-\tau} = E(S/S_T)$. Therefore,

$$-\tau \ln(1+r_a) = \ln\left[E\left(\frac{S}{S_T}\right)\right]. \tag{9.41}$$

So now we are working with $\ln[E(S/S_T)]$. Because S_T/S is lognormally distributed, it will also be true that the inverse of a lognormal distribution is lognormally distributed.[7] Thus, S/S_T is lognormally distributed. For any random variable x that is lognormally distributed, it will be the case that $\ln[E(x)] = E[\ln(x)] + \text{var}[\ln(x)]/2$. This result may not look correct, but follows from a widely used result in options: The log return is defined as $S_T/S = e^x$ so $\ln(S_T/S) = x$. It is known that $E(x) = \exp(\mu + \sigma^2/2)$.[8] Thus, $\ln E(x) = \mu + \sigma^2/2$, where μ is $E[\ln(x)]$ and σ is its standard deviation.

Because $-\tau \ln(1 + r_a) = \ln[E(S/S_T)]$ (see Equation (9.41)), then

$$-\tau \ln(1 + r_a) = E\left[\ln\left(\frac{S}{S_T}\right)\right] + \frac{\text{var}\left[\ln\left(\frac{S}{S_T}\right)\right]}{2}$$

$$= E\left[-\ln\left(\frac{S_T}{S}\right)\right] + \frac{\text{var}\left[-\ln\left(\frac{S_T}{S}\right)\right]}{2} = -E\left[\ln\left(\frac{S_T}{S}\right)\right] + \frac{\text{var}\left[\ln\left(\frac{S_T}{S}\right)\right]}{2}.$$

(9.42)

Note the origin of these results, $E[\ln(S/S_T)] = E[-\ln(S_T/S)] = -E[\ln(S_T/S)]$, because you can always pull a minus sign out from inside an expectations operator. Also, $\text{var}[\ln(S/S_T)] = \text{var}[-\ln(S_T/S)] = \text{var}[\ln(S_T/S)]$, because you can pull the constant (-1) out in front of the variance operator by squaring it, thereby obtaining one times the variance.

Now what we have is

$$E\left[\ln\left(\frac{S_T}{S}\right)\right] = \tau \ln(1 + r_a) + \frac{\text{var}\left[\ln\left(\frac{S_T}{S}\right)\right]}{2}.$$

(9.43)

Recall that we know that $\text{var}[\ln(S_T/S)]$ is $\sigma^2 \tau$. Thus, we have

$$E\left[\ln\left(\frac{S_T}{S}\right)\right] = \left[\ln(1 + r_a) + \frac{\sigma^2}{2}\right]\tau.$$

(9.44)

Because $\ln(1 + r) = r_c$, this is the result we want for B_1 to converge to $N(d_1)$.

For B_2 to converge to $N(d_2)$, recall the definition $\phi = (1 + r - d)/(u - d)$. Then $1 + r = [\phi u + (1 - \phi)d]$. Because $S_i = S_{i-1}u$ with probability ϕ and $S_i = S_{i-1}d$ with probability $1 - \phi$, then $E(S_i/S_{i-1}) = \phi u + (1 - \phi)d = 1 + r$. Because

$$E\left(\frac{S_T}{S}\right) = E\left[\prod_{i=1}^{n}\left(\frac{S_i}{S_{i-1}}\right)\right] = \prod_{i=1}^{n} E\left(\frac{S_i}{S_{i-1}}\right) = [\phi u + (1 - \phi)d]^n = (1 + r)^n = (1 + r_a)^\tau,$$

(9.45)

it follows that

$$\ln\left[E\left(\frac{S_T}{S}\right)\right] = \tau \ln(1 + r_a).$$

(9.46)

Again, recall

$$\ln\left[E\left(\frac{S_T}{S}\right)\right] = E\left[\ln\left(\frac{S_T}{S}\right)\right] + \frac{\text{var}\left[\ln\left(\frac{S_T}{S}\right)\right]}{2}.$$

Substituting our previous result, and note the switch to solving for $E[\ln(S_T/S)]$ and recall $\mathrm{var}[\ln(S_T/S)] = \sigma^2\tau$, we have

$$E\left[\ln\left(\frac{S_T}{S}\right)\right] = \left[\ln(1 + r_a) - \frac{\sigma^2}{2}\right]\tau. \qquad (9.47)$$

Because $\ln(1 + r_a) = r_c$, this is the result we needed to obtain convergence of B_2 to $N(d_2)$. Thus, we have shown that the binomial model for a call can be expressed as

$$c = SB_1 - Xe^{-r_c\tau}B_2, \qquad (9.48)$$

where

$$B_1 = e^{-r_c\tau}\sum_{j=a}^{n}\binom{n}{j}\phi^j(1 - \phi)^{n-j}u^j d^{n-j} \text{ and} \qquad (9.49)$$

$$B_2 = \sum_{j=a}^{n}\binom{n}{j}\phi^j(1 - \phi)^{n-j}, \qquad (9.50)$$

which will converge to the Black-Scholes-Merton model in the limit as $n \to \infty$, expressed as

$$c_t = S_t N(d_1) - Xe^{-r_c\tau}N(d_2), \qquad (9.51)$$

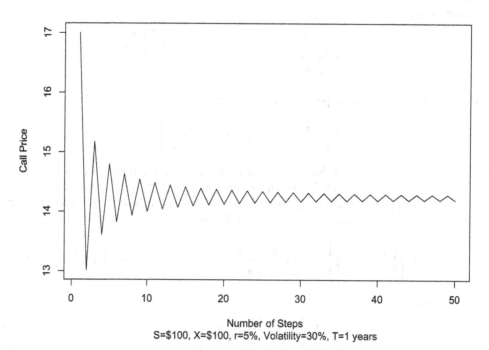

S=$100, X=$100, r=5%, Volatility=30%, T=1 years

FIGURE 9.1   Convergence of the Binomial Model for Calls to the Black-Scholes-Merton

where

$$d_1 = \frac{\ln(S_t/X) + (r_c + \sigma^2/2)\tau}{\sigma\sqrt{\tau}}, \text{ and} \qquad (9.52)$$

$$d_2 = \frac{\ln(S_t/X) + (r_c - \sigma^2/2)\tau}{\sigma\sqrt{\tau}}. \qquad (9.53)$$

For example, suppose $S = 100$, $X = 100$, $r = 5.0\%$, $\sigma = 30\%$, and $\tau = 1$ year. Figure 9.1 illustrates the call prices as the number of time steps increases. The binomial call price quickly converges to the Black-Scholes-Merton price of 14.23. The zig-zag pattern is well known. An odd number of steps produces values on the high side, and an even number of steps produces values on the low side.

9.3 PUT OPTIONS

Based on put-call parity, we have

$$p = SB_1 - Xe^{-r_c\tau}B_2 - S + Xe^{-r_c\tau} = Xe^{-r_c\tau}(1 - B_2) - S(1 - B_1). \qquad (9.54)$$

This binomial result will converge to the Black-Scholes-Merton put model in the limit as $n \to \infty$ or

$$p_t = c_t - S_t + Xe^{-r_c\tau} = Xe^{-r_c\tau}N(-d_2) - S_tN(-d_1). \qquad (9.55)$$

The convergence properties are the same as the call results as illustrated in Figure 9.2. This identical pattern is driven by put-call parity. Put-call parity can be expressed as $c_t =$

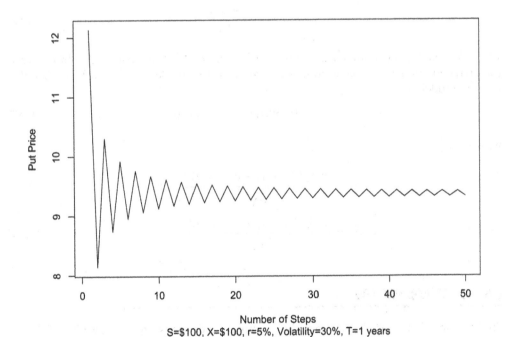

Number of Steps
S=$100, X=$100, r=5%, Volatility=30%, T=1 years

FIGURE 9.2 Convergence of the Binomial Model for Puts to the Black-Scholes-Merton

$p_t = S_t - X^{-r_c\tau}$. Thus, the binomial convergence of the put is simply shifted by the difference between the asset price and the present value of the strike price.

We now explore the important role of dividends, which can occur if the underlying asset is a stock or index.

9.4   DIVIDENDS

The payment of dividends has the effect of lowering the expected growth rate of the stock based on the risk-neutral probability measure. Hence, solely increasing dividends will lower the call price and raise the put price. Thus, we need an efficient way to incorporate dividends into the Black-Scholes-Merton model for European options.

As discussed in Chapter 7, dividends pose a significant problem within the binomial model. Based on the escrow method, we work backward through the tree checking whether the binomial model value is below the exercise value. Recall it may be optimal for the call buyer to exercise immediately before the ex-dividend date whereas the put buyer would exercise immediately after the ex-dividend date so as to enable the stock to drop, which benefits the put buyer. The escrow method works well with short dated options on individual stocks or other instruments with relatively few dividend payments over the life of the option.

Alternatively, we may assume a known continuous dividend yield. The dividend yield approach is useful for index options that contain a large number of relatively small dividends. As with discrete dividends, the continuous dividend yield will have the effect of lowering the expected growth rate of the stock. Thus, the risk-neutral probability of an up move occurring can be expressed as

$$\phi = \frac{e^{(r_c-\delta_c)\Delta t} - d}{u - d},\tag{9.56}$$

where δ_c denotes the annualized, continuously compounded dividend yield. In this case, the dividend adjusted binomial model converges to the dividend adjusted Black-Scholes-Merton model or

$$c_t = S_t e^{-\delta_c\tau}N(d_1) - Xe^{-r_c\tau}N(d_2),\tag{9.57}$$

where

$$d_1 = \frac{\ln(S_t/X) + (r_c - \delta_c + \sigma^2/2)\tau}{\sigma\sqrt{\tau}}, \text{and}\tag{9.58}$$

$$d_2 = \frac{\ln(S_t/X) + (r_c - \delta_c - \sigma^2/2)\tau}{\sigma\sqrt{\tau}}.\tag{9.59}$$

9.5   RECAP AND PREVIEW

We have shown in this chapter that the binomial model converges to the Black-Scholes-Merton model.[9] We made no assumptions regarding what the probabilities were, and we did not have to specify a particular formula for u and d. Hsia's proof requires only the condition of the DeMoivre-LaPlace limit theorem, which is that $np \to \infty$ as $n \to \infty$. This condition will not be met only if the probability per period approaches zero. This is clearly not the case. If the probability of an up move approaches zero, then the probability of a down move would approach one. The model would then be meaningless as there would be no uncertainty. In fact, it is well known that the probability value converges to $\frac{1}{2}$.[10]

This completes Part II. In Part III, we move into the continuous time world of Black, Scholes, and Merton. Chapter 10 introduces the statistical process that we shall use to describe movements in the underlying asset.

QUESTIONS AND PROBLEMS

1 When selecting the lowest integer within the binomial sum where the call option is in-the-money, we have the following relationships,

$$a^* = \frac{\ln(X/S) - n\ln d}{\ln(u/d)} \text{ and}$$

$$\iota = a - a^*.$$

If we assume that $d = 1/u$, explain the behavior of ϕ as n increases. Further, illustrate your results for $S = 100$, $X = 100$, $u = 1.25$, and $n = 1,2,3,4,5,100,$ and 101.

2 Suppose $S = 90$, $X = 100$, $u = 1.25$, $d = 0.80$, $n = 5$, and $r = 5.0\%$. Compute $B_1(j)$ and $B_2(j)$ for all six outcomes. With these results as well as the data provided here, calculate the call and put prices.

Counter	n choose j	S
0	1	29.49
1	5	46.08
2	10	72.00
3	10	112.50
4	5	175.78
5	1	274.66
Sum	1,024	114.87*

*Sum of the product of B_2S or the expected terminal stock price.

3 Prove: $Var\left[\ln\left(\frac{S_T}{S}\right)\right] = n\phi(1 - \phi)\left[\ln\left(\frac{u}{d}\right)\right]^2$.

4 Prove B_1 is always greater than or equal to B_2.

5 Within the vast literature related to the multiplicative binomial model, there has emerged what are known as coherent conditions. Coherent conditions address the conditions necessary to ensure that regardless of the number of time steps, the binomial model is consistent (coherent) with the parameters of the lognormal distribution. To be coherent, the following two conditions must hold at any point in time t, and any node j:

$$e^{r\Delta t} = \phi u + (1 - \phi)d \text{ and} \tag{1}$$

$$\sigma^2 \Delta t = \phi(1 - \phi) \left[\ln\left(\frac{u}{d}\right)\right]. \tag{2}$$

Based on these two conditions, identify the relationships among ϕ, u, and d.

NOTES

1. An alternative approach is provided by Jarrow and Rudd (1983), whose up and down factors are different from those of Cox, Ross, and Rubinstein, and these values hold for any number of time steps, not just in the limit. They then provide a general sketch of how the binomial model converges to the Black-Scholes-Merton model.
2. Chance (2008) provides a review of binomial option pricing models and the various formulas that have been used to convert volatility to the up and down factors. He shows that they all converge to a risk neutral probability of ½.
3. For ease of exposition, we suppress the time subscript t.
4. Note the symbol $\in$ is read as "is an element of." Thus, $(a \in n)$ is read "a is an element of n."
5. Although it is not particularly important, note that r is the harmonic mean return on the asset when the probability is ϕ^*.
6. Note that this expression is the inverse of the return. That is, it is the earlier price over the later price.
7. Let x be distributed lognormally meaning that $\ln x$ is distributed normally. Let $y = 1/x$. Is $\ln y$ distributed normally? $\ln y = \ln 1 - \ln x = -\ln x$. Given that $\ln x$ is distributed normally, changing its sign will not change its status as a normally distributed variable. It simply changes all positive outcomes to negative and vice versa. It does not change the normal distribution into some other type of distribution.
8. This result is obtained by evaluating the integral of the normally distributed random variable x using the normal density function.
9. Another convergence result is provided in the Rendleman and Bartter (1979) version of the model. Their proof seems somewhat more complex than the Cox-Ross-Rubinstein proof and requires the condition that the probabilities converge to ½. Although this condition is met, nonetheless, it is not necessary to impose it to obtain the proof.
10. For the Jarrow and Rudd proof, they simply set the probability to ½ and then do not alter it as time steps are added (p. 188). In other applications of the binomial model, such as elsewhere in the Jarrow-Rudd book, the probability is not arbitrarily set to ½, but simulations would show that it converges to ½, given their formulas for u and d, as the number of time steps increases. Of course, as noted, Chance (2008) demonstrated that in all of the well-known formulas the probability converges to ½.

Continuous Time Derivatives Pricing Theory

CHAPTER 10

The Basics of Brownian Motion and Wiener Processes

The prices of assets evolve in a random manner, meaning that stock prices, interest rates, exchange rates, and commodity prices are largely unpredictable. Thus, the financial landscape will be everchanging. Unpredictable, however, does not mean bizarre or meaningless. Indeed, it is important that we understand the probability process driving prices, because this helps us to develop good valuation models. Estimates of expected returns and volatilities and their effects on asset and derivative prices are essential elements in the financial decision-making process. Although there are many excellent references on the subject of this chapter, three we highly recommend are Baxter and Rennie (1996), Neftci (2000), and Malliaris and Brock (1982).

A *stochastic process* is a sequence of observations from a probability distribution. Rolling a pair of dice multiple times is a stochastic process. In this case, the distribution is stable because the possible outcomes do not change from one roll to the next. Rolling a 6-5 three times in a row, while highly unlikely, in no way changes the probability of rolling another 6-5. A changing distribution, however, would be the case if we drew a card from a deck and then drew another card without replacing the previously drawn card. Real-world asset prices probably come from changing distributions, though it is difficult to determine when a distribution has changed. Empirical analysis of past data can be useful in that context—not to predict the future, but to know when the numbers are coming out according to different bounds of probability. At this point, however, we shall focus on stochastic processes that are stable, which are sometimes called *homogeneous* or *identically distributed*.

10.1 BROWNIAN MOTION

In 1827, the Scottish botanist Robert Brown (1773–1858) observed the random behavior of pollen particles suspended in water. This phenomenon came to be known as *Brownian motion*. About 80 years passed before Albert Einstein developed the mathematical properties of Brownian motion. This gap in time, however, is not to suggest that no work was being done in the interim, but scientists did not always know what additional research was going on. For example, Louis Bachelier in his doctoral work on option pricing in 1900 made numerous advances of these mathematical properties several years before Einstein. It is not surprising that it was Einstein who received most of the credit, because he was certainly the most famous scientist of that era and possibly of all time. In this chapter and ensuing chapters, we shall borrow from the scientific theory of Brownian motion.

Let us start by assuming that a series of random numbers emanates from a standard normal or bell-shaped probability distribution. Let these variables be denoted as $\varepsilon_{t+\Delta t}$ where $t + \Delta t$ denotes the point in time when the variable is observed. Because the numbers are of the standard normal type, this means they on average equal zero and have a standard deviation of 1 (at the moment, we assume $\Delta t = 1$). Numbers like this have very limited properties, however, and in this form are not very useful for modeling asset prices, but we can take them further. Let us transform these numbers into another process identified with a W because it will eventually be called a Wiener process. Suppose we are currently at time t. This is our starting point. Now move ahead to time $t + \Delta t$ by drawing a number from the standard normal probability distribution. Call the drawn number $\varepsilon_{t+\Delta t}$, which has an expected value of zero and its variance is 1. A very simple transformation of the standard normal variable into the W variable would be to add $\varepsilon_{t+\Delta t}$ to W_t to get $W_{t+\Delta t}$. Another simple transformation would be to multiply $\varepsilon_{t+\Delta t}$ by a term we call dt, which is the length of time that elapses between t and $t + \Delta t$. Given that we typically measure financial variables in years, if that time interval happens to be one minute, dt would be $1/(60*24*265)$, or in other words, the fraction of a year that elapses between t and $t + \Delta t$.

One reason we like to multiply $\varepsilon_{t+\Delta t}$ by a time factor is that we would like our model to accommodate time intervals between t and $t + \Delta t$ of different lengths. These statistical shocks that are the source of randomness might be larger if they were spread out over a longer time period; hence, we need to scale them by a function of time. In fact, to model asset prices evolving continuously, we need the interval between t and $t + \Delta t$ to be as short as possible. Mathematicians use the terminology *in the limit*, roughly meaning that we are almost but not quite to a specific point. They also use dt instead of Δt. Here dt will approach zero in the limit but not actually get there. Unfortunately, the model $W_{t+\Delta t} = W_t + \varepsilon_{t+\Delta t} dt$ will become problematic when dt is nearly zero, because the variance of $W_{t+\Delta t}$ will be nearly zero. That is because dt is very small and to obtain the variance, we have to square it, which drives it even closer to zero. Thus, the variable W_t will have no variance, which takes away its randomness. Because it would not vary, we cannot even call it a *variable* anymore.

The problem is best solved by multiplying $\varepsilon_{t+\Delta t}$ by the square root of dt:

$$W_{t+1} = W_t + \varepsilon_{t+\Delta t} \sqrt{dt}. \tag{10.1}$$

Then when we square the term representing time to take the variance, we obtain dt, which is not zero.

This model has many convenient properties. Suppose we are interested in predicting a future value of W, say at time $t + \Delta t$. Then the expected value of $W_{t+\Delta t}$ is W_t because the expected random change in the process is zero. If you start off at W_t and keep incrementing it by values that average to zero, you can expect to get nowhere. The variance of $W_{t+\Delta t}$ is $t + \Delta t = \Delta t$, or, in other words, the variance equals how much time elapses between now, time t, and the future point, time $t + \Delta t$. This result is found intuitively by noting that the variance of each increment is dt. Integrating over all the increments, we obtain Δt. We shall do this in a more formal manner for the case in which the increment has a variance other than 1 later in this chapter.

Going forward, we seek to model a continuous process by allowing the time step, Δt, to get smaller, essentially trending to nearly zero. Thus, the point of observation of t or $t + \Delta t$ will be negligibly different. Going forward, we adopt the traditional notation where, for example, $\varepsilon_{t+\Delta t}$ is denoted simply as ε_t. The random variable, ε_t, is observed at time $t + \Delta t$, but generated between points in time t and $t + \Delta t$. Remember, the goal is to eventually migrate Δt to dt.

10.2   THE WIENER PROCESS

The process we have been working with is often called a standard Brownian motion or just Brownian motion. Now let us take the difference between $W_{t+\Delta t}$ and W_t and denote it as dW_t, which will be defined as

$$dW_t = \varepsilon_t \sqrt{dt}. \tag{10.2}$$

This process, the increment to the standard Brownian motion, is called a *Wiener process*, named after the American mathematician, Norbert Wiener (1894–1964), who did important work in this area. In option pricing, we are more interested in the process dW_t than in the process W_t. We shall transform dW_t into something more useful for modeling asset prices at a later point.[1]

It is important to note that the mathematics necessary to define the expected value and variance require the mathematical technique of integration. The ordinary rules of integration, however, do not automatically apply when the terms are stochastic. Fortunately, work by the Japanese mathematician Kiyoshi Itô (1915–2008), who did most of his work at the University of Kyoto, proved that the integral, defined as a *stochastic integral*, does exist though with a slightly different definition. This branch of mathematics is often called Itô calculus. Consequently, many of the rules of ordinary integration apply in similar or slightly different forms. We cover this material in Chapter 11.

First, let us look at some simple characteristics of the Wiener process. For starters, let us take the expected value,

$$E(dW_t) = E\left(\varepsilon_t \sqrt{dt}\right) = \sqrt{dt}E(\varepsilon_t) = 0. \tag{10.3}$$

Now we take the variance,

$$\text{var}\left(dW_t\right) = E\left(\varepsilon_t \sqrt{dt}\right)^2 - \left[E\left(\varepsilon_t \sqrt{dt}\right)\right]^2 = dtE(\varepsilon_t^2) - 0 = dt. \tag{10.4}$$

Note that $E(\varepsilon_t^2) = 1$ because $\text{var}(\varepsilon_t) = 1 = E(\varepsilon_t^2) - E(\varepsilon_t)^2$ and $E(\varepsilon_t) = 0$.

One interesting property of the Wiener process is that when you square it, it becomes perfectly predictable. This point seems very enigmatic. How can you generate random numbers, square them, and find them perfectly predictable? Let us see.

Suppose we draw a standard normal random variable, ε_t. Then we multiply it by the square root of the time interval dt. We know that this transformed value is unpredictable, but we know its expected value and its variance. The expected value is obviously zero. Using the rule that the variance of a constant times a random variable is the constant squared times the variance of the random variable, we see that its variance is dt.[2]

If, however, we define the variable of interest as dW_t^2, we get an important result. To determine dW_t^2 we simply draw the value of ε_t, multiply it by the square root of dt, and square the entire expression to obtain $\varepsilon_t^2 dt$. The variance of this expression is found by squaring dt and multiplying by the variance of ε_t^2. By definition, however, all values of dt^k

where $k > 1$ are zero. In other words, the length of the time interval is so short that raising it to a power more than one makes it shorter and effectively zero. The expected value of dW_t^2 is the expected value of $\varepsilon_t^2 dt$, which will be dt times the expected value of ε_t^2. So now we must evaluate $E(\varepsilon_t^2)$. We did that to get Equation (10.4). Thus, $E\left(dW_t^2\right) = 1^* dt = dt$. Because $dt^2 = 0$, then $\text{var}\left(dW_t^2\right) = 0$ and $E\left(dW_t^2\right) = dW_t^2 = dt$. In other words, any variable that has zero variance can be expressed as its expected value. We see that our intuitively unusual result is that although dW_t is random, its square is not. So, remember this important result:

$$dW_t^2 = dt. \tag{10.5}$$

We shall see it again.

Why do these things matter? They are the foundations of the most fundamental model used to price options. Let us look at how this process can be used to model movements in the price of an asset. For convenience, this asset could be a stock, currency, or commodity.

10.3   PROPERTIES OF A MODEL OF ASSET PRICE FLUCTUATIONS

We know that asset price fluctuations have several important characteristics. First, over the long run, asset prices typically go up, especially in the case of a stock, which is what we mostly model. They are said to "drift." This upward movement or drift represents the return from bearing risk. The Wiener process contains no drift, but as we show later, it is easy to make it drift either upward or even downward.

Second, asset prices are random. We know that Wiener processes are random, though we cannot use the basic Wiener process for assets because different assets have different volatilities. We can, however, transform the basic Wiener process in such a way to give it any volatility we desire.

Third, it should be more difficult to forecast asset prices further into the future than nearby. This point does not mean that asset prices are very predictable at all but that the margin of error, which is related to the variance of the future asset price, should be greater when predicting far into the future than when predicting into the near future. Imagine, for example, that the latest closing Dow Jones Industrial Average is 24,597.38. Suppose you can choose to make a prediction of the tomorrow's closing average or the average in six months. Which are you likely to be closer to? That answer should be obvious. There is certainly more uncertainty further out because the uncertainty of each day accumulates.

The final property is that an asset price should never be allowed to become negative. Corporate shareholders have limited liability, so the minimum value of their shares is zero.

We briefly illustrate these properties with an exchange-traded fund that seeks to track the S&P 500 index with dividends included with ticker symbol SPY. Figure 10.1 illustrates the value of $100 invested in SPY both with and without dividends. We see clearly the important role of dividends. Further, over this period, SPY moved up significantly, on average, whether including dividends or not. Although there is an upward drift, stocks often experience long periods of time without any substantial appreciation exhibiting the second property of randomness. Clearly, even with an ETF tracking the S&P 500 index, dividends are a significant consideration and cannot be ignored.

Figure 10.2 illustrates the notion that longer time periods have more uncertainty based on analyzing dividend adjusted prices. Panel A presents daily returns, Panel B presents

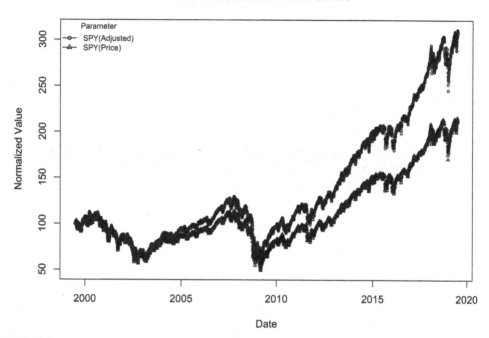

FIGURE 10.1 20-Year History of SPY, an Exchange-Traded Fund
Source: Yahoo Finance.

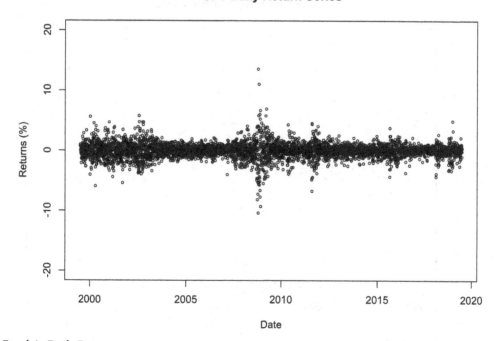

Panel A. Daily Returns

FIGURE 10.2 20-Year History of Holding Period Returns for Dividend Adjusted SPY

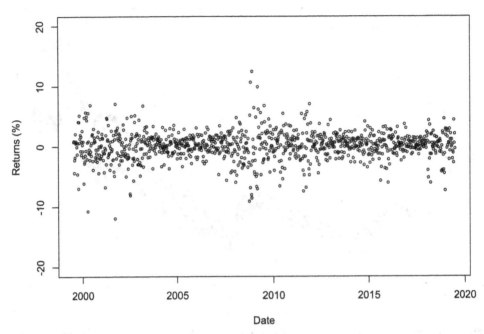

Panel B. Weekly Returns
Source: Yahoo Finance.

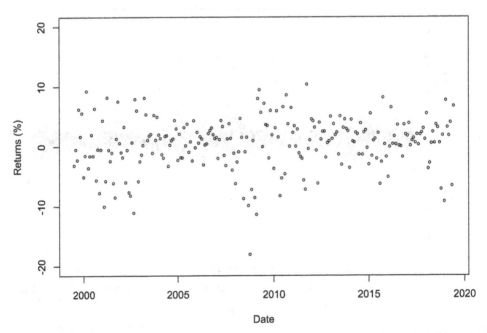

Panel C. Monthly Returns
Source: Yahoo Finance.

FIGURE 10.2 *Continued*

weekly returns, and Panel C presents monthly returns. The y-axis is fixed to ease observing the increased variability with longer holding period. The great recession of 2008–2009 is clearly visible in each time series.

Now our objective is to build a model that has these properties.

10.4 BUILDING A MODEL OF ASSET PRICE FLUCTUATIONS

The characteristics of the return on an asset can be described by the expected return and variance of return.[3] Let $E(R)$ be the expected return on the asset and σ^2 be the variance of the return, with both measures standardized to a common period of time, typically a year.[4] Thus, these are the annual expected return and variance. Then σ is the standard deviation, which is the square root of the variance. If we have an effective model, a sample of historical data will produce an average return and standard deviation equal to these values.

Let $R_t h$ be the return on the asset, between t and $t + h$, which is its price change divided by its base price minus one, over the holding period h. For example, R_t is stated as an annual rate and h, the holding period, is a fraction of a year. Thus, $R_t h$ is the return over the holding period. Now let us set $R_t h$ equal to $E(R)h + g_t$, where g_t represents the random component of the return, again generated between t and $t + h$. In other words, the return consists of the expected return plus a random component. We do not yet know what the random component, g_t, is, but we assume it is related to the variance. Now, let us force the expected value of our model to $E(R)h$. This requirement is easily fulfilled by letting $E(g_t) = 0$. Thus, we must keep this constraint in mind when we look for a suitable functional form for g_t.

Now let us make the model have the correct variance. If the variance over a year is σ^2, then the variance over the holding period h is $h\sigma^2$. Thus, the variance of our model should be such that it equals $h\sigma^2$. So far, our model has two terms, $E(R)h + g_t$. The term $E(R)h$ is a constant, so it has no variance. Thus, we need to make sure that var$(g_t) = h\sigma^2$.

One model that has the appropriate variance is $g_t = \sigma \varepsilon_t \sqrt{dt}$, which is simply σdW_t. This model is just a modified Wiener process with the transformation coming from multiplication by the volatility of the asset. Remember that we also have to have the expected value of g_t equal to zero, but that requirement is upheld because we already know that $g_t = \sigma dW_t$ and $E(dW_t) = 0$.

Now our model looks like this:

$$R_t = E(R)dt + \sigma \varepsilon_t \sqrt{dt}. \tag{10.6}$$

Let us check and see if this model has the other necessary properties. If S_t is the current asset price, then the asset price after the period dt is $S_{t+dt} = S_t \left[1 + E(R)dt + \sigma \varepsilon_t \sqrt{dt}\right]$. What is the variance of this future asset price? Over the next increment, the variance is

$$\text{var}\left[S_t + S_t E(R)dt + S_t \sigma \varepsilon_t \sqrt{dt}\right] = \text{var}(S_t) + \text{var}\left[S_t E(R)dt\right] + \text{var}\left(S_t \sigma \varepsilon_t \sqrt{dt}\right) = S_t^2 \sigma^2 dt. \tag{10.7}$$

Note that this result makes use of the fact that the constants, S_t, $E(R)$, and dt, have zero variance.

Now consider increments to a later period T. The variance is

$$\text{var}\left(S_T\right) = \int_t^T S_t^2\sigma^2 dj = S_t^2\sigma^2\int_t^T dj = S_t^2\sigma^2\left(T-t\right). \tag{10.8}$$

So, clearly the further out we look, that is, the larger T is, the higher the variance is.

Finally, the model must not permit the asset price to ever go to or below zero. Let us write the model in the following form:

$$\frac{dS_t}{S_t} = E\left(R\right)dt + \sigma\varepsilon_t\sqrt{dt}. \tag{10.9}$$

Multiplying through by S_t makes everything on the right-hand side (RHS) be multiplied by S_t. That is, the model can be written as

$$dS_t = E\left(R\right)S_t dt + \sigma S_t\varepsilon_t\sqrt{dt}. \tag{10.10}$$

The first term on the RHS is the drift term and the second is the noise term, both expressed in currency units. Note that as the asset price declines, both the drift term and the noise term are diminished. Thus, the smaller S_t is, the more the price changes are diminished. It is technically possible for a sufficiently negative shock to drive the price to zero, though this requires an incredibly large negative shock, which is an improbable event. But, from Equation (10.10), it is easily seen that if the price hits zero, any further price changes are all zero. Zero is said to be an absorbing barrier. Hence, our requirement that the price cannot be negative is met. That said, we cannot completely rule out a zero price. In Chapter 12, we shall see that the process we have described here is actually an approximation of a lognormal process. Given that it is not possible to take the log of zero, a zero price is ruled out. But for now, avoiding negative prices is our goal and that goal is achieved with this model.

One final adjustment is necessary. Most of the time the annualized expected return is written as either α or μ. We shall choose the former and will reserve μ for another term. Thus, the model is written as

$$\frac{dS_t}{S_t} = \alpha dt + \sigma dW_t, \tag{10.11}$$

where it is understood that $dW_t = \varepsilon_t\sqrt{dt}$. We see clearly here that S_t cannot obtain zero, otherwise the left-hand side (LHS) of this equation is undefined. A stochastic process of this type is called an *Itô process*. It is more generally stated in the form, $dS_t = \alpha\left(S,t\right)dt + \sigma\left(S,t\right)dW_t$, where the expected value and variance are allowed to change with S and t.

Equation (10.11) appears to imply that the LHS is normally distributed because dW_t is normally distributed. Recall multiplying a normally distributed variable by a constant and adding another constant does not change the distribution type, just its parameters. As we will see in Chapter 12, the LHS is in fact lognormally distributed. At this point, we simply note that this result is similar to discrete and continuous compounding covered in Chapter 3, Section 3.3.

To recap, the model allows us to replicate the behavior of the asset over a short holding period. We have taken the basic Brownian motion process and converted it into a form that models asset price movements. This model has many convenient and reasonable properties. We refer to the process as *geometric Brownian motion*. It is "geometric" in the sense that proportional changes, meaning percentage changes, in the asset price follow this stochastic process. Geometric Brownian motion can be formally stated as

$$dS_t = \alpha(S, t) S_t dt + \sigma S_t dW_t. \qquad (10.12)$$

It is important to emphasize that the driver of the geometric aspect is the presence of the underlying asset price in the noise term. Thus, as the asset price falls, the impact of the noise term also falls. Given the continuous nature of this process, the asset price never reaches zero. We can divide both sides by S_t and express in the traditional finance form or

$$\frac{dS_t}{S_t} = \alpha(S, t) dt + \sigma dW_t. \qquad (10.13)$$

We thus implicitly assume that the asset prices follow a lognormal distribution. This is not the normal or bell-shaped curve. A lognormal distribution is skewed toward positive returns in contrast to the normal distribution, which is symmetric. A lognormal distribution does imply, however, that the logarithm of the returns comes from the normal or bell-shaped distribution.[5] These properties are often desirable and fairly reasonable from an empirical standpoint.

An alternative process is known as *arithmetic Brownian motion*, where the asset prices follow a normal distribution.[6] It is "arithmetic" in the sense that absolute changes, meaning dollar changes, in the asset price follow this stochastic process and thus can easily be added together. Arithmetic Brownian motion can be formally stated as

$$dS_t = \alpha(S, t) dt + \sigma_\$ dW_t. \qquad (10.14)$$

It is important to emphasize that the driver of the arithmetic aspect is the *absence* of the underlying asset price in the noise term. Also, we introduce the subscript, $ (dollars), to highlight the unit of measure is whatever the unit of measure of the asset price (assumed $ here) and no longer expressed as a percentage. As discussed in later chapters, the drift term can be expressed in dollars or percentage depending on context. Thus, as the asset price falls, the noise term is not affected. Given the continuous nature of this process, the asset price can easily reach zero as well as go negative. One benefit of arithmetic Brownian motion is zero asset values are now possible—an unfortunate reality for many financial instruments.

Of course, no model will reproduce perfectly the process in which asset returns are generated. The real world can rarely be reduced to a set of mathematical equations. But as is nearly always the case, if a set of mathematical equations can reproduce the basic manner in which a real-world phenomenon occurs, it can have many uses. One of these uses is in pricing options on assets that follow the process described by the mathematical model covered here.

This model will be developed much further in Chapter 13.

10.5   SIMULATING BROWNIAN MOTION AND WIENER PROCESSES

To get a better understanding of the Brownian motion and Wiener processes, let us run some simulations in Excel. Recall that the Brownian motion process is Equation (10.11), and the Wiener process is Equation (10.2). Of course, in the model, time is continuous, so the time increments are close to zero. They are not, however, precisely equal to zero. Let us start by making them very short, say one hour. In a year, there are 365 days and 24 hours per day. Thus, each time increment is set at $1/(24*365) = 0.000114$. Let us assume a stock with an expected return of 0.12, a volatility of 0.50, and a starting stock price of 50.

The random component of the process is the Wiener process. The ε_t, which is a standard normal, is converted into the Wiener process. First, we need to simulate a standard normal. One reasonable way to do so is to use Excel's =rand() function. This function produces a uniformly distributed random number between zero and one. A uniform distribution is a distribution of continuous random variables between two endpoints a and b in which each outcome is equally likely. If $a = 0$ and $b = 1$, the distribution is called unit uniform, because the range of outcomes will be between 0 and 1. It is known that the expected value of the uniform distribution is $(1/2)(a+b)$, which is $1/2 = (0+1) = 0.5$ for the unit uniform. The variance is $(1/12)(b-a)^2$. For the unit uniform, the variance is $(1/12)(1-0)^2 = 1/12$. So, if we generate 12 unit uniform random variables and add them up, we obtain $u_1 + u_2 + \ldots + u_{12}$. Taking the expectation, we obtain $E(u_1) + E(u_2) + \ldots + E(u_{12}) = 0.5 + 0.5 + \ldots + 0.5 = 6$. We can then subtract 6, leaving an expected value of zero with no effect on the variance, which is easily obtained. These random drawings are independent so there is no covariance between them, and we can, therefore, simply add their variances to obtain the variance of the sum of the 12 unit uniform random variables. Thus, we have $\text{var}(u_1) + \text{var}(u_2) + \ldots + \text{var}(u_{12}) = 1/12 + 1/12 + \ldots + 1/12 = 1$. Thus, the sum of the 12 unit uniform random numbers minus 6 has expected value of zero and variance of 1, just like a standard normal. Recall that the central limit theorem tells us that the sum of a series of independent random variables from any distribution is normally distributed in the limit. We have used only 12 so the limiting condition is not likely to occur just yet, but we should be reasonably close to a normal distribution.[7]

Thus, the sum of 12 unit uniform random variables minus 6.0 will serve as our randomly generated value of ε_t. The rest is simple. We use Equation (10.2) to generate the Wiener process value and insert that value into Equation (10.11) to obtain the stock return. We then apply the return to the current stock price and obtain the next stock price.

Recall that the time increment is one hour. To avoid some complications, let us assume that prices are generated every hour, every day, for a full year so there is no market closing and no holidays. This assumption would mean $24*365 = 8,760$ prices in a year. Let us cut it down to a month, which would be $8,760/12 = 730$ prices. Then $dt = 1/8,760$, and we simulate 730 sequential prices or one month.

Table 10.1 shows the first five prices generated from the simulation.

Each row represents one hour. The column labeled ε_t is the sum of 12 unit uniform random variables minus 6.0. The column labeled dW_t is a transformation of the unit uniform to the Wiener process, Equation (10.2). The column labeled R_t is from Equation (10.11). In the first row, time 0, we simply set the starting price as 50. The first increment occurs

TABLE 10.1 Simulation of Geometric Brownian Motion

Time	ε_t	dW_t	R_t	S_t	dW_t^2
0				50.00	
1	−0.314189	−0.003357	−0.001665	49.92	0.000011
2	1.712835	0.018301	0.009164	50.37	0.000335
3	0.058170	0.000622	0.000324	50.39	0.000000
4	−0.697400	−0.007451	−0.003712	50.20	0.000056
5	−1.340360	−0.014321	−0.007147	49.84	0.000205

at increment 1 where the Brownian motion value is −0.314189. The remainder of the information for time increment 1 is found as follows. First find the Wiener process value,

$$dW_t = \varepsilon_t \sqrt{dt} = -0.314189\sqrt{1/8{,}760} = -0.003357.$$

Then convert this value to the stock return:

$$\frac{dS_t}{S_t} = \alpha dt + \sigma dW_t = 0.12\,(1/8{,}760) + 0.5\,(0.003357) = -0.001665.$$

Then the next stock price is

$$\$50\,(1 - 0.001665) = \$49.92.$$

We should also compute dW_t^2, because we shall show later that this value is essentially constant:

$$dW_t^2 = (-0.003357)^2 = 0.000011.$$

Figure 10.3 shows the simulation of the normal random variable, ε_t, with each dot representing one of the randomly drawn values. Recall that we created this variable by adding 12 unit uniform random variables and subtracting 6.0. The average value is 0.010416 and the standard deviation is 0.968623. These are close to but not exactly equal to their desired values, but that result would only hold with a lot more values.

Figure 10.4 shows the simulated values of dW_t and dW_t^2. The volatile and dots is dW_t, and the nearly straight line is dW_t^2. We noted that dW_t^2 should be constant, but that result occurs only in the limit, meaning with an extremely short time interval. Our time interval of one hour seems short, but as you probably know, there are many stock prices that generate within an hour under normal trading. If we increasingly shorten the time interval, then the process would become more stable. See Appendix 10A for details.

Figure 10.5 is the stock price simulated from the given results. It should be noted that the average return does not need to be anywhere near the expected return. Indeed, it is not, as the average return is 0.000069 per hour, which annualizes to almost 60%. The volatility, however, is likely to be close to the annual specified volatility of 0.50. Here it is 0.484312.

We turn now to summarize key characteristics and important properties of the Wiener process. These properties will prove useful in later chapters.

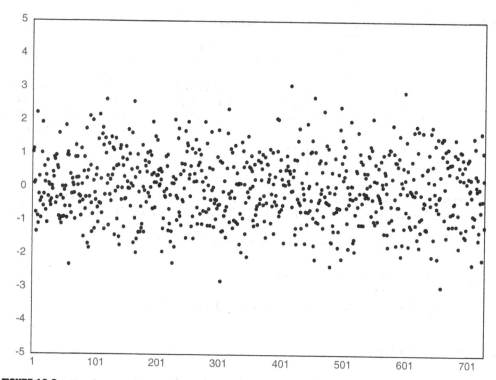

FIGURE 10.3 Simulation of Normal Distribution from Unit Uniform Distribution

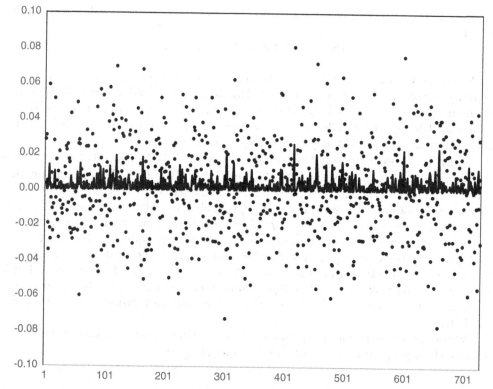

FIGURE 10.4 Simulation of dW_t and dW_t^2

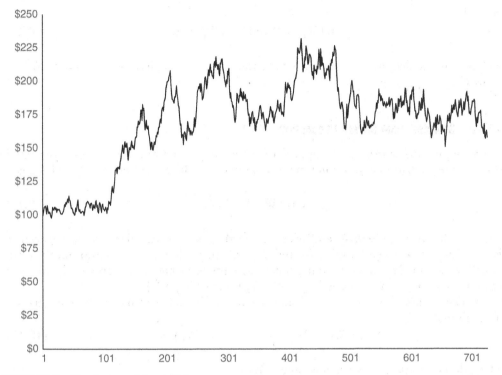

FIGURE 10.5 Simulation of Stock Price

10.8 FORMAL STATEMENT OF WIENER PROCESS PROPERTIES

We now bring together important properties of Wiener processes. We also now assume the initial starting time is $t = 0$. Assume constants $t \geq 0$ (e.g., calendar time expressed as fraction of year) and $0 < h < t$ (e.g., historical point in time compared to t assuming $t > 0$). As previously discussed, the standard Wiener process at time t or W_t can be expressed with four key attributes:

- Initial value of zero or $W_t = 0$ if $t = 0$.
- Independent increments. For every $t > 0$, increments $W_{t+\Delta t} - W_t$ are independent of any W_h, where Δt denotes an increment of calendar time.
- Normal distribution. For any h, $W_t - W_h$ is distributed normal with mean 0 and variance $t - h$.
- Continuity. For any t, W_t is continuous in t.

We now briefly review selected properties of the Wiener process as well as introduce other properties useful for addressing tasks using Wiener processes.

10.8.1 Univariate Properties

As previously discussed, the standard Wiener process follows a normal distribution where the probability density function can be expressed as (recall we assume the initial time is 0):

$$n_{W_t}(x) = \frac{e^{-\frac{x^2}{2t}}}{\sqrt{2\pi t}}. \tag{10.15}$$

Note that

$$E(W_t) = 0 \text{ and } \text{var}(W_t) = t. \tag{10.16}$$

Recall from the definition of variance and assuming $t > 0$, we have $\text{var}(W_t) = E(W_t^2) - E(W_t)^2 = t - 0 = t$.

10.6.2 Selected Time Series Properties

Again, assume constants $t > 0$ (e.g., calendar time expressed as a fraction of a year) and $0 < h < t$ (e.g., historical point in time compared to t). The covariance of W_t and W_h is

$$\text{cov}(W_t, W_h) = h. \tag{10.17}$$

Note that between 0 and h, W_t and W_h are perfectly positively correlated so the covariance is one. Further, between h and t, W_t and W_h are uncorrelated, so the covariance is zero. Recall we assumed $t > h$. Given independent increments as mentioned previously, we have $\text{cov}(W_t - W_h + W_h, W_h) = \text{cov}(W_t - W_h, W_h) + \text{cov}(W_h, W_h) = 0 + h = h$.

For constants c_1 and c_2, based on standard properties of covariance covered in Chapter 4, recall that

$$\text{cov}(c_1 W_t, c_2 W_h) = c_1 c_2 \text{cov}(W_t, W_h) = c_1 c_2 h. \tag{10.18}$$

Thus, the correlation between $c_1 W_t$ and $c_2 W_h$ is

$$\rho(c_1 W_t, c_2 W_h) \equiv \frac{\text{cov}(c_1 W_t, c_2 W_h)}{\sqrt{c_1^2 \text{var}(W_t)} \sqrt{c_2^2 \text{var}(W_h)}} = \frac{\text{cov}(W_t, W_h)}{\sqrt{\text{var}(W_t)} \sqrt{\text{var}(W_h)}} = \frac{h}{\sqrt{th}} = \sqrt{\frac{h}{t}}. \tag{10.19}$$

Now consider two generic points in time $j > 0$ and $k > 0$ and the same Wiener process; however, we do not know whether $j > k$, $j < k$, or $j = k$. In this case,

$$\text{cov}(W_j, W_k) = \min(j, k). \tag{10.20}$$

If we assume $j > k$, then based on Equation (10.17), we know $\text{cov}(W_j, W_k) = k$. Similarly, if $j < k$, we know $\text{cov}(W_j, W_k) = j$. Thus, Equation (10.20) holds for all j and k.

$$\rho(W_j, W_k) = \frac{\text{cov}(W_j, W_k)}{\sqrt{\text{var}(W_j)} \sqrt{\text{var}(W_k)}} = \frac{\min(j, k)}{\sqrt{jk}} = \sqrt{\frac{\min(j, k)}{\max(j, k)}}. \tag{10.21}$$

Now consider the same point in time $t > 0$ and two correlated Wiener processes ($W_{1,t}$ and $W_{2,t}$). That is, while $\varepsilon_{1,t}$ and $\varepsilon_{2,t}$ are independent, the Wiener processes are correlated. In this case, one of the Wiener processes (say $W_{2,t}$) can be expressed as a linear function of the other Wiener process ($W_{1,t}$) and an independent Wiener process, denoted z_t. That is,

$$W_{2,t} = \rho W_{1,t} + (1 - \rho) z_t. \tag{10.22}$$

In this case, we note

$$E\left(W_{2,t}\right) = \rho E\left(W_{1,t}\right) + (1-\rho) E\left(z_t\right) = \rho 0 + (1-\rho) 0 = 0, \tag{10.23}$$

$$\begin{aligned}
\text{var}\left(W_{2,t}\right) &= \rho^2 \text{var}\left(W_{1,t}\right) + (1-\rho)^2 \text{var}\left(z_t\right) + 2\text{cov}\left[\rho W_{1,t}, (1-\rho) z_t\right] \\
&= \rho^2 t + (1-\rho)^2 t + 2\rho(1-\rho)\text{cov}\left(W_{1,t}, z_t\right) \\
&= \rho^2 t + (1-\rho)^2 t + 2\rho(1-\rho) t = t, \text{and}
\end{aligned} \tag{10.24}$$

$$\text{cov}\left(W_{1,t}, W_{2,t}\right) = \text{cov}\left(W_{1,t}, \rho W_{1,t} + (1-\rho) z_t\right) = \rho t + (1-\rho)\text{cov}\left(W_{1,t}, z_t\right) = \rho t. \tag{10.25}$$

As a reminder, the last result occurs because z_t and $W_{1,t}$ are independent. Now, we have

$$\rho\left(W_{1,t}, W_{2,t}\right) = \frac{\text{cov}\left(W_{1,t}, W_{2,t}\right)}{\sqrt{\text{var}\left(W_{1,t}\right)}\sqrt{\text{var}\left(W_{2,t}\right)}} = \frac{\rho t}{\sqrt{t}\sqrt{t}} = \rho. \tag{10.26}$$

These Wiener process properties will be useful in future chapters.

10.7　RECAP AND PREVIEW

In this chapter, we introduced the basic Brownian motion and Wiener process. We did this on a very intuitive level. We identified certain desirable characteristics and, one by one, we added features that introduced those characteristics without affecting the characteristics we had already built into the model. We ultimately obtained the stochastic process that is widely used to model assets. We showed several of its properties, and we generated a simulation so you could get a feel for what these numbers look like.

In Chapter 11, we introduce the two important concepts of Itô's lemma and stochastic calculus, which will complete the knowledge base that we need to derive the Black-Scholes-Merton model.

APPENDIX 10A

Simulation of the Wiener Process and the Square of the Wiener Process for Successively Smaller Time Intervals

Suppose we run a second set of simulated values of the Wiener process for smaller and smaller time intervals. First, let us use a daily time interval ($dt = 1/365$), then an hourly time interval ($dt = 1/(24^*365)$), then a minute time interval ($dt = 1/(60^*24^*365)$), and then a second time interval ($dt = 1/(60^*60^*24^*365)$). We obtain the results in Table 10A.1 for dW_t and $d_t W^2$ with values quoted to six decimal places.

Note that the average and volatility shrink for both measures when the interval is shortened. Note in particular, that the volatility of dW_t^2 is virtually zero. Of course, in the limit, it is zero.

TABLE 10A.1 Values of dW_t and $dW_t{}^2$ for Various dt

	dW_t				$dW_t{}^2$			
	Daily	Hourly	Minute	Second	Daily	Hourly	Minute	Second
Average	−0.004201	−0.000858	−0.000111	−0.000014	0.002862	0.000119	0.000002	0.000000
Standard Deviation	0.053366	0.010893	0.001406	0.000182	0.004154	0.000173	0.000003	0.000000

QUESTIONS AND PROBLEMS

1 The Wiener process is defined as $dW_t = \varepsilon_t \sqrt{dt}$. Find the expected value and standard deviation of the Wiener process.

2 Geometric Brownian motion can be expressed as $dS_t = \alpha S_t dt + \sigma S_t dW_t$. Explain the behavior of changes in S_t for different values of S_t. In particular, explain why S_t never obtains the value of zero.

3 As discussed in this chapter, an alternative framework for modeling certain financial instruments is arithmetic Brownian motion with geometric drift or $dS_t = \alpha S_t dt + \sigma_\$ dW_t$. Explain the behavior of changes in S_t for different values of S_t. In particular, explain why S_t may obtain the value of zero or become negative.

4 The following two graphs were generated based on discretized versions of geometric Brownian motion ($dS_t = \alpha S_t dt + \sigma S_t dW_t$) and arithmetic Brownian motion ($dS_t = \alpha S_t dt + \sigma_\$ dW_t$). Identify the graphs and defend your answer.

Process: _____

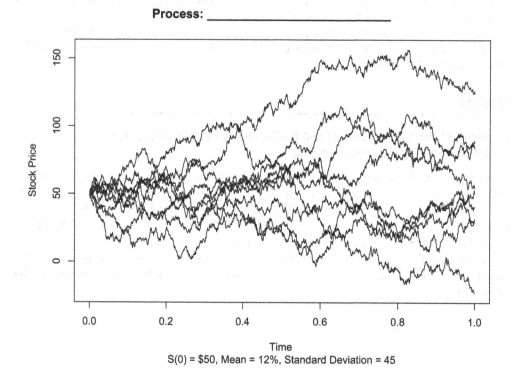

Time
S(0) = $50, Mean = 12%, Standard Deviation = 45

Process: _____

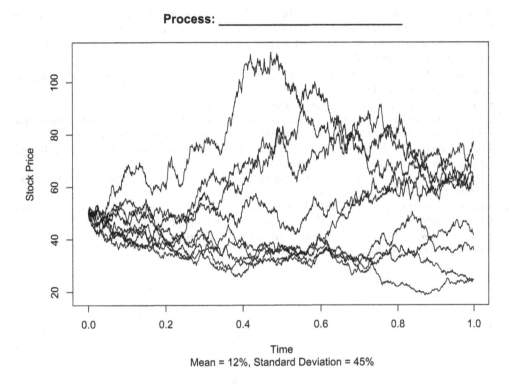

Time

Mean = 12%, Standard Deviation = 45%

5 Identify four important characteristics that a model of asset prices should encompass.

6 [Contributed by Brecklyn Groce] The following table gives five simulated values of a standard normal random variable. Convert these values to the increments of a Wiener process and find the simulated value after five days of a stock initially priced at 80 with an annual expected return of 10% and a volatility of 48% under the assumption that the time increment is one day, $1/365 = 0.00274$. You may wish to create a spreadsheet to work the problem.

Day	ε_t	ΔW_t	S_t
0			$80.00
1	0.808889		
2	0.480432		
3	0.695581		
4	−0.870413		
5	0.407389		

NOTES

1. The terms *Brownian motion*, *Wiener process*, and *Itô process* are often used interchangeably though there are technically some differences. Recall Robert Brown was a botanist, whereas

Wiener and Itô were mathematicians. Thus, mathematics-based authors lean toward Wiener or Itô processes, whereas physical and social scientists lean toward Brownian motion. In our context, we use standard Wiener process or just Wiener process to denote W_t alone, whereas Brownian motion denotes functions of W_t.

2. Remember that $dW_t = \varepsilon_t \sqrt{dt}$. Squaring the square root term gives dt. This term is then multiplied by the variance of ε_t, which is 1.0.

3. Other properties like skewness might be important but we ignore them for this model.

4. It is common but not required to express expected returns and variances on an annual basis. For example, banks quote CD rates for a year, though the CD may have some other maturity. The Federal Reserve may have a target Fed Funds rate, which is stated on an annual basis, though most borrowings at that rate are overnight.

5. This type of process is also sometimes called a *lognormal diffusion process*. We explore this result in more detail in Chapter 12, Section 12.1.

6. Arithmetic (accent on third syllable) Brownian motion offers an alternative approach to several financial problems. We will cover these processes in detail starting in Chapter 12.

7. Another method is to use the Excel function =normsinv(rand()), which directly gives a standard normal random variable.

CHAPTER 11

Stochastic Calculus and Itô's Lemma

Stochastic calculus is an important field of mathematics that works with stochastic processes. Much like ordinary calculus, stochastic calculus is based on several fundamental results. One of the most important stochastic calculus results used in options is Itô's lemma. Though this result was discovered about 1950, it did not get firmly established in the finance literature until 1973 when Black, Scholes, and Merton discovered that it could be used to model the price of a stock and ultimately to facilitate pricing an option. Some excellent references with applications in finance are Baxter and Rennie (1996), Neftci (2000), and Malliaris and Brock (1982). Good sources for more advanced treatments are Karatzas and Shreve (1991) and Karlin and Taylor (1981).

11.1 A RESULT FROM BASIC CALCULUS

Although there is a great deal of formal mathematical rigor to Itô's lemma, the essential elements are relatively simple. Let us begin, however, with a reminder of a few basic results from ordinary calculus. Recall that any differential in ordinary calculus, $dt < 1$, has a limit of zero if raised to a power greater than 1.0. In other words, $dt^k \to 0$ if $k > 1$. Consider a generic mathematical function such as $F(x, t)$ in which the first and second derivatives exist with respect to x and t. Using a Taylor series expansion, the change in value of the function can be expressed as

$$dF = \frac{\partial F}{\partial x}dx + \frac{\partial F}{\partial t}dt + \frac{1}{2}\frac{\partial^2 F}{\partial x^2}dx^2 + \frac{1}{2}\frac{\partial^2 F}{\partial t^2}dt^2 + \frac{\partial^2 F}{\partial x \partial t}dxdt + \cdots. \qquad (11.1)$$

Because $dx^2 \to 0$, $dt^2 \to 0$ and $dxdt \to 0$, we write Equation (11.1) as

$$dF = \frac{\partial F}{\partial x}dx + \frac{\partial F}{\partial t}dt. \qquad (11.2)$$

This statement means that the change in F is a function of the first-order changes in x and t. The change in x is multiplied by the partial derivative of F with respect to x and the change in t is multiplied by the partial derivative of F with respect to t. All of this is a formal way of stating that as x and t change, they induce a change in F. The changes in x and t are so small, however, that squared changes in x and t are zero in the limit, and their product is also zero in the limit.

11.2 INTRODUCING STOCHASTIC CALCULUS AND ITÔ'S LEMMA

In ordinary calculus, the variables are non-stochastic, which simply means that when we talk about a particular value of x, that value is known for certain. When x is stochastic, we leave the world of ordinary calculus and enter the world of *stochastic calculus*. There we cannot talk about a particular value of x. Instead, we must talk about a set of possible values of x that are generated according to a probability distribution. In stochastic calculus, results are proven by demonstrating what happens when squared values of a variable are multiplied by probabilities. A result is said to hold in "mean square limit."[1] A more formal statement of this concept is presented later in this chapter.

11.2.1 Generalized Itô Process

Let us now propose that x is stochastic and follows an Itô process, such as dW_t, or a more generalized process such as

$$dx = \mu(x,t)dt + \sigma(x,t)dW_t. \tag{11.3}$$

Recall that $dW_t = \varepsilon_t\sqrt{dt}$. With the expressions $\mu(x,t)$ and $\sigma(x,t)$, we are allowing the expectation and variance of x to be functions of the level of x and time t. Note that μ and σ depend on, at most, the stochastic variable x and time t.

Now suppose that we go back to the unspecified function $F(x,t)$ and examine the Taylor series expansion when x is stochastic. Again, we emphasize that the function F depends on at most variables x and t.[2] Although dt^2 is still zero because time is not stochastic, dx^2 is not zero. Recalling that $dW_t^2 = dt$, we square the generalized Itô process defined in Equation (11.3) and thus we have

$$dx^2 = \left[\mu(x,t)dt + \sigma(x,t)dW_t\right]^2$$
$$= \mu(x,t)^2 dt^2 + \sigma(x,t)^2 dW_t^2 + 2\mu(x,t)\sigma(x,t)dtdW_t. \tag{11.4}$$

Note that $dW_t dt$ is zero because

$$dtdW_t = dt\varepsilon_t\sqrt{dt} = \varepsilon_t dt^{3/2}. \tag{11.5}$$

And this expression goes to zero because of the power of dt higher than one. Hence,

$$dx^2 = \sigma^2(x,t)dW_t^2. \tag{11.6}$$

For Taylor series expansion purposes, we also examine $dxdt$ or

$$dxdt = \left[\mu(x,t)dt + \sigma(x,t)dW_t\right]dt = \mu(x,t)dt^2 + \sigma(x,t)dW_t dt = 0, \tag{11.7}$$

because, as just noted, $dW_t dt$ goes to zero in the limit.

So, now our Taylor series expansion when x is stochastic is

$$dF = \frac{\partial F}{\partial x}dx + \frac{\partial F}{\partial t}dt + \frac{1}{2}\frac{\partial^2 F}{\partial x^2}dx^2. \tag{11.8}$$

Note that we do not have to consider higher-order terms, as dx^3 would go to zero, because dx^2 is dt. With this setup, we are now ready to express Itô's lemma.

11.2.2 Itô Lemma

Assuming x follows a generalized Itô process expressed as

$$dx = \mu(x,t)dt + \sigma(x,t)dW_t,$$

where $F = F(x,t)$, subject to certain technical constraints,

$$dF = \frac{\partial F}{\partial x}\left[\mu(x,t)dt + \sigma(x,t)dW\right] + \frac{\partial F}{\partial t}dt + \frac{1}{2}\frac{\partial^2 F}{\partial x^2}\sigma(x,t)^2 dt. \tag{11.9}$$

This result is known as Itô's lemma, being named for the Japanese mathematician Kiyoshi Itô, whom we mentioned in Chapter 10 and who discovered this result. The equation describes the stochastic process of a function $F(x,t)$ that is driven by time t and a stochastic process for x of the form we previously referred to as a Wiener process.

Note that by substituting for dx (Equation (11.3)) and dx^2 (Equation (11.6)), we can express the stochastic process for F more commonly in finance as

$$dF = \left[\frac{\partial F}{\partial t} + \frac{\partial F}{\partial x}\mu(x,t) + \frac{1}{2}\frac{\partial^2 F}{\partial x^2}\sigma(x,t)^2\right]dt + \frac{\partial F}{\partial x}\sigma(x,t)dW_t. \tag{11.10}$$

In this manner, we see that the term within the square brackets is the expected change in F per unit of time and the variance is given as $(\partial F/\partial x)^2\sigma(x,t)^2$ per unit of time.[3] That is,

$$E(dF) = E\left\{\left[\frac{\partial F}{\partial t} + \frac{\partial F}{\partial x}\mu(x,t) + \frac{1}{2}\frac{\partial^2 F}{\partial x^2}\sigma(x,t)^2\right]dt + \frac{\partial F}{\partial x}\sigma_{x,t}dW\right\}$$

$$= E\left[\frac{\partial F}{\partial t} + \frac{\partial F}{\partial x}\mu(x,t) + \frac{1}{2}\frac{\partial^2 F}{\partial x^2}\sigma(x,t)^2\right]dt \text{ and} \tag{11.11}$$

$$\text{var}(dF) = \text{var}\left\{\left[\frac{\partial F}{\partial t} + \frac{\partial F}{\partial x}\mu(x,t) + \frac{1}{2}\frac{\partial^2 F}{\partial x^2}\sigma(x,t)^2\right]dt + \frac{\partial F}{\partial x}\sigma(x,t)dW_t\right\}$$

$$= \left(\frac{\partial F}{\partial x}\right)^2\sigma(x,t)^2 dt. \tag{11.12}$$

Note that the covariance terms in the variance formula are all zero and that all the uncertainty in F comes from the uncertainty in W_t.

Itô's lemma is widely used in pricing derivatives. The price of a derivative is said to be "derived" from the price of the underlying asset and time. Thus, $F(x,t)$ is a convenient specification of a derivative price, because its value is derived from x and t, with x known to be stochastic and t representing time. Suppose F is the price of an option or other derivative contract on an asset whose value is x. If we let that asset price evolve according to the Itô process, then we can characterize the change in the option price by Itô's lemma, as stated in the previous equation.

For example, consider an instrument that is presently valued as the future value of the spot price, where the rate applied, k, is known. The constant k denotes the carrying cost that generally includes financing charges and any other costs or benefits from holding the asset. That is,

$$F_t = S_t e^{k\tau},$$

where the time to expiration of this instrument is expressed as $\tau = T - t$.[4] If we assume that the underlying instrument follows geometric Brownian motion or $dS_t = \alpha S_t dt + \sigma S_t dW_t$, then based on Itô's Lemma as represented by Equation (11.10), we have

$$dF_t = \left(\frac{\partial F}{\partial t} + \frac{\partial F}{\partial S}\alpha S_t + \frac{1}{2}\frac{\partial^2 F}{\partial S^2}\sigma^2 S_t^2 \right) dt + \frac{\partial F}{\partial S}\sigma S_t dW_t.$$

With $F_t = S_t e^{k(T-t)}$, the three required derivatives are as follows:

$$\frac{\partial F}{\partial t} = -kS_t e^{k\tau},$$

$$\frac{\partial F}{\partial S} = e^{k\tau}, \text{and}$$

$$\frac{\partial^2 F}{\partial S^2} = 0.$$

Substituting these results, we have

$$dF_t = \left[-kS_t e^{k\tau} + e^{k\tau}\alpha S_t + \frac{1}{2}(0)\sigma^2 S_t^2 \right] dt + e^{k\tau}\sigma S_t dW_t$$

$$= (\alpha - k)F_t dt + \sigma F_t dW_t.$$

Therefore, if the underlying instrument price can be modeled with geometric Brownian motion, then this instrument is also geometric Brownian motion with an adjusted drift term and identical relative volatility.

Interestingly, if we modeled the underlying instrument as arithmetic Brownian motion with geometric drift or $dS_t = \alpha S_t dt + \sigma dW_t$, then we have[5]

$$dF_t = \left[-kS_t e^{k\tau} + e^{k\tau}\alpha S_t + \frac{1}{2}(0)\sigma^2 \right] dt + e^{k\tau}\sigma dW_t$$

$$= (\alpha - k)F_t dt + e^{k\tau}\sigma dW_t.$$

Note that the volatility term here is an absolute measure, such as currency units, rather than a relative measure such as percentage. In this case, the instrument price is also arithmetic Brownian motion with an adjusted drift term. The absolute volatility is growing when $k > 0$. We will return to explore arithmetic Brownian motion in more detail in Chapters 12 and 13.

11.3 ITÔ'S INTEGRAL

In addition, Itô's lemma can be expressed in integral form. Let us restate the problem. We are given a random variable x_t that follows the stochastic process:

$$dx_t = \mu(x,t)dt + \sigma(x,t)dW_t, \tag{11.13}$$

where we have subscripted t on x to reinforce the point that x can take on different values at different times t, depending on the evolution of W_t through time. Applying Itô's lemma to the function, $F(x,t)$, we obtain

$$dF(x_t,t) = \frac{\partial F}{\partial t}dt + \frac{\partial F}{\partial x_t}dx_t + \frac{1}{2}\frac{\partial^2 F}{\partial x_t^2}dx_t^2. \tag{11.14}$$

Now suppose we integrate over the period from time 0 to time t, an operation called *stochastic integration*, which is not the same as standard or non-stochastic integration. The latter is done by dividing the area under a curve into rectangles. The area under the curve is approximately the sum of the areas of each of the rectangles. As the number of rectangles goes to infinity, the sum of the areas of the rectangles approaches the area under the curve. Alternatively, standard integration can be viewed as adding up all of the infinitesimally small values dX/dt across values of t. Consequently, the derivative must be defined for all values of t. This type of standard integration is called a *Riemann integration*.

In a stochastic function, however, a derivative such as dx_t/dt is not defined. To put it somewhat roughly, the zig-zaggedness of x_t renders it impossible for the slope of a tangent line at any point to have a finite limit, which is a requirement for a derivative to exist. It is possible, however, to integrate a stochastic function by defining integration somewhat differently. Instead of being the limit of the area under each rectangle under the curve, a stochastic integral is defined as the *mean square limit*, meaning somewhat loosely that the integral is the expected value of the sum of the squared product of the volatility times the change in the stochastic variable. Such a limit will exist for the processes we typically encounter in finance. In this sense, a stochastic integral is much more like a volatility measure. More formally, the stochastic integral known as the Itô integral can be expressed as

$$\int_0^t \sigma(x_j,t)\,dx_j, \tag{11.15}$$

and in limit form is the following,

$$\lim_{n\to\infty} E\left[\sum_{k=1}^{n} \sigma(x_{k-1},k)\,(x_k - x_{k-1})\right]. \tag{11.16}$$

The mean squared difference is then zero,

$$\lim_{n\to\infty} E\left[\sum_{k=1}^{n} \sigma(x_{k-1},k)(x_k - x_{k-1}) - \int_0^t \sigma(x_j,j)\,dx_j\right]^2 = 0. \tag{11.17}$$

Remember that in the ordinary integral, the analogous expression is that the limit of the area of the rectangles equals the integral. For a stochastic integral, we must define this notion a little differently, in terms of expectations of squared differences.

Conveniently, some of the properties of ordinary integration are consistent with stochastic integration. For example, by definition,

$$\int_0^t dx_j = x_t - x_0. \tag{11.18}$$

In other words, whether x is stochastic or not, the sum of the changes in x from x_0 to x_t is, by definition, $x_t - x_0$. In the special case where the volatility is constant, that is, $\sigma_{x,t} = \sigma$ for all t, we can pull the constant out and obtain

$$\int_0^t \sigma dx_j = \sigma \left(x_t - x_0 \right). \tag{11.19}$$

11.4   THE INTEGRAL FORM OF ITÔ'S LEMMA

Now let us develop Itô's lemma in integral form. Using stochastic integration applied to Equation (11.14), we have

$$\int_0^t dF = \int_0^t \frac{\partial F}{\partial j} dj + \int_0^t \frac{\partial F}{\partial x_j} dx_j + \int_0^t \frac{1}{2} \frac{\partial^2 F}{\partial x_j^2} dx_j^2. \tag{11.20}$$

This equation looks like it is going to be a problem because of the term dx_j^2. In standard integration, we would see the term dx_j but not dx_j^2. So how do we handle this? We have previously noted that $dx_t^2 = \sigma(x,t)^2 dt$. Using that result and combining terms gives

$$F_t - F_0 = \int_0^t \left[\frac{\partial F}{\partial j} + \frac{1}{2} \frac{\partial^2 F}{\partial x_j^2} \sigma^2 \left(x_j, j \right) \right] dj + \int_0^t \frac{\partial F}{\partial x_j} dx_j, \tag{11.21}$$

which is Itô's lemma in integral formula and is sometimes called Itô's stochastic integral. Remember that either the differential or integral version of Itô's lemma automatically implies that the other exists, so either can be used, and in some cases, one is preferred over the other.

In Appendix 11A, we review several generalized stochastic integration results. Many of these results are useful when solving various financial problems with stochastic calculus. We turn now to additional applications of Itô's lemma.

11.5   SOME ADDITIONAL CASES OF ITÔ'S LEMMA

Let us look at two more cases of Itô's lemma. First, we look at the case where our random process is a function of not one but two random processes, x and y. An example of this situation might be a currency option, whereby the performance is determined by the exchange rate and an interest rate. Alternatively, there are options that pay off based on the greater or poorer performing of two assets that are stochastic.

So let x and y follow stochastic differential equations driven by Itô processes. For now, we do not have to specify those processes precisely. Consider a function F determined by x, y, and t. Applying Itô's lemma to F, we obtain

$$dF = \frac{\partial F}{\partial x}dx + \frac{\partial F}{\partial y}dy + \frac{\partial F}{\partial t}dt + \frac{1}{2}\frac{\partial^2 F}{\partial x^2}dx^2 + \frac{1}{2}\frac{\partial^2 F}{\partial y^2}dy^2 + \frac{\partial^2 F}{\partial x \partial y}dxdy. \tag{11.22}$$

Depending on the specifications of dx and dy, we may be able to simplify this expression further. Note that there is no interaction term with dx and dt or with dy and dt as these will go to zero by the product of dx and dW and the product of dy and dW. To take this equation further, we have to specify the processes describing x and y.

So now consider two processes, dx and dy, both with different but constant parameters and the same Brownian motion,

$$dx = \mu_x dt + \sigma_x dW_t$$
$$dy = \mu_y dt + \sigma_y dW_t. \tag{11.23}$$

Now consider another process, z, defined as the product of x and y. Products can occur in finance where the price of an asset is multiplied by an exchange rate or the quantity is multiplied by price:

$$z = xy. \tag{11.24}$$

We need to identify the stochastic process for dz. Applying Itô's lemma we obtain:[6]

$$dz = \frac{\partial z}{\partial x}dx + \frac{\partial z}{\partial y}dy + \frac{1}{2}\frac{\partial^2 z}{\partial x^2}dx^2 + \frac{1}{2}\frac{\partial^2 z}{\partial y^2}dy^2 + \frac{\partial^2 z}{\partial x \partial y}dxdy. \tag{11.25}$$

Because z is a very simple function of x and y, the partial derivatives in the above expression are easily obtained from Equation (11.24),

$$\frac{\partial z}{\partial x} = y, \quad \frac{\partial z}{\partial y} = x, \quad \frac{\partial^2 z}{\partial x^2} = 0,$$
$$\frac{\partial^2 z}{\partial y^2} = 0, \quad \frac{\partial^2 z}{\partial x \partial y} = \frac{\partial}{\partial x}\left(\frac{\partial z}{\partial y}\right) = \frac{\partial}{\partial y}\left(\frac{\partial z}{\partial x}\right) = 1. \tag{11.26}$$

Thus, simplifying equation (11.25) to

$$dz = ydx + xdy + dxdy.$$

Now let us examine $dxdy$. First, we show that the variance of $dxdy = 0$. Using the fact that $dt^k \to 0$ for $k > 1$, we have that

$$
\begin{aligned}
\text{var}(dxdy) &= \text{var}\left[(\mu_x dt + \sigma_x dW_t)(\mu_y dt + \sigma_y dW_t)\right] \\
&= \text{var}(\mu_x dt \mu_y dt + \mu_x dt \sigma_y dW_t + \mu_y dt \sigma_x dW_t + \sigma_x dW_t \sigma_y dW_t) \\
&= 0.
\end{aligned}
\tag{11.27}
$$

Working term by term, the first term has $dt^2 \to 0$. The next two terms involving $dt dW_t \to 0$, because dW_t has the square root of dt in it and that means that dt is raised to the power 3/2. Finally, recall that earlier we had $\text{var}\left(dW_t^2\right) = \text{var}\left(\varepsilon_t^2 dt\right) = dt^2 \text{var}\left(\varepsilon_t^2\right) = 0$. Now, if $\text{var}(dxdy) = 0$, then $dxdy = E(dxdy)$, which means that $dxdy$ is almost surely constant. Using the facts that $dW_t^2 = \varepsilon_t^2 dt$, $dt^k \to 0$ for $k > 1$, and $E(\varepsilon_t^2) = 1$, we show that $dxdy = \sigma_x \sigma_y dt$ as follows:

$$
\begin{aligned}
E(dxdy) &= E\left[(\mu_x dt + \sigma_x dW_t)(\mu_y dt + \sigma_y dW_t)\right] \\
&= E(\mu_x dt \mu_y dt + \mu_x dt \sigma_y dW_t + \mu_y dt \sigma_x dW_t + \sigma_x dW_t \sigma_y dW_t) \\
&= E(\sigma_x dW_t \sigma_y dW_t) = \sigma_x \sigma_y E\left(dW_t^2\right) = \sigma_x \sigma_y dt = dxdy.
\end{aligned}
\tag{11.28}
$$

Substituting this result into Equation (11.25), we obtain

$$
dz = ydx + xdy + \sigma_x \sigma_y dt.
\tag{11.29}
$$

Now let us assume the two processes, dx and dy, are driven by different Wiener processes, dW_t and dQ_t:

$$
\begin{aligned}
dW_t &= \varepsilon_{W,t}\sqrt{dt} \\
dQ_t &= \varepsilon_{Q,t}\sqrt{dt}.
\end{aligned}
\tag{11.30}
$$

Note the additional subscript for clarity. These processes could be independent, or they could be correlated. Let us start with the general case of a nonzero correlation, so let $\rho_{W,Q}$ be the correlation between the dW_t and dQ_t.

Now let us specify the processes for x and y, as driven by dW_t and dQ_t.

$$
\begin{aligned}
dx &= \mu_x dt + \sigma_x dW_t \\
dy &= \mu_y dt + \sigma_y dQ_t.
\end{aligned}
\tag{11.31}
$$

Recall that we are interested in z, the product of x and y. We found the total differential of z as Equation (11.25) and we obtained the partial derivatives in Equation (11.26). Therefore, dz is given as

$$
\begin{aligned}
dz &= xdy + ydz + dxdy = xdy + ydx + (\mu_x dt + \sigma_x dW_t)(\mu_y dt + \sigma_y dQ_t) \\
&= xdy + ydx + \mu_x \mu_y dt^2 + \mu_y \sigma_x dW_t dt + \mu_x \sigma_y dt dQ + \sigma_x \sigma_y dW_t dQ_t \\
&= xdy + ydx + \sigma_x \sigma_y dW_t dQ_t.
\end{aligned}
\tag{11.32}
$$

Notice in deriving this equation that the terms dt^2, $dW_t dt$, and $dQ_t dt$ go to zero because they all involve powers of dt greater than 1. Now let us examine the $dW_t dQ_t$ term. We start by taking the variance of $dW_t dQ_t$. We have

$$\text{var}(dW_t dQ_t) = \varepsilon_{W,t}\sqrt{dt}\,\varepsilon_{Q,t}\sqrt{dt} = \text{var}\left(dt\varepsilon_{W,t}\varepsilon_{Q,t}\right) = dt^2 \text{var}(\varepsilon_{W,t}\varepsilon_{Q,t}) = 0. \quad (11.33)$$

The zero variance means that $dW_t dQ_t = E(dW_t dQ_t)$, and so we evaluate $E(dW_t dQ_t)$. We can obtain this result by making use of the definition of covariance and correlation. Note the first line here is the correlation times the product of the standard deviations of the two processes (dt) and the second line is the definition of correlation. Thus,

$$
\begin{aligned}
\text{cov}(dW, dQ) &= dt\rho(dW, dQ) \\
&= E(dW_t dQ_t) - E(dW_t)\,E(dQ_t) \\
&= E(dW_t dQ_t).
\end{aligned}
\quad (11.34)
$$

Hence,

$$E(dW_t dQ_t) = dW_t dQ_t = \rho(dW, dQ)\,dt. \quad (11.35)$$

Thus,

$$dz = y\,dx + x\,dy + \sigma_x \sigma_y \rho(dW, dQ)\,dt. \quad (11.36)$$

We can do one more thing to make this result a bit cleaner. Note that the correlation is expressed in terms of dW and dQ but the other terms are expressed in terms of x and y. We know that $\rho_{dWdQ} = \text{cov}(dW, dQ)/\sigma_{dW}\sigma_{dQ}$. We also know from Equation (11.30) that the standard deviations of dW_t and dQ_t are each the square root of dt. Thus, $\sigma_{dW}\sigma_{dQ} = dt$, and, $\rho(dW, dQ)\,dt = \text{cov}(dW, dQ)$. We also know that $\rho(dx, dy) = \text{cov}(dx, dy)/\sigma_{dx}\sigma_{dy}$. Using Equation (11.31), $\text{cov}(dx, dy) = \sigma_{dx}\sigma_{dy}\text{cov}(dW, dQ) = \rho(dW, dQ)\,dt$. Then $\text{cov}(dW, dQ) = \rho(dW, dQ)\,dt/\sigma_{dW}\sigma_{dQ}$. Substituting into Equation (11.36),

$$dz = y\,dx + x\,dy + \rho(x, y)\,dt. \quad (11.37)$$

Of course, if these are independent Brownian motions, the correlation term disappears.

11.8 RECAP AND PREVIEW

In this chapter, we presented the concept of Itô's lemma, which is a specification roughly analogous to a Taylor series, but which applies to stochastic processes. Itô's lemma enables us to explain how the change in one variable is related to the changes in other variables. We also showed how Itô's lemma can be expressed in terms of a stochastic integral, and we showed how a stochastic integral differs from the standard integrals with which we are familiar from standard math. The information from this chapter will be needed to derive the Black-Scholes-Merton model.

In Chapter 12, we shall look at some other properties of the normal and lognormal diffusion process that will be necessary in order to complete the foundations of the

Black-Scholes-Merton model. After that point, we shall finally be able to dive into the derivation of the model.

APPENDIX 11A

Technical Stochastic Integral Results

11A.1 Selected Stochastic Integral Results

Before proceeding, it is important to recall from the definition of a Wiener process, we know

$$dW_t = \varepsilon_t \sqrt{dt}. \tag{11.38}$$

Recall the definition of an Itô integral found in Equation (11.16) can be represented here as

$$\int_0^t dW_t = \sum_{k=1}^n (W_k - W_{k-1}) : n \to \infty (h \to 0), \tag{11.39}$$

where $(0, t)$ is partitioned into intervals of equal length h, where $t = nh$.

Wiener Integration Theorem

If

$$f_t = \int_0^t dW_j \tag{11.40}$$

then

$$f_t = \int_0^t dW_j \stackrel{d}{=} \tilde{\varepsilon}\sqrt{dt} \text{ (equal in distribution)}, \tag{11.41}$$

where $\tilde{\varepsilon}$ denotes a standard normal random variable and $\stackrel{d}{=}$ denotes equal in distribution.

Proof: Within this proof, we demonstrate that f_t is equal in distribution to a standard normal random variable times the square root of time. We know dW is distributed normal and the sum of independent normals is normal. Thus, we focus on the first two moments of the distribution. Recall

$$f_t = \int_0^t dW_j \equiv \lim_{n \to \infty} \sum_{i=1}^n (W_i - W_{i-1}),$$

where $n = ht$ and h is equally spaced time steps. Taking the expected value,

$$E(f_t) = E\left(\lim_{n \to \infty} \sum_{i=1}^n (W_i - W_{i-1}) \right) = \lim_{n \to \infty} \sum_{i=1}^n E[(W_i - W_{i-1})] = 0.$$

Thus, the first moment is zero. We now turn to the second moment. The expected value of the square,

$$E(f_t^2) = E\left(\int\limits_{\tau=0}^{\tau=t} dW_\tau \int\limits_{\tau=0}^{\tau=t} dW_\tau\right)$$

$$= E\left[\left(\lim_{n\to\infty}\sum_{i=1}^{n}(W_i - W_{i-1})\right)\left(\lim_{n\to\infty}\sum_{j=1}^{n}(W_j - W_{j-1})\right)\right]$$

$$= \lim_{n\to\infty}\sum_{i=1}^{n}\sum_{j=1}^{n}E\left[(W_i - W_{i-1})\right](W_j - W_{j-1}).$$

Note that the covariance is zero when $i \neq j$,

$$E\left(f_t^2\right) = \lim_{n\to\infty}\left\{\begin{matrix}\sum_{i=1}^{n}E\left[(W_i - W_{i-1})^2\right]\\ +\sum_{i=1}^{n}\sum_{\substack{j=1\\j\neq1}}^{n}E[(W_i - W_{i-1})(W_j - W_{j-1})]\end{matrix}\right\}$$

$$= \lim_{n\to\infty}\left\{\sum_{i=1}^{n}E\left[(W_i - W_{i-1})^2\right]\right\}$$

$$= \lim_{n\to\infty}\left\{\sum_{i=1}^{n}E[(t_i - t_{i-1})]\right\} = \int\limits_{0}^{t} dj = t.$$

Therefore,

$$f_t = \int\limits_{0}^{t} dW_j \overset{d}{=} \tilde\varepsilon\sqrt{\int\limits_{0}^{t} dj} = \tilde\varepsilon\sqrt{t},$$

where $\tilde\varepsilon$ denotes a standard normal random variable and $\overset{d}{=}$ denotes equal in distribution.

We will apply this observation several times in the next few subsections. Throughout this material, we assume any required derivative of generic functions exists and is finite. Often, a financial application will require time-varying parameters. For example, an investor may believe that volatility will generally increase for a particular stock over the next year. The results presented next allow for time-varying parameters but not state-varying parameters. That is, the parameters are independent of the asset price.

11A.1.1   Time Functions of a Normally Distributed Process Lemma

If $g(t)$ is a function solely of calendar time and we have a function defined as

$$f_t \equiv \int\limits_{0}^{t} g(j)dW_j, \tag{11.42}$$

then

$$f_t = \sqrt{\frac{\int_0^t g^2(j)\,dj}{t}} \int_0^t dW_j = \tilde{\varepsilon} \sqrt{\int_0^t g^2(j)\,dj}. \qquad (11.43)$$

Proof: From the definition of stochastic integration, we have

$$f_t = \int_0^t g(j)\,dW_j \equiv \lim_{n\to\infty} \sum_{i=1}^n g(t_{i-1})(W_i - W_{i-1}), \qquad (11.44)$$

where $n = ht$ and h represents spaced time steps. Taking the expected value, we have

$$E(f_t) = E\left(\lim_{n\to\infty} \sum_{i=1}^n g(t_{i-1})(W_i - W_{i-1})\right) = \lim_{n\to\infty} \sum_{i=1}^n g(t_{i-1})\,E[(W_i - W_{i-1})] = 0. \qquad (11.45)$$

The expected value of the square,

$$E\left(f_t^2\right) = E\left(\int_0^t g(j)\,dW_j \int_0^t g(j)\,dW_j\right)$$

$$= E\left[\left(\lim_{n\to\infty} \sum_{i=1}^n g(t_{i-1})(W_i - W_{i-1})\right)\left(\lim_{n\to\infty} \sum_{j=1}^n g(t_{i-1})(W_j - W_{j-1})\right)\right]$$

$$= \lim_{n\to\infty} \sum_{i=1}^n \sum_{j=1}^n E[g(t_{i-1})(W_i - W_{i-1})\,g(t_{j-1})(W_j - W_{j-1})]. \qquad (11.46)$$

Note that the covariance is zero when $i \neq j$,

$$E(f_t^2) = \lim_{n\to\infty}\left[\begin{array}{l} \sum_{i=1}^n E\left[g^2(t_{i-1})(W_i - W_{i-1})^2\right] \\ + \sum_{i=1}^n \sum_{\substack{j=1 \\ j\neq 1}}^n E[g(t_{i-1})(W_i - W_{i-1})\,g(t_{j-1})(W_j - W_{j-1})] \end{array}\right]$$

$$= \lim_{n\to\infty}\left[\sum_{i=1}^n E\left[g^2(t_{i-1})(W_i - W_{i-1})^2\right]\right]$$

$$= \lim_{n\to\infty}\left[\sum_{i=1}^n g^2(t_{i-1})(t_i - t_{i-1})\right] = \int_0^t g^2(j)\,dj. \qquad (11.47)$$

Therefore,

$$f_t = \int_0^t g(j)\,dW_j = \sqrt{\frac{\int_0^t g^2(j)\,dj}{t}}\int_0^t dW_j = \tilde{\varepsilon}\sqrt{\int_0^t g^2(j)\,dj}. \qquad (11.48)$$

Based on this result, we note the mean and variance of this function are simply

$$E(f_t) = E(\tilde{\varepsilon})\sigma\sqrt{\int_0^t g^2(j)\,dj} = 0 \text{ and} \qquad (11.49)$$

$$\text{var}(f_t) = E(\tilde{\varepsilon}^2)\int_0^t g^2(j)\,dj = \int_0^t g^2(j)\,dj. \qquad (11.50)$$

We illustrate one application of this lemma. Suppose we wish to model the underlying instrument based on arithmetic Brownian motion with geometric drift. Specifically, suppose the underlying instrument is modeled as

$$dS_t = \alpha_t S_t dt + \sigma_t dW_t. \qquad (11.51)$$

In this case, we may have time varying geometric drift as well as time varying absolute volatility. Recall in a prior example where $F_t = S_t e^{k\tau}$, we found

$$dF_t = (\alpha - k)F_t dt + e^{k\tau}\sigma dW_t. \qquad (11.52)$$

Based on the previous lemma, let us focus solely on the noise term. Thus, $g(j) = \sigma e^{k(T-j)}$. The stochastic integral,

$$f_T = \int_0^T g(j)\,dW_j = \int_0^T \sigma e^{k(T-j)}\,dW_j,$$

can be represented as

$$f_T = \sigma\sqrt{\frac{\int_0^T e^{2k(T-j)}\,dj}{T}}\int_0^T dW_j = \tilde{\varepsilon}\sigma\sqrt{\int_0^T e^{2k(T-j)}\,dj}.$$

The solution to the standard integral is

$$\int_0^T e^{2k(T-j)}dj = e^{2k(T)}\int_0^T e^{-2k(j)}dj = e^{2k(T)}\left[\frac{e^{-2k(j)}}{-2k}\Big|_{j=0}^{j=T}\right]$$

$$= e^{2k(T)}\left(\frac{e^{-2k(T)}-1}{-2k}\right) = \frac{e^{2k(T)}-1}{2k}.$$

Thus, in this case, absolute volatility has an adjustment factor addressing the assumed underlying growth rate. Based on this result, we note the mean and variance of this function are

$$E\big(f_T\big) = E(\tilde{\varepsilon})\sigma\sqrt{\int_0^T e^{2k(T-j)}dj} = 0 \text{ and}$$

$$\text{var}\big(f_T\big) = E(\tilde{\varepsilon}^2)\sigma^2\int_0^T e^{2k(T-j)}dj = \sigma^2\left(\frac{e^{2k(T)}-1}{2k}\right).$$

11A.1.2   Time Functions of a Lognormally Distributed Process Lemma

Again, if $g(t)$ is a function solely of calendar time and we have a function defined as

$$h_t \equiv e^{\int_0^t g(j)dW_j}, \tag{11.53}$$

then

$$h_t = e^{f_t} = e^{\sqrt{\frac{\int_0^t g^2(j)dj}{t}}\int_0^t dW_j} = e^{\tilde{\varepsilon}\sqrt{\int_0^t g^2(j)dj}}. \tag{11.54}$$

Proof: Based directly on the previous lemma.

In this case, the mean and variance are based on properties of the lognormal distribution. Thus,

$$E\big(h_t\big) = e^{\frac{\int_0^t g^2(j)dj}{2}} \text{ and}$$

$$\text{var}\big(h_t\big) = E\big(h_t^2\big) - E\big(h_t\big)^2 = e^{2\int_0^t g^2(j)dj} - e^{\int_0^t g^2(j)dj} = e^{\int_0^t g^2(j)dj}\left(e^{\int_0^t g^2(j)dj} - 1\right).$$

11A.2   A General Linear Theorem

Mikosch (1998) provides an elegant introduction to stochastic calculus with a finance focus. On page 150 and following, he provides an extremely useful as well as powerful result. We review this result expressed here as a theorem. We follow the theorem with several useful finance applications that also rely on the stochastic integration results of the previous section.

11A.2.1   Mikosch's General Linear Theorem

Consider a generic linear stochastic integral of the form[7]

$$x_t = x_0 + \int_0^t \left[\mu_1(j)x_j + \mu_2(j) \right] dj + \int_0^t [\sigma_1(j)x_j + \sigma_2(j)]dW_j \qquad (11.55)$$

or expressed in stochastic differential form as

$$dx_t = \left[\mu_1(t)x_t + \mu_2(t) \right] dt + [\sigma_1(t)x_t + \sigma_2(t)]dW_t, \qquad (11.56)$$

where $\mu_1, \mu_2, \sigma_1, \sigma_2$ are deterministic continuous coefficient functions, $t \in [0, T]$ (bounded). Let

$$y_t \equiv e^{\int_0^t \left[\mu_1(j) - \frac{\sigma_1^2(j)}{2} \right] dj + \int_0^t \sigma_1(j) dW_j}, \qquad (11.57)$$

then

$$x_t = y_t \left(x_0 + \int_0^t \frac{\left[\mu_2(j) - \sigma_1(j)\sigma_2(j) \right]}{y_j} dj + \int_0^t \frac{\sigma_2(j)}{y_j} dW_j \right). \qquad (11.58)$$

Proof: See Mikosch (1998).

In most finance cases, either $\sigma_1(j) = 0$ or $\sigma_2(j) = 0$, but not both. When $\sigma_1(j) > 0$ and $\sigma_2(j) = 0$, the process is termed linear multiplicative noise. Recall that the binomial process was modeled through a multiplicative process $(S_0 u^j d^{n-j})$. When $\sigma_1(j) = 0$ and $\sigma_2(j) > 0$, the process is termed *linear additive noise*. In this case, the binomial process would be modeled through an additive process $[S_0 + ju + (n-j)d.]$

Although we would be interested in the $E(x_t)$ and $\text{var}(x_t)$, the solution for these parameters is a bit tedious for Mikosch's general linear theorem. Thus, we introduce two lemmas that are more straightforward. With the combination of the previous stochastic integration results as well as this theorem, we illustrate several important finance applications.

11A.2.2   Linear Additive Noise Lemma

Consider an underlying instrument that is linear with an additive noise term implying a terminal normal distribution. That is, the generic stochastic integral representation is

$$S_t = S_0 + \int_0^t \left[\mu_1(j)S_j + \mu_2(j) \right] dj + \int_0^t \sigma_2(j)dW_j, \qquad (11.59)$$

or the generic stochastic differential representation is

$$dS_t = \left[\mu_1(t)S_t + \mu_2(t) \right] dt + \sigma_2(t)dW_t. \qquad (11.60)$$

From the general linear theorem, we have

$$y_t \equiv e^{\int_0^t \mu_1(j)\,dj} \tag{11.61}$$

and

$$S_t = y_t \left[S_0 + \int_0^t \frac{\mu_2(j)}{y_j}\,dj + \int_0^t \frac{\sigma_2(j)}{y_j}\,dW_j \right]. \tag{11.62}$$

The mean and variance can be expressed as

$$E(S_t) = S_0 e^{\int_0^t \mu_1(j)\,dj} + \int_0^t \mu_2(j)\exp\left\{\int_j^t \mu_1(k)\,dk\right\}dj \text{ and} \tag{11.63}$$

$$\mathrm{var}(S_t) = \int_0^t \sigma_2^2(j)\exp\left\{2\int_j^t \mu_1(k)\,dk\right\}dj. \tag{11.64}$$

For example, consider arithmetic Brownian motion with additive drift. Thus, $\mu_1(j)=0$, $\mu_2(j)=\mu_2, \sigma_1(j)=0$, and $\sigma_2(j)=\sigma_2$. Thus, $dS_t = \mu_2 dt + \sigma_2 dW_t$. Based on the linear additive noise lemma, we have

$$y_t = e^{\int_0^t \mu_1(j)\,dj} = e^0 = 1 \tag{11.65}$$

and

$$S_t = y_t\left[S_0 + \int_0^t \frac{\mu_2(j)}{y_j}\,dj + \int_0^t \frac{\sigma_2(j)}{y_j}\,dW_j\right] = S_0 + \mu_2 t + \sigma_2 \int_0^t dW_j. \tag{11.66}$$

Thus, the expected value and variance for underlying instruments that follow arithmetic Brownian motion with additive drift is

$$E(S_t) = S_0 + \mu_2 t + \sigma_2 E\left(\int_0^t dW_j\right) = S_0 + \mu_2 t \text{ and} \tag{11.67}$$

$$\mathrm{var}(S_t) = \sigma_2^2 t. \tag{11.68}$$

Now consider arithmetic Brownian motion with geometric drift. Thus, $\mu_1(j)=\mu_1$, $\mu_2(j)=0, \sigma_1(j)=0$, and $\sigma_2(j)=\sigma_2$. Thus, $dS_t = \mu_1 S_t dt + \sigma_2 dW_t$. Based on the linear additive noise lemma, we have

$$y_t \equiv e^{\int_0^t \mu_1\,dj} = e^{\mu_1 t} \tag{11.69}$$

and

$$S_t = y_t\left(S_0 + \int_0^t \frac{\sigma_2}{y_j} dW_j\right) = e^{\mu_1 t}\left(S_0 + \sigma_2\int_0^t e^{-\mu_1 j} dW_j\right) = S_0 e^{\mu_1 t} + \sigma_2\int_0^t e^{\mu_1(t-j)} dW_j. \quad (11.70)$$

From the time functions of a normally distributed process lemma, we have the expected value and variance for underlying instruments that follow arithmetic Brownian motion with geometric drift is

$$E(S_t) = S_0 e^{\mu_1 t} + \sigma_2 E\left[\int_0^t e^{\mu_1(t-j)} dW_j\right] = S_0 e^{\mu_1 t} \text{ and} \quad (11.71)$$

$$\text{var}(S_t) = \sigma_2^2\int_0^t e^{2\mu_1(t-j)} dj = \sigma_2^2\frac{e^{2\mu_1 t} - 1}{2\mu_1}. \quad (11.72)$$

We turn now to the multiplicative case.

11A.2.3 Linear Multiplicative Noise Lemma

Consider an underlying instrument that is linear with a multiplicative noise term implying the terminal distribution is lognormal. That is, the generic stochastic integral representation is

$$S_t = S_0 + \int_0^t [\mu_1(j)S_j + \mu_2(j)]\, dj + \int_0^t \sigma_1(j)S_j dW_j, \quad (11.73)$$

or the generic stochastic differential representation is

$$dS_t = [\mu_1(t)S_t + \mu_2(t)]\, dt + \sigma_1(t)S_t dW_t. \quad (11.74)$$

Then

$$y_t = e^{\int_0^t\left[\mu_1(j)-\frac{\sigma_1^2(j)}{2}\right] dj + \int_0^t \sigma_1(j) dW_j} \quad (11.75)$$

and

$$S_t = y_t\left[S_0 + \int_0^t \frac{\mu_2(j)}{y_j} dj\right]. \quad (11.76)$$

The mean and variance are bit more involved for the multiplicative case. Let

$$A(a,b) \equiv e^{\int_a^b\left[\mu_1(j)-\frac{\sigma_1^2(j)}{2}\right] dj}. \quad (11.77)$$

Therefore, we have

$$E(S_t) = S_0 A(0,t) e^{\frac{\int_0^t \sigma_1^2(k)\,dk}{2}} + \int_0^t \mu_2(j) A(j,t) e^{\frac{\int_j^t \sigma_1^2(k)\,dk}{2}}\,dj \text{ and} \tag{11.78}$$

$$E(S_t^2) = S_0^2 A^2(0,t) e^{2\int_0^t \sigma_1^2(k)\,dk} + 2S_0 A(0,t)\int_0^t \mu_2(j) A(j,t) e^{2\int_j^t \sigma_1^2(k)\,dk + \frac{\int_0^j \sigma_1^2(k)\,dk}{2}}\,dj$$

$$+ \int_0^t \int_0^j \mu_2(j) A(j,t) \mu_2(k) A(k,t) e^{2\int_j^t \sigma_1^2(m)\,dm + \frac{\int_k^j \sigma_1^2(m)\,dm}{2}}\,dk\,dj$$

$$+ \int_0^t \int_j^t \mu_2(j) A(j,t) \mu_2(k) A(k,t) e^{2\int_k^t \sigma_1^2(m)\,dm + \frac{\int_j^k \sigma_1^2(m)\,dm}{2}}\,dk\,dj. \tag{11.79}$$

Thus, the variance is simply based on the definition of variance or

$$\text{var}(S_t) \equiv E(S_t^2) - [E(S_t)]^2, \tag{11.80}$$

where these two moments are based on Equations (11.78) and (11.79).

Now consider arithmetic Brownian motion with geometric drift. Thus, $\mu_1(j) = \mu_1$, $\mu_2(j) = 0$, $\sigma_1(j) = \sigma_1$, and $\sigma_2(j) = 0$. Thus, $dS_t = \mu_1 S_t dt + \sigma_1 S_t dW_t$. Based on the linear additive noise lemma, we have

$$y_t \equiv e^{\int_0^t \left(\mu_1 - \frac{\sigma_1^2}{2}\right) dj + \int_0^t \sigma_1 dW_j} = e^{\left(\mu_1 - \frac{\sigma_1^2}{2}\right) t + \sigma_1 \int_0^t dW_j} \tag{11.81}$$

and

$$S_t = y_t(S_0) = S_0 e^{\left(\mu_1 - \frac{\sigma_1^2}{2}\right) t + \sigma_1 \int_0^t dW_j}. \tag{11.82}$$

From the time functions of a lognormally distributed process lemma, we have

$$S_t = S_0 e^{\left(\mu_1 - \frac{\sigma_1^2}{2}\right) t + \sigma_1 \sqrt{t}\,\hat{\varepsilon}}. \tag{11.83}$$

Thus, the expected value and variance for underlying instruments that follow geometric Brownian motion with geometric drift, based on properties of a lognormal distribution, is

$$E(S_t) = S_0 E\left[e^{\left(\mu_1 - \frac{\sigma_1^2}{2}\right) t + \sigma_1 \sqrt{t}\,\hat{\varepsilon}} \right] = S_0 e^{\left(\mu_1 - \frac{\sigma_1^2}{2}\right) t} E\left(e^{\sigma_1 \sqrt{t}\,\hat{\varepsilon}} \right) = S_0 e^{\left(\mu_1 - \frac{\sigma_1^2}{2}\right) t} \left(e^{\frac{\sigma_1^2 t}{2}} \right) = S_0 e^{\mu_1 t},$$

$$\tag{11.84}$$

$$E(S_t^2) = S_0^2 E\left[e^{2\left(\mu_1 - \frac{\sigma_1^2}{2}\right)t + 2\sigma_1\sqrt{t}\tilde{\epsilon}}\right] = S_0^2 e^{2\left(\mu_1 - \frac{\sigma_1^2}{2}\right)t} E\left(e^{2\sigma_1\sqrt{t}\tilde{\epsilon}}\right)$$

$$= S_0^2 e^{2\left(\mu_1 - \frac{\sigma_1^2}{2}\right)t}\left(e^{\frac{4\sigma_1^2 t}{2}}\right) = S_0^2 e^{\left(2\mu_1 + \sigma_1^2\right)t}, \qquad (11.85)$$

and

$$\text{var}(S_t) = E(S_t^2) - E(S_t)^2 = S_0^2 e^{\left(2\mu_1 + \sigma_1^2\right)t} - \left(S_0 e^{\mu_1 t}\right)^2 = S_0^2 e^{2\mu_1 t}\left(e^{\sigma_1^2 t} - 1\right). \qquad (11.86)$$

QUESTIONS AND PROBLEMS

1 One model for the S&P 500 index futures price is represented as

$$F_t = S_t e^{(r-\delta)\tau},$$

where S_t denotes the S&P 500 index value, r denotes the risk free rate, δ denotes the dividend yield, and τ denotes the time to maturity, expressed in years. If S_t is assumed to follow geometric Brownian motion, $dS_t = \alpha S_t dt + \sigma S_t dW_t$, where α denotes the expected return on the index and σ denotes the standard deviation (both annualized, continuously compounded). What is the expected return and standard deviation of the continuously compounded percentage change in the future price?

2 One model for the S&P 500 index futures price is represented as

$$F_t = S_t e^{(r-\delta)\tau},$$

where S_t denotes the S&P 500 index value, r denotes the risk-free rate, δ denotes the dividend yield, and τ denotes the time to expiration, expressed in years. Now if S_t is assumed to follow arithmetic Brownian motion with geometric drift, $dS_t = \alpha S_t dt + \sigma dW_t$, where α denotes the expected return on the index (annualized, continuously compounded) and σ denotes the absolute standard deviation (annualized unit changes). What is the expected change and standard deviation of the change in the future price?

3 Prove the time functions of a normally distributed process lemma stated as:
If $g(t)$ is a function solely of calendar time and we have a function defined as

$$f_t \equiv \int_0^t g(j) dW_j,$$

then

$$f_t = \sqrt{\frac{\int_0^t g^2(j)\,dj}{t}} \int_0^t dW_j = \tilde{\varepsilon}\sqrt{\int_0^t g^2(j)\,dj}.$$

4 Suppose a stochastic integral can be expressed as

$$f_t \equiv \int_0^t e^{c(j)j}\,dW_j.$$

Derive the mean and variance of f_t.

5 Suppose the US dollar (USD) to Narnian lokum (NNL) is $S_t = 1.25$ USD for each NNL. Based on an analysis of the two countries, the exchange rate is assumed to follow geometric Brownian motion. Specifically, $dS_t = \alpha S_t dt + \sigma S_t dW_t$, where $\alpha = 5.0\%$ and $\sigma = 35\%$. Compare the expected change in the exchange rate and standard deviation of the expected change from the US perspective as well as the Narnian perspective assuming a five-year horizon. (Note: If the spot exchange rate is $S_t = \$1.25/N$, then the reciprocal exchange rate is $R_t = N0.8/\$\ [= N(1/1.25)/\$]$.

NOTES

1. The term *mean square limit* can be thought of somewhat like the concept of variance, which is the mean squared deviation around the expected value. It is approximately correct to say that a result in stochastic calculus holds when the variance converges to a finite value.
2. The variable F can depend on other parameters, but they must be constant across x and t.
3. Remember that the expected value of dW_t is zero. Hence, the expected value of dF comes from the first term on the right-hand side; the term in parentheses is the expected value, which is a constant. Multiplying by dt scales it by the length of the time interval. The variance in the stochastic process comes from the second term on the right-hand side. The variance is the square of whatever term is multiplied by dW_t times the variance of dW_t, which is dt.
4. This instrument closely resembles a forward contract under the carry arbitrage model covered in Chapter 22, but the details of that instrument are not relevant for this application of Itô's lemma. Also, note that τ is in years and $T - t$ is typically just an index, that is, a type of counter that can be used to identify points in time; there is no harm done in letting $T - t$ represent days/365 to make it completely consistent with the definition of τ.
5. In arithmetic Brownian motion with arithmetic drift, the asset follows the stochastic process, $dS_t = \alpha dt + \sigma dW_t$, meaning that the price change and not the return is modeled as a function of its drift and volatility. In such a case, the asset value can go below zero.
6. There is no term related to dt because z is not directly determined by t.
7. Generic refers to μ_1, μ_2, σ_1, and σ_2 being a function of, at most, calendar time. Many different functional forms are possible.

CHAPTER 12

Properties of the Lognormal and Normal Diffusion Processes for Modeling Assets

In this chapter, we will take what we have already learned about stochastic processes and adapt it to the case of an asset. We will show how certain properties of the asset's return behavior are derived. Finally, we will show how to obtain the formula for the future asset price in terms of a base price and information on what happened in the intervening period. In examining the diffusion processes of assets, let us start by noting that there are many excellent treatments of the subject of this chapter and those in this entire unit. For further study, we recommend several that are particularly well written such as Shimko (1992), Neftci (2006), Ross (1999), and Malliaris and Brock (1982). In this chapter, we use the term *asset* to represent any of a variety of potential exposures, such as stocks, exchange rates, or combinations of assets, one of which we will introduce later in this chapter.

Recall that in Chapter 10, we obtained the stochastic process for the asset price in a heuristic manner. We started with four desirable properties for such a process summarized succinctly here:

1. The asset price may have a nonzero drift.
2. Changes in the asset price are random.
3. Over time, the asset price becomes more unpredictable.
4. Over time, the asset price is nonnegative.

Based on these properties, we proposed geometric Brownian motion (GBM) with geometric drift. The GBM process has all four properties and can be expressed in two ways,

$$dS_t = \alpha S_t dt + \sigma S_t dW_t \text{ and}$$

$$\frac{dS_t}{S_t} = \alpha dt + \sigma dW_t. \tag{12.1}$$

One clear undesirable property evident in Equation (12.1) is the assumption that the underlying instrument is greater than zero or $S_t > 0$. That is, property (4) holds, but there is no chance for the underlying to be zero in the future. For numerous finance applications, this is simply inappropriate. For example, there is always a chance that a particular stock's price goes permanently to zero. Interest rates can and do go to zero as well as negative. Many financial risks involve differences in prices, such as profits, calendar-related spreads, or product-related spreads.[1] Thus, we also explore properties of the normal diffusion process for modeling various financial applications because it admits zero and negative values.

Recall the random variable dW_t is the transformed Brownian motion, which we called a Wiener process. Recall that dW_t is normally distributed with $E(dW_t) = 0$, $\text{var}(dW_t) = dt$. Also, we noted that dW_t^2 is non-stochastic and equal to dt. Using these results, the expectation and variance of Equation (12.1) are

$$E\left(\frac{dS_t}{S_t}\right) = \alpha dt \text{ and}$$

$$\text{var}\left(\frac{dS_t}{S_t}\right) = \sigma^2 dt. \tag{12.2}$$

Given the fact that dS_t/S_t is just a linear transformation of a normally distributed random variable dW_t, then it is also normally distributed.[2] In this chapter, we formally derive this stochastic process and some important results related to it.

Alternatively, we can model some underlying exposure with arithmetic Brownian motion (ABM). There are two forms of ABM, either with geometric drift or arithmetic drift. Recall from Chapter 10, we introduced the $ symbol to emphasize the unit of measure is the same as the asset price and not percentage. Thus, arithmetic Brownian motion with geometric drift (ABMGD) can be expressed as

$$dS_t = \alpha S_t dt + \sigma_\$ dW_t, \tag{12.3}$$

or, depending on context, arithmetic Brownian motion with arithmetic drift (ABMAD) can be expressed as

$$dS_t = \alpha_\$ dt + \sigma_\$ dW_t. \tag{12.4}$$

These ABM processes resolve the undesirable property evident in Equation (12.1) because the underlying exposure can be positive, zero, or negative. Also, at times we may wish to have the underlying instrument growing geometrically (e.g., stocks) and at other times arithmetically (e.g., spreads such as refined petroleum and crude oil). Note that in this case relative return may be undefined if $S_t = 0$ or lacking financial interpretation if $S_t < 0$. For the ABM stochastic process, we are interested in dS_t and not dS_t/S_t. Again, volatility, and perhaps mean, is measured in currency units and not percentage.

12.1 A STOCHASTIC PROCESS FOR THE ASSET RELATIVE RETURN

The relative return on the asset from a starting point of time 0 to a point of time dt is[3]

$$\frac{S_{dt}}{S_0}. \tag{12.5}$$

The relative return from time dt to time $2dt$ is

$$\frac{S_{2dt}}{S_{dt}}. \tag{12.6}$$

This pattern continues so that at a given future time T, the relative return is

$$\frac{S_T}{S_{T-dt}}.\tag{12.7}$$

Backing up to time 0, the relative return on the asset to time T is

$$\frac{S_T}{S_0}.\tag{12.8}$$

This return can be expressed by linking the successive relative returns,

$$\frac{S_T}{S_0} = \left(\frac{S_{dt}}{S_0}\right)\left(\frac{S_{2dt}}{S_{dt}}\right)\cdots\left(\frac{S_{T-dt}}{S_{T-2dt}}\right)\left(\frac{S_T}{S_{T-dt}}\right).\tag{12.9}$$

Now let us convert Equation (12.9) into the log or continuously compounded return,

$$\ln\left(\frac{S_T}{S_0}\right) = \ln\left(\frac{S_{dt}}{S_0}\right) + \ln\left(\frac{S_{2dt}}{S_{dt}}\right) + \cdots + \ln\left(\frac{S_{T-dt}}{S_{T-2dt}}\right) + \ln\left(\frac{S_T}{S_{T-dt}}\right).\tag{12.10}$$

We see that the log return for the period of time 0 to time T is the sum of the log returns of the subperiods from time 0 to time T. Now recall that the central limit theorem says that a random variable that is defined as the sum of a series of other random variables from any distribution that is constant approaches a normal distribution. Thus, we know that the return from time 0 to time T is normally distributed, provided the distribution of the intermediate returns is constant. We can also propose that each subperiod is infinitesimally small such that it, too, is made up of a series of component returns over infinitesimally small subperiods. Hence, it is reasonable to propose that the return over any arbitrary "short" period from t to $t + dt$ is normally distributed with dt period expectation of μ_p and variance of σ_p^2, which is formally written as

$$\ln\left(\frac{S_{t+dt}}{S_t}\right) \sim N\left(\mu_p, \sigma_p^2\right).\tag{12.11}$$

It is important to emphasize that when the price is lognormally distributed, the log of the relative return is normally distributed. Thus, the asset price is lognormally distributed. By definition, the log return is defined in the following manner:

$$\ln\left(\frac{S_{t+dt}}{S_t}\right) = \ln S_{t+dt} - \ln S_t = d\ln S_t.\tag{12.12}$$

In other words, the log return is the change in the log values of the asset price at the beginning and ending of the holding period. We then propose that the log return follows the GBM stochastic process

$$d\ln S_t = \mu dt + \sigma dW.\tag{12.13}$$

where the expectation and variance are, therefore,

$$E(d \ln S_t) = \mu dt$$

$$\text{var}(d \ln S_t) = \sigma^2 dt. \tag{12.14}$$

Now, however, we want the return dS_t/S_t. Let us use the following transformation:

$$G_t = \ln S_t, \tag{12.15}$$

so that

$$S_t = e^G. \tag{12.16}$$

We wish to find the stochastic process for S_t, so we can use Itô's lemma. Temporarily dropping the time subscript, we obtain

$$dS = \frac{\partial S}{\partial G} dG + \frac{1}{2} \frac{\partial^2 S}{\partial G^2} dG^2. \tag{12.17}$$

The partial derivatives are easily obtained as

$$\frac{\partial S}{\partial G} = e^G = S$$

$$\frac{\partial^2 S}{\partial G^2} = e^G = S. \tag{12.18}$$

Substituting these results, we get

$$dS = SdG + \frac{1}{2}SdG^2. \tag{12.19}$$

Because $dG = d \ln S$, the differentials, dG and dG^2, are

$$dG = \mu dt + \sigma dW$$

$$dG^2 = \sigma^2 dt. \tag{12.20}$$

Substituting these results, we obtain

$$dS = S(\mu dt + \sigma dW_t) + \frac{1}{2}S\sigma^2 dt. \tag{12.21}$$

Dividing both sides by S_t and adding the time subscript, we now have the stochastic process for dS_t/S_t,

$$\frac{dS_t}{S_t} = (\mu + \sigma^2/2)dt + \sigma dW_t. \tag{12.22}$$

Defining $\alpha = \mu + \sigma^2/2$, we have

$$\frac{dS_t}{S_t} = \alpha dt + \sigma dW_t. \tag{12.23}$$

And this is the equation we proposed heuristically in Chapter 10. We have now provided a formal derivation of its existence. The expectation and variance are

$$E\left(\frac{dS_t}{S_t}\right) = \alpha dt$$

$$\text{var}\left(\frac{dS_t}{S_t}\right) = \sigma^2 dt. \tag{12.24}$$

Thus, we now have the stochastic differential equations for the return (Equation (12.23)) and the log return (Equation (12.13)). The return over the longer horizon is dS_T/S_0, and the log return over the long horizon is normally distributed, which means that dS_T/S_0 is lognormally distributed.[4] Both the infinitesimal return, dS_t/S_t, and the infinitesimal log return, $d\ln S_t$, are normally distributed.

For those students who remember their calculus, they might more easily make the link that because $d\ln S(t) = S'(t)/S'(t)$, then $d\ln S(t) = dS(t)/S(t) = \alpha dt + \sigma dW_t$.

12.2 A STOCHASTIC PROCESS FOR THE ASSET PRICE CHANGE

The price change on the asset from a starting point of time 0 to a point of time dt is simply the dollar change or[5]

$$dS_{dt} = S_{dt} - S_0. \tag{12.25}$$

The price change from time dt to time $2dt$ is

$$dS_{2dt} = S_{2dt} - S_{dt}. \tag{12.26}$$

This pattern continues so that at a given future time T, the price change is

$$dS_T = S_T - S_{T-dt}. \tag{12.27}$$

Backing up to time 0, the price change on the asset to time T is

$$\Delta S_T = S_T - S_0. \tag{12.28}$$

Thus a price change can be expressed by linking the successive returns,

$$S_T - S_0 = (S_{dt} - S_0) + (S_{2dt} - S_{dt}) + \cdots + (S_{T-dt} - S_T) + (S_T - S_{T-dt}). \tag{12.29}$$

Note that due to the potential for nonpositive values, log transformations are not pursued with ABM. With price changes, we assume the change in asset value is normally distributed.

For ABM with arithmetic drift, we have the expected price change and price change variance are[6]

$$E(dS_t) = \alpha_\$ dt \text{ and} \tag{12.30}$$

$$\text{var}(dS_t) = \sigma_\$^2 dt. \tag{12.31}$$

With ABM with geometric drift, we have the expected price change and price change variance as follows:[7]

$$E(dS_t) = S_0 e^{\alpha dt} \tag{12.32}$$

and

$$\text{var}(dS_t) = \begin{bmatrix} \sigma_\$^2 \left(\frac{e^{2\alpha dt} - 1}{2\alpha} \right) & \text{if } \alpha \neq 0 \\ \sigma_\$^2 dt & \text{if } \alpha = 0. \end{bmatrix} \tag{12.33}$$

12.3 SOLVING THE STOCHASTIC DIFFERENTIAL EQUATION

The equations for the relative return and log return are stochastic processes, as well as stochastic differential equations. A differential equation has a potential solution, which is a function such that the derivatives conform to the differential equation. In this context, a solution would be the asset price at some future time t, expressed in terms of the asset price at a previous time such as time 0.

To solve the stochastic differential equation, (12.23), we want to obtain a future price S_t in terms of a current price, say S_0. We take the equation for the log return, Equation (12.13) using dG_t as the LHS, and integrate over the interval 0 to t,

$$\int_0^t dG_j = \int_0^t \mu dj + \int_0^t \sigma dW_j. \tag{12.34}$$

The left-hand side is clearly $G_t - G_0$. The first integral on the right-hand side of Equation (12.34) is a standard Riemann integral and equals

$$\int_0^t \mu dj = \mu \int_0^t dj = \mu t. \tag{12.35}$$

The second integral on the right-hand side of Equation (12.34) is a stochastic integral and, fortunately, one of the simplest of stochastic integrals. It is obtained as

$$\int_0^t \sigma dW_j = \sigma \int_0^t dW_j = \sigma(W_t - W_0). \tag{12.36}$$

In fact, in this case, the stochastic integral is so simple, it is the same as the Riemann integral. The variable W_t is the value of the Brownian motion process at time t. It is quite common that W_0 is set at zero, so let us do it. So now we have σW_t as the solution to Equation (12.36). Then $G_t - G_0 = \mu t + \sigma W_t$ is the solution to Equation (12.34). To get this solution in terms of S_t and S_0, we exponentiate this result,

$$e^{G_t - G_0} = e^{\mu t + \sigma W_t}. \tag{12.37}$$

Because $S_t = e^{G_t}$, and $S_0 = e^{G_0}$,

$$S_t = S_0 e^{\mu t + \sigma W_t}. \tag{12.38}$$

Equation (12.38) is presumably the solution to the stochastic differential equation, (12.23). Now, to be absolutely certain that we have the correct solution, we need to check it by applying Itô's lemma to S_t, inserting the derivatives, and determining if we obtain (12.23). Our stochastic differential equation for S_t obtained from Itô's lemma is

$$dS_t = \frac{\partial S_t}{\partial W_t} dW_t + \frac{\partial S}{\partial t} dt + \frac{1}{2} \frac{\partial^2 S_t}{\partial W_t^2} dW_t^2. \tag{12.39}$$

We then insert the partial derivatives that we obtain by differentiating the solution, Equation (12.38). The partial derivatives are

$$\frac{\partial S_t}{\partial W_t} = S_t \sigma$$

$$\frac{\partial^2 S_t}{\partial W_t^2} = S_t \sigma^2$$

$$\frac{\partial S_t}{\partial t} = S_t \mu. \tag{12.40}$$

Now recall that $dW_t^2 = dt$. Substituting these results into Equation (12.39) and rearranging, we obtain:

$$\frac{dS_t}{S_t} = \alpha dt + \sigma dW_t. \tag{12.41}$$

This is the original stochastic process, Equation (12.23). Thus, our solution is correct.

12.4 SOLUTIONS TO STOCHASTIC DIFFERENTIAL EQUATIONS ARE NOT ALWAYS THE SAME AS SOLUTIONS TO CORRESPONDING ORDINARY DIFFERENTIAL EQUATIONS

Let us see how solving a stochastic differential equation (SDE) is different from solving an ordinary differential equation (ODE). Consider the ordinary differential equation:

$$dy_t = y_t dW_t, \tag{12.42}$$

where W_t is non-stochastic. Remember that this means that we are absolutely certain of the value of W_t. This is a fairly simple ODE. Assuming $y_t > 0$, we start by rewriting the equation as

$$\frac{1}{y_t} \frac{dy_t}{dW_t} dW_t = dW_t. \tag{12.43}$$

We now integrate over 0 to t:

$$\int_0^t \frac{1}{y_s} \frac{dy_s}{dW_s} dW_s = \int_0^t dW_s. \tag{12.44}$$

With $W_0 = 0$, the solution is $\ln y_t = W_t$ or alternatively, $y_t = e^{W_t}$.[8]

Now we let W_t be stochastic. We start by proposing a general form for the solution. Specifically, we shall say that $y_t = e^{x_t}$. In other words, x_t is some function that solves the equation and in which x_t is a function of W_t. In the special case $x_t = 0$, we have $y_0 = 1$. In the ODE case, $x_t = W_t$. First, we use Itô's lemma on x_t and obtain

$$dx_t = \frac{\partial x_t}{\partial y_t} dy_t + \frac{1}{2} \frac{\partial^2 x_t}{\partial y_t^2} dy_t^{\,2}. \tag{12.45}$$

The partial derivatives are $\partial x_t / \partial y_t = 1/y_t$ and $\partial^2 x_t / \partial y_t^2 = -(1/y_t^2)$. We also have $dy_t = y_t dW_t$ and $dy_t^2 = y_t^2 dt$, due to the properties of dW_t. Substituting these results, we obtain

$$dx_t = dW_t - \frac{1}{2} dt. \tag{12.46}$$

Now we perform the integration,

$$\int_0^t dx_s = \int_0^t dW_s - \int_0^t \frac{1}{2} ds$$

$$x_t - x_0 = W_t - t/2. \tag{12.47}$$

With $x_t = \ln y_t$, then

$$y_t = e^{W_t - t/2}. \tag{12.48}$$

Notice that now we have an additional term $t/2$. Thus, at least in this common situation, and quite often otherwise, the solution to an SDE is not the same as a solution to an ODE.

12.5 FINDING THE EXPECTED FUTURE ASSET PRICE

Given the solution,

$$S_t = S_0 e^{\mu t + \sigma W_t}, \tag{12.49}$$

to the stochastic differential equation, we shall now use it to obtain the expected asset price at t. Such a concept is commonly seen in real markets, such as when someone asks an analyst, "Where do you think the Dow will be in a year?" Using Equation (12.49), we express the problem as follows:

$$E(S_t) = S_0 E\left(e^{\mu t + \sigma W_t}\right) = S_0 e^{\mu t} E\left(e^{\sigma W_t}\right). \tag{12.50}$$

This expectation is evaluated by recognizing that W_t is normally distributed. We are reminded that the probability density for a normally distributed random variable W_t, which has mean zero and variance t, is

$$f(W_t) = \frac{1}{\sqrt{2\pi t}} e^{-W_t^2/2t}. \tag{12.51}$$

Thus, we can find the expected value of S_t by evaluating the following expression:

$$E(e^{\sigma W_t}) = \int_{-\infty}^{\infty} e^{\sigma W_t} \frac{1}{\sqrt{2\pi t}} e^{-W_t^2/2t} dW_t. \tag{12.52}$$

We express the right-hand side as

$$\int_{-\infty}^{\infty} \frac{1}{\sqrt{2\pi t}} e^{\sigma W_t} e^{-W_t^2/2t} dW_t. \tag{12.53}$$

Working on the exponent, we obtain

$$\sigma W_t - W_t^2/2t = \frac{2\sigma W_t t - W_t^2}{2t}$$

$$= \frac{2\sigma W_t t - W_t^2}{2t} + \frac{\sigma^2 t}{2} - \frac{\sigma^2 t}{2}$$

$$= \frac{2\sigma W_t t - W_t^2 - \sigma^2 t^2}{2t} + \frac{\sigma^2 t}{2}$$

$$= -\frac{1}{2} \frac{(W_t - \sigma t)^2}{t} + \frac{\sigma^2 t}{2}. \tag{12.54}$$

So now we have

$$e^{\sigma^2 t/2} \int_{-\infty}^{\infty} \frac{1}{\sqrt{2\pi t}} e^{-\frac{1}{2}\left(\frac{W_t - \sigma t}{\sqrt{t}}\right)^2} dW_t. \tag{12.55}$$

The integrand is the probability density function for a normally distributed random variable with mean σt and variance t and, by definition, integrates to a value of 1.0. Thus,

$$E(e^{\sigma W_t}) = e^{\sigma^2 t/2}. \tag{12.56}$$

So our expectation is,

$$E(S_t) = S_0 e^{\mu t + \sigma^2 t/2} = S_0 e^{(\mu + \sigma^2/2)t}. \tag{12.57}$$

Note that this result is also equal to $E(S_t) = S_0 e^{\alpha t}$ because $\alpha = \mu + \sigma^2/2$. This is an intuitively simple result. It says that the expected future asset price is the current asset price compounded at the expected rate of return.

We now explore the choice of modeling financial problems with either GBM or ABM.

12.6   GEOMETRIC BROWNIAN MOTION OR ARITHMETIC BROWNIAN MOTION?

The key focus of this book is addressing approaches to valuing financial derivatives. Based on materials covered thus far, there appears to be a key decision to be made in our journey. Do we go with GBM or ABM?

In their exploration of this question, Brooks and Brooks (2017) explore the genesis of the decision to pursue GBM over ABM. Historically, the first approach was ABM introduced by Bachelier (1900). In France, during Bachelier's time, option contracts technically could admit negative strike prices. Further, it was only a few decades prior to Bachelier's work that France introduced the Limited Liability Acts on July 24, 1867. For a host of reasons, French companies were slow in adopting a limited liability charter.[9] Thus, it appears possible that Bachelier addressed options on company stock that had unlimited liability or a potential for negative prices. Although we will never know if these features motivated Bachelier to pursue ABM over GBM, it is an interesting artifact.

Osborne (1959) provided one of the original analyses of continuously compounded stock returns. Based on the observation that for percentage changes less than 15% there is not a material difference between modeling with the normal or lognormal distributions, he opted for the lognormal distribution based on the lognormal appearance of the *cross-sectional* distribution of stock prices.[10] Sprenkle (1961) argued in favor of the lognormal distribution due to it having no chance of negative values. Alexander (1961) raised some concerns regarding this choice due to time series financial data exhibiting fat tails.

Samuelson (1965) seems to have created the phrase "geometric Brownian motion." He opted for GBM over ABM primarily due to the nonnegativity of the lognormal distribution and some technical problems with Bachelier's (1900) model.[11] Following Samuelson (1965), Black and Scholes (1973) as well as Merton (1973) chose GBM.

Brooks and Brooks (2017) identify several issues related to the choice between modeling the underlying with either the lognormal or normal distribution. We briefly highlight a few here:

- **Positive values, $S_t > 0$.** With GBM, there is a zero probability of the asset price being zero in the future. ABM solves this problem. Based on US Census data, approximately 0.7% of businesses file for some form of bankruptcy each year. Thus, some percentage of these bankruptcy filing companies result in equity prices being permanently zero. Note there is a fundamental difference between legal bankruptcy, such as a court filing, and economic bankruptcy, such as the stock being deemed worthless. Many legal bankruptcies do not result in economic bankruptcy.

- **Portfolios, $\Pi_t = \sum_{i=1}^{n} N_i S_{i,t}$.** With GBM, a simple portfolio of assets follows no known statistical distribution. Thus, internal coherence of risk measures is not feasible due to the lack of the additive property of the lognormal distribution. ABM solves this problem.

- **Time-series independence.** With GBM, changes in the asset returns are assumed to be independent of the asset price itself. ABM suffers from dependence particularly over long periods of time. For example, if a stock price doubles in value, the absolute volatility of price changes does not automatically double with arithmetic Brownian motion.

- **Factor models,** $S_t = f(x_1, x_2, \ldots, x_n)$. An asset can have multiple risk factors. ABM can easily handle multiple factors, whereas with GBM it is a difficult challenge. Mathematically, the two approaches can be expressed as

$$dS_t = \mu(S, t)dt + \sum_{i=1}^{n} \sigma_{\$,i}(t)dW_i \text{ (ABM) and}$$

$$dS_t = \mu(S, t)dt + \sum_{i=1}^{n} \sigma_i(t)x_i dW_i. \text{ (GBM)}$$

Clearly, the ABM version results in the terminal distributions being normally distributed, whereas the GBM version follows no known tractable distribution.

- **Evidence.** Finance is an empirical science; hence, the model choice depends heavily on financial market behavior. For example, asset prices often behave significantly different in other countries when compared to the US. Thus, access to alternative frameworks is justified.

- **Risk.** Risk can be represented based on relative measures (%) or absolute measures ($ or other currency). Stock splits require ABM to manually adjust absolute volatility, but there is a leverage effect that is implicitly handled by ABM. The leverage effect is the intuitive observation that as a stock price declines it becomes more highly leveraged. Thus, by definition the stock price would be expected to be more volatile in a relative sense. GBM automatically adjusts for stock splits, but the leverage effect is ignored.

- **Extreme volatilities.** ABM can handle extreme volatilities better than GBM. For example, with simulations, the lognormal distribution is numerically unstable for high volatilities. In practice, option volatilities in excess of 100% are common for some options. For example, relative volatilities have been found to be above 1,000% for electricity options. Shocking, indeed!

- **Binomial convergence.** GBM can be derived from a multiplicative binomial process. ABM implies an additive binomial process. Both lattices can be modeled to converge to the appropriate terminal distributions.

- **Homogeneity of degree 1.** GBM-based models typically are homogeneous of degree 1, whereas ABM-based models are not.[12] For example, within an ABM framework stock splits will result in option pricing models that do not properly adjust. Scaling absolute volatility, however, can easily solve this problem.

- **Geometric drift.** Both ABM and GBM can handle geometric drift.

Ultimately, the choice between ABM-based models or GBM-based models should be based on which model provides the most useful information for improving the financial decision-making process. The better approach for one type of financial process may not be the better approach for another. Thus, there is good reason to study both.

12.7 RECAP AND PREVIEW

In this chapter, we took what we have already learned about stochastic processes and adapted it to the case of an asset. We showed how certain properties of the asset's return behavior are derived out of the model. We showed how to obtain the formula for the future asset price in terms of a base price and information on what happened in the intervening period. We also examined and compared both arithmetic Brownian motion and GBM.

In the following chapter, we shall do what we have been alluding to many times: derive the Black-Scholes-Merton model.

QUESTIONS AND PROBLEMS

1 Geometric Brownian motion with geometric drift is often represented in one of two ways:

$$dS_t = \alpha S_t dt + \sigma S_t dW_t \text{ and}$$

$$dS_t = (\mu + \sigma^2/2)S_t dt + \sigma S_t dW_t.$$

In the finance context, explain the economic relationship between α and μ.

2 Mathematically, there are significant differences between GBM with geometric drift and ABM with geometric drift. In early empirical work on stock prices, Osborne (1959) asserted, "Percentage changes of less than 15 per cent, expressed as fractions from unity, are very nearly natural logarithms of the same ratio" (p. 146). Identify five important issues related to the choice of GBM or ABM.

3 Volatility is often reported as a standard deviation, σ. If we assume the standard deviation itself follows GBM with geometric drift, what is the stochastic process of the variance as well as its mean and variance?

4 Assuming S_t follows geometric Brownian motion, explain the relationship between $E[\ln(S_T/S_0)]$ and $\ln[E(S_T)/S_0]$.

5 Based on a historical analysis of a tech company, the annualized average arithmetic return was 29.4% and the annualized average geometric return was 25.95%. The annualized standard deviation of arithmetic returns was 26.2%. Based on the analysis contained in this chapter, how can you explain the relationship between these two annualized returns?

NOTES

1. Profits are simply revenues minus expenses. An example of a calendar spread is the difference between the wholesale price of a good purchased in June and the retail price of the same good sold in December. An example of a product spread is the cost of crude oil purchased by a refinery and the subsequent price of refined products sold.
2. A linear transformation of a normally distributed variable x to a variable y by multiplying x by a constant and adding a constant preserves the normality.
3. Technically, we may consider using Δt in this explanation rather than dt. Given the context here of diffusion processes, we simply use dt.
4. In a lognormal distribution, the log return is normally distributed.
5. Here we use dS to mean changes over short time periods and ΔS to mean long time period.
6. Recall we use $ symbol to emphasize the different unit of measure with ABM.
7. The proof of this result is covered in Chapter 11. See Section 11.5.1.

8. The reason we know this is the solution is that the derivative of the natural log function is $(1/y_t)$ times the derivative of y_t with respect to W_t.

9. See, for example, Antoin E. Murphy, "Corporate Ownership in France: The Importance of History," NBER, Working Paper 10716, 2004, http://www.nber.org/papers/w10716.

10. A cross-sectional distribution is taken at a point in time but examines different instruments, such as stock prices. Time series distribution is taken across time for selected instruments, again such as stock prices.

11. Without limited liability, a call option can be higher than the corresponding stock price because the call enjoys limited liability, whereas the underlying stock does not.

12. Homogeneity is covered in detail in Chapter 17 but mentioned here just to have an exhaustive list of the issues related to GBM and ABM.

CHAPTER 13

Deriving the Black-Scholes-Merton Model

From the basic principles associated with the standard geometric Brownian motion stochastic process used in modeling asset prices and with an understanding of Itô's lemma, we can now derive the Black-Scholes-Merton (Black and Scholes 1973; Merton 1973b) model for pricing European options. Let the asset price follow the standard lognormal diffusion process given by the stochastic differential equation known as geometric Brownian motion,

$$\frac{dS_t}{S_t} = \alpha dt + \sigma dW_t, \tag{13.1}$$

where dS_t is the change in the asset price per unit of time dt, S_t is the asset price at time t, α is the drift or expected rate of return on the asset, σ is the volatility of the return on the asset, and dW_t is the Wiener process that represents the uncertainty. Equation (13.1) describes the rate of return on the asset. Recall that $dW_t = \varepsilon_t \sqrt{dt}$, where ε_t is a standard normal random variable (mean 0, variance 1) and that the key properties of dW_t are that $E(dW_t) = 0$, $\text{var}(dW_t) = dt$, and that $dW_t^2 = dt$.

13.1 DERIVATION OF THE EUROPEAN CALL OPTION PRICING FORMULA

Consider a European call option with exercise price X. The option price is assumed to be a function of only two variables: the asset price and time.[1] Thus, we write the option price function in its general form as $c(S_t, t)$ and more loosely as c_t, which denotes the option price at time t and where the option price is implicitly a function of the asset price at time t, S_t. The option's time to expiration is $\tau = T - t$. At expiration, the option price is $c_T = \max(0, S_T - X)$.

Let us construct a hedge portfolio consisting of h units of the asset and one short call option. The value of this portfolio at time t is

$$V_t = hS_t - c_t. \tag{13.2}$$

The value h could be subscripted with a t, as its value is set and known at t. We hold h_t units of the asset per one short call. As we move forward, the value of our holdings changes due to changes in the value of the asset and the call and the increment of time. We are still holding h_t units of the asset and one short call. As noted later, we shall have to reset h_t

according to the new market conditions, such as at time $t + dt$ to obtain h_{t+dt}. Thus, h_t is set at t and remains constant until we rebalance, which occurs after we determine how our portfolio has performed. Consequently, for the purpose of this derivation, we can treat h_t as a constant and denote it as h.

Our objective is to make this portfolio risk free. Because the call price is a function of S_t, which follows the stochastic differential equation in (13.1), and t, we use Itô's lemma to express the call price change in terms of the changes in S_t and t,[2]

$$dc = \frac{\partial c}{\partial S} dS + \frac{\partial c}{\partial t} dt + \frac{1}{2} \frac{\partial^2 c}{\partial S^2} dS^2. \tag{13.3}$$

Given Equation (13.1) and the properties of the Wiener process, dW_t, we know that $dS^2 = S^2 \sigma^2 dt$. Henceforth, we shall make this substitution whenever we use the result from Itô's lemma.

We know that the value of the portfolio, Equation (13.2), is a function of the asset price and the option price. Hence, using the total differential rule, we can express the change in the value of the portfolio as

$$dV = \frac{\partial V}{\partial S} dS + \frac{\partial V}{\partial c} dc. \tag{13.4}$$

From Equation (13.2), we can determine the partial derivatives as

$$\frac{\partial V}{\partial S} = h$$

$$\frac{\partial V}{\partial c} = -1. \tag{13.5}$$

Thus, the change in the value of the portfolio is

$$dV = h dS - dc. \tag{13.6}$$

Substituting from Itô's lemma, Equation (13.3), for dc in (13.6) gives

$$dV_t = h dS - \left(\frac{\partial c}{\partial S} dS + \frac{\partial c}{\partial t} dt + \frac{1}{2} \frac{\partial^2 c}{\partial S^2} S^2 \sigma^2 dt \right). \tag{13.7}$$

Equation (13.7) is a stochastic partial differential equation that defines the change in the value of the portfolio in terms of several expressions. Observe that the stochastic terms are the dS terms but note that they amount to the simple expression $h dS - (\partial c / \partial S) dS$. We are free to set h to whatever we want as long as its value can be determined before the asset price changes. It should be apparent that by setting h to $\partial c / \partial S$, the two dS terms cancel, leaving the following expression for the change in the value of the portfolio:

$$dV = - \left(\frac{\partial c}{\partial t} dt + \frac{1}{2} \frac{\partial^2 c}{\partial S^2} S^2 \sigma^2 dt \right). \tag{13.8}$$

Notice in Equation (13.8) that there are no stochastic terms, so this portfolio is perfectly hedged. Thus, to avoid arbitrage the portfolio value, V, must increase at the risk-free rate. This specification is made by the requirement that

$$dV = Vr_c dt. \tag{13.9}$$

We now substitute $hS - c = (\partial c/\partial S)S - c$ for V in Equation (13.9) to obtain

$$dV = \left(\frac{\partial c}{\partial S}S - c\right)r_c dt. \tag{13.10}$$

We now have two expressions for dV, Equations (13.8) and (13.10). Equating these and canceling the dt terms gives

$$r_c S\frac{\partial c}{\partial S} + \frac{\partial c}{\partial t} + \frac{1}{2}\frac{\partial^2 c}{\partial S^2}\sigma^2 S^2 = r_c c. \tag{13.11}$$

Equation (13.11) is a partial differential equation of which the unique solution, c, is the call option price. The unique solution is determined by the boundary conditions, which refers to the value of c at the option expiration. In the case of a European call, the condition is that $c_T = \max(0, S_T - X)$.[3] Equation (13.11), or a variation of it, appears frequently in derivative pricing theory and is often termed the *Black-Scholes-Merton partial differential equation*.

Equation (13.11), formally a second-order parabolic partial differential equation, is actually a variation of a famous equation in physics that models the transfer of heat. The solution procedure is well known to physicists and one version of it is presented later in this chapter. Black and Scholes (1973) first obtained the solution by taking advantage of previous research on option pricing that produced an idea of what the solution would look like. Bachelier (1900) had derived the solution under arithmetic Brownian motion.[4] Several other researchers derived similar solutions under diverse assumptions, but the key complication was that none of this earlier work on option pricing used the risk-free hedge.[5] As such, they all require the use of a discount rate that reflected the risk premium that an investor would demand to hold an option. Risk premia are not observable in the market, and they require that the model incorporate assumptions about the risk preferences of investors. These requirements are fairly standard in general equilibrium models, such as the capital asset pricing model and most models of macroeconomic equilibrium. These models typically aggregate the demand and supply of assets and produce an equilibrium price, as well as the optimal holdings of assets by individuals. These models are powerful but demanding ones that impose many restrictions on their users and can be very difficult to empirically estimate and to implement in practice.

Partial equilibrium models, however, derive prices and certain other pieces of information only for sectors within a market or economy. For example, a partial equilibrium option pricing model derives the price of an option as well as the hedge ratio and the option sensitivities to changes in its inputs. It does not, however, tell us how many, if any, options a market participant would hold. Nonetheless, partial equilibrium models are valuable for pricing and trading instruments such as options, keeping in mind, however, that they apply only to this sector of the market.

The early research on option pricing produced partial equilibrium models, but these models imposed certain general equilibrium requirements, meaning that we needed to know something about investor risk preferences and the rates at which they would discount risky future cash flows. These requirements are quite strong. Hence, these models never gained much acceptance.

The key insight provided by Black, Scholes, and Merton of the risk-free hedge was a giant leap forward in economic and financial theory. Interestingly, there is some indication that a mathematically oriented practitioner, Ed Thorp, working with Sheen Kassouf (1967), developed the model before Black, Scholes, and Merton did. Thorp and Kassouf, however, did the wise thing and kept it to themselves, apparently making a considerable amount of money. Black, Scholes, and Merton did credit Thorp and Kassouf for the idea of risk-free hedging.[6] For a history of the development of the ideas behind the model, see Szpiro (2011).

Despite their limitations, each of the option pricing formulas that preceded Black-Scholes-Merton resembles the correct solution as found by Black, Scholes, and Merton. Black and Scholes originally derived the equation by using the capital asset pricing model, which provides the equation for the expected return on a risky asset as a function of its risk. Though Black had a PhD in applied mathematics from Harvard, he was not aware that the differential equation was the heat transfer equation. It is rumored that someone else noticed and showed him how to obtain the solution. At approximately the same time, Robert Merton (1973b) wrote a paper developing the model with virtually the full mathematical solution. It is not clear whether Merton or Black and Scholes came up with the result first, but Black and Scholes have without doubt received the most credit, because the model is more often known as the Black-Scholes model.[7] The model should probably be called the Black-Scholes-Merton model, and we shall do so, but Merton has never really received full credit. Black died in 1995, and in 1997, Merton and Scholes received the Nobel Prize in Economics for this work.

The solution to the equation is the Black-Scholes-Merton option pricing model,

$$c = SN(d_1) - Xe^{-r_c\tau}N(d_2),\qquad(13.12)$$

where

$$d_1 = \frac{\ln(S/X) + (r_c + \sigma^2/2)\,\tau}{\sigma\sqrt{\tau}}$$

$$d_2 = \frac{\ln(S/X) + (r_c - \sigma^2/2)\,\tau}{\sigma\sqrt{\tau}} = d_1 - \sigma\sqrt{\tau}.\qquad(13.13)$$

The value $N(d_i)$, $i = 1, 2$, is the cumulative normal probability,

$$N(d_i) = \int_{-\infty}^{d_i} \frac{1}{\sqrt{2\pi}} e^{-x^2/2} dx.\qquad(13.14)$$

Note that the expected return on the asset is not a factor in pricing the option. All of the information needed to capture investor risk preferences is embedded in the price of the underlying, S. If we attempted to incorporate risk preferences by, say, using a discount rate greater than the risk-free rate, we would be double counting, that is, penalizing the option twice for the effect of risk.

Of the five inputs into the model, four are easily observable. We know the price of the asset, S. The option contract specifies the exercise price, X, and when the option expires. The time to expiration, τ, is simply the number of remaining days divided by 365. The risk-free rate, r_c, would be the continuously compounded rate on a risk-free security that matures at the option expiration. The only remaining factor is the volatility, and it is indeed an unobservable factor. Investors will have different opinions on the volatility, and this diversity of opinions will generate trading.[8]

13.2 THE EUROPEAN PUT OPTION PRICING FORMULA

The model for pricing a European put is easily derived from put-call parity. We know that

$$p = c - S + Xe^{-r_c\tau}. \tag{13.15}$$

We can then substitute into Equation (13.15) for c from the Black-Scholes-Merton formula, Equation (13.12) to obtain

$$p = Xe^{-r_c\tau}N(-d_2) - SN(-d_1). \tag{13.16}$$

Alternatively, the put option pricing model can be derived by setting up a risk-free hedge involving the holding of h units of the asset while being long one put. The derivation would proceed exactly as in the call option pricing model derivation, except that the boundary condition would be $p_T = \max(0, X - S_T)$, and the hedge ratio h would be found to be $\partial p / \partial S$. From put-call parity, Equation (13.15), we can see that $\partial p / \partial S = \partial c / \partial S - 1$.

Interestingly, Rubinstein (1991b) has shown that the Black-Scholes-Merton call and put models can be written compactly as

$$v = \iota SN(\iota d) - \iota Xe^{-r_c\tau}N\left[\iota\left(d - \sigma\sqrt{\tau}\right)\right]$$

where

$$d = \frac{\ln\left(\frac{S}{X}\right) + \left(r_c + \frac{\sigma^2}{2}\right)\tau}{\sigma\sqrt{\tau}}$$

$$\iota = \begin{cases} +1 \text{ for call} \\ -1 \text{ for put,} \end{cases} \tag{13.17}$$

where v is the value of the option, whether a call or a put.

Once the Black-Scholes-Merton equation is found, one can determine the partial derivatives, which can then be substituted back into the partial differential equation to verify that the solution is correct. In Chapter 14, we examine the comparative statics of the Black-Scholes-Merton option pricing model and use these derivatives to verify the solution.

13.3 DERIVING THE BLACK-SCHOLES-MERTON MODEL AS AN EXPECTED VALUE

Another approach to deriving the Black-Scholes-Merton model is to treat the option as if it were valued in a world of risk-neutral investors. The Black-Scholes-Merton price is then found as the present value of the expected call price at expiration. We term this solution methodology as the *expectations approach.*

In such a world, investors are not concerned about risk, and the risk-free rate is an appropriate discount rate. Thus,

$$c = e^{-r_c\tau}E_t\left(c_T\right). \tag{13.18}$$

The use of risk neutrality may appear to be unpalatable, but it is entirely appropriate in option pricing, as we discussed in previous chapters. We have already noted that the expected return on the asset is not a factor in the Black-Scholes-Merton model. The risk preferences of investors are reflected in an asset's expected return; thus, the Black-Scholes-Merton model does not require an adjustment for investor risk preferences. We are not assuming that investors are actually risk neutral. On the contrary, virtually all investors are risk averse, but their feelings about risk are reflected in the asset price. Risk is priced in the market for the underlying asset, and the price of risk, what we call the *risk premium*, shows up in the price of the asset. All other things equal, the price of the asset is lower if there is more risk or if the degree of risk aversion of investors increases, but the risk aversion is embedded in the asset price, and we do not have to take it into account in pricing the option. Of course, these are points we have already made, but they must be recalled and accepted to proceed further.

The process of pricing an option via risk neutrality amounts to taking discounted expectations under an *equivalent martingale probability distribution.* The price of any asset that trades in a no-arbitrage world can be obtained as the discounted expectation of its future payoff, where the probabilities that determine its expectation are called *equivalent martingale probabilities* and the discount rate is the risk-free rate. A martingale is a stochastic process in which the expectation of an asset's future value is its current value. As has been shown elsewhere, a world of no-arbitrage is equivalent to a world in which the asset price can be obtained using the probabilities that would exist if investors were risk neutral; discounting then proceeds at the risk-free rate. In option pricing theory, we should be confident that no arbitrage opportunities exist and we can take investors' feelings about risk as given and reflected already in the asset price. No further adjustment for risk is required and indeed any further adjustment would be a double penalty. Let us state more formally that our option price is

$$c = e^{-r_c\tau}E_t^Q\left(c_T\right), \tag{13.19}$$

which means that the expectation of the option payoff, c_T, is taken at time t using equivalent martingale probabilities and discounted to the present at the risk-free rate. Now let us proceed to evaluate this expression.

Recall from our model that $dW = \varepsilon\sqrt{dt}$.[9] The time increment over the life of the option is $\Delta t = \tau$. Now let us define the stochastic increment $\Delta W = \varepsilon^*\sqrt{\Delta t} = \varepsilon^*\sqrt{\tau}$. The term ε is not the instantaneous increment. It represents the accumulated value of all the instantaneous increments. Recall the original stochastic process, $dS/S = \alpha dt + \sigma dW$. Given that ε^* is normally distributed and is the source of all the uncertainty, we should be able to use normal probability theory to evaluate the expectation. We shall need to express the stochastic process in such a manner that the return on the asset is normally distributed. In geometric Brownian motion, the log return on the asset is normally distributed, so we will need the stochastic process for the log of the asset return.

Define $dS + S$ as the asset price at an instant, dt, later. Thus, we can write the stochastic process as $dS + S = S[1 + \alpha dt + \sigma dW]$. Working with the term in brackets, note that we can write it in the following way:

$$1 + \alpha dt + \sigma dW = 1 + \left[(\alpha - \sigma^2/2)\, dt + \sigma dW + ((\alpha - \sigma^2/2)\, dt + \sigma dW)^2/2 \right]. \quad (13.20)$$

By multiplying out the terms on the right-hand side, it is easy to verify that the statement is true. Now define $\mu = \alpha - \sigma^2/2$ and the statement can be written as $1 + \left[\mu dt + \sigma dW + (\mu dt + \sigma dW)^2/2 \right]$, which is equivalent to a second-order Taylor series expansion of the function $e^{\mu dt + \sigma dW}$. A second-order expansion is sufficient because all terms higher than second order will involve powers of dt greater than 1.0.

Now we can write out the stochastic process as $dS + S = Se^{\mu dt + \sigma dW}$. Dividing by S we obtain $dS/S + 1 = e^{\mu dt + \sigma dW}$. Taking natural logs, we have the stochastic process of the log return on the asset,

$$\ln\left(\frac{dS + S}{S} \right) = \mu dt + \sigma dW. \quad (13.21)$$

This result confirms that the log return is normally distributed with mean μ and volatility σ. For our purposes here, we use the following version,

$$dS + S = Se^{\mu dt + \sigma dW}. \quad (13.22)$$

Noting that the time increment until expiration is τ, we can state that the asset price at expiration is $S_{t+\tau} = S_T$ and the stochastic process for W is $\Delta W_T = \varepsilon^* \sqrt{\tau}$. Thus,

$$S_T = S_t e^{\mu\tau + \sigma\varepsilon^* \sqrt{\tau}}, \quad (13.23)$$

with ΔW_T normally distributed with mean zero and variance τ, per the central limit theorem.[10]

Our objective is to evaluate the expectation of c_T. By definition,

$$E_t^Q(c_T) = \int_X^\infty (S_T - X)\, f(S_T)\, dS_T, \quad (13.24)$$

where $f(S_T)$ is the density for S_T. Note that we need integrate only over the interval (X,∞) because $c_T = 0$ for $0 \le S_T \le X$. Thus,

$$E_t^Q(c_T) = \int_X^\infty S_T f(S_T)\, dS_T - \int_X^\infty X f(S_T)\, dS_T$$

$$= \int_X^\infty S_T f(S_T)\, dS_T - X \Pr^Q(S_T > X), \tag{13.25}$$

where $\Pr^Q$ is the probability under the risk neutral or equivalent martingale measure that $S_T > X$. The overall expression is the expected value of S_T given that $S_T > X$ minus the expected payout of the exercise price, that is, the exercise price times the probability that the option will be exercised. That is, for the entire expression,

$$E_t^Q(c_T) = E_t^Q(S_T|S_T > X)\Pr^Q(S_T > X) - X\Pr^Q(S_T > X). \tag{13.26}$$

Starting with the simpler case, let us first evaluate the second term on the right-hand side. By definition,

$$\Pr^Q(S_T > X) = \Pr^Q\left(Se^{\mu\tau+\sigma\varepsilon^*\sqrt{\tau}} > X\right). \tag{13.27}$$

Note that $Se^{\mu\tau+\sigma\varepsilon^*\sqrt{\tau}} > X$ is equivalent to $\ln(S/X) + \mu\tau + \sigma\varepsilon^*\sqrt{\tau} > 0$, or

$$\varepsilon^* > -\frac{[\ln(S/X) + \mu\tau]}{\sigma\sqrt{\tau}}. \tag{13.28}$$

Recall that α is the expected simple return on the asset and μ is the expected logarithmic return on the asset where $\mu = \alpha - \sigma^2/2$. Under the equivalent martingale/risk neutrality approach, we can let $\alpha = r_c$ so that $\mu = r_c - \sigma^2/2$ so that

$$\varepsilon^* > -\frac{[\ln(S/X) + (r_c - \sigma^2/2)\,\tau]}{\sigma\sqrt{\tau}}. \tag{13.29}$$

You should recognize this as $\varepsilon^* > -d_2$ or $\varepsilon^* < d_2$. So our second term in (13.25) is

$$\varepsilon^* > -\frac{[\ln(S/X) + (r_c - \sigma^2/2)\tau]}{\sigma\sqrt{\tau}} \Rightarrow N(d_2). \tag{13.30}$$

Now we look at the first term on the right-hand side of (13.25). It can be written as

$$\int_X^\infty S_T f(S_T)\, dS_T = \int_X^\infty Se^{\mu\tau+\sigma\varepsilon^*\sqrt{\tau}} f(S_T)\, dS_T$$

$$= Se^{\mu\tau}\int_X^\infty e^{\sigma\varepsilon^*\sqrt{\tau}} f(S_T)\, dS_T. \tag{13.31}$$

We showed that $S_T > X$ implies that $\varepsilon^* > d_2$ and $\varepsilon^* > -d_2$ so we can change the variable of integration giving us the equivalent statement,

$$Se^{\mu\tau} \int_{-d_2}^{\infty} e^{\sigma\varepsilon^*\sqrt{\tau}} f(\varepsilon^*) \, d\varepsilon^*. \tag{13.32}$$

Because ε^* is normally distributed, we can substitute the density for the normal distribution,

$$Se^{\mu\tau} \int_{-d_2}^{\infty} e^{\sigma\varepsilon^*\sqrt{\tau}} \left(e^{-\varepsilon^{*2}/2} / \sqrt{2\pi} \right) d\varepsilon^*$$

$$= Se^{\mu\tau} \int_{-d_2}^{\infty} \left(e^{-\varepsilon^{*2}/2 + \sigma\varepsilon^*\sqrt{\tau}} / \sqrt{2\pi} \right) d\varepsilon^*. \tag{13.33}$$

By completing the square in the exponent, we obtain

$$Se^{\mu\tau} \int_{-d_2}^{\infty} \left(e^{-\left(\varepsilon^* - \sigma\sqrt{\tau}\right)^2/2 + \sigma^2\tau/2} / \sqrt{2\pi} \right) d\varepsilon^*. \tag{13.34}$$

Now we make a change of variables. Let $y = \varepsilon^* - \sigma\sqrt{\tau}$ so that $\varepsilon^* = \sigma\sqrt{\tau} + y$. This changes the lower limit of integration. Formerly, we were interested in the value of ε^* over the range $(-d_2, \infty)$. Because $d_2 = d_1 - \sigma\sqrt{\tau}$, this means that $\varepsilon^* > -\left(d_1 - \sigma\sqrt{\tau}\right)$ or $\varepsilon^* > \sigma\sqrt{\tau} - d_1$. Substituting $y + \sigma\sqrt{\tau}$ for ε^* means that $y + \sigma\sqrt{\tau} > \sigma\sqrt{\tau} - d_1$; thus, $y < d_1$. Now our expression can be written as

$$Se^{\mu\tau + \sigma^2\tau/2} \int_{-\infty}^{d_1} \left(e^{-y^2/2} / \sqrt{2\pi} \right) dy. \tag{13.35}$$

The integral is simply the value $N(d_1)$. Because $\mu + \sigma^2/2 = r_c$, our first term can be written as $Se^{r_c\tau} N(d_1)$. Thus,

$$E_T^Q(c_T) = Se^{r_c\tau} N(d_1) - XN(d_2). \tag{13.36}$$

And the value of the call today is the present value of $E_T^Q[c_T]$ obtained using the factor $e^{-r_c\tau}$. Thus, we obtain the model,

$$c = SN(d_1) - Xe^{-r_c\tau}N(d_2), \tag{13.37}$$

which is the Black-Scholes-Merton formula.

This derivation makes it somewhat easier to see the proper interpretation of the normal probabilities, $N(d_1)$ and $N(d_2)$. $N(d_2)$ is the probability that the call will be exercised provided one assumes that the asset expected return is the risk-free rate. Of course, any risky asset will command a risk premium so the actual drift is α, which is higher than r by the risk premium. Consequently, the actual probability of exercise is higher than $N(d_2)$. $N(d_1)$, however, does not lend itself to a simple probability interpretation. $SN(d_1)$ is interpreted as the expected asset price at expiration conditional on the asset price exceeding the exercise price, times the probability that the asset price exceeds the exercise price, discounted at the risk-free rate.

13.4 DERIVING THE BLACK-SCHOLES-MERTON MODEL AS THE SOLUTION OF A PARTIAL DIFFERENTIAL EQUATION

The mathematical details that lead from the partial differential equation (PDE), Equation (13.11), to the Black-Scholes-Merton formula, Equation (13.12), are seldom seen in print. One exception is Kutner (1988). This section is adapted from that article. There are a few other sources of derivations, which are listed in the references, that take a slightly different approach. We refer to this general solution methodology as the PDE approach.

Solving a differential equation frequently involves guessing at the form of the solution. Some clues as to the correct form were provided in the research on options that preceded Black, Scholes, and Merton. Assuming we do not yet know the solution, let us propose that it can be written as $c_t = e^{-r_c \tau} y$ where $y = f(u, q)$. You can see here that y seems to play the role of the payoff, which is a function of two variables. The variables u and q are proposed to be of the form,[11]

$$u = (2/\sigma^2)(r_c - \sigma^2/2)[\ln(S/X) - (r_c - \sigma^2/2)(-\tau)]$$

$$q = -(2/\sigma^2)(r_c - \sigma^2/2)^2(-\tau) . \tag{13.38}$$

The partials of $c_t = e^{-r_c \tau} y(u, q)$ are found as

$$\frac{\partial c}{\partial S} = e^{-r_c \tau} y_u (2/\sigma^2)(r_c - \sigma^2/2)/S$$

$$\frac{\partial c}{\partial t} = e^{-r_c \tau}\left[-y_u(2/\sigma^2)(r_c - \sigma^2/2)^2 - y_q(2/\sigma^2)(r_c - \sigma^2/2)^2 + r_c y\right]$$

$$\frac{\partial^2 c}{\partial S^2} = e^{-r_c \tau}\left[y_u(2/\sigma^2)(r_c - \sigma^2/2)(1/S^2) - y_q(2/\sigma^2)^2(r_c - \sigma^2/2)^2(1/S^2)y_{uu}\right], \tag{13.39}$$

where y_u, y_q, and y_{uu} are shorthand notation for $\partial y/\partial u$, $\partial y/\partial t$ and $\partial^2 y/\partial u^2$. Note that $\partial t = -\partial \tau$. Substituting into the Black-Scholes-Merton PDE gives

$$rSe^{-r\tau}y_u(2/\sigma^2)(r_c - \sigma^2/2)/S_t - r_c c_t$$

$$+ e^{-r\tau}\left[y_u(2/\sigma^2)(r_c - \sigma^2/2)^2 - y_q(2/\sigma^2)(r_c - \sigma^2/2)^2 + r_c y\right]$$

$$+ \frac{1}{2}e^{-r\tau}\left[y_u(2/\sigma^2)(r_c - \sigma^2/2)(1/S^2) + (2/\sigma^2)^2(r_c - \sigma^2/2)^2(1/S^2)y_{uu}\right] = 0, \tag{13.40}$$

which simplifies to

$$y_q = -y_{uu}.$$

(13.41)

This is the heat exchange equation of thermodynamics. The solution is well known, and Black and Scholes cite a contemporary math book. If V is a function of x and t and $V_t = KV_{xx}$ for $-\infty < x < \infty$, $t > 0$ and $V = f(x)$ when $t = 0$ for $-\infty < x < \infty$, the solution is of the form

$$V = \frac{1}{\sqrt{\pi}} \int_{-\infty}^{\infty} f\left(x + 2\eta\sqrt{kt}\right) e^{-\eta^2} d\eta.$$

(13.42)

In our problem, V is the option price, x is u, t is q, $k = 1$, and $f(x)$ is the boundary condition. The variable q can equal zero only at expiration ($\tau = 0$). We need to find the function $f(x)$ consistent with the option's expiration value:

$$y(u, 0) = c(S_T, 0) = S_T - X \quad if \ S_T \geq X$$
$$= 0 \qquad\qquad if \ S_T < X.$$

(13.43)

Let

$$y(u, 0) = X\left(e^{u\left(\sigma^2/2\right)/\left(r - \sigma^2/2\right)} - 1\right) \quad if \ u \geq 0$$
$$= 0 \qquad\qquad\qquad\qquad\qquad if \ u < 0.$$

(13.44)

At $\tau = 0$, u will equal $(2/\sigma^2)\left(r_c - \sigma^2/2\right)\ln\left(S_T/X\right)$. So at $\tau = 0$,

$$y(u, 0 | \tau = 0) = X\left(e^{\ln(S_T/X)} - 1\right) = S_T - X \quad if \ u \geq 0$$
$$= 0 \qquad\qquad\qquad\qquad\qquad\qquad if \ u < 0.$$

(13.45)

Thus, $y(u, 0) \mid \tau = 0$ is consistent with $c(S_T, 0)$. A function $f(x)$ that meets this condition is

$$f(u) = X\left(e^{u\left(\sigma^2/2\right)/\left(r - \sigma^2/2\right)} - 1\right) \quad if \ u \geq 0$$
$$= 0 \qquad\qquad\qquad\qquad\qquad if \ u < 0.$$

(13.46)

The solution is, thus,

$$y = \frac{1}{\sqrt{\pi}} \int_{-u/2\sqrt{q}}^{\infty} X\left[e^{(u+2\eta\sqrt{q})\left(\sigma^2/2\right)/\left(r_c - \sigma^2/2\right)} - 1\right] e^{-\eta^2} d\eta.$$

(13.47)

The lower limit was changed from $-\infty$ to $-u/2\sqrt{q}$, because we require the integrand to be nonnegative to get a positive value of y. Thus, the first exponential function must be greater than or equal to one. So the exponent must be restricted to $(u + 2\eta\sqrt{q}) \geq 0$, meaning that

$\eta \geq -u/2\sqrt{q}$. Now let $\eta = a\sqrt{2}/2$. We will have $-\eta^2 = -a^2/2$, $d\eta = da/\sqrt{2}$, and the lower limit of integration will change to $-u/\sqrt{2q}$. Then the equation will become

$$y = \int_{-u/2\sqrt{q}}^{\infty} \frac{1}{\sqrt{\pi}} X \left[e^{\left(u+a\sqrt{2q}\right)\left(\sigma^2/2\right)/\left(r_c-\sigma^2/2\right)} - 1 \right] e^{-a^2/2} da. \tag{13.48}$$

Now note that $-u/\sqrt{2q} = (1/\sigma)\left[\ln(S/X) - \left(r_c - \sigma^2/2\right)(-\tau)\right]/\sqrt{\tau} = -d_2$. Next, we transform the equation back into the equation for c_t by multiplying by $e^{-r_c\tau}$, and then we separate the integrals:

$$c = \int_{-d_2}^{\infty} \frac{1}{\sqrt{2\pi}} X e^{-r\tau} \left[e^{\left(u+2a\sqrt{q}\right)\left(\sigma^2/2\right)/\left(r_c-\sigma^2/2\right)} - 1 \right] e^{-a^2/2} da$$

$$= \int_{-d_2}^{\infty} \frac{1}{\sqrt{2\pi}} X e^{-r\tau} e^{\left[\left(u+2a\sqrt{q}\right)\left(\sigma^2/2\right)/\left(r_c-\sigma^2/2\right)\right]-a^2/2} da - X e^{-r\tau} \int_{-d_2}^{\infty} \frac{1}{\sqrt{2\pi}} e^{-a^2/2} da. \tag{13.49}$$

The first term can be simplified by using the definitions of u and q. After rearranging those definitions, we have

$$u(2/\sigma^2)\left(r_c - \sigma^2/2\right) = \left[\ln(S/X) - \left(r_c - \sigma^2/2\right)(-\tau)\right]$$

$$a\sqrt{2q}(1/\sigma^2)\left(r_c - \sigma^2/2\right) = -a\sigma\sqrt{\tau}. \tag{13.50}$$

Thus, the first term in our solution is

$$X e^{-r_c\tau} e^{\ln(S/X)} \int_{-d_2}^{\infty} \frac{1}{\sqrt{2\pi}} e^{r\tau} e^{-\left[a^2/2-a\sigma\sqrt{\tau}+\sigma^2\tau\right]} da$$

$$= S \int_{-d_2}^{\infty} \frac{1}{\sqrt{2\pi}} e^{-(1/2)\left[a^2-2a\sigma\sqrt{\tau}+\sigma^2\tau\right]} da. \tag{13.51}$$

The integrand is $e^{-(1/2)(a-\sigma\sqrt{\tau})^2}$. Now define $b = a - \sigma\sqrt{\tau}$ so $db = da$ and the lower limit of integration becomes $-d_2 - \sigma\sqrt{\tau}$. Let $d_1 = d_2 + \sigma\sqrt{\tau}$ so we have

$$S \int_{-d_1}^{\infty} \frac{1}{\sqrt{2\pi}} e^{-b^2/2} db, \tag{13.52}$$

which is simply $S[1 - N(-d_1)] = SN(d_1)$.

The second term in the solution is simply $-Xe^{-r_c\tau}\left[1 - N(-d_2)\right] = -Xe^{-r_c\tau}N(d_2)$. Thus,

$$c = SN(d_1) - Xe^{-r_c\tau}N(d_2), \tag{13.53}$$

which is the Black-Scholes-Merton formula.

An alternative route to the solution uses the LaPlace transform on the partial differential equation. This converts it to an ordinary differential equation, for which standard solution techniques can be applied.

Wilmott, Howison, and DeWynne (1995) have an excellent illustration of the steps in converting the Black-Scholes-Merton PDE into the heat transfer equation of physics and then solving that equation to obtain the option pricing formula.

Thus, we have two approaches to arriving at the Black-Scholes-Merton model, the expectations approach and the PDE approach. The Feynman-Kac theorem provides a convenient methodology for moving between these two approaches. Although we have solution methodologies for both the expectations approach as well as the PDE approach with the Black-Scholes-Merton model, there are other more complicated models where one approach may be less complicated than the other. Once, say the PDE is derived based on arbitrage arguments, we can easily transform the problem to taking an expectation via the Feynman-Kac theorem.

The generic univariate Feynman-Kac theorem is briefly introduced here. Recall the Itô process parameters can depend on at most the underlying variable and time, represented here as x and t. Thus, the PDE parameters have the same constraint. The PDE parameters also depend on at most x and t. The three parameters associated with the PDE are expressed as $\mu(x,t)$, $\hat{\mu}(x,t)$, and $\hat{\sigma}(x,t)$. Formally, the Feynman-Kac theorem can be expressed as follows:

Assume a univariate Itô process expressed generically as

$$dx = \mu(x,t)\,dt + \sigma(x,t)\,dz. \tag{13.54}$$

Subject to certain technical conditions, we have the following statement:

$$\mu(x,t)f = \frac{\partial f}{\partial t} + \hat{\mu}(x,t)\frac{\partial f}{\partial x} + \frac{1}{2}\hat{\sigma}^2(x,t)\frac{\partial^2 f}{\partial x^2}, \tag{13.55}$$

subject to

$$f(x,T) = f(x_T), \tag{13.56}$$

if and only if

$$f(x,t) = E_{\hat{\mu},\hat{\sigma}^2}\left[f(x_T)\,e^{-\int_t^T \mu(x_y,y)dy}\right]. \tag{13.57}$$

Thus, the Feynman-Kac theorem provides a straightforward linkage between the expectations approach (Equation (13.57)) and the PDE approach (Equation (13.55) with boundary condition Equation (13.56)) to various option valuation problems, including

the Black-Scholes-Merton model. The discounting $\left(e^{-\int_t^T \mu(x_y,y)dy}\right)$ in Equation (13.57) is based on the parameter on the left-hand side of Equation (13.55), $\mu(x, t)$. In the Black-Scholes-Merton model, it is simply r_c. The expectation is taken based on the remaining two parameters from the PDF. In the Black-Scholes-Merton model, for example, $\hat{\mu}(x, t) = \mu S$ and $\hat{\sigma}(x, t) = \sigma S$.

We now illustrate the application of the Feynman-Kac theorem within our context here. Based on assuming geometric Brownian motion for the underlying asset, we have the following result. The Black-Scholes-Merton PDE of

$$r_c S \frac{\partial c}{\partial S} + \frac{\partial c}{\partial t} + \frac{1}{2} \frac{\partial^2 c}{\partial S^2} \sigma^2 S^2 = r_c c,$$

subject to

$$c(S, T) = \max(0, S_T - X),$$

if and only if

$$c(S, t) = e^{-r_c(T-t)} E_{r_c, \sigma} \left[\max(0, S_T - X) \right].$$

Due to the lognormal distribution assumption, we know

$$S_T = S_t e^{\left(r_c - \frac{\sigma^2}{2} \right)(T-t) + \sigma \sqrt{T-t} \varepsilon}, \tag{13.58}$$

where ε is distributed normal with mean 0 and standard deviation 1.

Based on the previous two sections, we know that both approaches result in the Black-Scholes-Merton model.

13.5 DECOMPOSING THE BLACK-SCHOLES-MERTON MODEL INTO BINARY OPTIONS

Recall in Chapter 7, we introduced binary options, asset-or-nothing options and cash-or-nothing options. We can obtain the value of an asset-or-nothing call by using the method of expectations, as covered in Section 13.4. Recall that we derived the Black-Scholes-Merton model and found that it has two components, $SN(d_1)$ and $Xe^{-r_c \tau} N(d_2)$. The first represents the expected value of the underlying asset at expiration conditional on the value of the underlying asset exceeding the exercise price times the probability that the value of the underlying asset exceeds the exercise price, discounted to the present at the risk-free rate. The second equals the discounted value of a payoff of X at expiration provided that the price of the underlying asset exceeds the exercise price X. The payoff of the first component is precisely that of an asset-or-nothing call, and the payoff of the second is precisely the payoff of a cash-or-nothing call. Therefore, as explained in Chapter 7, a standard European option can be viewed as a long position in an asset-or-nothing call and a short position in a cash-or-nothing call both struck at X and the latter paying X.[12] Similarly, a standard European put can be decomposed into a long position in a cash-or-nothing put and a short position in an asset-or-nothing put.

13.6 BLACK-SCHOLES-MERTON OPTION PRICING WHEN THERE ARE DIVIDENDS

Recall in Chapter 2 that we examined the boundary conditions for option pricing with and without dividends. In almost every situation, we subtracted the present value of the dividends, D_t, from the asset price. We also discussed how this consideration comes into play in the binomial model, as covered in Chapter 7. Whenever we subtract the present value of the dividends over the life of the option, we are assuming that the dividends over the life of the option are known. Hence, when we are interested in pricing an option in a continuous time world, the stochastic process followed by the asset price, Equation (13.1), can apply only to the value of the asset price minus the present value of the dividends. Without going through the derivation again, we will just say that the Black-Scholes-Merton formula for calls and puts requires a small adjustment, which is that the asset price, S_t, must represent the asset price minus the present value of the dividends.

There is one other common adjustment for dividends, which is to assume that dividends are paid continuously and that the aggregate annual rate is expressed as δ, which is interpreted as the continuously compounded annual dividend yield. This adjustment is, more or less, an approximation, because no stock pays dividends continuously, but index options in which the underlying index contains mostly dividend-paying stocks could be reasonably approximated with this method. This is because some stocks in the index pay dividends on one day and some on other days. On a given day if a single stock pays dividends, then the index is paying a dividend. Though there does tend to be some dividend clustering in months and days of the week, there is still a modest degree of dispersion of dividends, and the method is widely used for index options. In that case, the Black-Scholes-Merton model and all of the dividend-adjusted boundary conditions that we covered in Chapter 2 will have $S_t e^{-\delta \tau}$ used instead of S_t.

Finally, we should add that some options are on physical assets that incur storage costs, which are sometimes called holding or carrying costs. For example, oil is very expensive to store. Thus, while oil is being held, there is not only the loss of revenue from interest tied up in the oil, which is captured by the risk-free rate, but there is also an additional cost that represents out-of-pocket expense of storing the oil. In this case, there should be a further adjustment downward of the asset price by the present value of these costs.

We now turn to exploring selected limiting results for the Black-Scholes-Merton model.

13.7 SELECTED BLACK-SCHOLES-MERTON MODEL LIMITING RESULTS

In this section, we explore the Black-Scholes-Merton model for calls when the asset price tends to zero or positive infinity, time tends to zero, volatility tends to zero or to positive infinity, and the exercise price tends to zero. We assume the dividend yield is zero.

From the Black-Scholes-Merton model, we note that d_1 can be written as

$$d_1 = \frac{\ln(S)}{\sigma \sqrt{\tau}} - \frac{\ln(X) - \left(r_c + \sigma^2/2 \right) \tau}{\sigma \sqrt{\tau}}. \tag{13.59}$$

Only the first term on the right-hand side is affected by changes in S. Thus, as S tends to $+\infty$, then that implies ($\Rightarrow$) $\ln(S)$ tends to $+\infty$ (denoted $\ln(S) \to +\infty$), which implies $d_1, d_2 \to +\infty$. Based on the properties of $N(d)$, we have the following result,

$$S \to +\infty \Rightarrow \ln(S) \to +\infty \Rightarrow d_1, d_2 \to +\infty \Rightarrow N(d_1), N(d_2) \to 1.0 \Rightarrow c = S - Xe^{-r_c\tau}.$$
(13.60)

Thus, as the underlying asset price increases, then the option value tends to infinity.

As S tends to zero, $\ln(S)$ tends to $-\infty$, then $d_1, d_2 \to -\infty$. Based on properties of the $N(d)$, we have the following result,

$$S \to 0 \Rightarrow \ln(S) \to -\infty \Rightarrow d_1, d_2 \to -\infty \Rightarrow N(d_1), N(d_2) \to 0.0 \Rightarrow c = 0.$$
(13.61)

In this case, as the underlying asset price decreases, then the option value tends toward zero.

We now turn to the case where time to expiration tends to positive infinity. We note that d_1 can be written as

$$d_1 = \frac{\ln(S/X)}{\sigma\sqrt{\tau}} + \frac{(r_c + \sigma^2/2)\sqrt{\tau}}{\sigma}.$$
(13.62)

The first term on the right-hand side tends toward zero as τ tends to $+\infty$ and the second term on the right-hand side tends toward $+\infty$ as τ tends to $+\infty$. Thus, assuming positive interest rates, $N(d_1) \to 1.0$. Note that the second term of the Black-Scholes-Merton model goes to zero due to the present value factor ($e^{-r_c\tau}$). That is, the present value of the exercise price is zero. Thus,

$$\tau \to +\infty \Rightarrow d_1 \to +\infty, Xe^{-r_c\tau}N(d_2) \to 0 \Rightarrow N(d_1) \to 1.0 \Rightarrow c = S.$$
(13.63)

We now turn to the case where time to maturity tends to zero. In other words, we are approaching very closely to maturity. Recall, for the model to be coherent, then it must converge to $\max(0, S_T - X)$ at expiration. We must address three cases: (1) $S_T > X$, (2) $S_T < X$, and (3) $S_T = X$. When $S_T > X$, we note that d_1 can be written as

$$d_1 = \frac{\ln(S/X)}{\sigma\sqrt{\tau}} + \frac{(r_c + \sigma^2/2)\sqrt{\tau}}{\sigma}.$$
(13.64)

We sketch the results as time to expiration approaches zero for these three cases:

1. In-the-money ($S_T > X$):

$$\tau \to 0 \ \& \ \ln(S/X) > 0 \Rightarrow d_1, d_2 \to +\infty \Rightarrow N(d_1), N(d_2) \to 1.0 \Rightarrow c = S_T - X.$$
(13.65)

2. Out-of-the-money ($S_T < X$):

$$\tau \to 0 \,\& \ln(S/X) < 0 \Rightarrow d_1, d_2 \to -\infty \Rightarrow N(d_1), N(d_2) \to 0.0 \Rightarrow c = 0.$$
(13.66)

3. At-the-money $(S_T = X)$:

$$\tau \to 0 \,\& \ln(S/X) = 0 \Rightarrow d_1, d_2 \to 0 \Rightarrow N(d_1), N(d_2) \to 0.5 \Rightarrow c = 0. \qquad (13.67)$$

Thus, as time to expiration approaches zero, we have the call price converging to $\max(0, S_T - X)$.

We now turn to the case where volatility tends to positive infinity. In this case, we note that d_1 and d_2 can be written as

$$d_1 = \frac{\ln\left(Se^{r_c\tau}/X\right)}{\sigma\sqrt{\tau}} + \frac{\sigma\sqrt{\tau}}{2} \text{ and} \qquad (13.68)$$

$$d_2 = \frac{\ln\left(Se^{r_c\tau}/X\right)}{\sigma\sqrt{\tau}} - \frac{\sigma\sqrt{\tau}}{2}. \qquad (13.69)$$

Thus, as volatility tends to positive infinity, then the first term on the right-hand side tends to zero and we have the following results:

$$\sigma \to +\infty, d_1 \to +\infty, d_2 \to -\infty \Rightarrow N(d_1) \to 1.0, N(d_2) \to 0.0 \Rightarrow c = S. \qquad (13.70)$$

We now turn to the case where volatility tends to zero. In this case, we note that d_1 can be written as

$$d_1 = \frac{\ln\left(Se^{r_c\tau}/X\right)}{\sigma\sqrt{\tau}} + \frac{\sigma\sqrt{\tau}}{2}. \qquad (13.71)$$

For volatility, we must address three cases, (1) $Se^{r_c\tau} > X$, (2) $Se^{r_c\tau} < X$, and (3) $Se^{r_c\tau} = X$. We sketch the results as volatility tends to zero for these three cases. Note that the expected terminal asset value under the risk neutral probability measure has the asset growing at the risk-free rate or $Se^{r_c\tau}$.

1. Expected terminal value in-the-money $(Se^{r_c\tau} > X)$:

$$\sigma \to 0 \,\& \ln\left(Se^{r_c\tau}/X\right) > 0 \Rightarrow d_1, d_2 \to +\infty \Rightarrow N(d_1), N(d_2) \to 1.0 \Rightarrow c = S - Xe^{-r_c\tau}. \qquad (13.72)$$

2. Expected terminal value out-of-the-money $(Se^{r_c\tau} < X)$:

$$\sigma \to 0 \,\& \ln\left(Se^{r_c\tau}/X\right) < 0 \Rightarrow d_1, d_2 \to -\infty \Rightarrow N(d_1), N(d_2) \to 0.0 \Rightarrow c = 0. \qquad (13.73)$$

3. Expected terminal value at-the-money $(Se^{r_c\tau} = X)$:

$$\sigma \to 0 \,\& \ln\left(Se^{r_c\tau}/X\right) = 0 \Rightarrow d_1, d_2 \to 0 \Rightarrow N(d_1), N(d_2) \to 0.5$$
$$\Rightarrow c = S(0.5) - Xe^{-r_c\tau}(0.5) = 0.5e^{-r_c\tau}(Se^{r_c\tau} - X) = 0. \qquad (13.74)$$

Again, as volatility tends to zero, we have the call price converges to the lower boundary.

We now turn to the case where the exercise price tends to zero. In this case, we note that d_1 can be written as

$$d_1 = -\frac{\ln(X)}{\sigma\sqrt{\tau}} + \frac{\ln(S) + \left(r_c + \frac{\sigma^2}{2}\right)\tau}{\sigma\sqrt{\tau}}. \tag{13.75}$$

For the exercise price, we simply note

$$X \to 0 \Rightarrow \ln(X) \to -\infty \Rightarrow d_1, d_2 \to +\infty \Rightarrow N(d_1), N(d_2) \to 1.0 \Rightarrow c = S. \tag{13.76}$$

Thus, as the exercise price approaches zero, then the call price approaches the dividend yield adjusted asset price.

Note that using a similar approach, we can find the limits for puts. Alternatively, we can apply the limiting results with put-call parity and arrive at the same put limits. For completeness, we conclude this chapter by manually computing Black-Scholes-Merton option pricing model values.

13.8 COMPUTING THE BLACK-SCHOLES-MERTON OPTION PRICING MODEL VALUES

We provide a few illustrations of the manual option value computation. The key to manually calculating option values is the standard normal cumulative distribution table given in Chapter 5 (Table 5.1). We will refer back to Table 5.1 during these illustrations.

The dividend-adjusted Black-Scholes-Merton option pricing model for calls and puts can be expressed as

$$c = Se^{-\delta\tau}N(d_1) - Xe^{-r_c\tau}N(d_2), \tag{13.77}$$

$$p = Xe^{-r_c\tau}N(-d_2) - Se^{-\delta\tau}N(-d_1), \tag{13.78}$$

where

$$d_1 = \frac{\ln(S/X) + \left(r_c - \delta + \sigma^2/2\right)\tau}{\sigma\sqrt{\tau}}$$

$$d_2 = d_1 - \sigma\sqrt{\tau}. \tag{13.79}$$

We start with a non-dividend example with an at-the-money call option where the stock price is 50, the volatility is 40%, the interest rate is 2%, and the time to maturity is 0.25. The first step in the calculations requires the computation of d_1 and d_2. Recall at-the-money implies $S = X$, thus we have

$$d_1 = \frac{\ln(S/X) + \left(r_c - \delta + \sigma^2/2\right)\tau}{\sigma\sqrt{\tau}}$$

$$= \frac{\ln(50/50) + \left(0.02 - 0.0 + 0.4^2/2\right)0.25}{0.4\sqrt{0.25}} = \frac{0.025}{0.2} = 0.125$$

$$d_2 = d_1 - \sigma\sqrt{\tau} = 0.125 - 0.4\sqrt{0.25} = -0.075.$$

The second step is to look up the values of $N(d_1)$ and $N(d_2)$ from Table 5.1. Recall the values are given such that the first column provides the value of the first decimal and the first row provides the values of the second and third decimals incrementing by 0.005. Thus, the value for $N(d_1)$ is found by locating the row starting with 0.1. From that row locate the column with a heading of 0.025. At the intersection of this row and column is the value 0.549738 and therefore $N(d_1) = 0.549738$. The value for $N(d_2)$ is found by locating the row starting with −0.0. From that row locate the column with a heading of 0.075. At the intersection of this row and column is the value 0.470107 and therefore $N(d_2) = 0.470107$.

The final step is to substitute these values into the call equation or

$$c = Se^{-\delta\tau}N(d_1) - Xe^{-r_c\tau}N(d_2)$$

$$= 50e^{-0(0.25)}0.549738 - 50e^{-0.02(0.25)}0.470107$$

$$= 27.4869 - 23.3881 = 4.0988.$$

There are several ways to compute the put value. We first note the put-call parity approach. From the symmetry of the standard normal cumulative distribution function, we know that $N(-d) = 1 - N(d)$. Thus, the put equation can be rearranged to one expression of put-call parity or

$$p = Xe^{-r_c\tau}\left(1 - N(d_2)\right) - Se^{-\delta\tau}\left(1 - N(d_1)\right)$$

$$= Xe^{-r_c\tau} - Se^{-\delta\tau} + \left[Se^{-\delta\tau}N(d_1) - Xe^{-r_c\tau}N(d_2)\right]$$

$$= Xe^{-r_c\tau} - Se^{-\delta\tau} + c.$$

With call value calculated, we can quickly compute the put value as

$$p = Xe^{-r_c\tau} - Se^{-\delta\tau} + c$$

$$= 50e^{-0.02(0.25)} - 50e^{-0(0.25)} + 4.0988$$

$$= 49.7506 - 50 + 4.0988 = 3.8494.$$

Alternatively, we can work through the same three steps for computing the call option. The first step in manual calculations for a put requires the computation of $-d_1$ and $-d_2$. Thus, given we have already computed the values of $-d_1$ and $-d_2$, we simply have

$$-d_1 = -\frac{\ln(S/X) + \left(r_c - \delta + \sigma^2/2\right)\tau}{\sigma\sqrt{\tau}} = -0.125$$

$$-d_2 = 0.075.$$

The second step is to look up the values of $N(-d_1)$ and $N(-d_2)$ from Table 5.1. The value for $N(-d_1)$ is found by locating the row starting with -0.1. From that row locate the column with heading of 0.025. At the intersection of this row and column is the value 0.450262 and therefore $N(-d_1) = 0.450262$. The value for $N(d_2)$ is found by locating the row starting with 0.0. From that row locate the column with heading of 0.075. At the intersection of this row and column is the value 0.529893 and therefore $N(d_2) = 0.529893$.

The final step is to substitute these values into the put equation or

$$p = Xe^{-r_c\tau}N(-d_2) - Se^{-\delta\tau}N(-d_1)$$

$$= 50e^{-0.02(0.25)}0.529893 - 50e^{-0(0.25)}0.450262$$

$$= 26.3625 - 22.5131 = 3.8494.$$

Thus, both put values are identical. The key to successfully using Table 5.1 is knowing where to round. The purpose of such a dense table is to avoid significant rounding error with manual calculations. With Table 5.1, we round up or down on the quarters, that is, 0.XX25 and 0.XX75. For example, if $d_1 = 0.3224$, then we round down to $d_1 = 0.320$ and use $N(d_1) = 0.625516$. Alternatively, if $d_1 = 0.3226$, then we round up to $d_1 = 0.325$ and use $N(d_1) = 0.627409$. For most applications, Table 5.1 is granular enough to provide option values without too much estimation error.

Now let us consider the influence of a 2% dividend yield ($\delta = 0.02$) on our calculations given in the previous example. This will enable us to see firsthand the influence of dividends on call and put values. As before, the first step in manual calculations requires the computation of d_1 and d_2 or

$$d_1 = \frac{\ln(S/X) + \left(r_c - \delta + \sigma^2/2\right)\tau}{\sigma\sqrt{\tau}}$$

$$= \frac{\ln(50/50) + \left(0.02 - 0.02 + 0.4^2/2\right)0.25}{0.4\sqrt{0.25}} = 0.1$$

$$d_2 = d_1 - \sigma\sqrt{\tau} = 0.1 - 0.4\sqrt{0.25} = -0.1.$$

The value for $N(d_1)$ is found by locating the row starting with 0.1 and selecting the first value or $N(d_1) = 0.539828$. Similarly, we find the value for $N(d_2) = 0.460172$. The final step is to substitute these values into the call equation or

$$c = Se^{-\delta\tau}N(d_1) - Xe^{-r_c\tau}N(d_2)$$

$$= 50e^{-0.02(0.25)}0.539828 - 50e^{-0.02(0.25)}0.460172$$

$$= 26.8568 - 22.8938 = 3.9630.$$

Applying put-call parity, we have

$$p = Xe^{-r_c\tau} - Se^{-\delta\tau} + c$$
$$= 50e^{-0.02(0.25)} - 50e^{-0.02(0.25)} + 3.9630 = 3.9630.$$

Thus, both call and put values are identical because the underlying growth rate under the equivalent martingale measure is zero. Clearly, we see that the call value declined and the put value increased when the dividend yield was changed from 0% to 2%.

13.9   RECAP AND PREVIEW

In this chapter, we showed how the Black-Scholes-Merton model is derived. We illustrated the construction of the dynamic hedge that eliminates the effect of movements in the underlying asset. With the risk eliminated, the hedge portfolio should earn the risk-free rate. There is one and only one call option value that ensures this will be true. If it is not true, there is an opportunity for a dynamic arbitrage, which we assume cannot exist under these conditions. We also derive the put version of the model.

In addition, we showed how the model can be derived as the discounted expected payoff of the option under the risk-neutral measure. Further, we established the link between the expectations approach and the PDE approach for deriving the Black-Scholes-Merton model. We also showed how the Black-Scholes-Merton model can be decomposed into two models: price asset-or-nothing options and cash-or-nothing options. Hence, a standard European option is—in effect—a combination of a long position in an asset-or-nothing option and short X cash-or-nothing options. The latter are also referred to as *binary options* or *digital options*.

We conclude this chapter by exploring some of the Black-Scholes-Merton model limits as well as illustrating a few calculations of option prices.

In Chapter 14, we show how the derivatives of the model are obtained, and we provide the Greeks, which as you should recall from the binomial model, are the sensitivities to the factors that go into the model.

APPENDIX 13.A

Deriving the Arithmetic Brownian Motion Option Pricing Model

In this appendix, we derive the call and put prices based on arithmetic Brownian motion (ABM) incorporating a dividend yield. We assume ABMGD or

$$dS_t = \alpha S_t dt + \sigma_S dW_t. \tag{13.80}$$

As given in this chapter, we construct a hedge portfolio consisting of h units of the asset and one short call option. The value of this portfolio at time t is

$$V_t = hS_t - c_t. \tag{13.81}$$

Given that we seek to make this portfolio risk free, we use Itô's lemma to express the call price change in terms of the changes in S_t and t,

$$dc = \frac{\partial c}{\partial S}dS + \frac{\partial c}{\partial t}dt + \frac{1}{2}\frac{\partial^2 c}{\partial S^2}dS^2. \tag{13.82}$$

Given the properties of the Wiener process, dW_t, we know that $dS^2 = \sigma_\$^2 dt$ for ABM. We know that the value of the portfolio is a function of the asset price, the dividend yield, and the option price. Hence, using the total differential rule, we can express the change in the value of the portfolio as

$$dV = \frac{\partial V}{\partial S}(dS + \delta S dt) + \frac{\partial V}{\partial c}dc, \tag{13.83}$$

where $\frac{\partial V}{\partial S} = h$ and $\frac{\partial V}{\partial c} = -1$. Note that the dividend yield results in a positive cash flow to the portfolio over time. Thus, the change in the value of the portfolio is

$$dV = h(dS + \delta S dt) - dc. \tag{13.84}$$

Substituting from Itô's lemma for dc gives

$$dV_t = h(dS + \delta S dt) - \left(\frac{\partial c}{\partial S}dS + \frac{\partial c}{\partial t}dt + \frac{1}{2}\frac{\partial^2 c}{\partial S^2}\sigma_\$^2 dt\right), \tag{13.85}$$

which is a stochastic partial differential equation that defines the change in the value of the portfolio in terms of several expressions. Observe that the stochastic terms are the dS terms but note that they amount to the simple expression $hdS - (\partial c/\partial S)\,dS$. As with GBM, we are free to set h to whatever we want as long as its value can be determined before the asset price changes. It should be apparent that by setting h to $\partial c/\partial S$, the two dS terms cancel, leaving the following expression for the change in the value of the portfolio:

$$dV = \frac{\partial c}{\partial t}\delta S dt - \left(\frac{\partial c}{\partial t}dt + \frac{1}{2}\frac{\partial^2 c}{\partial S^2}\sigma_\$^2 dt\right). \tag{13.86}$$

Notice in this case again that there are no stochastic terms, so this portfolio is perfectly hedged. Thus, to avoid arbitrage the portfolio value, V, must increase at the risk-free rate. This specification is made by the requirement that

$$dV = Vr_c dt. \tag{13.87}$$

We now substitute $hS - c = (\partial c/\partial S)\,S - c$ for V to obtain

$$dV = \left(\frac{\partial c}{\partial S}S - c\right)r_c dt. \tag{13.88}$$

We then have two expressions for dV. Equating these two expressions and canceling the dt terms gives

$$(r_c - \delta)\,S\frac{\partial c}{\partial S} + \frac{\partial c}{\partial t} + \frac{1}{2}\frac{\partial^2 c}{\partial S^2}\sigma_\$^2 = r_c c. \tag{13.89}$$

This equation is a partial differential equation of which the solution, c, is the call option price. As before, the solution is subject to the boundary conditions, which refers to the value of c at the option expiration. In the case of a European call, the condition is that $c_T = \max(0, S_T - X)$.

The ABM model is known to be represented as[13]

$$c = e^{-r_c\tau}\left[\left(Se^{(r_c-\delta)\tau} - X\right)N(d_n) + \sigma_\$\sqrt{\frac{e^{2(r_c-\delta)\tau} - 1}{2(r_c - \delta)}}n(d_n)\right], \qquad (13.90)$$

where $N(d_n)$ is the area under the standard cumulative normal distribution up to d_n and $n(d_n)$ is the value for the standard normal probability density function where

$$d_n = \frac{Se^{(r_c-\delta)\tau} - X}{\sigma_\$\sqrt{\frac{e^{2(r_c-\delta)\tau}-1}{2(r_c-\delta)}}}. \qquad (13.91)$$

Note that the put formula follows directly from put-call parity or

$$p = Xe^{-r_c\tau} - Se^{-\delta\tau} + c$$

$$= Xe^{-r_c\tau} - Se^{-\delta\tau} + e^{-r_c\tau}\left[\left(Se^{(r_c-\delta)\tau} - X\right)N(d_n) + \sigma_\$\sqrt{\frac{e^{2(r_c-\delta)\tau} - 1}{2(r_c - \delta)}}n(d_n)\right]$$

$$= e^{-r_c\tau}\left[\left(X - Se^{(r_c-\delta)\tau}\right)N(-d_n) + \sigma_\$\sqrt{\frac{e^{2(r_c-\delta)\tau} - 1}{2(r_c - \delta)}}n(d_n)\right]. \qquad (13.92)$$

We now proceed with illustrating the manual calculations of call and put values based on arithmetic Brownian motion. We again start with a non-dividend-paying example with an at-the-money call option where the stock price is 50, the volatility is 19.967, the interest rate is 2%, and the time to maturity is 0.25. The first step in manual calculations requires the computation of d_n. Recall that at-the-money implies $S = X$, thus we have

$$d_n = \frac{Se^{(r_c-\delta)\tau} - X}{\sigma_\$\sqrt{\frac{e^{2(r_c-\delta)\tau}-1}{2(r_c-\delta)}}} = \frac{50e^{(0.02-0)0.25} - 50}{19.967\sqrt{\frac{e^{2(0.02-0)0.25}-1}{2(0.02-0)}}} = 0.025.$$

The second step is to look up the values of $N(d_n)$ from Table 5.1. As before, the value for $N(d_n)$ is found by locating the row starting with 0.0. From that row locate the column with heading of 0.025. Thus, $N(d_n) = 0.509973$. The final step is to substitute these values into the call equation or

$$c = e^{-r_c\tau}\left[\left(Se^{(r_c-\delta)\tau} - X\right)N(d_n) + \sigma_\$\sqrt{\frac{e^{2(r_c-\delta)\tau} - 1}{2(r_c - \delta)}}n(d_n)\right]$$

$$= e^{-0.02(0.25)}\left[\left(50e^{(0.02-0)0.25} - 50\right)0.025 + 19.967\sqrt{\frac{e^{2(0.02-0)\tau} - 1}{2(0.02 - 0)}}\frac{e^{-0.025^2/2}}{\sqrt{2\pi}}\right] = 4.0988.$$

Put-call parity is the most efficient way to compute the put value when the call value is known. Thus,

$$p = Xe^{-r_c\tau} - Se^{-\delta\tau} + c = 50e^{-0.02(0.25)} - 50e^{-0(0.25)} + 4.0988 = \$3.8495.$$

But for a slight rounding error, these option values are identical to those found in the chapter. The only input difference was a price change volatility of 19.967 rather than a relative volatility of 40%. Notice that 40% of 50 stock is 20. Thus, relative volatility of 40% or a $50 stock implies a one standard deviation move would be $20. This $20 is closely aligned with our assumed $19.967. Though not precise, we see that with a slight reduction to 19.967 we have the same option values. This slight difference is driven by the different distributional assumptions made where the Black-Scholes-Merton model is based on the lognormal distribution of terminal asset prices, whereas the arithmetic Brownian motion model is based on the normal distribution of terminal asset prices.

As in the chapter, let us consider the influence of a 2% dividend yield ($\delta = 0.02$) on our calculations given in the previous example. In this case, d_n or[14]

$$d_n = \frac{Se^{(r_c-\delta)\tau} - X}{\sigma_\$\sqrt{\frac{e^{2(r_c-\delta)\tau}-1}{2(r_c-\delta)}}} = \frac{50e^{(0.02-0.02)0.25} - 50}{19.967\sqrt{\frac{e^{2(0.02-0.02)0.25}-1}{2(0.02-0.02)}}} = 0.0.$$

The value for $N(d_n)$ is 0.5 and thus we have

$$c = e^{-r_c\tau}\left[\left(Se^{(r_c-\delta)\tau} - X\right)N(d_n) + \sigma_\$\sqrt{\frac{e^{2(r_c-\delta)\tau} - 1}{2(r_c - \delta)}}n(d_n)\right]$$

$$= e^{-0.02(0.25)}\left[\left(50e^{(0.02-0.02)0.25} - 50\right)0.025 + 19.967\sqrt{\frac{e^{2(0.02-0.02)\tau} - 1}{2(0.02 - 0.02)}}\frac{e^{-0.0^2/2}}{\sqrt{2\pi}}\right]$$

$$= 3.9630.$$

Applying put-call parity, we have

$$p = Xe^{-r_c\tau} - Se^{-\delta\tau} + c$$

$$= 50e^{-0.02(0.25)} - 50e^{-0.02(0.25)} + 3.9630 = 3.9630.$$

Again, both call and put values are identical because the underlying growth rate under the equivalent martingale measure is zero. Also, we see that the call value declined and the put value increased when the dividend yield was changed from 0% to 2%. Therefore, the Black-Scholes-Merton model and arithmetic Brownian motion model have similar properties.

QUESTIONS AND PROBLEMS

1 Assuming a stock price is modeled with arithmetic Brownian motion with geometric drift, $dS_t = \alpha S_t dt + \sigma_\$ dW_t$, derive the partial differential equation based on the hedging approach for call options given in this chapter.

2 Assuming a stock price is modeled with geometric Brownian motion with geometric drift, $dS_t = \alpha S_t dt + \sigma S_t dW_t$, derive the partial differential equation based on the hedging approach for put options given in this chapter. Do not rely on put-call parity.

3 Prove Equation (13.20) is true:

$$1 + \alpha dt + \sigma dW = 1 + \left[(\alpha - \sigma^2/2)\, dt + \sigma dW + ((\alpha - \sigma^2/2)\, dt + \sigma dW)^2/2 \right].$$

4 Suppose a particular asset price follows geometric Brownian motion or $dS_t = \mu S_t dt + \sigma S_t dW_t$. Prove the probability that S_t is greater than X, $\Pr(S_t > X)$, is $N(d_2)$ as defined in the Black-Scholes-Merton model except that the risk-free interest rate, r_c, is replaced with the investor's expected rate of return, μ.

5 Suppose we have a stock trading at 100 per share. If the stock price follows $dS_t = \mu S_t dt + \sigma S_t dW_t$, compute the probability of the stock exceeding its current value with a particular risk-averse investor's view with $\mu = 8.0\%$ and $\sigma = 31.6\%$. Compare your answer to the case of a risk-neutral investor's view with the risk-free interest rate of $r_c = 5.0\%$ and $\sigma = 40\%$. Explain your results.

6 Suppose we have a stock trading at 100 per share. Further suppose that the stock's volatility is 30% and the risk-free rate is 5%. Using Table 5.1, compute the Black-Scholes-Merton option value of an at-the-money call and put option with one year to expiration. (Assume the dividend yield is zero.)

7 Suppose we have a stock trading at 100 per share, the stock's volatility is 30%, and the risk-free rate is 5%. In this case, assume the dividend yield is 5%. Using Table 5.1, compute the Black-Scholes-Merton option value of at-the-money call and put options with one year time to expiration.

8 Given the arithmetic Brownian motion with geometric drift model, $dS = \mu S dt + \sigma dW$, derive the corresponding partial differential equation. Further, given the partial derivatives below, verify that the call model given here is valid:

$$c = (S - Xe^{-r\tau})\, N(d_n) + \sigma_\$ \sqrt{\frac{1 - e^{-2r\tau}}{2r}}\, n(d_n),\, d_n = \frac{Se^{r\tau} - X}{\sigma_\$ \sqrt{\frac{e^{2r\tau}-1}{2r}}} = \frac{S_0 - Xe^{-r\tau}}{\sigma_\$ \sqrt{\frac{1-e^{-2r\tau}}{2r}}},$$

$$n(d_n) = \frac{e^{-d_n^2/2}}{\sqrt{2\pi}},\, N(d_n) = \int_{-\infty}^{d_n} \frac{e^{-x^2/2}}{\sqrt{2\pi}}\, dx,\, \frac{\partial c}{\partial S} = N(d_n),\, \frac{\partial^2 c}{\partial S^2} = \frac{n(d_n)}{\sigma_\$ \sqrt{\frac{1-e^{-2r\tau}}{2r}}},\, \text{and}$$

$$\frac{\partial c}{\partial t} = -rXe^{-r\tau} N(d_n) - \frac{\sigma e^{-2r\tau}}{2\sqrt{\frac{1-e^{-2r\tau}}{2r}}}\, n(d_n).$$

NOTES

1. Naturally the option price is also a function of the exercise price, X, the continuously compounded risk-free rate, r_c, and the volatility, σ, but these values are static and, thus, they are parameters rather than variables. They can, of course, vary from option to option, but they are not allowed to vary internally, that is, within the dynamics of the evolution of the underlying asset and the price of the specific option over the life of the option.

2. To reduce notational clutter, where feasible, we eliminate the t subscripts on c_t, V_t and S_t.

3. Differential equations are also sometimes associated with initial conditions, such as when the value of the variable at the start is known. In the case of an option, the initial condition is that the price is nonnegative and bounded from above by the asset price.

4. Bachelier assumes arithmetic Brownian motion with arithmetic drift where the asset follows the stochastic process, $dS = \alpha_S dt + \sigma_S dW$. The asset value can go below zero.

5. See for example, Sprenkle (1961), Boness (1964), and Samuelson and Merton (1965).

6. Rubinstein (1976) has shown that the Black-Scholes-Merton model can be derived an alternative way, which requires no assumptions about the ability to continuously adjust a hedge.

7. Full details of this interesting side story are not clear, but it seems to be the case that Merton developed the mathematical insights, but in deference to Black and Scholes, he withheld the publication of his paper until the Black and Scholes paper appeared in print. Interestingly, Black and Scholes were working on two papers at the time, with the second paper being an empirical test of the model using over-the-counter option price data. That paper was presented at the annual meeting of the American Finance Association in 1971 and appeared in print in *The Journal of Finance* in 1972 (Black and Scholes 1972). Meanwhile, the original Black and Scholes paper, which derived the model, was rejected by the University of Chicago's renowned *Journal of Political Economy (JPE)*. Black and Scholes then submitted the paper to Harvard's *Review of Economics and Statistics*, which also rejected it. Then the University of Chicago's Merton Miller suggested that the *Journal of Political Economy* reconsider the paper, because Miller believed it was a major breakthrough. The *JPE* then agreed to publish the paper and it appeared in print in 1973, about the same time as the Merton paper but after the model's empirical tests had already appeared in the *Journal of Finance*. Other interesting stories about the development and use of the model are in Black (1989a, 1989b) and Mehrling (2005), and an excellent treatment of its impact on Wall Street is Bernstein (1996).

8. We illustrate the derivation of the arithmetic Brownian motion alternative in Appendix 13A. Although the ultimate formula is different, the approach to finding it is very similar.

9. For notational simplicity, we have dropped the time subscript on W_t and ε_t.

10. See Chapter 10, Section 10.3, if you need a review of the properties of the Wiener process.

11. The terms u and q certainly look remarkably like terms in the Black-Scholes-Merton model, leading one to wonder how the equation would be solved if one did not already know the solution. As noted, earlier models had provided some guidance here. Moreover, solving partial differential equations often proceeds based on guesses of general forms of solutions.

12. A cash-or-nothing call can be structured to pay a different amount than X. In that case, X would be replaced by the amount to be paid, but d_2 would be calculated using X or whatever is the strike.

13. Note if the dividend yield equals the interest rate, then the square root term equals the square root of time to maturity. This solution is proven in Chapter 14.

14. Recall when the interest rate equals the dividend yield then the denominator is simply the volatility times the square root of time to maturity.

The Greeks in the
Black-Scholes-Merton Model

Recall the Greeks refer to the partial derivatives of a pricing model with respect to various input parameters. In this chapter, we derive and explain the Greeks related to the Black-Scholes-Merton model. We focus first on the call option results and then turn to put options, relying on put-call parity for a smooth transition. After incorporating dividends, we will examine selected sensitivities of the Greeks as well as some extensions.

In Chapter 13, we derived the Black-Scholes-Merton model, as given here:

$$c = SN(d_1) - Xe^{-r_c\tau}N(d_2)$$

$$d_1 = \frac{\ln(S/X) + (r_c + \sigma^2/2)\,\tau}{\sigma\sqrt{\tau}}$$

$$d_2 = d_1 - \sigma\sqrt{\tau}$$

$$N(d_i) = \int_{-\infty}^{d_i} \frac{1}{\sqrt{2\pi}} e^{-x^2/2} dx, i = 1,2. \qquad (14.1)$$

We noted that this formula is the solution to the following partial differential equation,

$$r_c S \frac{\partial c}{\partial S} + \frac{\partial c}{\partial t} + \frac{1}{2}\frac{\partial^2 c}{\partial S^2}\sigma^2 S^2 = r_c c. \qquad (14.2)$$

The only way to be absolutely certain that the solution of a differential equation is correct is to take the derivatives of the solution and insert them into the differential equation. If the solution is correct, the differential equation will turn into the solution equation. So, let us do that here. As is indicated in Equation (14.2), we shall need to take the first and second derivatives with respect to the asset price, which are called the *delta* and *gamma*, respectively, and the first derivative with respect to time, which is called the *theta*. We approximated these sensitivity measures in the binomial model. Now, we shall more formally derive them in the continuous time world.

Before starting, however, there are some interim results that will prove useful in this process. Recall that $N(d)$ is the value of the cumulative density function, also called the distribution function, of a standard normal random variable, d. Thus, the integrand is simply the probability density function denoted $n(d)$. We first take the partial derivatives of $N(d_1)$ and $N(d_2)$ with respect to d_1 and d_2, respectively,

$$\frac{\partial N(d_1)}{\partial d_1} = \frac{1}{\sqrt{2\pi}}e^{-d_1^2/2} = n(d_1) \tag{14.3}$$

and

$$\frac{\partial N(d_2)}{\partial d_2} = \frac{1}{\sqrt{2\pi}}e^{-d_2^2/2} = \frac{1}{\sqrt{2\pi}}e^{-(d_1-\sigma\sqrt{\tau})^2/2} = n(d_2). \tag{14.4}$$

Further, we can establish the relationship between $n(d_1)$ and $n(d_2)$. Focusing first on $\left(d_1 - \sigma\sqrt{\tau}\right)^2$, we note

$$\left(d_1 - \sigma\sqrt{\tau}\right)^2 = d_1^2 - 2\sigma\sqrt{\tau}d_1 + \sigma^2\tau$$

$$= d_1^2 - 2\sigma\sqrt{\tau}\left[\frac{\ln(S/X) + \left(r_c + \sigma^2/2\right)\tau}{\sigma\sqrt{\tau}}\right] + \sigma^2\tau$$

$$= d_1^2 - 2\ln\left(S/Xe^{-r_c\tau}\right).$$

Substituting this result into Equation (14.4), we have either

$$n(d_1) = \frac{Xe^{-r\tau}n(d_2)}{S} \text{ or} \tag{14.5}$$

$$n(d_2) = \frac{Sn(d_1)}{Xe^{-r\tau}}. \tag{14.6}$$

With these results, we note

$$\frac{S}{Xe^{-r\tau}} = \frac{n(d_2)}{n(d_1)} = \frac{\frac{e^{-d_2^2/2}}{\sqrt{2\pi}}}{\frac{e^{-d_1^2/2}}{\sqrt{2\pi}}} = e^{\left(d_1^2-d_2^2\right)/2}. \tag{14.7}$$

Focusing on the exponent,

$$\frac{d_1^2 - d_2^2}{2} = \frac{1}{2}\left[d_1^2 - \left(d_1 - \sigma\sqrt{\tau}\right)^2\right] = \frac{1}{2}\left(d_1^2 - d_1^2 + 2d_1\sigma\sqrt{\tau} - \sigma^2\tau\right) = d_1\sigma\sqrt{\tau} - \frac{\sigma^2\tau}{2}$$

$$= \left[\frac{\ln\left(\frac{S}{X}\right) + \left(r + \frac{\sigma^2}{2}\right)\tau}{\sigma\sqrt{T}}\right]\sigma\sqrt{\tau} - \frac{\sigma^2\tau}{2} = \ln\left(\frac{S}{X}\right) + r\tau. \tag{14.8}$$

Therefore,

$$e^{\left(d_1^2 - d_2^2\right)/2} = e^{\ln\left(\frac{S}{X}\right) + r\tau} = \frac{S}{Xe^{-r\tau}}. \tag{14.9}$$

We shall also need the derivatives of d_1 and d_2 with respect to S,

$$\frac{\partial d_1}{\partial S} = \frac{\partial}{\partial S}\left(\frac{[\ln(S/X) + (r_c + (\sigma^2/2))\,\tau]}{\sigma\sqrt{\tau}}\right) = \frac{1}{S\sigma\sqrt{\tau}}$$

$$\frac{\partial d_2}{\partial S} = \frac{\partial\left(d_1 - \sigma\sqrt{\tau}\right)}{\partial S} = \frac{\partial d_1}{\partial S} = \frac{1}{S\sigma\sqrt{\tau}}. \tag{14.10}$$

We also need the derivatives of d_1 and d_2 with respect to time to expiration, τ. Recall the quotient rule from Chapter 3:

$$y = \frac{b}{v} \text{ where } b = f(x), v = g(x), \text{ and } v \neq 0,$$

$$\frac{dy}{dx} = \frac{v\frac{db}{dx} - b\frac{dv}{dx}}{v^2} \text{ (the quotient rule).} \tag{14.11}$$

Note for d_1, $b = \ln(S/X) + (r_c + \sigma^2/2)\tau$ and $v = \sigma\sqrt{\tau}$. Thus, $\partial b/\partial\tau = (r_c + \sigma^2/2)$ and $\partial v/\partial\tau = \sigma/(2\sqrt{\tau})$. Further, we assume $\sigma > 0$. Thus, applying the quotient rule, we have

$$\frac{\partial d_1}{\partial\tau} = \frac{v\frac{db}{dx} - b\frac{dv}{dx}}{v^2} = \frac{\sigma\sqrt{\tau}\left(r_c + \frac{\sigma^2}{2}\right) - [\ln(S/X) + (r_c + \sigma^2/2)\,\tau]\left[\sigma/(2\sqrt{\tau})\right]}{\sigma^2\tau} \text{ and}$$

$$= \frac{\left(r_c + \frac{\sigma^2}{2}\right) - [\ln(S/X) + (r_c + \sigma^2/2)\,\tau]\,[1/(2\tau)]}{\sigma\sqrt{\tau}} = \frac{\left(r_c + \frac{\sigma^2}{2}\right)}{\sigma\sqrt{\tau}} - \frac{d_1}{2\tau}. \tag{14.12}$$

$$\frac{\partial d_2}{\partial\tau} = \frac{\partial d_1}{\partial\tau} - \sigma\frac{\partial\sqrt{\tau}}{\partial\tau} = \frac{\left(r_c + \frac{\sigma^2}{2}\right)}{\sigma\sqrt{\tau}} - \frac{d_1}{2\tau} - \frac{\sigma}{2\sqrt{\tau}}$$

$$= \frac{\left(r_c + \frac{\sigma^2}{2}\right)}{\sigma\sqrt{\tau}} - \frac{d_1 - \sigma\sqrt{\tau}}{2\tau} = \frac{\left(r_c + \frac{\sigma^2}{2}\right)}{\sigma\sqrt{\tau}} - \frac{d_2}{2\tau}. \tag{14.13}$$

From the formula for d_1, we will also use the following:

$$S = Xe^{d_1\sigma\sqrt{\tau} - \left(r + \sigma^2/2\right)\tau} \text{ and} \tag{14.14}$$

$$X = Se^{-d_1\sigma\sqrt{\tau} + \left(r + \sigma^2/2\right)\tau}. \tag{14.15}$$

14.1   DELTA: THE FIRST DERIVATIVE WITH RESPECT TO THE UNDERLYING PRICE

This measure is called the *delta*. We take the derivative of the call price with respect to the underlying price,

$$
\begin{aligned}
\frac{\partial c}{\partial S} &= S\frac{\partial N(d_1)}{\partial S} + N(d_1) - Xe^{-r_c\tau}\frac{\partial N(d_2)}{\partial S} \\
&= S\frac{\partial N(d_1)}{\partial d_1}\frac{\partial d_1}{\partial S} + N(d_1) - Xe^{-r_c\tau}\frac{\partial N(d_2)}{\partial d_2}\frac{\partial d_2}{\partial S}.
\end{aligned} \tag{14.16}
$$

Substituting from Equations (14.3), (14.4), and (14.10), we obtain

$$
\frac{\partial c}{\partial S} = N(d_1) + Sn(d_1)\frac{1}{S\sigma\sqrt{\tau}} - Xe^{-r_c\tau}n(d_2)\frac{1}{S\sigma\sqrt{\tau}}. \tag{14.17}
$$

Finally, based on Equation (14.6) we note that the last two terms cancel, thus the call option delta is

$$
\Delta_c = \frac{\partial c}{\partial S} = N(d_1). \tag{14.18}
$$

The delta expresses the instantaneous change in the option price for a change in the underlying price. Because $N(d_1)$ is a probability, its value is between 0 and 1. In the derivation of the Black-Scholes-Merton model, delta is also the hedge ratio, representing the number of units of the asset required to offset one unit of the option when the underlying price changes in continuous time.

14.2   GAMMA: THE SECOND DERIVATIVE WITH RESPECT TO THE UNDERLYING PRICE

The second derivative with respect to the underlying price is the gamma and is obtained as follows:

$$
\begin{aligned}
\Gamma_c &= \frac{\partial\left(\frac{\partial c}{\partial S}\right)}{\partial S} = \frac{\partial^2 c}{\partial S^2} = \frac{\partial N(d_1)}{\partial S} \\
&= \frac{\partial N(d_1)}{\partial d_1}\frac{\partial d_1}{\partial S} = \frac{n(d_1)}{S\sigma\sqrt{\tau}}.
\end{aligned} \tag{14.19}
$$

This value is called the *gamma*. We can indeed think of it as the second derivative with respect to the asset price or we can think of it as the first derivative of the delta. Thus, it represents the instantaneous rate of change of the delta as the underlying price changes.

14.3 THETA: THE FIRST DERIVATIVE WITH RESPECT TO TIME

Now we need the derivative with respect to time to expiration.

$$\frac{\partial c}{\partial \tau} = S\frac{\partial N(d_1)}{\partial d_1}\frac{\partial d_1}{\partial \tau} - Xe^{-r_c\tau}\frac{\partial N(d_2)}{\partial d_2}\frac{\partial d_2}{\partial \tau} + r_cN(d_2)\,Xe^{-r_c\tau}$$

$$= Sn\,(d_1)\frac{\partial d_1}{\partial \tau} - Xe^{-r_c\tau}n(d_2)\frac{(\partial d_1 - \sigma\sqrt{\tau})}{\partial \tau} + r_cN(d_2)\,Xe^{-r_c\tau}$$

$$= Sn\,(d_1)\frac{\partial d_1}{\partial \tau} - Xe^{-r_c\tau}n(d_2)\frac{\partial d_1}{\partial \tau} + Xe^{-r_c\tau}n(d_2)\frac{\sigma}{2\sqrt{\tau}} + r_cN(d_2)\,Xe^{-r_c\tau}. \quad (14.20)$$

Based on Equation (14.6), we know the first two terms cancel and we obtain

$$\frac{\partial c}{\partial \tau} = \frac{\sigma Xe^{-r_c\tau}}{2\sqrt{\tau}}n(d_2) + r_cXe^{-r_c\tau}N(d_2). \quad (14.21)$$

Based again on Equation (14.6), we can express this result as

$$\frac{\partial c}{\partial \tau} = \frac{\sigma S}{2\sqrt{\tau}}n(d_1) + r_cXe^{-r_c\tau}N(d_2). \quad (14.22)$$

Thus, the derivative with respect to time to expiration is clearly positive, which is intuitively consistent. A call option with a longer time to expiration is worth more. Generally speaking, however, we define theta as the derivative with respect to the point in time, that is, $\partial c/\partial t$. Recall that we defined the time to expiration as $\tau = T - t$. What we shall need in order to check the solution to the partial differential equation is $\partial c/\partial t$, but it is easy to see that $\partial c/\partial t = -\partial c/\partial \tau$. Hence,

$$\Theta_c = \frac{\partial c}{\partial t} = -\frac{\sigma S}{2\sqrt{\tau}}n(d_1) - r_cXe^{-r_c\tau}N(d_2). \quad (14.23)$$

This formula is one way to express theta. Alternatively, one could compute the partial with respect to time to expiration.

14.4 VERIFYING THE SOLUTION OF THE PARTIAL DIFFERENTIAL EQUATION

Now that we have the three necessary partial derivatives—Equations (14.18), (14.19), and (14.23)—we can substitute them into Equation (14.2):

$$r_cSN\,(d_1) - \frac{\sigma Sn\,(d_1)}{2\sqrt{\tau}} - r_cXe^{-r_c\tau}N(d_2) + \frac{1}{2}n(d_1)\frac{1}{S\sigma\sqrt{\tau}}\sigma^2 S^2 = r_cc$$

$$r_cc = r_cSN\,(d_1) - r_cXe^{-r_c\tau}N(d_2) - \frac{\sigma S}{2\sqrt{\tau}}n(d_1) + \frac{\sigma S}{2\sqrt{\tau}}n(d_1)$$

$$c = SN\,(d_1) - Xe^{-r_c\tau}N(d_2). \quad (14.24)$$

And the end result is the Black-Scholes-Merton formula.

Recall for arithmetic Brownian motion with geometric drift, the PDE can be expressed as

$$r_c S \frac{\partial c}{\partial S} + \frac{\partial c}{\partial t} + \frac{1}{2} \frac{\partial^2 c}{\partial S^2} \sigma^2 = r_c c. \tag{14.25}$$

Further, recall that the resultant call value is

$$c_t = (S_t - X e^{-r_c(T-t)}) N(d_n) + \sigma \sqrt{\frac{1 - e^{-2r_c(T-t)}}{2r_c}} n(d_n), \tag{14.26}$$

where

$$d_n = \frac{S_0 - X e^{-r_c(T-t)}}{\sigma \sqrt{\frac{1 - e^{-2r_c(T-t)}}{2r_c}}}. \tag{14.27}$$

It can be shown, based on the ABM model, that

$$\frac{\partial c}{\partial S} = N(d_n) \text{ (delta)} \tag{14.28}$$

$$\frac{\partial^2 c}{\partial S^2} = \frac{n(d_n)}{\sigma \sqrt{\frac{1 - e^{-2r_c(T-t)}}{2r_c}}} \text{ (gamma)} \tag{14.29}$$

$$\frac{\partial C}{\partial t} = -r_c X e^{-r_c(T-t)} N(d_n) - \frac{\sigma e^{-2r_c(T-t)}}{2\sqrt{\frac{1 - e^{-2r_c(T-t)}}{2r_c}}} n(d_n) \text{ (theta)}. \tag{14.30}$$

Substituting delta, gamma, and theta into the ABM PDE will result in Equation (14.26), and we have

$$r_c c_t = -r_c X e^{-r(T-t)} N(d_n) - \frac{\sigma e^{-2r_c(T-t)}}{2\sqrt{\frac{1 - e^{-2r_c(T-t)}}{2r_c}}} n(d_n) + r_c [N(d_n)] S + \frac{1}{2} \sigma^2 \left[\frac{n(d_n)}{\sigma \sqrt{\frac{1 - e^{-2r_c(T-t)}}{2r_c}}} \right]$$

$$= r_c S N(d_n) - r_c X e^{-r(T-t)} N(d_n) - \frac{\sigma e^{-2r_c(T-t)} n(d_n)}{2\sqrt{\frac{1 - e^{-2r_c(T-t)}}{2r}}} + \frac{\sigma n(d_n)}{2\sqrt{\frac{1 - e^{-2r_c(T-t)}}{2r_c}}}$$

$$= r_c(S - X e^{-r_c(T-t)}) N(d_n) + \frac{\sigma n(d_n)}{\sqrt{\frac{1 - e^{-2r_c(T-t)}}{2r_c}}} \frac{(1 - e^{-2r_c(T-t)})}{2}$$

$$= r_c \left(S - X e^{-r_c(T-t)} \right) N(d_n) + r_c \sigma \sqrt{\frac{1 - e^{-2r_c(T-t)}}{2r_c}} n(d_n). \tag{14.31}$$

Therefore, we have verified that both the Black-Scholes-Merton and ABM models satisfy their respective partial differential equations. To further validate these models, one must confirm that boundary conditions and nonnegativity constraints are satisfied. See Chapter 13, Section 7, for the limiting argument for the Black-Scholes-Merton model.

14.5   SELECTED OTHER PARTIAL DERIVATIVES OF THE BLACK-SCHOLES-MERTON MODEL

In addition, we may be interested in how the option value varies with the volatility and risk-free rate. Note that the Black-Scholes-Merton model assumes that the volatility and the risk-free rate are constant. Thus, there is an element of incoherence to these calculations. Remember, however, that this is a financial model and as such will only ever be an approximation to the behavior of the underlying asset. We also know from empirical observation that both the volatility and risk-free rate often vary significantly over time. Thus, we wish to take their derivatives. As previously noted, the volatility and risk-free rate are assumed to not vary, but their partial derivatives show the sensitivity of the option price to different values of the volatility and interest rate. This is not the same as allowing them to vary within the model, but it is better than making no such adjustment. Models in which these input values vary do exist, but they are far more complex and face the difficult challenge of finding a way to hedge away their risk so that the final solution meets the risk-neutrality condition.

14.5.1   Vega: The Partial Derivative with Respect to the Volatility

If we differentiate the call price with respect to the volatility, σ, we get

$$\frac{\partial c}{\partial \sigma} = S \frac{\partial N(d_1)}{\partial d_1} \frac{\partial d_1}{\partial \sigma} - X e^{-r_c \tau} \frac{\partial N(d_2)}{\partial d_2} \frac{\partial d_2}{\partial \sigma}. \tag{14.32}$$

Note that $\partial d_2 / \partial \sigma = \partial d_1 / \partial \sigma - \sqrt{\tau}$ and substituting from Equations (14.3) and (14.4), we have

$$\frac{\partial c}{\partial \sigma} = S n(d_1) \frac{\partial d_1}{\partial \sigma} - X e^{-r_c \tau} n(d_2) \left(\frac{\partial d_1}{\partial \sigma} - \sqrt{\tau} \right). \tag{14.33}$$

Substituting from Equation (14.6) for $n(d_2)$ and cancelling terms, we obtain

$$v_c = \frac{\partial c}{\partial \sigma} = S \sqrt{\tau} n(d_1). \tag{14.34}$$

This value is clearly positive and is known as the *vega*.[1] Interestingly, the traditional interpretation of vega in the BSM model is that it helps one on the upside and does not hurt on the downside, but Chance (1994) shows that the effect is technically from the downside only and is strongly affected by the static and somewhat questionable assumption that the underlying is unaffected by the volatility change.

14.5.2   Rho: The Partial Derivative with Respect to the Risk-Free Rate

If we differentiate the call price with respect to the risk-free rate, we obtain

$$\frac{\partial c}{\partial r} = S\frac{\partial N(d_1)}{\partial d_1}\frac{\partial d_1}{\partial r} - \left[Xe^{-r_c\tau}\frac{\partial N(d_2)}{\partial d_2}\frac{\partial d_2}{\partial r} - \tau N(d_2)Xe^{-r_c\tau}\right]. \tag{14.35}$$

Making various substitutions as in the previous examples, we simplify this equation to

$$\rho_c = \frac{\partial c}{\partial r} = Xe^{-r_c\tau}\tau N(d_2). \tag{14.36}$$

This expression, called *rho*, is clearly positive.[2]

14.6   PARTIAL DERIVATIVES OF THE BLACK-SCHOLES-MERTON EUROPEAN PUT OPTION PRICING MODEL

The Black-Scholes-Merton European put option pricing model is repeated here as

$$p = Xe^{-r_c\tau}N(-d_2) - SN(-d_1). \tag{14.37}$$

Recall that put-call parity expresses the relationship between the put and call prices. To be consistent with the continuous time framework of Black-Scholes-Merton, we restate put-call parity using continuous discounting:

$$p = c - S + Xe^{-r_c\tau}. \tag{14.38}$$

We can obtain the derivatives of the put model by differentiating Equation (14.38) as follows:

$$\frac{\partial p}{\partial S} = \frac{\partial c}{\partial S} - 1$$
$$\frac{\partial^2 p}{\partial S^2} = \frac{\partial^2 c}{\partial S^2}$$
$$\frac{\partial p}{\partial \tau} = \frac{\partial c}{\partial \tau} - r_c Xe^{-r_c\tau}$$
$$\frac{\partial p}{\partial \sigma} = \frac{\partial c}{\partial \sigma}$$
$$\frac{\partial p}{\partial r_c} = \frac{\partial c}{\partial r_c} - \tau Xe^{-r_c\tau}. \tag{14.39}$$

Substituting the partial derivatives of the call and simplifying, we obtain the put delta as

$$\Delta_p = \frac{\partial p}{\partial S} = N(d_1) - 1 = -N(-d_1). \tag{14.40}$$

Thus, the put delta is clearly negative, as it should be. The gamma is

$$\Gamma_p = \frac{\partial^2 p}{\partial S^2} = \frac{1}{S\sigma\sqrt{\tau}}n(d_1) = \Gamma_c. \tag{14.41}$$

The put gamma, which is the same as the call gamma, is clearly positive. The theta is obtained as

$$\frac{\partial p}{\partial \tau} = \frac{S\sigma}{2\sqrt{\tau}}n(d_1) - r_c X e^{-r_c\tau}N(-d_2)$$

$$\Theta_p = \frac{\partial p}{\partial t} = -\frac{S\sigma}{2\sqrt{\tau}}n(d_1) + r_c X e^{-r_c\tau}N(-d_2). \tag{14.42}$$

The put theta cannot be definitively signed. Indeed, it is well known that a European put can have a negative or positive theta. That is, the European put can have a lower or higher value with longer expiration. The negative relationship of put value to time to expiration occurs when the put is deep in-the-money. The upper limit of the put value, which exists by the fact that the put cannot achieve a higher value than the exercise price, can make a long-term put less valuable than a short-term put, the latter being closer to the time of expiration and the payoff from exercise. Because exercise of a put pays the exercise price, waiting a long time for that payoff penalizes the put holder. This unsigned theta on a European put is precisely the reason why if the put were American, there is a possibility that it would be optimal to exercise it early.

The vega is

$$v_p = \frac{\partial p}{\partial \sigma} = S\sqrt{\tau}n(d_1). \tag{14.43}$$

The put vega is the same as the call vega and is clearly positive. Finally, the rho is

$$\rho_p = \frac{\partial p}{\partial r} = -\tau X e^{-r_c\tau}N(-d_2). \tag{14.44}$$

The put rho is negative, because a higher interest rate reflects the loss of greater interest from waiting to exercise the option.

14.7 INCORPORATING DIVIDENDS

In the case of continuous dividends, recall the Black-Scholes-Merton model is adjusted where $S' = e^{-\delta\tau}S$ and can be expressed as

$$c = S'N(d_1) - Xe^{-r_c\tau}N(d_2)$$

$$p = Xe^{-r_c\tau}N(-d_2) - S'N(-d_1)$$

$$d_1 = \frac{\ln(S'/X) + (r_c + \sigma^2/2)\,\tau}{\sigma\sqrt{\tau}}$$

$$d_2 = d_1 - \sigma\sqrt{\tau}. \tag{14.45}$$

TABLE 14.1   Dividend-Adjusted Greeks

Greek	Calls	Puts
Delta	$e^{-\delta\tau}N(d_1)$	$-e^{-\delta\tau}N(-d_1)$
Gamma	$\dfrac{e^{-\delta\tau}n(d_1)}{S\sigma\sqrt{\tau}}$	Same as call
Theta	$-\dfrac{\sigma S'n(d_1)}{2\sqrt{\tau}}-rXe^{-r\tau}N(d_2)$ $+\delta S'N(d_1)$	$-\dfrac{\sigma S'n(d_1)}{2\sqrt{\tau}}+rXe^{-r\tau}N(-d_2)$ $-\delta S'N(-d_1)$
Vega	$S'\sqrt{\tau}n(d_1)$	Same as call
Rho	$\tau Xe^{-r\tau}N(d_2)$	$-\tau Xe^{-r\tau}N(-d_2)$

For example, the dividend-adjusted deltas are simply

$$\Delta_c = \frac{\partial c}{\partial S} = \frac{\partial c}{\partial S'}\frac{\partial S'}{\partial S} = N(d_1)e^{-\delta\tau} \text{ and} \qquad (14.46)$$

$$\Delta_p = \frac{\partial p}{\partial S} = \frac{\partial p}{\partial S'}\frac{\partial S'}{\partial S} = -N(-d_1)e^{-\delta\tau}. \qquad (14.47)$$

Table 14.1 presents the five dividend-adjusted Greeks for calls and puts.

The values of these Greeks are sensitive to changes in other parameters. We explore some of these sensitivities in Section 14.8.

14.8   GREEK SENSITIVITIES

We now consider the sensitivities of option values and Greeks to changes in the underlying. For this illustration we start with options that have the following parameters, $X = 100$, $r_c = 5.0\%$, $\sigma = 30\%$, $T - t = 1.0$ year, and $\delta = 0.0\%$. Figure 14.1 illustrates option values and Greeks over the stock price range from 0 to 200.

Several observations can be made from Figure 14.1. First, the call delta moves toward zero as the asset price falls and the call delta moves to one as the asset price rises (with the dividend yield assumed zero). Second, both gamma and vega directly reflect the lognormal distribution assumption with its positive skewness. Third, the difference between call and put deltas, thetas, and rhos are constants and relatively close to one based on the put-call parity relationship.

We now consider the sensitivities of option values and Greeks to changes in volatility. For this illustration, we again start with options that are at-the-money with $S = X = 100$, $r_c = 5.0\%$, $T - t = 1.0$ year, and $\delta = 0.0\%$. Figure 14.2 illustrates option values and Greeks over the volatility range from 0.0% to 60%.

Several observations can be made from Figure 14.2. First, based on the value graph, we observe that the difference between the call and the put is simply $S - Xe^{-r_c\tau}$. Second,

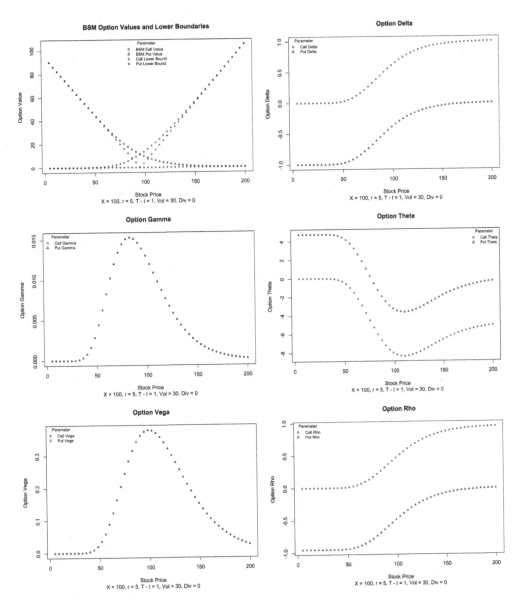

FIGURE 14.1 Sensitivity of Option Values and Greeks to Changes in Stock Price

both gamma and vega are very sensitive as volatility approaches zero. Third, the difference between call and put thetas and rhos are constant across volatilities.

We now consider the sensitivities of option values and Greeks to changes in time to expiration. For this illustration, we again start with options that are at-the-money with $S = X = 100$, $r_c = 5.0\%$, $\sigma = 30\%$, and $\delta = 0.0\%$. Figure 14.3 illustrates option values and Greeks over the time to expiration range from zero to two years.

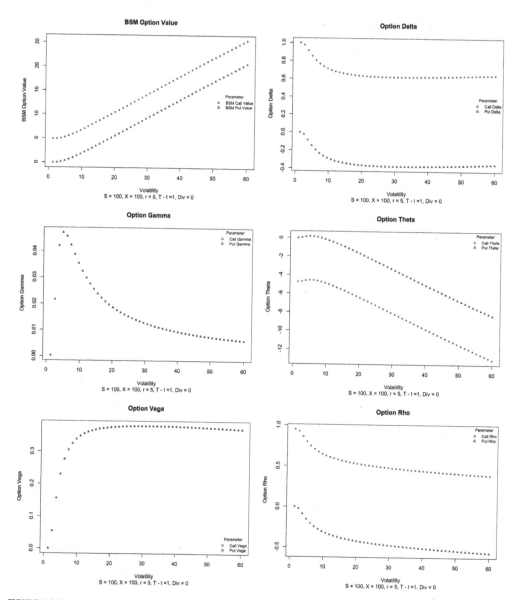

FIGURE 14.2 Sensitivity of Option Values and Greeks to Changes in Volatility

Several observations can be made from Figure 14.3. First, based on the value graph, we observe that the at-the-money call is worth more than the at-the-money put, where the value difference reflects the magnitude of $S - Xe^{-r_c\tau}$. Second, both gamma and vega are very sensitive as time to expiration approaches zero. Third, the difference between call and put thetas shrinks as the time to expiration shortens. Finally, the rhos have opposite signs and increase in absolute value as time to expiration increases.

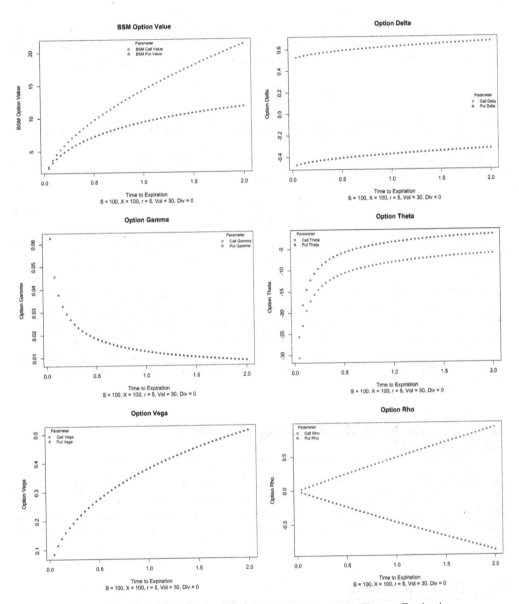

FIGURE 14.3 Sensitivity of Option Values and Greeks to Changes in Time to Expiration

14.9 ELASTICITIES

The Greeks presented here are simply partial derivatives. Thus, a call delta answers the question of how much a call price will increase for a small dollar change in the underlying asset. There is no adjustment for the absolute price level. In response to this problem, some have argued that various elasticities would be more informative.

For example, rather than report the call delta, it may be more informative to report the percentage change in the call price with respect to the percentage change in the underlying asset. Thus, the elasticity of the call price to the asset price, known as omega, is

$$\Omega_c = \frac{\partial c/c}{\partial S/S} = \frac{\partial c}{\partial S}\frac{S}{c} = \Delta_c \frac{S}{c}. \tag{14.48}$$

Note that elasticity is simply the first derivative, or delta, adjusted by what we call a leverage factor, S/c.

Similar calculations could be performed on all the reported Greeks . For example, the elasticity of a call price with respect to volatility is

$$\frac{\partial c/c}{\partial \sigma/\sigma} = \frac{\partial c}{\partial \sigma}\frac{\sigma}{c} = \Theta_c \frac{\sigma}{c}. \tag{14.49}$$

We see that the reported elasticities could be significantly different from reported Greeks. Clearly, with the Greek formulas, we can easily compute the corresponding elasticities.

One well-known weakness of the Greeks and related elasticities is the lack of information regarding the different likelihoods of various parameter changes. For example, beyond the asset price, option prices are generally very sensitive to changes in volatilities. The Greeks do not address the likelihood of volatility increasing from 30% to 33% or a 10% increase. Moreover, as partial derivatives, the Greeks are technically accurate only for infinitesimal changes. One way to address this issue is through simulation, a topic we take up in Chapter 23.

14.10 EXTENDED GREEKS OF THE BLACK-SCHOLES-MERTON OPTION PRICING MODEL

There are many more Greeks, corresponding to the many different partial and cross-partial derivatives. For example, one can differentiate the vega with respect to time, so as to determine how vega changes as the option approaches expiration. There are numerous possible combinations. Table 14.2 provides selected Greek names where known or otherwise brief syntax is given.

14.11 RECAP AND PREVIEW

In this chapter, we derived the calculus derivatives of the Black-Scholes-Merton model to obtain the Greeks, which are the sensitivities of the option value to the factors that go into the models. We further illustrate selected sensitivities of Greeks to changes in the asset price, volatility, and time to expiration. Finally, we introduce elasticities as well as selected extended Greeks.

In Chapter 15, we examine a mathematical result known as Girsanov's theorem and show how it establishes our ability to derive the model under the risk-neutral or martingale measure.

TABLE 14.2   Extended Greek Names or Definitions

Parameter	S	t	σ	r	X	δ
First-Order Derivatives						
Value (O)	$\partial O \partial S$ Delta (Δ)	$\partial O \partial t$ Theta (θ)	$\partial O \partial \sigma$ Vega (v)	$\partial O \partial r$ Rho (ρ)	$\partial O \partial X$	$\partial O \partial \delta$
Elasticity	$O\%S\%$ Omega (Ω)	$O\%t\%$	$O\%\sigma\%$	$O\%r\%$	$O\%X\%$	$O\%\delta\%$
Selected Second-Order Derivatives						
Delta (Δ)	$\partial\Delta\partial S$ Gamma (Γ)	$\partial\Delta\partial t$ Charm	$\partial\Delta\partial\sigma$ Vanna	$\partial\Delta\partial r$	$\partial\Delta\partial X$	$\partial\Delta\partial$
Theta (θ)		$\partial\theta\partial t$	$\partial\theta\partial\sigma$ Veta	$\partial\theta\partial r$	$\partial\theta\partial X$	$\partial\theta\partial\delta$
Vega (v)			$\partial v\partial\sigma$ Vomma	$\partial v\partial r$ Ver	$\partial v\partial X$	$\partial v\partial\delta$
Rho (ρ)				$\partial\rho\partial r$	$\partial\rho\partial X$	$\partial\rho\partial\delta$
$\partial O\partial X$					$(\partial O\partial X)\partial X$	$(\partial O\partial X)\partial\delta$
$\partial O\partial\delta$						$(\partial O\partial\delta)\partial\delta$
Selected Third-Order Derivatives						
Gamma (Γ)	$\partial\Gamma\partial S$ Speed	$\partial\Gamma\partial t$ Color	$\partial\Gamma\partial\sigma$ Zomma	$\partial\Gamma\partial r$	$\partial\Gamma\partial X$	$\partial\Gamma\partial\delta$
$\partial\theta\partial t$		$(\partial\theta\partial t)\,\partial t$	$(\partial\theta\partial t)\partial\sigma$	$(\partial\theta\partial t)\partial r$	$(\partial\theta\partial t)\partial X$	$(\partial\theta\partial t)\partial\delta$
Vomma			∂Vomma$\partial\sigma$ Ultima	∂Vomma∂r	∂Vomma∂X	∂Vomma$\partial\delta$
$\partial\rho\partial r$				$(\partial\rho\partial r)\partial r$	$(\partial\rho\partial r)\partial X$	$(\partial\rho\partial r)\partial\delta$
...						

QUESTIONS AND PROBLEMS

1   Derive the call and put theta for a stock that pays a continuous dividend yield, δ.

2   Prove the following relationships for stocks with a dividend yield, δ:

$$n(d_1) = \frac{Xe^{-rT}n(d_2)}{Se^{-\delta T}} \text{ or}$$

$$n(d_2) = \frac{Se^{-\delta T}n(d_1)}{Xe^{-rT}}.$$

3   Derive the partial derivative of the call price with respect to the dividend yield, δ.

4 One of the most important cross-partial derivatives is vanna, the partial derivative of delta with respect to the volatility, because delta is key to risk management and volatility is not constant in practice. Derive the vanna for both calls and puts on a dividend paying stock with yield, δ.

$$Vanna_c = \frac{\partial \Delta_c}{\partial \sigma} = \frac{\partial}{\partial \sigma} e^{-\delta \tau} N(d_1)$$

5 Derive $\partial c / \partial X$ for a stock paying a dividend yield, δ.

NOTES

1. *Vega* is not a Greek word and the symbol typically used is the Greek nu because it looks similar to ν for volatility. Thus, it has come to be used as one of the option Greeks.
2. It is also possible to differentiate with respect to the exercise price. One would obtain $\partial c / \partial X = -e^{-r_c \tau} N(d_2)$, which is negative. Because the exercise price does not change, however, this value has no name and is rarely needed. One case in which it has been used is in extracting the risk-neutral density implied by the option price. It is fairly easy to see how this is done. Note that the cumulative density is in the formula for the partial with respect to X. This implies that the density itself can be extracted. By using a sample of the prices of options that vary only by slightly different exercise prices, one can infer the density that implies the option price.

CHAPTER 15

Girsanov's Theorem in Option Pricing

At this point we have already derived the Black-Scholes-Merton model. Therefore, it would seem that little else needs to be done. Indeed, that is partially correct, but there are some additional results that are useful to understand to gain a better perspective on this model as well as give you the tools to create your own models. In this chapter, we cover one of those results, which is called Girsanov's theorem. Good references on this material are Karatzas and Shreve (1991) and Neftci (2000).

Girsanov's theorem appears to be a somewhat complex rule that at best provides limited marginal value if one already understands derivative pricing. Indeed, it is quite possible to obtain a solid understanding of derivative pricing without encountering Girsanov's theorem. Nowhere did it appear in the literature on derivatives for many years, despite many exceptionally powerful advances developed by financial economists. Yet Girsanov's theorem provides the foundation for the mathematical rigor that underlies derivative pricing. Unfortunately, Girsanov's theorem in its raw form provides virtually no intuition and is rarely, if ever, presented with any economic insights. In short, mathematicians often approach the derivative pricing problem in their own ways, one being the application of Girsanov's theorem, and financial economists approach it in other ways. That these two groups arrive at the same answer is not surprising, but they speak different languages and travel different routes. Here we cause them to converge. The material is admittedly more mathematical than financial, but that is because the readership of this book is likely to be more financial than mathematical and is more in need of seeing the mathematics and being reminded of when the mathematicians are saying something that a financial economist recognizes.

15.1 THE MARTINGALE REPRESENTATION THEOREM

In applying Girsanov's theorem, what we do is alter the probability distribution of the asset return such that it follows a stochastic process known as a *martingale*. A martingale is a process without a drift, that is, it has a zero price change.[1] In that case, we can obtain the value today of the derivative by determining the expected future value of the derivative, where the expectation is arrived at by using the altered probability distribution. In doing it this way, we avoid the problem of having to solve a differential equation.

First, however, let us identify why we are doing this. One of the important results of mathematical finance is that any two stochastic processes that are both martingales can be related to each other via a simple transformation. For example, given one martingale, x_t, and another, y_t, the relationship,

$$y_t = y_0 + \int_0^t \lambda_j dx_j,$$ (15.1)

represents one martingale in terms of the other. Accordingly, this specification is referred to as the *martingale representation theorem*. It is important to note, however, and often overlooked in the mathematical finance literature that these two processes cannot simply be any two martingales. They must be rationally connected as we explain next.

Suppose we consider x and y, which we specify to be two independent stochastic variables with expected returns of zero. The variable x may be the stock price of a gas company in Warsaw, Poland, whereas y might be the price of corn in Madison County, Iowa. There is no amount of mathematical wizardry that can convert the price of a gas company in Warsaw to the price of corn in Iowa, even though both may be martingales given the probability distribution of states of the global economy. The martingale representation theorem seems to say that these variables can be expressed in terms of each other, but what is missing can be easily seen by recalling the binomial model.

Suppose x can go up to x^+ or down to x^-. Now we want to model y. Intuitively we might specify that y can go up to y^+ or down to y^-. If we allow only two states of nature and jointly consider x and y in the same model, then whenever x is x^+, y is y^+, and whenever x is x^-, y is y^-. In that case, the variables x and y are perfectly related to each other. If you know the value of one, you know the value of the other. In a very loose sense x and y can be thought of as perfectly correlated, though correlation is a specific and rather strong type of relationship that does not have to hold to make this point. In our example, it is obvious that when the price of corn is up, the price of the Polish gas company can be up or down. There is probably little relation between our x and y. Furthermore, we do not require such an extreme example to make this point. The variable x could be the price of Microsoft and y is the price of Exxon. Though both share a common relationship as driven by general stock market movements, there is no way that we can completely determine the value of one from the other. So, if Microsoft and Exxon were martingales, we could not relate one to the other as the martingale representation theorem seems to suggest. Even two stocks in the same industry, such as Ford and General Motors, cannot be related by the martingale representation theorem.

But if x is the price of Microsoft stock and y is the price of a call option on Microsoft, we can connect the two, because the price of the latter is completely determined by the price of the former. In the binomial sense, we can easily see this point. The option payoff at expiration for a given outcome is completely identified by the underlying asset price in that outcome.[2] Thus, the martingale representation theorem requires that one variable be completely determined by the other. Well, that was the binomial case, which applies in a discrete time world. In a continuous time world, Girsanov's theorem will provide the connection.

How we use the martingale representation theorem is in the following manner. If the price of an asset, as indicated by the x variable, is a martingale, then by the described transformation, we can turn it into the y variable, if y is completely determined by x. If y is the option on x, then we have replicated y with x. It should be apparent that λ is likely to represent a certain number of shares of x held to replicate y.

Those who have studied option pricing theory from a financial economics perspective, however, may find this result a bit disconcerting because they know that one must hold bonds as well as shares to replicate an option. The martingale representation theorem gives us only the condition under which the uncertainty in y can be captured by the uncertainty in x. More formally, if we differentiate y_t with respect to x_t from Equation (15.1) we obtain

$$dy_t = \lambda_t dx_t. \tag{15.2}$$

This is the result we need and the one that clearly indicates that the uncertainty in y is driven by the uncertainty in x. The bonds we must hold for replication introduce no element of uncertainty.

So the martingale representation theorem tells us that we must find a stochastic process λ_t such that we are holding just the right number of units of x. In addition, we must invest a certain amount in risk-free bonds such that we replicate y. Formally,

$$y_t = \lambda_t x_t + \theta_t B_t, \tag{15.3}$$

where B_t is the value of the bonds, and which accrues value by the factor $e^{r_c dt}$. Note that λ and θ are variables. They are indexed by t and are determined as the asset evolves along its stochastic paths. We need to know their values at t.[3] They change as we move through time, but they are completely determined once we know the new price of the asset. Financial economists know this as the dynamic nature of the option delta. Indeed delta is what we are currently designating as λ_t.

Mathematical finance formally tells us that if the asset can be transformed into a martingale, we can find a stochastic process such that the asset can be transformed into another martingale that replicates the option. Once we have replicated the option, we can price it using the asset price, the number of units of the asset we hold, the investment in risk-free bonds, and the interest rate, and so on, or in other words, all known values. We can obtain the option value by finding the expectation of the future value of this process.

Now let us proceed to examine the mathematics of how derivative prices are found. Our first step is to learn how to change the drift, that is, the expected return of a random variable.

15.2 INTRODUCING THE RADON-NIKODYM DERIVATIVE BY CHANGING THE DRIFT FOR A SINGLE RANDOM VARIABLE

Let us begin by examining the process of changing a probability distribution for a general random variable. We start with a random variable x, which is simply a single unknown outcome and not a stochastic process. We shall take x as distributed normally with expected value μ and variance σ^2. The probability density of x is

$$f(x) = \frac{1}{\sigma\sqrt{2\pi}} e^{-\frac{\left(\frac{x-\mu}{\sigma}\right)^2}{2}} = \frac{d\Pr(x)}{dx}, \tag{15.4}$$

where the probability distribution function or cumulative probability is $\Pr(x)$. Now suppose that we wanted to change the location of this probability distribution. Specifically, we wish to shift the expected value by an amount γ, such that the expected value is $\mu + \gamma$. Then the density we want is

$$f(x) = \frac{1}{\sigma\sqrt{2\pi}} e^{-\frac{\left(\frac{x-\mu-\gamma}{\sigma}\right)^2}{2}}. \tag{15.5}$$

Note that the numerator of the exponent takes x and subtracts our shifted expected value $\mu + \gamma$.

We have changed the expected value but not the variance, meaning that we have changed the location but not the shape of the distribution. We need not specify if γ is positive or negative, because we could shift the expected value upward or downward. Let us call this new density $f^Q(x)$ and the new distribution function $\mathrm{Pr}^Q(x)$, so $f^Q(x) = d\mathrm{Pr}^Q(x)/dx$. Let us see how we can make this distribution change so that $\mathrm{Pr}(x)$ is replaced by $\mathrm{Pr}^Q(x)$.

Look at the expression in the previous exponent, $-[(x - \mu - \gamma)/\sigma]^2/2$, and note that it equals $-(1/2\sigma^2)(x^2 - 2x\mu - 2\gamma x + 2\mu\gamma + \mu^2 - \gamma^2)$. If we compare this expression to what we started with, $-[(x - \mu)/\sigma]^2/2 = -(1/2\sigma^2)(x^2 - 2x\mu + \mu^2)$, we see that we can get the new expression by multiplying the original expression by a certain factor. That is,

$$\frac{1}{\sigma\sqrt{2\pi}} e^{-\frac{\left(\frac{x-\mu-\gamma}{\sigma}\right)^2}{2}} = \left(\frac{1}{\sigma\sqrt{2\pi}} e^{-\frac{\left(\frac{x-\mu}{\sigma}\right)^2}{2}}\right)\left(e^{-\frac{(2\gamma x - 2\gamma\mu - \gamma^2)}{2\sigma^2}}\right). \tag{15.6}$$

Let us now designate this new multiplier as

$$\phi(x) = e^{-\frac{(2\gamma x - 2\gamma\mu - \gamma^2)}{2\sigma^2}}. \tag{15.7}$$

Because $f(x) = d\,\mathrm{Pr}(x)/dx$, then $f(x)dx = d\,\mathrm{Pr}(x)$. What we want is $f^Q(x)dx = d\mathrm{Pr}^Q(x)$, where we have a new probability measure $\mathrm{Pr}^Q(x)$. Because $f(x)\phi(x) = f^Q(x)$, then $f(x)dx\phi(x) = f^Q(x)dx$. Then

$$\frac{f^Q(x)}{f(x)} = \phi(x). \tag{15.8}$$

It follows that

$$\frac{d\mathrm{Pr}^Q(x)}{d\mathrm{Pr}(x)} = \frac{f^Q(x)dx}{f(x)dx} = \phi(x). \tag{15.9}$$

Our multiplier $\phi(x)$ can be thought of as an adjustment that converts one probability measure, $\mathrm{Pr}(x)$, into another, $\mathrm{Pr}^Q(x)$. We must be careful, however, in that one cannot just arbitrarily multiply one measure by some other factor because the resulting measure should be of the same type of distribution as the one we started with. In this case, we started with the normal distribution, and we end with the normal distribution but with a different expected value. In some cases, our random variable will be standard normal, meaning that $\mu = 0$ and $\sigma = 1$. In that case,

$$\phi(x) = e^{\frac{x\gamma - \gamma^2}{2}}. \tag{15.10}$$

This special function $\phi(x)$, which we noted can be expressed as $d\text{Pr}^Q(x)/d\text{Pr}(x)$, is a derivative itself and is called the *Radon-Nikodym derivative*. For this derivative to exist, it is necessary that the function $\text{Pr}^Q(x)$ and the function $\text{Pr}(x)$ be considered *equivalent probability measures*. What this means is that if an event is possible under one measure, then it is possible under the other measure. In other words, events that cannot occur in the first place cannot be made possible by simply changing the probability measure. Likewise events that can occur in the first place cannot be made impossible by changing the probability measure. The probability of a possible event, however, can and will be changed by this change of measure.

Before providing a formal statement of Girsanov's theorem, we introduce a complete probability space with a particular interest in how it is applicable to financial modeling.

15.3 A COMPLETE PROBABILITY SPACE

The complete probability space is a very formal mathematical representation of our perceptions of the possible uncertain outcomes and their probabilities, such as the random movements in asset prices. In financial practice, one always faces uncertainty where both likelihood and outcome are unknown. In finance, uncertainty means we do not know the potential outcomes and/or their corresponding likelihoods. Risk means we can assert both the potential outcomes and their corresponding likelihoods. Most quantitative tasks seek to model uncertainty with some risk-based framework. With the complete probability space, uncertainty is reduced to risk, again where both likelihood and outcome are known. For a thorough treatment of this topic, see Harrison and Kreps (1979) and Harrison and Pliska (1981).

Girsanov's theorem relies on a formal mathematical structure. Hence, we introduce some of the formalities here. We assume a finite time horizon; that is, time is modeled as a real number over $(0, \widehat{T})$, for fixed $0 \leq t \leq \widehat{T}$, where $\widehat{T} < \infty$.

The uncertainty is characterized by a complete probability space expressed as (Ω, I, P), where the state space Ω is the set of all possible realizations of the stochastic economy between time 0 and time $\widehat{T}$ and has a typical element ω representing a single sample path, I is called the sigma field of distinguishable events denoted I for information at time $\widehat{T}$, and P is a probability measure defined on the elements of I. P is short for $\text{Pr}(x)$. A sigma field is a means of keeping track of what is known and unknown as illustrated next.

To illustrate a complete probability space, consider a three-period binomial example where each period is one year and the likelihood of an up move is 60%. Thus, the state space can be represented as:

$$\Omega = \left\{ \begin{array}{l} \varnothing, \{d\}, \{u\}, \{d,d\}, \{d,u\}, \{u,d\}, \{u,u\}, \{d,d,d\}, \{d,d,u\}, \\ \{d,u,d\}, \{d,u,u\}, \{u,d,d\}, \{u,d,u\}, \{u,u,d\}, \{u,u,u\} \end{array} \right\},$$

where $\varnothing$ represents the null set. The initial time period, 0, in this illustration is $\widehat{T} - 3$. For example, one element of the state space is $\omega = \{u, d, u\}$. The sigma field of distinguishable events, I, keeps track of the information along the complete past sample path, where the percentages in parentheses here are probabilities:

$$t=0 \qquad I_0 = \left\{ \begin{array}{l} \varnothing, \{d\}, \{u\}, \{d,d\}, \{d,u\}, \{u,d\}, \{u,u\}, \\ \{d,d,d\}, \{d,d,u\}, \{d,u,d\}, \{d,u,u\}, \\ \{u,d,d\}, \{u,d,u\}, \{u,u,d\}, \{u,u,u\} \end{array} \right\} (100\%)$$

$$t=1: \qquad I_1 = \left\{ \begin{array}{l} \{d\}, \{d,d\}, \{d,u\}, \{d,d,d\}, \\ \{d,d,u\}, \{d,u,d\}, \{d,u,u\} \end{array} \right\} (40\%)$$

$$I_0 = \left\{ \begin{array}{l} \{u\}, \{u,d\}, \{u,u\}, \{u,d,d\}, \\ \{u,d,u\}, \{u,u,d\}, \{u,u,u\} \end{array} \right\} (60\%)$$

$$t=2: \qquad I_2 = \{\{dd\}, \{ddd\}, \{ddu\}\} \ (16\%)$$
$$ I_2 = \{\{du\}, \{dud\}, \{duu\}\} \ (24\%)$$
$$ I_2 = \{\{ud\}, \{udd\}, \{udu\}\} \ (24\%)$$
$$ I_2 = \{\{uu\}, \{uud\}, \{uuu\}\} \ (36\%)$$

$$t=3: \qquad I_3 = \{\{d,d,d\}\} \ (6.4\%)$$
$$ I_3 = \{\{d,d,u\}\} \ (9.6\%)$$
$$ I_3 = \{\{d,u,d\}\} \ (9.6\%)$$
$$ I_3 = \{\{d,u,u\}\} \ (14.4\%)$$
$$ I_3 = \{\{u,d,d\}\} \ (9.6\%)$$
$$ I_3 = \{\{u,d,u\}\} \ (14.4\%)$$
$$ I_3 = \{\{u,u,d\}\} \ (14.4\%)$$
$$ I_3 = \{\{u,u,u\}\} \ (21.6\%)$$

An adapted process on a complete probability space is non-anticipating. Generally, process x is said to be adapted if and only if for every outcome x_t, x_t is known at time t. Itô processes are adapted processes.

There are several other highly technical assumptions, but we leave them for further study in the mentioned references. With this foundation, we are now ready for a formal statement of Girsanov's theorem.

15.4 FORMAL STATEMENT OF GIRSANOV'S THEOREM

Define the complete probability space as (Ω, I, P). Let Q be a probability measure defined on I. Q is equivalent to P if for every A that is an element of I, $P(A) = 0$ if and only if $Q(A) = 0$.

Assuming equivalent probability measures for P and Q, the Radon-Nikodym derivative of Q with respect to P is defined as

$$Z \equiv \frac{dQ}{dP} = \frac{\Pr^Q(x)}{\Pr(x)}. \tag{15.11}$$

For our purposes here, recall the properties of the standard Wiener processes based on the notation used here:

- Let W and Z be standard Wiener processes on probability measures P and Q, respectively. Then $W_0 = 0$ and $Z_0 = 0$.
- $W_u - W_t$ and $W_s - W_t$, $Z_u - Z_t$ $Z_s - Z_t$ are P-independent and Q-independent if $0 \leq s < t < u$.
- $W_u - W_t$, $Z_u - Z_t$ are $N(0, u - t)$-distributed under P and Q, respectively.

We are now ready for a formal statement of Girsanov's theorem. We will then proceed to explain various aspects of it.

Girsanov's theorem: Let W_t be a standard Wiener process with respect to probability measure P on I_t. Let (γ_t) be an adapted process such that $\int_0^t \gamma_j dW_j$ is defined. Define a martingale, M_t, as $M_t \equiv e^{\int_0^t \gamma_j dW_j - \frac{1}{2} \int_0^t \gamma_j^2 dj}$. Define a new probability measure Q on I_t as $Q(A) \equiv E(1_A M_t); A \in I_t$. Then $\int_0^t d\widehat{W}_j \equiv \int_0^t dW_j - \int_0^t \gamma_j dj$ is a standard Wiener process with respect to Q.

15.5 CHANGING THE DRIFT IN A CONTINUOUS TIME STOCHASTIC PROCESS

For applications of Girsanov's theorem in finance, the random variable we deal with is often a stochastic process, and in many cases, the random variable will be a Brownian motion, W_t, such that

$$W_t = \int_0^t dW_j, \tag{15.12}$$

where we set $W_0 = 0$ and the increments are distributed with expected value zero and variance dt.[4] The density of W_t is[5]

$$f(W_t) = \frac{1}{\sqrt{2\pi t}} e^{-\frac{1}{2} \frac{W_t^2}{t}}. \tag{15.13}$$

We wish to change this Brownian motion into another process that has a new probability measure Q. We shall shift the distribution by the amount γ_t. In later applications, we shall see that γ_t will become an extremely simple function of t, but for now let us leave it unspecified.

So what we want is a new Brownian motion that has expected value of γ_t. What will accomplish this trick is to designate

$$\phi_t = e^{\int_0^t \gamma_u dW_u - \frac{1}{2} \int_0^t \gamma_u^2 du}. \tag{15.14}$$

For this transformation to be possible, we must impose a constraint on the behavior of γ_u. Specifically, we require that

$$E\left(e^{\int_0^t \gamma_u^2 du}\right) < \infty. \tag{15.15}$$

This statement is called the *Novikov condition*. In simple terms it means that the variation in γ_u must be finite. For all our applications, the Novikov condition will be met.[6]

If these requirements are met, then γ_t can be shown to be a martingale. Let us first apply Itô's lemma on ϕ_t:

$$d\phi_t = \frac{\partial \phi_t}{\partial W_t} dW_t + \frac{1}{2}\frac{\partial^2 \phi_t}{\partial W_t^2} dW_t^2. \tag{15.16}$$

First, we find the partials[7]

$$\frac{\partial \phi_t}{\partial W_t} = \frac{\partial \left(e^{\int_0^t \gamma_u dW_v - \frac{1}{2}\int_0^t \gamma_u^2 du}\right)}{\partial W_t}$$

$$= \phi_t \frac{\partial \left(\int_0^t \gamma_u dW_v - \frac{1}{2}\int_0^t \gamma_u^2 du\right)}{\partial W_t}$$

$$= \phi_t \gamma_t. \tag{15.17}$$

And, because γ_t is not determined by W_t, the second derivative is

$$\frac{\partial^2 \phi_t}{\partial W_t^2} = 0. \tag{15.18}$$

So

$$d\phi_t = \phi_t \gamma_t dW_t = e^{\int_0^t \gamma_u dW_u - \frac{1}{2}\int_0^t \gamma_u^2 du} \gamma_t dW_t. \tag{15.19}$$

Now let us consider the value at $t = 0$:

$$\phi_0 = e^{\int_0^0 \gamma_u dW_u - \frac{1}{2}\int_0^0 \gamma_u^2 du} \gamma_0 dW_0 = e^0 = 1. \tag{15.20}$$

Now we have

$$\int_0^t d\phi_u = \int_0^t \phi_u \gamma_u dW_u = \phi_t - \phi_0, \text{ so} \tag{15.21}$$

$$\phi_t - 1 = \int_0^t \phi_u \gamma_u dW_u,$$

$$\text{and } \phi_t = 1 + \int_0^t \phi_u \gamma_u dW_u. \tag{15.22}$$

Because dW_u on the right-hand side is known to be a martingale, its expectation is zero and, therefore, $E(\phi_t) = 1 = \phi_0$. Consequently, ϕ_t is a martingale.

Thus, we can now be certain that Girsanov's theorem applies. Our Brownian motion can be transformed as follows:

$$W_t^Q = W_t - \int_0^t \gamma_u du, \tag{15.23}$$

where W_t^Q is a Wiener process under the new probability measure $\Pr^Q$ such that

$$\frac{d \Pr Q}{d \Pr} = \phi_Q. \tag{15.24}$$

Remember that ϕ is the Radon-Nikodym derivative.

We shall ultimately need the Wiener differential, dW_t^Q, which is obtained as follows:

$$W_t^Q = W_t - \int_0^t \gamma_u du,$$

$$W_t^Q - W_t = -\int_0^t \gamma_u du \quad \text{implies}$$

$$dW_t^Q - dW_t = -\gamma_t dt, \quad \text{so}$$

$$dW_t^Q = dW_t - \gamma_t dt. \tag{15.25}$$

Now let us step back and think about how this result is important for our purposes. We shall want to convert our asset price process to a martingale. This will remove the drift and permit us to price the option by evaluating its expected future value under the new probability measure. When we remove the drift, what we are doing is removing the risk premium and the risk-free rate. If we know the value of the risk premium, it would be no problem: We would simply subtract it out. But we do not know what the risk premium is. We do not know how much of the asset's expected return to remove. We do know, however, that if we remove just enough that the asset return is a martingale, then we require no discounting whatsoever. So the trick is to change the probability distribution so that the asset return is a martingale. Here is where the finance ends and the math takes over. Girsanov's theorem tells us how to change a probability distribution to leave it the same type of distribution with the same variance but with a different drift. What we have just seen is that the Brownian motion process, W_t, can be changed such that it is still a martingale. Because the asset price process is a simple transformation of the Brownian motion process, it should be easy to transform it as well into a martingale.

We have seen that we are subtracting a function γ_t. This means that γ can potentially change with t. We are somewhat lucky here, because for our purposes γ_t is a very simple function of t: $\gamma_t = \gamma t$.[8] Now notice what we obtain for our Radon-Nikodym derivative:

$$\phi_t = e^{\int_0^t \gamma dW_u - \frac{1}{2}\int_0^t \gamma^2 du} = e^{\gamma \int_0^t dW_u - \frac{1}{2}\gamma^2 \int_0^t du}. \tag{15.26}$$

In other words, if we multiply the density function of W_t by ϕ_t as specified, we should obtain the density function for a new Brownian motion, which we shall call W_t^Q, and in which the expected value has been shifted by γt. Let us see. Given,

$$f(W_t) = \frac{1}{\sqrt{2\pi t}} e^{-(W_t^2)/2t}, \quad \text{and}$$

$$\frac{d\mathrm{Pr}^Q}{d\,\mathrm{Pr}} = e^{(\gamma W_t - \gamma^2 t)/2}, \tag{15.27}$$

we obtain by multiplication

$$f(W_t) = \frac{d\mathrm{Pr}^Q}{d\,\mathrm{Pr}} = \frac{1}{\sqrt{2\pi t}} e^{-(W_t^2/2t + \gamma W_t - \gamma^2 t/2)}$$

$$= \frac{1}{\sqrt{2\pi t}} e^{-(1/2)\left(\frac{W_t - \gamma t}{t}\right)^2}. \tag{15.28}$$

Equation (15.28) is the density for a Brownian motion with its zero expected value shifted by $-\gamma t$. So

$$W_t^Q = W_t - \gamma t. \tag{15.29}$$

To recap, we have that W_t is a Brownian motion under the probability measure Pr, such that

$$E^P(W_t) = W_0 = 0$$

$$E^P(W_t^Q) = E^P(W_t - \gamma t) = E^P(W_t) - \gamma t$$

$$= 0 - \gamma t. \tag{15.30}$$

The first statement defines that W_t is a Brownian motion under Pr. The second statement says that under Pr, W_t^Q is not a Brownian motion. Its expectation, $-\gamma t$, is not zero, except at $t = 0$, and varies with t. But W_t^Q is a Brownian motion under Q:

$$E^Q(W_t^Q) = W_0^Q = 0. \tag{15.31}$$

This statement follows because W_t and W_t^Q both start at a value of zero. Under Q, W_t, is not a Brownian motion because

$$E^Q(W_t) = E^Q(W_t^Q + \gamma t) = E^Q(W_t^Q) + \gamma t = \gamma t. \tag{15.32}$$

When we say that some random process, such as W_t or W_t^Q, is a Brownian motion under a given measure, we are saying that the probabilities of its possible paths are assigned such that its central property, a constant expectation of zero, is preserved. When the probabilities are changed such that the process no longer has a zero expected value, it is no longer the same thing. But another process can and in this case does have the property of a Brownian motion under the new probability measure.

15.8 CHANGING THE DRIFT OF AN ASSET PRICE PROCESS

We previously obtained the familiar stochastic process for an asset:

$$\frac{dS_t}{S_t} = \alpha dt + \sigma dW_t. \tag{15.33}$$

If we change W_t such that now $W_t = W_t^Q + \gamma t$, from Equation (15.29), then we can substitute its differential, $dW_t = dW_t^Q + \gamma dt$, into Equation (15.33) to obtain

$$\frac{dS_t^Q}{S_t^Q} = \alpha dt + \sigma(dW_t^Q + \gamma dt)$$

$$= (\alpha + \sigma\gamma)dt + \sigma dW_t^Q. \tag{15.34}$$

So if we change the probability measure for W_t, which is the same probability that drives S_t, we are now working with this new stochastic process and a new and different set of probabilities. But have we converted S_t^Q to a martingale? Well, not yet. If, however, we specify that $\gamma = -\alpha/\sigma$, the drift becomes zero, leaving us with

$$\frac{dS_t^Q}{S_t^Q} = \sigma dW_t^Q, \tag{15.35}$$

which is clearly a martingale. From this result, we can assign an obvious interpretation to $\gamma = -\alpha/\sigma$, an interpretation at which we have already arrived. First, ignoring the minus sign, the expression α in the numerator is the expected return. The denominator is clearly the risk. Thus, γ is the return over the risk, which makes it a measure of a risk-return trade-off. It is somewhat more natural, however, to specify γ in a slightly different manner:

$$\gamma = -\frac{\alpha - r_c}{\sigma}, \tag{15.36}$$

which, when substituted back into the stochastic differential equation, gives us

$$\frac{dS_t^Q}{S_t^Q} = r_c dt + \sigma dW_t^Q. \tag{15.37}$$

Now we have a more natural interpretation of γ. Again, ignoring the minus sign, the numerator is the expected return minus the risk-free rate or the risk premium. The denominator is the risk.[9] In financial economics, this ratio is the risk premium per unit of risk and is sometimes called the *market price of risk*. It reflects the relative risk-return trade-off, that is, the additional expected return necessary to induce investors to assume risk.[10] Now the adjustment is, more or less, just a subtraction of the risk premium per unit of risk, but the beauty of it all is that we never have to obtain the risk premium. By converting the process to a martingale, we automatically remove the risk premium.

For the log process, recall that its drift, μ, is equal to $\alpha - \sigma^2/2$, so then[11]

$$\gamma = -\frac{\mu + \sigma^2/2 - r_c}{\sigma}. \tag{15.38}$$

But if you have been paying close enough attention, you should note that it appears we no longer have S_t in the form of a martingale. After all, its new expected return is

$$E\left(\frac{dS_t^Q}{S_t}\right) = r_c dt, \tag{15.39}$$

which is typically not zero. With positive interest rates, we have now specified that the asset drifts upward at the risk-free rate. Now we seem to have a problem, but a slightly different spin on things saves the day.

First, we should be comforted in knowing that by removing the risk premium, we have taken out the most difficult part of the problem: the estimation of the risk premium for applied use or empirical research. There is an entire subdiscipline of financial economics called *empirical asset pricing* that is devoted to estimating risk premiums. We should be able to avoid this concern and solve the option pricing problem from what we now know, and indeed that is the case. Financial economists have long known that if we change the asset's expected return to the risk-free rate, we can then evaluate the expected option payoff under the assumption that the asset price is randomly generated by the standard stochastic differential equation with a drift set at the risk-free rate. Economists then go on to explain that everything we need to know about how investors feel about risk is impounded into the asset price. It is not necessary to reflect any effect of investors' risk preferences on the option price. Consequently, we can proceed to evaluate the option as if the expected return on the asset *is* the risk-free rate. This approach is often called *risk-neutral pricing*, and we have covered it previously. What we have done is equivalent to the well-known procedure of taking a short position in an option, hedging it with a long position in units of the underlying asset, thereby eliminating the risk, followed by setting the return on this hedged portfolio to the risk-free rate. From there we obtain a partial differential equation whose solution is the option pricing model. But how do we salvage our approach, which now leaves us with S_t no longer a martingale?

The trick lies in recognizing that we can work with the discounted value of S_t. In other words, say we start off at time 0 with a value of S_0. Then at time t, we have a new value, S_t. But suppose we transform our asset price into its discounted value, $S_t e^{-r_c t}$. Now let us look at some of our previous results. Recall that the solution to the stochastic differential equation that gives us S_t in terms of S_0 was

$$S_t = S_0 e^{\mu t + \sigma W_t}. \tag{15.40}$$

Suppose that instead of working with S_t, we work with its discounted value, $S_t e^{-r_c t}$. Then multiplying the previous equation by $e^{-r_c t}$, we obtain

$$S_t e^{-r_c t} = S_0 e^{(\mu - r_c)t + \sigma W_t}. \tag{15.41}$$

Right now, however, we are under the original probability measure. Substituting $W_t^Q + \gamma t$ for W_t, we obtain

$$S_t e^{-r_c t} = S_0 e^{(\mu - r_c)t + \sigma(W_t^Q + \gamma t)}. \tag{15.42}$$

Noting that we defined γ as $-(\mu - \sigma^2/2 - r_c)/\sigma$ and substituting this result, we obtain

$$S_t e^{-r_c t} = S_0 e^{-\sigma^2 t/2 + \sigma W_t^Q}.$$
(15.43)

You may wish to look back to Chapter 12, Section 5, where we used this result for S_t, along with the density function for a normally distributed W_t to obtain the expected future asset price:

$$E(S_t) = S_0 e^{(\mu + \sigma^2/2)t}.$$
(15.44)

If we follow that same procedure here for $S_t e^{-r_c t}$, we obtain

$$E(S_t e^{-r_c t}) = S_0 e^{-\sigma^2 t/2} e^{\sigma^2 t/2} = S_0.$$
(15.45)

The absence of a positive expected return shows that the discounted price is a martingale.

Let us take one more step. We shall apply Itô's lemma to the discounted asset price. First, for simplicity of notation, let us use S_t^* as $S_t e^{-r_c t}$. Now applying Itô's lemma to S_t^*, we obtain

$$dS_t^* = \frac{\partial S_t^*}{\partial W_t} dW_t + \frac{\partial S_t^*}{\partial t} dt + \frac{1}{2} \frac{\partial^2 S_t^*}{\partial W_t^2} dW_t^2.$$
(15.46)

Using Equation (15.41), we can obtain the partial derivatives:

$$\frac{\partial S_t^*}{\partial W_t} = S_t^* \sigma$$

$$\frac{\partial^2 S_t^*}{\partial W_t^2} = S_t^* \sigma^2$$

$$\frac{\partial S_t^*}{\partial t} = S_t^* (\mu - r_c).$$
(15.47)

Substituting these results into Equation (15.46), and with $dW_t^2 = dt$, we obtain the following stochastic differential equation for S_t^*:

$$\frac{dS_t^*}{S_t^*} = (\mu - r_c + \sigma^2/2)dt + \sigma dW_t$$

$$= (\alpha - r_c)dt + \sigma dW_t.$$
(15.48)

We see that once we have taken the risk-free rate into effect in specifying the underlying variable, we no longer account for the risk-free rate in the drift. But we still do not have a martingale. Recall that to obtain a martingale, we need to substitute $dW_t^Q + \gamma t$ where $\gamma = -(\alpha - r_c)/\sigma$ into Equation (15.48), giving us

$$\frac{dS_t^*}{S_t^*} = \sigma dW_t^Q,$$
(15.49)

which is clearly a martingale, because there is no drift.

To summarize, we adjust the drift of the asset price process by changing the probabilities such that we obtain a martingale. What we also must do is to discount the asset price at the risk-free rate and then work with the discounted asset price to change the probability measure, leaving us with a martingale. Then we easily evaluate the option by applying the probability distribution of the discounted asset price to the option payoff. In this way the option price is found as the expected payoff of the option at expiration, without any discounting, because it has already been done.

If we do the one—discounting the asset price—without the other—changing the measure—we have technically not completed the process, but as it turns out, we can get away with changing the measure without discounting the asset price. Recall that $dS_t/S_t = r_c dt + \sigma dW_t$. So, let us substitute $dW_t^Q + \gamma dt$ and $\gamma = -(\alpha - r_c)/\sigma$ to obtain

$$\frac{dS_t}{S_t} = r_c dt + \sigma dW_t^Q. \tag{15.50}$$

We can work with this model and do the discounting later, that is, after we have evaluated the expected option payoff at expiration. That is because whether we discount before we have derived the expectation or after, we have not altered the fundamental process or the results of taking expectations other than by the simple linear adjustment, $e^{-r_c t}$. Although mathematicians would probably prefer that we convert the asset price to a martingale, requiring that we do the discounting beforehand, financial economists would probably prefer to do the discounting afterwards. That is because the latter approach is more in line with the intuition provided by economic theory: The price of any asset is its expected future value, discounted to the present at an appropriate rate. One typically finds the expectation first and then does the discounting. The fact that in this case the appropriate rate is the risk-free rate is quite intuitive. The risk has been removed via the risk-free hedge, or alternatively, the risk can be viewed as being fully imbedded into the price of the underlying asset and, therefore, cannot legitimately be incorporated again without double counting. Moreover, if the risk is either not present or removed, investors' risk preferences are irrelevant to the valuation process. In that case, one might just as well use the simplest form of risk preferences—risk neutrality—wherein investors discount future values at the risk-free rate.

Finally, we might just simply say that if the price of the underlying asset is given, any two investors regardless of their feelings about risk will value the option in the same manner. Consequently, we can treat both investors as though they had the simplest risk preferences of all, risk neutrality.

15.7 RECAP AND PREVIEW

Girsanov's theorem shows that a martingale can often be represented by another martingale with a change in the location or drift of the process. This result is a pure mathematical concept, but it has great relevance to finance in that it shows that the stochastic process for an asset can be changed into another stochastic process with the expected return equal to the risk-free rate. This result, which we called the change of measure, plays a subtle but important role in derivatives pricing. In this chapter, we showed how this result is derived, and we applied it to the return on an asset on which we would be interested in valuing a derivative.

In Chapter 16, we illustrate how the discrete and continuous stochastic processes we have been working with are related. In doing so, we will show how the discrete change of measure relates to the continuous change of measure.

QUESTIONS AND PROBLEMS

1 Assume a given stock price follows geometric Brownian motion with geometric drift along with known constant mean and known constant standard deviation. Find the equivalent measure with a known risk-free rate and known constant standard deviation.

2 Assume an underlying instrument follows arithmetic Brownian motion with arithmetic drift with a known constant mean and known constant standard deviation. Find the equivalent measure with a known risk-free rate having geometric drift and known constant standard deviation.

3 Suppose x is distributed independent, multivariate normal with each variable's standard deviation of 1, $N(\mu_i, 1)$. Then the PDF is

$$f_{\mathbf{x}}^{\mu,1}(x_1, \ldots, x_n : \mu, \Sigma = I) = \frac{1}{\sqrt{(2\pi)^n}} e^{-\frac{1}{2}\sum_{i=1}^{n}(x_i - \mu_i)^2}.$$

Suppose we want x to be distributed independent, multivariate normal with each variable having mean 0 and standard deviation 1, $N(0,1)$. Then the PDF is

$$f_{\mathbf{x}}^{0,1}(x_1, \ldots, x_n : \mu = 0, \Sigma = I) = \frac{1}{\sqrt{(2\pi)^n}} e^{-\frac{1}{2}\sum_{i=1}^{n}x_i^2}.$$

Find the appropriate Radon-Nikodym derivative to make this transformation.

4 Based on Girsanov's theorem, is it possible to convert arithmetic Brownian motion with geometric drift to geometric Brownian motion with geometric drift?

5 Evaluate whether the following statement related to the martingale representation theorem is true:
Given any two martingales, x_t and y_t, the relationship between them can simply be stated as

$$y_t = y_0 + \int_0^t \lambda_j dx_j.$$

NOTES

1. In addition, a martingale must be finite, and each realization must be independent of the previous one.
2. Of course, the exercise price also determines the option payoff, but it is not random, so it causes us no problems.
3. Mathematicians call this property *previsibility*.
4. Recall that specifically, $dW_t = \varepsilon_t \sqrt{dt}$ where ε_t is a standard normal random variable.
5. The following specification is because W_t has expected value of zero, as it starts off at $W_0 = 0$, and has a variance of t. The term in parentheses is the value of the random variable minus its expected value divided by its variance.
6. As we shall ultimately see, γ_u will be a very simple constraint that in any rational financial market the variation is finite.
7. The result on the third line is obtained from the result on the second line by differentiating the stochastic integral at the end point.
8. If this were not the case, we would see that our method would not work later on.
9. It is even more natural that instead of defining γ as $-(\alpha - r_c)/\sigma$, we define it as $(\alpha - r_c)/\sigma$ and then change the sign such that instead of adding γ to α, we are subtracting it. In that way, we appear to be subtracting a risk premium, a more sensible way to describe what is happening.
10. In the study of general market equilibrium, such as the capital asset pricing model, the appropriate measure of risk is not the standard deviation, but the systematic risk, also known as β. The derivatives pricing framework, however, takes the general equilibrium as given and as fully reflected in the price of the underlying asset. Risk is measured in isolation and σ is the only measure of risk recognized within option pricing theory. You may also recognize that this return/risk ratio is the Sharpe ratio, which is widely used in investment performance evaluation.
11. Let us emphasize that this is not a new definition for γ. It simply expresses γ in terms of the log return.

CHAPTER 16

Connecting Discrete and Continuous Brownian Motions

We have now almost fully developed the binomial and Black-Scholes-Merton models. In this chapter, however, we will circle back and tie their stochastic processes together. Some of this material is repetitive, but it will be helpful in cleaning up some of the loose ends.

To model an asset price, we must start with a framework that reflects the noise produced by random information generated in such a manner that its expectation and volatility are the same through time. As we have described, the model is typically called a *Brownian motion*, named for the Scottish scientist, Robert Brown, who supposedly observed it about 1827 in pollen suspended in water. Much of the scientific work for this model was done by Einstein and Norbert Wiener, the latter for whom a form of Brownian motion, the Wiener process, was named. This process is also called a random walk, though technically a random walk is slightly different.[1]

16.1 BROWNIAN MOTION IN A DISCRETE WORLD

An extremely simple form of Brownian motion is a binomial process in which a random variable W, starting off at a value of $W_0 = 0$, can take on a value of $W_1^+ = +1$ or $W_1^- = -1$ in the next time period or sometimes termed *time step*. We denote the change from the point in time 0 to the point in time 1 as ΔW_0. Thus, the change notation refers to the initial point in time (0 here) and not the final point in time (1 here). Figure 16.1 illustrates this process.

Although some versions of this type of model permit unequal probabilities, many desirable properties are associated with the case where the probabilities of the up and down moves are ½. If, however, one is attempting to model a process with a given expectation and volatility, then there are formulas that specify the probabilities, which will not generally be ½. In the limiting case, that is, when the number of time steps over which a fixed period of time is large, however, it is well known that the formulas converge to ½.

Now let us examine some of the properties of the model. First consider the increment, ΔW_0. The expected value and variance of the increment are

$$E\left(\Delta W_0\right) = \left(\frac{1}{2}\right)(+1) + \left(\frac{1}{2}\right)(-1) = 0$$

$$\text{var}\left(\Delta W_0\right) = \left(\frac{1}{2}\right)(+1 - 0)^2 + \left(\frac{1}{2}\right)(-1 - 0)^2 = 1. \tag{16.1}$$

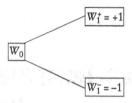

FIGURE 16.1   One-Period Binomial Wiener Process

Thus, because $W_1 = W_0 + \Delta W_0$, its expectation is

$$E(W_1) = E(W_0 + \Delta W_0) = E(W_0) + E(\Delta W_0)$$
$$= 0 + 0 = 0. \tag{16.2}$$

The variance of W_1 is, by definition,

$$\text{var}(W_1) = E(W_1^2) - [E(W_1)]^2. \tag{16.3}$$

Focusing first on the expected value of the square of W_1,

$$E(W_1^2) = \left(\frac{1}{2}\right)(+1)^2 + \left(\frac{1}{2}\right)(-1)^2 = 1. \tag{16.4}$$

Because $E(W_1) = 0$, we have

$$\text{var}(W_1) = 1 - 0^2 = 1. \tag{16.5}$$

Another way to obtain this result is to recall that we want $\text{var}(W_1)$, which is var $(W_0 + \Delta W_0)$. The variance of a sum is the sum of the variances and twice all pairwise covariances, but W_0 is a constant so its variance is zero and, therefore, there is no covariance with ΔW_0. Thus, $\text{var}(W_1) = \text{var}(\Delta W_0)$, which we already found as 1.

In this form, however, the model is too simple and has limited use. We can extend it somewhat by adding time periods. Note, however, that if we just let the process move from +1 to +2 or 0 or from −1 to 0 or −2, the variance would obviously increase.[2] With a large number of time periods, we might find ourselves with an unreasonable variance. One way to scale the variance is to establish that we are trying to model a random process over a fixed period of time. We might say, for example, that we wish to model movements over a period $t = 1$, which might be one year, and we could capture these movements with a model of $n = 2$ periods. We are, therefore, establishing that the time step is $\Delta t = 1/2$. We might be inclined, therefore, to adjust the model so that $\Delta W = \pm 1 \Delta t$, but this adjustment will cause a problem. Intuitively, we might expect that we can better capture reality if we shrink the time interval such that t is fixed but n is large. In that case, $\Delta t \to 0$. What does this leave us with? No motion at all, as it is easy to see that the variance will approach zero.[3]

Alternatively, let us try the model

$$\Delta W = \pm \sqrt{\Delta t}. \tag{16.6}$$

Now let us observe what we have in Figure 16.2.

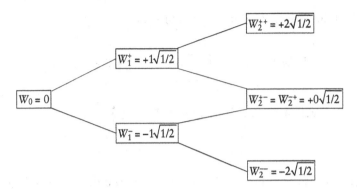

FIGURE 16.2 Two-Period Binomial Wiener Process

Now let us find the expectation and variance of ΔW_0 and ΔW_1. For the first increment,

$$E(\Delta W_0) = \left(\frac{1}{2}\right)\left(+1\sqrt{1/2}\right) + \left(\frac{1}{2}\right)\left(-1\sqrt{1/2}\right) = 0$$

$$\text{var}(\Delta W_0) = \left(\frac{1}{2}\right)\left(+1\sqrt{1/2} - 0\right)^2 + \left(\frac{1}{2}\right)\left(-1\sqrt{1/2} - 0\right)^2 = 1/2. \qquad (16.7)$$

For W_1, note again that the variance is the sum of the variance of W_0 and ΔW_1. W_0 is known so it has no variance nor a covariance with ΔW_1, and we can ignore any covariance between W_0 and ΔW_1. Therefore,

$$E(W_1) = E(W_0) + E(\Delta W_0) = 0 + 0 = 0$$

$$\text{var}(W_1) = \text{var}(W_0) + \text{var}(\Delta W_0) = 0 + 1/2 = 1/2. \qquad (16.8)$$

Now we need the expectation and variance of W_2:[4]

$$E(W_2) = W_0 + E(\Delta W_0) + E(\Delta W_1) = 0 + 0 + 0 = 0$$

$$\text{var}(W_2) = \text{var}(W_0) + \text{var}(\Delta W_0) + \text{var}(\Delta W_1) = 0 + 1/2 + 1/2 = 1. \qquad (16.9)$$

Thus, we observe again a zero mean but a variance that is growing linearly in time. Recall $\Delta t = 1/2$, so after two time steps, we have a variance of one as well as $t = 1$.

Now let us consider the general case of a process W_t, where t can be any future point in time:[5]

$$E(\Delta W_t) = \left(\frac{1}{2}\right)\left(+1\sqrt{\Delta t}\right) + \left(\frac{1}{2}\right)\left(-1\sqrt{\Delta t}\right) = 0$$

$$\text{var}(\Delta W_t) = \left(\frac{1}{2}\right)(+1\sqrt{\Delta t} - 0)^2 + \left(\frac{1}{2}\right)\left(-1\sqrt{\Delta t} - 0\right)^2 = \Delta t. \qquad (16.10)$$

Now consider the expectation and variance of W_t. First let us set the following relationship,

$$W_t = W_0 + \sum_{i=0}^{n-1} \Delta W_i. \qquad (16.11)$$

Then we can find the expected value and variance as

$$E(W_t) = W_0 + E\left(\sum_{i=1}^{n-1} \Delta W_i\right) = 0 + 0 + \ldots 0 = 0$$

$$\text{var}(W_t) = \text{var}(W_0) + \text{var}\left(\sum_{i=1}^{n-1} \Delta W_i\right)$$

$$= 0 + \Delta t + \Delta t + \ldots \Delta t = t. \tag{16.12}$$

The latter result comes from the fact that summing from 0 to $n-1$ is n items. So n times Δt is t because $\Delta t = t/n$.

16.2   MOVING FROM A DISCRETE TO A CONTINUOUS WORLD

It is possible to demonstrate quite formally that if we hold t fixed and increase n, in the limit, the random component, ± 1, will converge to a standard normal random variable, which of course has expected value of 0 and variance of 1.[6] In that case, our model is as follows:

$$dW_t = \varepsilon_t \sqrt{dt}$$

$$W_t = \int_0^t dW_s. \tag{16.13}$$

If we substitute the first equation into the second, the integral looks somewhat strange due to the square root term. This integral is no longer a standard Riemann integral but is instead a stochastic integral. We treated these integrals in Chapter 11, and as it turns out, with constant variance, a stochastic integral of this form is more or less the same as a Riemann integral. Therefore, we can say quite comfortably that with $W_0 = 0$,

$$W_t = \int_0^t dW_s = W_t - W_0 = W_t. \tag{16.14}$$

This result would appear to hold by definition, but a formal proof requires the tools of stochastic integration, which requires defining the integral, not in terms of a limit of an absolute difference but rather the expectation of a mean squared difference. Of course, we covered this in Chapter 11.

The variable W_t is a Brownian motion. Recall the requirements for a Brownian motion are as follows:

- The process commences at zero ($W_0 = 0$) and is continuous.
- The random variable is normally distributed with an expectation of zero and variance of t over time t.
- The increments, $W_{t+dt} - W_t = \Delta W_t$, are independent and normally distributed.

As noted, this process is also sometimes called a *Wiener process*, hence the W notation. Some experts make slight distinctions between Brownian motion and Wiener processes. Some even refer to W_t as the Brownian motion and dW_t as the Wiener process. Technically, our discrete time model is not a Brownian motion; it is often just referred to as a random walk.

We covered these points in Chapter 10 but let us review them here. Remember that $\varepsilon_t \sim N(0, 1)$, so

$$E\left(dW_t\right) = E\left(\varepsilon_t \sqrt{dt}\right) = \sqrt{dt}E\left(\varepsilon_t\right) = 0$$

$$\text{var}\left(dW_t\right) = \text{var}\left(\varepsilon_t \sqrt{dt}\right) = dt\,\text{var}\left(\varepsilon_t\right) = dt$$

$$E(W_t) = E\left(\int_0^t dW_j\right) = \int_0^t E\left(dW_j\right) = 0$$

$$\text{var}(W_t) = \text{var}\left(\int_0^t dW_j\right) = \int_0^t \text{var}\left(dW_j\right) = \int_0^t dj = t. \qquad (16.15)$$

For the last line, note that because the increments are independent, all covariances between dW_j and dW_k are zero, $j \neq k$.

Now suppose we are interested in the covariance between overlapping Brownian motions, W_s and W_t, where $t > s$. Recall due to time series independence, we have

$$\text{cov}\left(W_s, W_t\right) = \text{cov}\left(W_0 + \int_0^s dW_j, W_0 + \int_0^t dW_j\right)$$

$$= \text{cov}\left(W_0, W_0\right) + \text{cov}\left(W_0, \int_0^t dW_j\right) + \text{cov}\left(\int_0^s dW_j, W_0\right) + \text{Cov}\left(\int_0^s dW_j, \int_0^t dW_j\right)$$

$$= 0 + 0 + 0 + \text{cov}\left(\int_0^s dW_j, \int_0^t dW_j\right) = \text{cov}\left(\int_0^s dW_j, \int_0^s dW_j + \int_s^t dW_j\right)$$

$$= \text{cov}\left(\int_0^s dW_j, \int_0^s dW_j\right) + \text{cov}\left(\int_0^s dW_j \int_s^t dW_j\right) = \text{cov}\left(\int_0^s dW_j, \int_0^s dW_j\right) + 0$$

$$= s. \qquad (16.16)$$

Because the correlation is defined as the covariance divided by the product of the standard deviations, the correlation between W_s and W_t would be

$$\rho\left(W_s, W_t\right) = \frac{s}{\sqrt{s}\sqrt{t}} = \sqrt{s/t}. \qquad (16.17)$$

This seemingly minor result turns out to play a major role in compound option pricing models where it is necessary to determine a correlation for the underlying asset price on two dates, with the asset price driven by the value of W on the two dates. We shall cover compound options in Chapter 18.

Now let us look at the probability density for W_t. We know that in general, a normally distributed random variable x with mean μ and variance σ^2 has a density of

$$f(x) = \frac{1}{\sqrt{2\pi\sigma^2}}e^{-\frac{1}{2}\left(\frac{x-\mu}{\sigma}\right)^2}. \tag{16.18}$$

What we know about W_t is that it is normally distributed with $\mu = 0$ and $\sigma^2 = t$. Thus, its density is

$$f(W_t) = \frac{1}{\sqrt{2\pi t}}e^{-\frac{1}{2}\left(\frac{W_t^2}{t}\right)}. \tag{16.19}$$

We also know that dW_t is normally distributed with $\mu = 0$ and $\sigma^2 = dt$. Thus, its density is

$$f(dW_t) = \frac{1}{\sqrt{2\pi dt}}e^{-\frac{1}{2}\left(\frac{dW_t}{\sqrt{dt}}\right)^2}. \tag{16.20}$$

Because we know that $dW_t = \varepsilon_t\sqrt{dt}$, we can substitute into Equation (16.20) and obtain

$$f(dW_t) = \frac{1}{\sqrt{2\pi dt}}e^{-\frac{1}{2}\varepsilon_t^2}. \tag{16.21}$$

Now consider two distinct Brownian motions, W_x and W_y. Let us examine a new process that is a product of these two processes, specifically, $dW_x dW_y$. We know that the two increments are defined as

$$dW_x = \varepsilon_x\sqrt{dt}$$
$$dW_y = \varepsilon_y\sqrt{dt}. \tag{16.22}$$

Now let us find the variance of their product. By definition $\text{var}(dW_x dW_y) = E[(dW_x dW_y)^2] - [E(dW_x dW_y)]^2$. Then $E(dW_x dW_y) = E\left(\varepsilon_x\sqrt{dt}\varepsilon_y\sqrt{dt}\right) = dtE(\varepsilon_x\varepsilon_y)$. Now use the definition of covariance, $\text{cov}(\varepsilon_x, \varepsilon_x) = E(\varepsilon_x\varepsilon_x) - E(\varepsilon_x)E(\varepsilon_y)$, which here reduces to $E(\varepsilon_x\varepsilon_x)$ because the expectation of each ε is zero. Now consider the correlation between the processes ε_x and ε_y, which shall be denoted as ρ_{xy}. Because correlation is the covariance divided by the product of the standard deviations, then $\text{cov}(\varepsilon_x, \varepsilon_x) = \rho_{xy}$ because the two standard deviations are each 1.0. So, $E(\varepsilon_x\varepsilon_x) = \rho_{xy}$.[7] So $E(dW_x dW_y) = \rho_{xy}dt$. But we want the square of this value. Obviously, this square is zero because we have $\rho_{xy}^2 dt^2$ and dt^2 goes to zero in the limit. It follows that the second term in the variance definition, $[E(dW_x dW_y)]^2$, goes to zero because it is $(\rho_{xy}dt)^2$. Thus,

$$\text{var}(dW_x dW_y) = 0. \tag{16.23}$$

If the variance is zero, then

$$E\left(dW_x dW_y\right) = dW_x dW_y = \rho_{xy}dt. \tag{16.24}$$

So we see that the product of two Brownian motions is non-stochastic.

Mathematicians often refer to the derivative with respect to time, in this case dW_t/dt, as the *velocity*.[8] For Brownian motion, however, the velocity does not exist, as shown here:

$$\frac{dW_t}{dt} = \lim_{\Delta t \to 0} \frac{\Delta W_t}{\Delta t} = \lim_{\Delta t \to 0} \frac{\varepsilon_t \sqrt{\Delta t}}{\Delta t} = \lim_{\Delta t \to 0} \frac{\varepsilon_t}{\sqrt{\Delta t}} \to \pm\infty. \tag{16.25}$$

The intuition is that Brownian motion is characterized by infinitesimally small zig-zags. We cannot take limits while permitting the time increment to shrink. For a derivative to exist, we must be able to take a limit, meaning that the line drawn tangent to the point where we are taking the derivative must converge to a stable value, and this simply does not occur here. This result is, however, not a problem. We shall never need the derivative dW_t/dt.

Perhaps the most important characteristic of the process W_t is the property that its squared increment is no longer stochastic. We covered this point previously but let us review it here. Consider the squared variable dW_t^2. Let us take its expectation,

$$E\left(dW_t^2\right) = E\left[(\varepsilon_t \sqrt{dt})^2\right] = dt E(\varepsilon_t^2) = dt, \tag{16.26}$$

which results from the fact that $\text{var}\left(\varepsilon_t\right) = E\left(\varepsilon_t^2\right) - [E\left(\varepsilon_t\right)]^2 = 1$. But $E\left(\varepsilon_t\right)$ is zero, so $E\left(\varepsilon_t^2\right) = \text{var}\left(\varepsilon_t\right) = 1$.

Now let us take the variance of dW_t^2

$$
\begin{aligned}
\text{var}\left(dW_t^2\right) &= E\left[(dW_t^2)^2\right] - \left[E\left(dW_t^2\right)\right]^2 \\
&= E\left[(\varepsilon_t^2 dt)^2\right] - \left[E\left(\varepsilon_t^2 dt\right)\right]^2 \\
&= dt^2 E\left(\varepsilon_t^4\right) - dt^2 \\
&= 0. \tag{16.27}
\end{aligned}
$$

A key element of this result is remembering that in the limit $dt^k \to 0$ for all $k > 1$.

We shall eventually be interested in generalizing our Brownian motion so that it has a nonzero mean and a variance other than t. Technically, this would no longer be a Brownian motion, but it is often still referred to in this manner. Suppose we wish to create a stochastic process, x_t, in which the increment, dx_t, has mean μ and variance σ. Then we simply do the following: Let the process be defined as

$$dx_t = \mu dt + \sigma dW_t. \tag{16.28}$$

The properties of this process are

$$E\left(dx_t\right) = \mu dt \quad \text{and} \quad \text{var}\left(dx_t\right) = \sigma^2 dt. \tag{16.29}$$

The latter result arises because the variance of σdW_t is the constant σ^2 times the variance of dW_t, which is dt. The process, x_t, defined by the stochastic integral

$$x_t = x_0 + \int_0^t dx_j, \tag{16.30}$$

would have the properties

$$E\left(x_t\right) = x_0 + E\left(\int_0^t dx_j\right) = x_0 + \int_0^t E\left(dx_j\right) = x_0 + \int_0^t \mu dj = x_0 + \mu t$$

$$\operatorname{var}\left(x_t\right) = \operatorname{var}\left(\int_0^t dx_j\right) = \int_0^t \operatorname{var}\left(dx_t\right) = \operatorname{var}\left(\int_0^t \sigma^2 dj\right) = \sigma^2 t. \tag{16.31}$$

Note that there are no covariance terms for dx_t in the variance expression in the last line because the dx_t values are independent.

These results give x_t more general characteristics and enable us to use it to model more realistic phenomena, such as stock prices, exchange rates, and so on.

16.3   CHANGING THE PROBABILITY MEASURE WITH THE RADON-NIKODYM DERIVATIVE IN DISCRETE TIME

We examined the Radon-Nikodym derivative for a continuous time process in Chapter 15. In the current chapter, we started with a discrete time Brownian motion model. We then extended it to the continuous time case. Before leaving, we need to return to the discrete time case and examine how to change the probability measure. This procedure is extremely important but is much more difficult in the continuous time case. This material draws heavily on the excellent treatment in Baxter and Rennie (1996).

So let us again go back to the simple world with two outcomes. We can generalize it a little bit without any problems by having our variable W_0 move to W_1^+ or W_1^- with probabilities ϕ and $1 - \phi$, respectively. Now let us suppose that we are interested in changing the probabilities to q and $1 - q$. Without further study, one may wonder why we would want to do this or even whether we could do this without creating a problem. Probabilities are, after all, usually given by external phenomena, and we cannot often change them.[9] But assigned probabilities can actually sometimes be changed. We have already seen that we can create artificial probabilities that lead to correct prices of derivatives but do not require knowledge of the true probabilities, the expected returns of assets, or the utility preferences of investors. This procedure in continuous time, however, is quite complex, as we previously showed. We now illustrate it in the discrete time setting.

So what we have is illustrated in Figure 16.3.

What we want is illustrated in Figure 16.4.

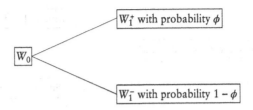

FIGURE 16.3   One-Period Wiener Process with Actual Probability ϕ

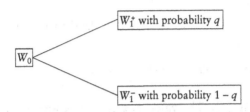

FIGURE 16.4   One-Period Wiener Process with Artificial Probability q

Let us define the ratios of probabilities, q/ϕ and $(1-q)/(1-\phi)$, as γ^s where s indicates the $+$ or $-$ state. That is,

$$\gamma^+ = q/\phi$$
$$\gamma^- = (1-q)/(1-\phi). \tag{16.32}$$

The probabilities q and ϕ represent different probability measures. The probabilities ϕ and $1-\phi$ are said to be measure P and q and $1-q$ are measure Q.[10] We can take the expected value of W_1 under either measure P or measure Q. Under measure Q, we have, by definition,

$$E^Q(W_1) = qW_1{}^+ + (1-q)W_1{}^-. \tag{16.33}$$

This statement can be rewritten as follows:

$$E^Q(W_1) = \phi(q/\phi)W_1{}^+ + (1-\phi)\left[(1-q)/(1-\phi)\right]W_1{}^-. \tag{16.34}$$

We can write this expectation compactly as

$$E^Q(W_1) = E^P(\gamma^s W_1). \tag{16.35}$$

In other words, we can take the expectation under the Q measure by taking the expectation under the P measure, provided we adjust the random process by a new specific stochastic process involving a ratio of probabilities. So γ^s, which is a ratio of probabilities in state s, is a stochastic process itself. We can easily see this in Equation (16.32). It takes on a value of q/ϕ or $(1-q)/(1-\phi)$ in the two states respectively.

Let us extend this result one more period. Figure 16.5 illustrates the process we now have.

Now we must index our ratio of probabilities by time, that is, γ_t^s. First note that the process starts at $\gamma_0 = 1.0$.[11] The stochastic process of γ_t^s can be derived as follows. First consider

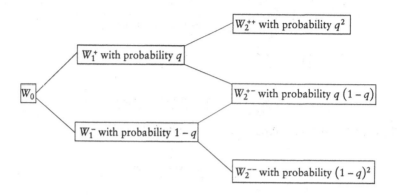

FIGURE 16.5   Two-Period Wiener Process with Probability q

γ_t^+, which as we know is q/ϕ. The value q/ϕ can be expressed as a weighted average of the next two values of the process as shown here:

$$\gamma_1^+ = (\phi)\gamma_2^{++} + (1-\phi)\gamma_2^{+-}$$

$$= (\phi)\left(\frac{q^2}{\phi^2}\right) + (1-\phi)\left(\frac{q(1-q)}{\phi(1-\phi)}\right)$$

$$= q^2/\phi + q(1-q)/\phi$$

$$= \left(q^2 + q - q^2\right)/\phi = q/\phi. \tag{16.36}$$

Thus, in the time 1+ state, we weight the ratio of the probabilities of the next two values of the process, q^2/ϕ^2 and $q(1-q)/\phi(1-\phi)$ by ϕ and $1-\phi$, respectively. Similarly in the time 1− state, we have for γ_1^-,

$$\gamma_1^- = (\phi)\gamma_2^{+-} + (1-\phi)\gamma_2^{--}$$

$$= (\phi)\left[\frac{(1-q)q}{(1-\phi)\phi}\right] + (1-\phi)\frac{(1-q)^2}{(1-\phi)^2}$$

$$= (1-q)q/(1-\phi) + (1-q)^2/(1-\phi)$$

$$= (1-q)/(1-\phi). \tag{16.37}$$

And we see that γ_1^- is a probability-weighted average of the next two values of the process. This exercise shows how γ_t is a stochastic process. Ordinarily, one would not think that a probability is a stochastic process itself, but in this case, the ratio of probabilities is stochastic. At the time 1+ state, there are two upcoming outcomes: γ will equal either q^2/ϕ^2 with probability ϕ or $q(1-q)/\phi(1-\phi)$ with probability $(1-\phi)$. A similar statement applies in the time 1− state. So we see that γ_1^s is a stochastic process itself with values as shown in Figure 16.6.

And as we saw, each value is the expectation of what it will be in the next period. So, in general

$$\gamma_t = E^P\left(\gamma_{t+1}^s\right). \tag{16.38}$$

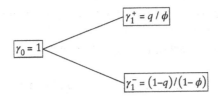

FIGURE 16.6 Stochastic Process of γ

This process γ^s_{t+1} is the discrete time analog to the Radon-Nikodym derivative and represents a relationship between two equivalent probability measures. Equivalent probability measures are two probability measures that meet certain requirements, the main one being that any event that is possible under one measure must be possible under the other and vice versa. In other words, the impossible cannot be created from the possible and the impossible cannot be destroyed from the possible. In continuous time, this concept is much more difficult, except that it is more easily seen for what it is: a derivative of one probability distribution with respect to another. In discrete time, it is merely a ratio of probabilities, but calculus derivatives are technically not defined in discrete time. We covered the continuous case in Chapter 15.

16.4   THE KOLMOGOROV EQUATIONS

We now take a look at a set of equations that appears in the analysis of movements of a stochastic process. These equations are called the *Kolmogorov equations*. They are slight variations of the partial differential equation that is well known in derivative pricing theory in which the option price change is related to its derivatives and changes in time and the underlying. But instead of relating prices to their derivatives, the Kolmogorov equations relate probabilities to their derivatives. There are two such equations: the *forward equation* and *backward equation*.

Consider the following setup in a two-state discrete time world. In other words, the random process can move up or down in a given time and then up or down again and again. Suppose the random variable of interest starts off in state j at time 0. We are interested in the probability that the process is in state k at time n. In the two-state discrete world, the only way the process could get to state k at time n is to be at state $k+1$ at time $n-1$ and move down with probability $1-\phi$ or to be in state $k-1$ at time $n-1$ and move up with probability ϕ. If the probability of moving from state j at time 0 to state $k+1$ at time $n-1$ is $\phi_{j,k+1,n-1}$ and the probability of moving from state j at time 0 to state $k-1$ at time $n-1$ is $\phi_{j,k+1,n-1}$, then the probability of being in state k at time n is

$$\phi_{j,k,n} = (1-\phi)\phi_{j;k+1;n-1} + \phi\phi_{j;k-1;n-1}. \tag{16.39}$$

Note that the parameter ϕ with no subscript is the probability of a move upward, and its complement, $1-\phi$, is the probability of a move downward. We are interested in moving from one state to another, not moving up or down. The variable ϕ with a subscript represents a compound probability, reflecting movements over time and states.

Consider a process x_t that changes by the value φx over the interval Δt. Let us express the probability in Equation (16.39) more generally as $\phi_{x;t;x_0}$, which is the probability of being in state x at time t, given that we were in state x_0 at time 0. We would write this probability as

$$\phi_{x;t;x_0} = \phi\phi_{x-\Delta x;t-\Delta t;x_0} + (1-\phi)\phi_{x+\Delta x;t-\Delta t;x_0}. \tag{16.40}$$

If we expand $\phi_{x-\Delta x;t-\Delta t;x_0}$ and $\phi_{x+\Delta x;t-\Delta t;x_0}$ in a second-order Taylor series, we obtain

$$\phi_{x+\Delta x;t-\Delta t;x_0} = \phi_{x;t;x_0} - \frac{\partial\phi}{\partial t}\Delta t + \frac{\partial\phi}{\partial x}\Delta x + \left(\frac{1}{2}\right)\frac{\partial^2\phi}{\partial x^2}\Delta x^2$$

$$\phi_{x-\Delta x;t-\Delta t;x_0} = \phi_{x;t;x_0} - \frac{\partial\phi}{\partial t}\Delta t - \frac{\partial\phi}{\partial x}\Delta x + \left(\frac{1}{2}\right)\frac{\partial^2\phi}{\partial x^2}\Delta x^2, \tag{16.41}$$

where we have suppressed some of the arguments on the joint probabilities for notational simplicity. We shall bring them back later. We now use (16.41) and (16.40) to obtain

$$\phi_{x;t;x_0} = \phi\left[\phi_{x;t;x_0} - \frac{\partial\phi}{\partial t}\Delta t - \frac{\partial\phi}{\partial x}\Delta x + \left(\frac{1}{2}\right)\frac{\partial^2\phi}{\partial x^2}\Delta x^2\right]$$

$$+ (1-\phi)\left[\phi_{x;t;x_0} - \frac{\partial\phi}{\partial t}\Delta t + \frac{\partial\phi}{\partial x}\Delta x + \left(\frac{1}{2}\right)\frac{\partial^2\phi}{\partial x^2}\Delta x^2\right]. \tag{16.42}$$

Rearranging, we have

$$\phi_{x;t;x_0} = \phi\left[\phi_{x;t;x_0} - \frac{\partial\phi}{\partial t}\Delta t + \left(\frac{1}{2}\right)\frac{\partial^2\phi}{\partial x^2}\Delta x^2\right] - \phi\frac{\partial\phi}{\partial x}\Delta x$$

$$+ \phi_{x;t;x_0} - \frac{\partial\phi}{\partial t}\Delta t + \left(\frac{1}{2}\right)\frac{\partial^2\phi}{\partial x^2}\Delta x^2 + \frac{\partial\phi}{\partial x}\Delta x$$

$$- \phi\left[\phi_{x;t;x_0} - \frac{\partial\phi}{\partial t}\Delta t + \left(\frac{1}{2}\right)\frac{\partial^2\phi}{\partial x^2}\Delta x^2\right] - \phi\frac{\partial\phi}{\partial x}\Delta x$$

$$0 = -\frac{\partial\phi}{\partial t}\Delta t + \frac{\partial\phi}{\partial x}\Delta x + \left(\frac{1}{2}\right)\frac{\partial^2\phi}{\partial x^2}\Delta x^2 - 2\phi\frac{\partial\phi}{\partial x}\Delta x$$

$$\frac{\partial\phi}{\partial t}\Delta t = \frac{\partial\phi}{\partial x}\Delta x(1-2\phi) + \left(\frac{1}{2}\right)\frac{\partial^2\phi}{\partial x^2}\Delta x^2. \tag{16.43}$$

It turns out that an appropriate set of values for ϕ and Δx is[12]

$$\Delta x = \sigma\sqrt{\Delta t}$$

$$\phi = \frac{1}{2}\left(1 + \frac{\mu\sqrt{\Delta t}}{\sigma}\right)$$

$$1-\phi = \frac{1}{2}\left(1 - \frac{\mu\sqrt{\Delta t}}{\sigma}\right). \tag{16.44}$$

Substituting from (16.44) into (16.43), we obtain

$$\frac{\partial \phi}{\partial t}\Delta t = \frac{\partial \phi}{\partial x}\sigma\sqrt{\Delta t}\left(1 - 1 - \frac{\mu\sqrt{\Delta t}}{\sigma}\right) + \left(\frac{1}{2}\right)\frac{\partial^2 \phi}{\partial x^2}\Delta x^2$$

$$= -\frac{\partial \phi}{\partial x}\mu\Delta t + \left(\frac{1}{2}\right)\frac{\partial^2 \phi}{\partial x^2}\sigma^2\Delta t$$

$$\frac{\partial \phi}{\partial t} = -\frac{\partial \phi}{\partial x}\mu + \left(\frac{1}{2}\right)\frac{\partial^2 \phi}{\partial x^2}\sigma^2. \tag{16.45}$$

Reinserting the omitted notation, the result is the forward equation,

$$\frac{\partial \phi_{x;t;x_0}}{\partial t} = -\frac{\partial \phi_{x;t;x_0}}{\partial x}\mu + \left(\frac{1}{2}\right)\frac{\partial^2 \phi\left(x;t;x_0\right)}{\partial x^2}\sigma^2. \tag{16.46}$$

Equation (16.46) is a continuous time representation of a partial differential equation for the probability density of being in a future state at a particular time and is called the *forward equation*, also known as the *Fokker-Planck equation*. It relates the probability at a forward time to the prior paths that take it to that point. The forward equation is a partial differential equation with a solution of the probability density,

$$\phi_{x;t;x_0} = \frac{1}{\sigma\sqrt{2\pi t}}e^{-\frac{1}{2}\frac{(x-x_0-\mu t)^2}{\sigma^2 t}}. \tag{16.47}$$

Equation (16.47) is also recognized as the density for a variable x with mean $x_0 + \Delta t$ and variance $\Delta^2 t$.[13] The special case of $\phi = 0$ is the one-dimensional equation for heat transfer in which $\phi_{x;t;x_0}$ is the temperature.

Now let us look briefly at the backward equation. In this case, we want to know the probability that we started at a particular point, given that we ended up at a particular point. As an example, and to distinguish the backward problem from the forward problem, note Figure 16.7.

Notice how we have labeled the states. For example, the lowest state at time 4 is referred to as state 4,1. The next highest state is labeled 4,2, and so on, with the highest state in time 4 being state 4,5. A similar pattern follows for times 3, 2, and 1. Of course, there is only one state at time 0.

Now suppose we are in state 1,2. Find this state in the Figure 16.7. Suppose we are interested in knowing the probability of being in state 4,4. Note that to get from 1,2 to 4,4, we can go up two periods and down one or down one period and up two. We can also go up one period, then down one period, and then up one period. Thus, there is more than one path that can take us to 4,4 from 1,2. We do not need to consider the path from state 1,1 to 2,2 to 3,3 to 4,4, because it did not start off in state 1,2. The probability of getting to state 4,4 from state 1,2 is the forward probability. Alternatively, we might be in state 4,4 and want to know what is the probability that we came from state 1,2. The paths are the same as in the forward probability, but the perspective is different. We can think of the forward probability as $\Pr(k = 4, 4|j = 1, 2)$ and the backward probability as $\Pr(j = 1, 2|k = 4, 4)$.

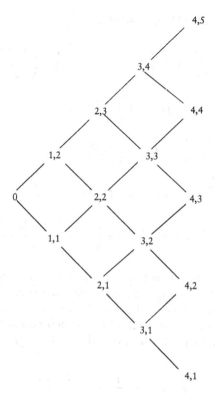

For illustration of the forward probability, let state j be state 1,2; state k be state 4,4; then $n = 3$. Recall the forward probability stated in Equation (16.39):

$$\phi_{j;k;n} = (1 - \phi)\phi_{j;k+1;n-1} + \phi\phi_{j;k-1;n-1}.$$

In this example, to go from state 1,2 to state 4,4, the up state one state before state 4,4 is state 3,4 and the down state one state before state 4,4 is state 3,3; thus, state $k + 1$ is equivalent to state 3,4 and state $k - 1$ is equivalent to state 3,3. We can write the forward looking probability as

$$\Pr(k = 4, 4 \,|\, j = 1, 2) = \phi_{(1,2);(4,4);3} = (1 - \phi)\phi_{(1,2);(3,4);2} + \phi\phi_{(1,2);(3,3);2}.$$

Now, we can take this further by finding expressions for $\phi_{(1,2);(3,4);2}$ and $\phi_{(1,2);(3,3);2}$. The probability $\phi_{(1,2);(3,4);2}$ represents the path of moving from state 1,2 to state 3,4 in two steps. This is accomplished in the same way as in the binomial world in exactly one way, two up movements or ϕ^2. The probability $\phi_{(1,2);(3,3);2}$ represents the paths of moving from state 1,2 to state 3,3, which is accomplished in two different ways, both having exactly one up and one down movement or $2\phi(1 - \phi)$. Thus, we can write the probability further as

$$\Pr(k = 4, 4 | j = 1, 2) = \phi_{(1,2);(4,4);3}$$

$$= (1 - \phi)\phi_{(1,2);(3,4);2} + \phi\phi_{(1,2);(3,3);2}$$

$$= (1 - \phi)\phi^2 + \phi 2\phi(1 - \phi)$$

$$= 3\phi^2(1 - \phi).$$

Now, more generally for the backward probability, let $\phi_{j;k;n}$ be the probability of going from state j at time 0 to state k in n steps. Let $\phi_{j-1;k;m}$ be the probability of going from state $j - 1$ to k in m steps and $\phi_{j+1;k;m}$ be the probability of going from state $j + 1$ to state k in m steps. Given that the probability is ϕ that we went up to state $j + 1$ at time $n - m$ and $1 - \phi$ that we went down to state $j - 1$ at time $n - m$, then the probability of going from state j to state k in $m + 1$ steps is

$$\phi_{j;k;m+1} = \phi\phi_{j+1;k;m} + (1 - \phi)\phi_{j-1;k;m}. \tag{16.48}$$

For illustration of the backward probability, let state k be state 4,4; state j be state 1,2; then $n = 3$, $m = 2$, *and* $n - m = 1$. In this example, to go from state 4,4 to state 1,2, the upstate one state before reaching state 1,2 from state 4,4 is state 2,3 and the down state one state before state 1,2 is state 2,2; thus, state $j + 1$ is equivalent to state 2,3 and state $j - 1$ is equivalent to state 2,2. We can write the backward looking probability as

$$\phi_{j;k;m+1} = \phi\phi_{j+1;k;m} + (1 - \phi)\phi_{j-1;k;m}$$

$$\Pr(j = 1, 2 | k = 4, 4) = \phi_{(1,2);(4,4);3} = \phi\phi_{(2,3);(4,4);2} + (1 - \phi)\phi_{(2,2);(4,4);2}.$$

Because we are asking what the probability is that we came from state 1,2, then we should note that it is impossible to have ever moved through some states. In Figure 16.8, the nodes where it would be impossible to have traveled from to get to the current state 4,4 are denoted in bold.

Now, suppose for a minute that state 4,4 has a lattice of its own where the lattice arrives to state 4,4 as the final point. Figure 16.9 illustrates this particular backward lattice.

This backward lattice consists of states that do not actually exist. The impossible states are expressed as bold. The possible states correspond to the nonbold states shown in Figure 16.8. If we treat this backward lattice as a usual lattice found in the binomial world, then we can use the usual binomial probability of each state, with a probability (down movement) = $1 - \phi$ and probability (up movement) = ϕ.

For the columns 0, 1, and 2, we can see that there exists states in each column that are not real. If we are coming from a state in one of these columns, then we must consider that we came from only a possible state in the column (not a fake state created during the creation of the backward lattice). Conditional probabilities are used to update the probabilities for a given column. Column 1, for example, has only two possible states—1,1

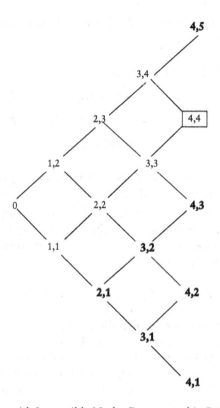

FIGURE 16.8 Forward Lattice with Impossible Nodes Demarcated in Bold

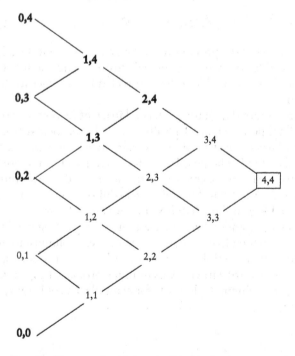

FIGURE 16.9 Backward Lattice Illustrated with Impossible Nodes Demarcated in Bold

and 1,2; thus, if we are looking for the probability that we came from a state in column 1 given that we are currently in state 4,4, we must update the probability to be conditional that we have to be in either state 1,1 or 1,2 only. Thus,

$$\Pr(j = 1,2 \cup j = 1,1|k = 4,4 \cap j = 1, x) = \phi_{(1,1);(4,4);3} + \phi_{(1,2);(4,4);4} = 1.$$

Following a similar procedure as we did for the forward equation, we obtain the continuous time backward equation

$$\frac{\partial \phi(x,t;x_0)}{\partial t} = \mu \frac{\partial \phi(x,t;x_0)}{\partial x_0} + \frac{1}{2}\frac{\partial^2 \phi(x,t;x_0)}{\partial x_0^2}\sigma^2, \qquad (16.49)$$

where we note that the difference from the forward equation is that here we are viewing the problem from the current position x_0. In the forward equation, we are viewing the problem from the future position, x. The solution to the backward equation, (16.49), is the same as the solution to the forward equation, (16.46), namely, the density (16.47).

The Kolmogorov equations for continuous time express the probability densities of looking from one state to another, either looking forward in time or looking back. The general idea behind such equations is to specify the likelihoods of various paths that a random variable can take in the future or has taken in the past. As is often the case, the lesson can oftentimes be more easily seen in discrete time. As such, we introduce a discrete time example.

Figure 16.10 shows a three-period binomial tree for an asset priced at 100 that can go up by 26% or down by 20% each period. The probability of an up move is 0.54, so the probability of a down move is 0.46. There are three values in each cell: the asset price, a cell identifier, and the probability of moving to that cell from the initial cell, 0,0.

For the discrete time analog of the forward equations, we present Table 16.1, which indicates the probability of being in a future state, given that one is in a current state. The

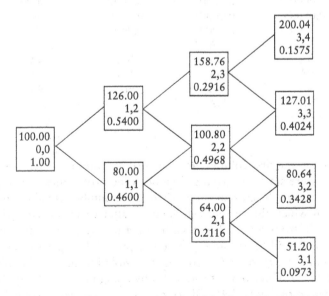

FIGURE 16.10 Binomial Tree for Illustrating Forward and Backward Probabilities

TABLE 16.1 Forward Probabilities in a Binomial World

Panel A. Points in Time 0–3

Current State	Future States									
	0,0	1,1	1,2	2,1	2,2	2,3	3,1	3,2	3,3	3,4
0,0	NA	0.4600	0.5400	0.2116	0.4968	0.2916	0.0973	0.3428	0.4024	0.1575
1,1	NA	NA	NA	0.4600	0.5400	NA	0.2116	0.4968	0.2916	NA
1,2	NA	NA	NA	NA	0.4600	0.5400	0.0000	0.2116	0.4968	0.2916
2,1	NA	NA	NA	NA	NA	NA	0.4600	0.5400	0.0000	0.0000
2,2	NA	NA	NA	NA	NA	NA	0.0000	0.4600	0.5400	0.0000
2,3	NA	NA	NA	NA	NA	NA	0.0000	0.0000	0.4600	0.5400
3,1	NA	NA	NA	NA	NA	NA	NA	NA	NA	NA
3,2	NA	NA	NA	NA	NA	NA	NA	NA	NA	NA
3,3	NA	NA	NA	NA	NA	NA	NA	NA	NA	NA
3,4	NA	NA	NA	NA	NA	NA	NA	NA	NA	NA

Panel B. Point in Time 4

Current State	Future States				
	4,1	4,2	4,3	4,4	4,5
0,0	0.0448	0.2102	0.3702	0.2897	0.0850
1,1	0.0973	0.3428	0.4024	0.1575	0.0000
1,2	0.0000	0.0973	0.3428	0.4024	0.1575
2,1	0.2116	0.4968	0.2916	0.0000	0.0000
2,2	0.0000	0.2116	0.4968	0.2916	0.0000
2,3	0.0000	0.00000	0.2116	0.4968	0.2916
3,1	0.4600	0.5400	0.0000	0.0000	0.0000
3,2	0.0000	0.4600	0.5400	0.0000	0.0000
3,3	0.0000	0.0000	0.4600	0.5400	0.0000
3,4	0.0000	0.0000	0.0000	0.4600	0.5400
4,1	NA	NA	NA	NA	NA
4,2	NA	NA	NA	NA	NA
4,3	NA	NA	NA	NA	NA
4,4	NA	NA	NA	NA	NA
4,5	NA	NA	NA	NA	NA

rows, as described in the first column, are the current states in which one could be. The columns are the future states. A given cell value is the probability of being in the future state, given that one is in the current state. There are a number of values indicated as NA, which are cases in which the current cell is either after or in the same time step as the future state. That is, from state 2,2, you cannot get to any prior states, nor can you get to state 2,1 or 2,3. For all future states, there is a probability given when feasible. From state 2,2, you cannot get to state 3,1, so it is infeasible but that probability is shown as 0.0000. From state 2,2, you can get to state 3,3 by going up, probability 0.54, or to state 3,2 by going down, probability 0.46. Of course, you cannot get to state 3,4 from 2,2, so that probability is 0.0000.

TABLE 16.2   Backward Probabilities in a Binomial World

Current State	State You Came From									
	0,0	1,1	1,2	2,1	2,2	2,3	3,1	3,2	3,3	3,4
0,0	NA	NA	NA	NA	NA	NA	NA	NA	NA	NA
1,1	1.0000	NA	NA	NA	NA	NA	NA	NA	NA	NA
1,2	1.0000	NA	NA	NA	NA	NA	NA	NA	NA	NA
2,1	1.0000	1.0000	0.00	NA	NA	NA	NA	NA	NA	NA
2,2	1.0000	0.5400	0.4600	NA	NA	NA	NA	NA	NA	NA
2,3	1.0000	0.0000	1.000	NA	NA	NA	NA	NA	NA	NA
3,1	1.0000	1.0000	0.0000	1.0000	0.0	0.0	NA	NA	NA	NA
3,2	1.0000	0.7013	0.2987	0.5400	0.4600	0.0	NA	NA	NA	NA
3,3	1.0000	0.3699	0.6301	0.0000	0.5400	0.4600	NA	NA	NA	NA
3,4	1.0000	0.0000	1.0000	0.0000	0.0000	1.0000	NA	NA	NA	NA
4,1	1.0000	1.0000	0.0000	1.0000	0.0000	0.0000	1.0	0.00	0.00	0.00
4,2	1.0000	0.7788	0.2212	0.7013	0.2987	0.0000	0.54	0.46	0.00	0.00
4,3	1.0000	0.5400	0.4600	0.2916	0.4968	0.2116	0.0	0.54	0.46	0.00
4,4	1.0000	0.28125	0.71875	0.0000	0.3699	0.6301	0.0	0.00	0.54	0.46
4,5	1.0000	0.0000	1.0000	0.0000	0.0000	1.0000	0.0	0.00	0.00	1.00

Referring back to our first forward probability example, we can see that this table matches what we had previously said with

$$\Pr(k = 4, 4 \mid j = 1, 2) = \phi_{(1,2);(4,4);3} = 3\phi^2(1 - \phi) = 3(0.54^2)(1 - 0.54) = 0.4024.$$

The backward probabilities are slightly more difficult to understand. A backward probability asks the question of, given you are in a particular state, what is the probability you got here from a specific past state. Table 16.2 shows the backward probabilities for all possible current states. The values of NA indicate states from which you could not have come, such as the same state or a contemporaneous state. Inasmuch as there is no time travel, you also could not have come from a future state. You can, however, have come from a prior state. For example, in state 2,2, there is a 0.46 chance you came from state 1,2, having gone down, and a 0.54 chance you came from state 1,1, having gone up. Notice that the probability is 1.0000 that you were in state 0,0.

Understanding these types of measures is important for path-dependent options. The forward measures are clearly more important in finance, because we nearly always look ahead rather than back, but understanding how something came about is helpful in modeling. Estimates of the future nearly always start with what has happened in the past. The historic paths taken by asset prices can be helpful in developing and validating models. Thus, they are an important element in the study of stochastic processes.

16.5   RECAP AND PREVIEW

In this chapter, we showed how the change of probability measure in the discrete time stochastic process that we use is related to the change of probability measure in the continuous time stochastic process we use. We also show the Kolmogorov equations, which define how movements at one point in time are related to movement at a future or prior point in time.

This chapter completes Part III. In Part IV, we return to a more direct application of the Black-Scholes-Merton model. Specifically, we examine types of options other than the standard European option, and we derive the relevant pricing models under the Black-Scholes-Merton assumptions.

QUESTIONS AND PROBLEMS

1 Assume we wish to model movements over the next calendar month, $t = 1/12$, with a two-period model, $n = 2$ periods. Propose a Wiener process with an expected value of 0 and standard deviation of $\sqrt{t} = \sqrt{1/12}$.

2 Assume we wish to model movements over some generic time period, t, with an n-period model. Propose a Wiener process with an expected value of 0 and standard deviation of $\sqrt{t}$.

3 Suppose we have two points in time, s and u. Unfortunately, we do not know which point in time is larger than the other point in time. Derive the covariance and correlation between two overlapping Brownian motions, W_s and W_u.

4 Assume that the actual probability of an up move is $\phi = 60\%$ and the artificial probability of an up move is $q = 50\%$ for each time step. Build a two-period binomial model of the Radon-Nikodym derivative and show that this process is a martingale under the actual probability measure but not the artificial probability measure.

5 Assuming arithmetic Brownian motion with arithmetic drift with constant coefficients, derive the expected value and standard deviation.

NOTES

1. A random walk is a stochastic process consisting of a series of independent and identically distributed random variables. A Brownian motion is a random walk, but all random walks are not Brownian motions. Technically, Brownian motions are required to be continuous. We relax this requirement in these introductory sections as we eventually make the limiting argument.
2. Now there would be three outcomes at time step 2: +2 with probability ¼, 0 with probability ½, and –2 with probability ¼. The expected value is still zero, but the variance is now $(1/4)(2 - 0)^2 + (1/2)(0 - 0)^2 + (1/4)(-2 - 0)^2 = 2$.
3. To obtain the variance we would end up squaring Δt, which would take it to zero when Δt is small.
4. We could also find the variance of W_2 the traditional way, by probability weighting the three squares of the outcomes less the expected value squared.
5. Recall we follow standard notation where the subscript is simple t even though it is not observed until $t + \Delta t$ because we eventually focus on the limiting result.
6. The formal proof of this result is a variation of the DeMoivre-LaPlace limit theorem, which proves that a binomial distribution converges to a normal distribution in the limit.

7. The covariance is actually the product of the correlation and the two standard deviations of the εs, which are both 1.0.
8. The term *velocity* refers to the speed of something. Hence, a derivative such as dW_t/dt would refer to the rate at which W_t changes with time, which, of course, is the speed.
9. Consider, however, that blackjack dealers frequently change decks to alter the probabilities.
10. It would be convenient if the probabilities under measure P were p and $1 - p$, or the value of a put.
11. At time 0, there is only one state, so we do not need to index it by state.
12. It can be proven by using the moment-generating function that in the limit these values produce the desired periodic mean, μ, and volatility, σ.
13. That this density solves the differential equation can be verified by taking the partial derivatives with respect to t and x and substituting back into the differential equation.

Extensions and Generalizations of Derivative Pricing

CHAPTER 17

Applying Linear Homogeneity to Option Pricing

In this chapter, we shall illustrate how the mathematical principle of linear homogeneity and its associated Euler rule can be used in pricing options. Our objective is to illustrate the pricing of exchange options and forward start options. We shall, however, illustrate how the Black-Scholes-Merton model can be priced using this procedure. We then show how the exchange option model is a more general model than Black-Scholes-Merton, which is but a special case of an exchange option. The analysis provided here relies on the concepts of homogeneous functions, linear homogeneity, and Euler's rule. But first let us introduce exchange options.

17.1 INTRODUCTION TO EXCHANGE OPTIONS

We will use the concept of linear homogeneity to price exchange options but we first have to define an exchange option. The exchange option, first covered by Margrabe (1978), has proven to be an extremely powerful generalization of the Black-Scholes-Merton model. In addition, the intuitive insights in the derivation of the exchange option model are very useful in other applications in option pricing.

Consider a European call option in which at expiration the holder can exchange one unit of asset 2 and receive one unit of asset 1. Let $c_t(S_1, S_2)$ denote the price of this call option. Its payoff at expiration is

$$c_T(S_1, S_2) = \max(0, S_{1T} - S_{2T}). \tag{17.1}$$

With this option, asset 2 plays the role of the exercise price, but the difference from the standard case is that the value of asset 2 is stochastic. Thus, an exchange option is an option with a stochastic exercise price. Alternatively, one can view this option as a put in which asset 2 can be exchanged for asset 1. In the context of a put, asset 1 plays the role of the exercise price. Let this option price be denoted as $p_t(S_2, S_1)$ and the payoff at expiration be

$$p_T(S_2, S_1) = \max(0, S_{1T} - S_{2T}). \tag{17.2}$$

Thus, the current prices of the options are equal, $p_t(S_2, S_1) = c_t(S_1, S_2)$.

TABLE 17.1   Put-Call Parity of Exchange Options

			Value at Expiration	
	Instrument	Current Value	$S_{1T} \leq S_{2T}$	$S_{1T} > S_{2T}$
Portfolio A	Long exch. call to exchange asset 2 for asset 1	$c_t(S_1, S_2)$	0	$S_{1T} - S_{2T}$
	Short exch. put to exchange asset 1 for asset 2	$-p_t(S_1, S_2)$	$-(S_{2T} - S_{1T})$	0
	Total	$c_t(S_1, S_2)$ $-p_t(S_1, S_2)$	$S_{1T} - S_{2T}$	$S_{1T} - S_{2T}$
Portfolio B	Long asset 1	S_{1t}	$+S_{1T}$	$+S_{1T}$
	Short asset 2	$-S_{2t}$	$-S_{2T}$	$-S_{2T}$
	Total	$S_{1t} - S_{2t}$	$S_{1T} - S_{2T}$	$S_{1T} - S_{2T}$

First let us establish put-call parity between exchange options. In the previous example, we argued that $p_t(S_2, S_1) = c_t(S_1, S_2)$. But what about the relationship of $c_t(S_2, S_1)$ to $p_t(S_2, S_1)$? Keep in mind that in options notation, the first term in parentheses is the underlying, and the second is the strike. Table 17.1 illustrates equivalent portfolios A and B, with the former being long the call and short the put and the latter being long asset 1 and short asset 2. We see that the payoffs of the two portfolios are equivalent.

Because the portfolios have the same payoff, their current values must be equal. Therefore,

$$c(S_1, S_2) - p(S_1, S_2) = S_1 - S_2. \tag{17.3}$$

In short, the difference in the value of a call option to exchange asset 2 for asset 1 and the value of a put option to exchange asset 1 for asset 2 is the difference in the values of asset 2 and asset 1. This difference will be negative when asset 2 is priced higher than asset 1.

With this introduction to exchange options, we turn to explaining homogeneous functions, linear homogeneity, and Euler's rule, all useful concepts related to valuing exchange options.

17.2   HOMOGENEOUS FUNCTIONS

A function $f(x_1, x_2, \ldots, x_n)$ is said to be homogeneous of degree k with respect to each variable, $x_1, x_2, \ldots, x_n$ if the following condition holds

$$f(\lambda x_1, \lambda x_2, \ldots, \lambda x_n) = \lambda^k f(x_1, x_2, \ldots, x_n). \tag{17.4}$$

An example of such a function is

$$g(x, y, z) = 2x^2 + 3yz - z^2.$$

Here we see that it is homogeneous of degree two.

$$g(\lambda x, \lambda y, \lambda z) = 2\lambda^2 x^2 + 3\lambda^2 yz - \lambda^2 z^2$$
$$= \lambda^2 \left(2x^2 + 3yz - z^2\right).$$

What we did is multiply each variable by λ, and the result is the original function multiplied by λ^2. In that case, the function is homogeneous of degree two with respect to x, y, and z.

A function can be homogeneous with respect to fewer than the full number of variables in it. For example, from Equation (17.4), a function $f(x, y, z)$ is said to be homogeneous of degree k with respect to the variables x and y if the following condition holds:

$$f(\lambda x, \lambda y, z) = \lambda^k f(x, y, z).$$

In other words, if variables x and y are each increased by the factor λ, and it leads to the function increasing by the factor λ^k, then the function is said to be homogeneous of degree k with respect to x and y. Note that we did not increase z by the factor λ. Homogeneity is always expressed with respect to one or more variables.

The most fundamental case of homogeneity is degree zero. That is, if we multiply one or more variables by λ^0, we naturally obtain the original function times $\lambda^0 = 1$. In other words, the function does not change at all. Consider the simple function $f(x, y, z) = xz/y$. Then note what happens when we multiply x and y by λ:

$$f(\lambda x, \lambda y, z) = \frac{\lambda xz}{\lambda y} = \frac{xz}{y}.$$

This function is unaffected by changing x and y by the factor λ^0 and is, thus, homogeneous of degree zero with respect to x and y. Note, however, that if we tried just z, we would get the result $x\lambda z/y$, which is $\lambda xz/y$, and the function is homogeneous of degree 1 with respect to z, but it is not homogeneous of degree 0.

For a case of linear homogeneity, or homogeneity of degree 1, consider the function

$$f(x, y, z) = z\sqrt{xy}.$$

It is homogeneous of degree one with respect to x and y because

$$f(\lambda x, \lambda y, z) = z\sqrt{\lambda x \lambda y} = \lambda z\sqrt{xy} = \lambda f(x, y, z).$$

With respect to x, y, and z, however, it is homogeneous of degree two:

$$f(\lambda x, \lambda y, \lambda z) = \lambda z\sqrt{\lambda x \lambda y} = \lambda^2 z\sqrt{xy} = \lambda^2 f(x, y, z).$$

17.3 EULER'S RULE

Functions that are homogeneous of degree one are referred to as *linearly homogeneous*, and this property is called *linear homogeneity*. The Swiss mathematician Leonhard Euler proved that linearly homogeneous functions have the property that

$$x\frac{\partial f}{\partial x} + y\frac{\partial f}{\partial y} = f(x,y), \tag{17.5}$$

which is called *Euler's rule*.

In economics, this special characteristic is used to describe production functions that have constant returns to scale. That is, if an economic output changes by a factor λ when the inputs change by the factor λ, then we say that the production function has constant returns to scale. This type of production function is relatively simple and convenient to use, and so it appears often in microeconomic models. Typically the inputs are capital (K) and labor (L).

For example, consider the economic production function

$$Q = AK^{\alpha}L^{1-\alpha}; A > 0; 0 < \alpha < 1,$$

where A is a scalar. Clearly, this economic production function is homogeneous of degree one with respect to capital and labor because

$$\hat{Q} = A(\lambda K)^{\alpha}(\lambda L)^{1-\alpha} = A\lambda K^{\alpha}L^{1-\alpha}.$$

Further, note that if we define $f(K,L) \equiv AK^{\alpha}L^{1-\alpha}$, then

$$K\frac{\partial f}{\partial K} + L\frac{\partial f}{\partial L} = K\alpha\frac{AK^{\alpha}L^{1-\alpha}}{K} + L(1-\alpha)\frac{AK^{\alpha}L^{1-\alpha}}{L} = AK^{\alpha}L^{1-\alpha} = f(K,L).$$

Thus, we see that this production function is linearly homogeneous and therefore satisfies Euler's rule.

17.4 USING LINEAR HOMOGENEITY AND EULER'S RULE TO DERIVE THE BLACK-SCHOLES-MERTON MODEL

In Merton's (1973b) classic article on option pricing theory, he demonstrated that the price of a European call option is linearly homogeneous with respect to the asset price and exercise price. Merton derived this result by defining the problem in terms of the rate of return on the option. In this chapter, we shall take a slightly different but economically equivalent approach.

As we previously covered, the price of an option can be obtained by appealing to the risk-neutral or equivalent martingale approach, which changes the probabilities of the expiration value of the underlying asset such that the current asset price is the expected future asset price discounted at the risk-free rate. As such, the expected expiration value of the option is taken using the adjusted probabilities and discounted at the risk-free rate to obtain the well-known Black-Scholes-Merton formula.

An option is an asset, so by definition the price at time t of an option is the expected future value of that option discounted at a rate appropriate for the time value of money and the risk. Consequently, without changing the probability distribution, we can obtain the option pricing formula by taking the expected value of the option at expiration using the actual probabilities and discounting at a rate suitable for the option's risk. In other words,

$$c_t = e^{-k\tau} E(c_T),\qquad(17.6)$$

where k is the risk-adjusted discount rate appropriate for the option and $\tau = T - t$ is the time to expiration. Again, this statement holds by definition for any asset, whether an option or not.

Of course, the value of a European call option at expiration T is

$$c_T = \max(0, S_T - X),\qquad(17.7)$$

where S_T is the underlying asset price at expiration and X is the exercise price. Now observe that this function is linearly homogeneous with respect to the asset price and exercise price, because

$$\max(0, \lambda S_T - \lambda X) = \lambda \max(0, S_T - X).\qquad(17.8)$$

Thus, we know that the expiration value of the option is linearly homogeneous, but we are interested in whether the current value of the option is linearly homogeneous. Here we appeal to the aforementioned result that the current value of any asset is the expected future value discounted back to the current time. Determining an expected value is a linear mathematical operation. That is, taking an expectation is a linear operation and discounting merely involves multiplying an expected payoff by a discount factor. Note that a linear operation performed on a linearly homogeneous function preserves its property of linear homogeneity. Discounting an expected payoff, such as by the factor $e^{-k\tau}$, involves simply multiplying it by a constant, so we preserve the linear homogeneity of the function. Thus, the current price of the option is linearly homogeneous with respect to the asset price and the exercise price. Hence, we can employ Euler's rule, and it will be quite useful.

Applying Euler's rule to the Black-Scholes-Merton model call price, c_t, focused on S_t and X, we have

$$S_t \frac{\partial c}{\partial S} + X \frac{\partial c}{\partial X} = c(S_t, X).\qquad(17.9)$$

From Chapter 14, we know

$$\Delta_c = \frac{\partial c}{\partial S} = N(d_1) \text{ and}\qquad(17.10)$$

$$\frac{\partial c}{\partial X} = -e^{-r_c \tau} N(d_2).\qquad(17.11)$$

By substituting these derivatives, we attain a statement of the Black-Scholes-Merton model that is consistent with linear homogeneity. We introduce the current price of a zero-coupon bond with par value of X as $B_t \equiv e^{-r_c \tau} X$. Hence,

$$S_t \frac{\partial c}{\partial S} + B_t \frac{\partial c}{\partial B_t} = c(S_t, X),\qquad(17.12)$$

where

$$\frac{\partial c}{\partial B_t} = \frac{\partial c}{\partial X}\frac{\partial X}{\partial B_t} = \frac{\partial c/\partial X}{\partial B_t/\partial X} = -\frac{e^{-r_c\tau}N(d_2)}{e^{-r_c\tau}} = -N(d_2). \tag{17.13}$$

Alternatively, let us simply start by assuming the option price is linearly homogeneous with respect to S_t and B_t, two assets that can be easily traded. Note that we are not starting with any specified stochastic process. Using Euler's rule, we know that the current option price, which is a function of S_t and t, can be expressed as

$$c_t = \gamma S_t + \omega B_t, \tag{17.14}$$

where γ is $\partial c_t/\partial S_t$ and ω is $\partial c_t/\partial B_t$. An interpretation of Equation (17.14) is that one call option can be replicated by holding γ units of the asset, each worth S_t, and ω units of a risk-free bond with current value $B_t = Xe^{-r_c\tau}$.

The total differential of the option price is

$$dc_t = \gamma dS_t + \omega dB_t. \tag{17.15}$$

The differential dS_t can be left in this form. We can, however, obtain the exact differential for the risk-free bond, dB_t. We simply take the derivative with respect to t,

$$\frac{dB_t}{dt} = r_c Xe^{-r_c\tau} = r_c B_t. \tag{17.16}$$

We then obtain the desired differential as

$$dB_t = r_c B_t dt. \tag{17.17}$$

Consequently, the call price differential is

$$dc_t = \gamma dS_t + \omega r_c B_t dt. \tag{17.18}$$

At this point, it is helpful to substitute $c_t - \gamma S_t$ for ωB_t from Equation (17.14), leaving us

$$dc_t = \gamma dS_t + (c_t - \gamma S_t)\, r_c dt. \tag{17.19}$$

Now, we assume only that the call price depends on S_t and t and that the underlying follows an Itô process expressed as $dS_t = \alpha(S_t, t)\, dt + \sigma(S_t, t)\, dW_t$. Itô's lemma can be used to obtain an equivalent expression for the change in the price of the call,

$$dc_t = \frac{\partial c_t}{\partial S_t}dS_t + \frac{\partial c_t}{\partial t}dt + \frac{1}{2}\frac{\partial^2 c_t}{\partial S_t^2}\sigma^2(S_t, t)\, dt, \tag{17.20}$$

from which we can now set Equation (17.20) equal to Equation (17.19). And by choosing γ to equal $\partial c_t/\partial S_t$, we eliminate the risky term dS_t, leaving the following non-stochastic partial differential equation,

$$c_t r_c = \frac{\partial c_t}{\partial S_t}r_c S_t + \frac{\partial c_t}{\partial t} + \frac{1}{2}\frac{\partial^2 c_t}{\partial S_t^2}\sigma^2(S_t, t), \tag{17.21}$$

which is a second-order partial differential equation. This partial differential equation solution, subject to the boundary condition $c_T = \max(0, S_T - X)$, depends on the specification of the volatility term.

Several important insights can be gained from linear homogeneity. First, no assumptions were made regarding risk-free hedging. Second, the only assumption made regarding the underlying asset stochastic process is that it follows an Itô process. Third, the drift term of the Itô process has no influence on the resulting partial differential equation. Fourth, if we assume a generic geometric Brownian motion of the form $dS_t = \alpha(S_t, t)\,dt + \sigma S_t dW_t$, we obtain the Black-Scholes-Merton model. In Appendix 17A, we explore linear homogeneity applied to the arithmetic Brownian motion model.

17.5 EXCHANGE OPTION PRICING

Now let us use the concept of linear homogeneity to price exchange options. We assume each asset follows the lognormal diffusion process,

$$\frac{dS_i}{S_i} = \alpha_i dt + \sigma_i dW_i$$

$$i = 1,2, \tag{17.22}$$

and we specify that the correlation between the two Wiener processes driving the asset prices is ρ_{12}.[1] Let $c(S_1, S_2)$ be the value today of the exchange call option, which gives the right to tender asset 2 for asset 1 at expiration, T. As noted, the payoff of this option is $c_T(S_1, S_2) = \max(0, S_1 - S_2)$. It is easy to see that this terminal payoff is linearly homogeneous with respect to the two asset values. Because the value of the option today is a simple discounted expectation of the payoff at expiration, the current value of the option must also be linearly homogeneous.

Using Euler's theorem, we can express the value of the option as

$$c(S_1, S_2) - \frac{\partial c(S_1, S_2)}{\partial S_1} S_1 - \frac{\partial c(S_1, S_2)}{\partial S_2} S_2 = 0. \tag{17.23}$$

Although this statement, viewed from the context of Euler's theorem, is purely mathematical, it also has a natural interpretation in finance. It says that a portfolio consisting of the purchase of one unit of the exchange call option and short positions in $\partial c / \partial S_1$ units of asset 1 and $\partial c / \partial S_2$ units of asset 2 would require no initial investment. Therefore, to avoid profitable arbitrage, such a portfolio must generate an instantaneous return of zero.

The differential of Equation (17.23) is

$$dc(S_1, S_2) - \frac{\partial c(S_1, S_2)}{\partial S_1} dS_1 - \frac{\partial c(S_1, S_2)}{\partial S_2} dS_2 = 0. \tag{17.24}$$

Now we apply the multivariate version of Itô's lemma to the value of the call:[2]

$$dc(S_1, S_2) = \frac{\partial c(S_1, S_2)}{\partial S_1} dS_1 + \frac{\partial c(S_1, S_2)}{\partial S_2} dS_2 + \frac{\partial c(S_1, S_2)}{\partial t} dt$$

$$+ \frac{1}{2} \left[\frac{\partial^2 c(S_1, S_2)}{\partial S_1^{\,2}} \sigma_1^{\,2} S_1^{\,2} + 2 \frac{\partial^2 c(S_1, S_2)}{\partial S_1 \partial S_2} \sigma_1 \sigma_2 \rho_{12} S_1 S_2 + \frac{\partial^2 c(S_1, S_2)}{\partial S_2^{\,2}} \sigma_2^{\,2} S_2^{\,2} \right] dt.$$

$$(17.25)$$

Equating Equation (17.25) with (17.24), and after canceling, we obtain

$$\frac{\partial c(S_1, S_2)}{\partial t} + \frac{1}{2} \left[\frac{\partial^2 c(S_1, S_2)}{\partial S_1^{\,2}} \sigma_1^{\,2} S_1^{\,2} + 2 \frac{\partial^2 c(S_1, S_2)}{\partial S_1 \partial S_2} \sigma_1 \sigma_2 \rho_{12} S_1 S_2 + \frac{\partial^2 c(S_1, S_2)}{\partial S_2^{\,2}} \sigma_2^{\,2} S_2^{\,2} \right] = 0.$$

$$(17.26)$$

The solution is the exchange option price,

$$c(S_1, S_2) = S_1 N(d_1) - S_2 N(d_2),$$

where

$$d_1 = \frac{\ln(S_1/S_2) + (\sigma^2/2)\tau}{\sigma \sqrt{\tau}}$$

$$d_2 = d_1 - \sigma \sqrt{\tau}$$

$$\sigma = \sqrt{\sigma_1^{\,2} + \sigma_2^{\,2} - 2\rho_{12}\sigma_1\sigma_2}.$$

$$(17.27)$$

Note that σ is the volatility of a new variable, defined as the proportional change in the log of the ratio S_1/S_2. It is obtained as follows:

$$\text{var}\left[\ln\left(\frac{S_1}{S_2} \right) \right] = \text{var}(\ln S_1 - \ln S_2)$$

$$= \text{var}(\ln S_1) + \text{var}(\ln S_2) - 2\text{cov}(\ln S_1, \ln S_2)$$

$$= \sigma_1^{\,2} + \sigma_2^{\,2} - 2\rho_{12}\sigma_1\sigma_2.$$

$$(17.28)$$

We can check the solution by taking the partial derivatives of Equation (17.27) and inserting them into Equation (17.26).

Because we know that c is linearly homogeneous with respect to the prices of assets 1 and 2, we can say that $ac(S_1, S_2) = c(aS_1, aS_2)$, where a is a constant. Define a as $1/S_2$, which gives us $(1/S_2) c(S_1, S_2) = c(S, 1)$ or $c(S_1, S_2) = S_2 c(S, 1)$ where $S = S_1/S_2$. In effect we have created a new artificial asset, the ratio of the value of asset 1 to the value of asset 2. Because it is not possible to deliver such an asset, this instrument can perhaps be more easily expressed as a cash-settled call option that enables one to exchange one unit of currency and receive the value $S = S_1/S_2$ in cash. Thus, the exchange call is an ordinary European call on S with an exercise price of 1. We can, therefore, differentiate the exchange option price by differentiating its equivalent value, $S_2 c(S, 1)$. We will also need second partial derivatives with respect to S_1, S_2, the cross partial of S_1 and S_2, and the first derivative with respect to t.

Thus, the model can be expressed as

$$c(S_1, S_2) = S_2 c(S, 1)$$
$$= S_2 \left[SN(d_1) - N(d_2) \right],$$

where

$$S = S_1/S_2. \tag{17.29}$$

Equation (17.29) is an ordinary Black-Scholes-Merton option on the artificial asset S with exercise price of 1. The d_1 and d_2 variables in this variation are the same as shown in Equation (17.27).

By converting S_1 and S_2 to S and the strike to 1, we have normalized these assets. In asset management and asset allocation in particular, we are often interested in the relative performance of asset 1 in terms of asset 2. In other words, we want to know which asset class outperforms the other and by how much. In practice, the actual prices of asset 1 and 2 are often significantly different and hence some form of normalization is necessary. Investors decide the dollar amount to invest in each asset class independent of the quoted price of each asset. We see here clearly that an exchange option can capture in option value in terms of the relative behavior of the two assets.

Now, for completeness, we need to verify the solution to the PDE in the traditional manner. That is, we must show that Equation (17.27) solves Equation (17.26). Our approach is to transform the PDE into one that is isomorphic to the Black-Scholes-Merton and infer the solution via the Black-Scholes-Merton model.

Let us initially examine the first derivatives with respect to S_1 and S_2, commonly referred to as the option delta. We will make use of the artificial asset S to simplify the math, thereby enabling us to use many of the results we already know from the Black-Scholes-Merton model. We shall need certain partial derivatives. The first partial derivatives with respect to the underlying prices are

$$\frac{\partial c(S_1, S_2)}{\partial S_1} = \frac{\partial \left[S_2 c(S, 1) \right]}{\partial S_1}$$
$$= S_2 \frac{\partial c(S, 1)}{\partial S_1} = S_2 \frac{\partial c(S, 1)}{\partial S} \frac{\partial S}{\partial S_1}$$
$$= \frac{\partial c(S, 1)}{\partial S}$$

$$\frac{\partial \left[S_2 c(S, 1) \right]}{\partial S_2} = S_2 \frac{\partial c(S, 1)}{\partial S} \frac{\partial S}{\partial S_2} + c(S, 1)$$
$$= S_2 \frac{\partial c(S, 1)}{\partial S} \left(-\frac{S}{S_2} \right) + c(S, 1)$$
$$= c(S, 1) - S \frac{\partial c(S, 1)}{\partial S}. \tag{17.30}$$

The second partial derivatives with respect to the asset prices are

$$\frac{\partial^2 c(S_1, S_2)}{\partial S_1^2} = \frac{\partial}{\partial S_1}\left[\frac{\partial c(S_1, S_2)}{\partial S_1}\right] = \frac{\partial}{\partial S_1}\left[\frac{\partial c(S, 1)}{\partial S}\right] = \frac{\partial}{\partial S}\left[\frac{\partial c(S, 1)}{\partial S}\frac{\partial S}{\partial S_1}\right] = \frac{\partial^2 c(S, 1)}{\partial S^2}\left(\frac{1}{S_2}\right)$$

$$\frac{\partial^2 c(S_1, S_2)}{\partial S_2^2} = \frac{\partial}{\partial S_2}\left[\frac{\partial c(S_1, S_2)}{\partial S_2}\right] = \frac{\partial}{\partial S_2}\left[c(S, 1) - S\frac{\partial c(S, 1)}{\partial S}\right] = \frac{\partial c(S, 1)}{\partial S_2} - S\frac{\partial}{\partial S_2}\left[\frac{\partial c(S, 1)}{\partial S}\right]$$

$$= \frac{\partial c(S, 1)}{\partial S_2} - \frac{\partial c(S, 1)}{\partial S}\frac{\partial S}{\partial S_2} - S\frac{\partial}{\partial S_2}\left[\frac{\partial c(S, 1)}{\partial S}\right] - S\frac{\partial}{\partial S_2}\left[\frac{\partial c(S, 1)}{\partial S}\right]$$

$$= \frac{\partial c(S, 1)}{\partial S}\frac{\partial S}{\partial S_2} - \frac{\partial c(S, 1)}{\partial S}\frac{\partial S}{\partial S_2} - S\frac{\partial^2 c(S, 1)}{\partial S^2}\frac{\partial S}{\partial S_2}$$

$$= -S\frac{\partial^2 c(S, 1)}{\partial S^2}\left(-\frac{S}{S_2}\right) = \frac{\partial^2 c(S, 1)}{\partial S^2}\left(\frac{S^2}{S_2}\right)$$

$$\frac{\partial^2 c(S_1, S_2)}{\partial S_1 \partial S_2} = \frac{\partial}{\partial S_2}\left[\frac{\partial c(S_1, S_2)}{\partial S_1}\right]$$

$$= \frac{\partial}{\partial S_2}\left[\frac{\partial c(S, 1)}{\partial S}\right] = \frac{\partial^2 c(S, 1)}{\partial S^2}\frac{\partial S}{\partial S_2}$$

$$= \frac{\partial^2 c(S, 1)}{\partial S^2}\left(-\frac{S}{S_2}\right). \tag{17.31}$$

We also need the first partial derivative with respect to time:

$$\frac{\partial c(S_1, S_2)}{\partial t} = \frac{\partial S_2 c(S, 1)}{\partial t} = S_2\frac{\partial c(S, 1)}{\partial t}. \tag{17.32}$$

Expressing Equation (17.26) based on the derivatives above in terms of S and σ, we find

$$S_2\frac{\partial c(S, 1)}{\partial t} + \frac{1}{2}\frac{\partial^2 c(S, 1)}{\partial S^2}\left(\frac{1}{S_2}\right)\sigma_1^2(SS_2)^2 + \frac{\partial^2 c(S, 1)}{\partial S^2}\left(-\frac{S}{S_2}\right)\sigma_1\sigma_2\rho_{12}(SS_2)S_2$$

$$+ \frac{1}{2}\frac{\partial^2 c(S, 1)}{\partial S^2}\left(\frac{S^2}{S_2}\right)\sigma_2^2 S_2^2 = 0$$

$$S_2\left[\frac{\partial c(S, 1)}{\partial t} + \frac{1}{2}\frac{\partial^2 c(S, 1)}{\partial S^2}\sigma_1^2 S^2 - \frac{\partial^2 c(S, 1)}{\partial S^2}\sigma_1\sigma_2\rho_{12}S^2 + \frac{1}{2}\frac{\partial^2 c(S, 1)}{\partial S^2}\left(\frac{S^2}{S_2}\right)\sigma_2^2 S^2\right] = 0$$

$$\frac{\partial c(S, 1)}{\partial t} + \frac{1}{2}\frac{\partial^2 c(S, 1)}{\partial S^2}\sigma^2 S^2 = 0. \tag{17.33}$$

And this is the PDE that was presented as Equation (17.26).

We can also see that the first two lines in the equation show that this partial differential equation is the same as the Black-Scholes-Merton partial differential equation when the interest rate, r_c, is set to zero, the underlying asset price is S, and the volatility is σ. Recall the Black-Scholes-Merton PDE as given in Equation (13.11):

$$r_c S \frac{\partial c}{\partial S} + \frac{\partial c}{\partial t} + \frac{1}{2}\frac{\partial^2 c}{\partial S^2}\sigma^2 S^2 = r_c c.$$

Substituting $r_c = 0$, then we have Equation (17.33).

Consequently, we can say that the exchange option is equivalent to S_2 units of an ordinary European call when the underlying asset is S, the strike is one, the interest rate is zero, and the volatility is $\sigma^2 = \sigma_1^2 + \sigma_2^2 - 2\rho_{12}\sigma_1\sigma_2$. The last two lines above verify that this PDE is the same as the one we previously obtained. Thus, we note that Equation (17.27) solves Equation (17.26).

This result is useful in better understanding not only the exchange option but also the ordinary European option. The latter can be viewed as an exchange option where the asset exchanged is cash. The exchange option implies a zero interest rate because it can be replicated by holding asset 1 and shorting asset 2. The shorting of asset 2 would not have an expected return of r_c as it would if it were risk free. Rather the holder of asset 2 would demand its expected return, α_2, as compensation. Consequently, the short seller of asset 2, who is trying to replicate the exchange option, would not have an expected return of $-r_c$ but rather of $-\alpha_2$. In any ordinary European call, the second term in the pricing equation is the present value of the exercise price. In the exchange option, the second term is also the present value of the exercise price. The current price of asset 2 is its present value.

Interestingly, in the same issue of *The Journal of Finance* directly preceding the Margrabe article, there is an article by Fischer (1978), in which he modeled bonds indexed to inflation. He showed that to price such a bond one needs the formula for an option where the exercise price is stochastic. Such an option is equivalent to an exchange option, and naturally Fisher derives the same formula as Margrabe.

The traditional exchange option Greeks are provided in Appendix 17C. We now turn to explore spread options.

17.6 SPREAD OPTIONS[3]

We now consider spread options where there is a nonzero exercise price. In this case, suppose we have a European spread call option in which at expiration the holder can exchange α_2 units of asset 2 and receive α_1 units of asset 1 but also must pay a prespecified exercise price, X. We assume α_1 and α_2 are both positive. A European spread put option in which at expiration the holder can exchange α_1 units of asset 1 and receive α_2 units of asset 2 but again has to pay a prespecified exercise price, X. Note that X can be positive or negative. The payoffs at expiration are

$$c_T = \max(0, \alpha_1 S_{1T} - \alpha_2 S_{2T} - X) \text{ and} \tag{17.34}$$

$$p_T = \max(0, X - \alpha_1 S_{1T} + \alpha_2 S_{2T}). \tag{17.35}$$

With spread options, neither asset plays the role of the exercise price. Thus, a spread option is a more complex exchange option with a known exercise price.

Pearson (1995), Carmona and Durrleman (2003, 2006), Li, Deng, and Zhou (2008), and others have offered approximations when solving for the spread option price when each asset follows geometric Brownian motion. Brooks and Cline (2015) provide a closed-form solution that also incorporates dividends. Based on the results in the previous section as well as the more complex terminal boundary condition, the spread option model based on geometric Brownian motion can be expressed as

$$
c_t = \alpha_1 S_{1t} e^{-\delta_1 \tau} \int_{-\infty}^{\infty} N[d_{1,1}(j)]\, n(j) dj - \alpha_2 S_{2t} e^{-\delta_2 \tau}
$$

$$
\int_{-\infty}^{\infty} N[d_{1,2}(j)]\, n(j) dj - X e^{-r_c \tau} \int_{-\infty}^{\infty} N[d_2(j)]\, n(j) dj, \tag{17.36}
$$

$$
p_t = c_t - \alpha_1 S_{1t} e^{-\delta_1 \tau} + \alpha_2 S_{2t} e^{-\delta_2 \tau} + X e^{-r_c \tau}, \tag{17.37}
$$

where

$$
n(j) = \frac{e^{-\frac{j^2}{2}}}{\sqrt{2\pi}}, \tag{17.38}
$$

$$
N[d_i(j)] = \int_{-\infty}^{d_i(j)} \frac{e^{-\frac{k^2}{2}}}{\sqrt{2\pi}} dk, \tag{17.39}
$$

$$
d_{1,1}(j) \equiv \frac{\ln\left[\dfrac{\alpha_1 S_{1t} e^{\left(r_c - \delta_1 + \frac{\sigma_1^2}{2}\right)\tau + \rho\sigma_1\sqrt{\tau}j}}{X + \alpha_2 S_{2t} e^{\left(r_c - \delta_1 - \frac{\sigma_1^2}{2} + \rho\sigma_1\sigma_2\right)\tau + \sigma_2\sqrt{\tau}j}}\right]}{\sigma_1\sqrt{\tau(1-\rho^2)}}, \tag{17.40}
$$

$$
d_{1,2}(j) \equiv \frac{\ln\left[\dfrac{\alpha_1 S_{1t} e^{\left(r_c - \delta_1 - \frac{\sigma_1^2}{2} + \rho\sigma_1\sigma_2\right)\tau + \rho\sigma_1\sqrt{\tau}j}}{X + \alpha_2 S_{2t} e^{\left(r_c - \delta_2 + \frac{\sigma_2^2}{2}\right)\tau + \sigma_2\sqrt{\tau}j}}\right]}{\sigma_1\sqrt{\tau(1-\rho^2)}}, \text{ and} \tag{17.41}
$$

$$
d_2(j) = \frac{\ln\left[\dfrac{\alpha_1 S_{1t} e^{\left(r_c - \delta_1 - \frac{\sigma_1^2}{2}\right)\tau + \rho\sigma_1\sqrt{\tau}j}}{X + \alpha_2 S_{2t} e^{\left(r_c - \delta_2 - \frac{\sigma_2^2}{2}\right)\tau + \sigma_2\sqrt{\tau}j}}\right]}{\sigma_1\sqrt{\tau(1-\rho^2)}}. \tag{17.42}
$$

This solution is not an approximation like Pearson (1995), Carmona and Durrleman (2003, 2006), Li, Deng, and Zhou (2008) and others; rather, it is an exact result. It is still technically a double integral as the standard $N(d)$ function is itself an integral. In practice, however, it is a single integral because of the existence of very accurate numerical approximations available to compute the standard $N(d)$ function. Fast solutions to single integral problems are widely available, particularly with fast and efficient computer languages such as R or C++.

The solution in the case of arithmetic Brownian motion with geometric drift is much more straightforward. Note that the spread is normally distributed and is denoted as

$$SP_T = \alpha_1 S_{1T} - \alpha_2 S_{2T}. \tag{17.43}$$

Under risk-neutral valuation, we have

$$E(SP_T) = \alpha_1 S_{1t} e^{(r_c - \delta_1)\tau} - \alpha_2 S_{2t} e^{(r_c - \delta_2)\tau} = SP_t e^{r_c \tau}, \tag{17.44}$$

where one measure of the current spread at t is

$$SP_t \equiv \alpha_1 S_{1t} e^{-\delta_1 \tau} - \alpha_2 S_{2t} e^{-\delta_2 \tau}. \tag{17.45}$$

Let σ_{SP} denote the standard deviation of changes in the spread. Thus, based on the Brownian motion (ABM) model, the spread option prices are

$$c_t = \left(SP_t - Xe^{-r\tau}\right) N(d_n) + \sigma_{SP} \sqrt{\frac{e^{r_c \tau} - 1}{2r_c}} n(d_n), \tag{17.46}$$

$$p_t = c_t - \alpha_1 S_{1t} e^{-\delta_1 \tau} + \alpha_2 S_{2t} e^{-\delta_2 \tau} + Xe^{-r\tau}, \tag{17.47}$$

where

$$n(d_n) = \frac{e^{\frac{-d_n^2}{2}}}{\sqrt{2\pi}}, \tag{17.48}$$

$$N(d_n) = \int_{-\infty}^{d_n} n(x)dx, \text{ and} \tag{17.49}$$

$$d_n = \frac{SP_t e^{r_c \tau} - X}{\sigma_{SP} \sqrt{\frac{e^{r_c \tau} - 1}{2r_c}}}. \tag{17.50}$$

17.7 FORWARD START OPTIONS

Another useful application of Euler's rule is that of a forward start option, which is an option that is purchased today but does not begin until a later date. When the premium is paid today, the purchaser specifies that they wish to acquire an option at a designated later date that has a particular degree of moneyness.

Figure 17.1 illustrates the timeline. Let time t be the day the forward start option is initiated and time T_1 be the date on which the underlying option is received and T_2 be

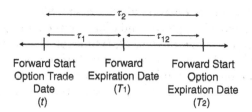

FIGURE 17.1   Forward Start Option Timeline

the date when the underlying option expires. That is, at time T_1 $(t < T_1 < T_2)$ when the asset price is S_1 the option will be granted with an exercise price of αS_1. For example, an at-the-money option has $\alpha = 1$. An option with exercise price 5% higher than the asset price has $\alpha = 1.05$. Because we do not know today what the asset price will be at T_1, we cannot prespecify the exercise price. This would appear to make it difficult to price the option, but in fact it is easy to price the option because we can use the principle of linear homogeneity. The original reference on this is Rubinstein (1991a).

First let us define $c(S_t, \alpha, T_1, T_2)$ as the value of a forward start option at time t in which the exercise price is determined at time T_1 and equals α times the value of the underlying at time T_1, and the option expires at time T_2. Although the option is created and paid for at time t, at time T_1, the option formally comes into existence, that is, its exercise price is now declared. At this point, it has the property of a standard European option, which has a time to expiration of τ_{12}. Consequently, we know that it is linearly homogeneous with respect to the asset price and exercise price. Writing its value as $c(S_1, \alpha S_1, \tau_{12})$ where subscript 1 denotes $t = 1$, we can state that this option's value is equal to

$$c(S_1, \alpha S_1, \tau_{12}) = S_1 c(1, \alpha, \tau_{12}). \tag{17.51}$$

The value $c(1, \alpha, \tau_{12})$ is known at all times. It is simply the value of an option where the underlying is the value of one unit of the currency, the exercise price is α, and the time to expiration is τ_{12}. One could easily plug these values into the Black-Scholes-Merton model and obtain the value. Even at time t, we know this value. Equation (17.51), however, says that the forward start option value at time T_1 is the product of S_1 and this option. At time t, we do not know what S_1 will be, but we can still price the option at time t.

To do so, we need to find a combination of instruments that replicates the option, or, in other words, produces a value at T_1 equal to that of the option. It should be obvious that holding $c(1, \alpha, \tau_{12})$ units of the asset will produce a value of $c(1, \alpha, \tau_{12})$ S_1 at time T_1. Consequently, the value of the forward start option today is

$$c(S_t, \alpha, \tau_1, \tau_2) = S_t c(1, \alpha, \tau_{12}). \tag{17.52}$$

In other words, holding $c(1, \alpha, \tau_{12})$ units of the asset valued at S_1 will reproduce the option value at time t. Do not, however, confuse this statement with any suggestion of full replication. Buying $c(1, \alpha, \tau_{12})$ units of the asset at time valued at S_t will not replicate the final payoff of the option. The replication strategy is dynamic. We hold a static amount, $c(1, \alpha, \tau_{12})$, of the underlying until time T_1. At that point, we have an option with a fixed exercise price of $S_1 \alpha$ on an underlying whose value is fluctuating. This option is now an ordinary European call and is replicated using the appropriate number of units of the asset as given by the dynamic delta.

If instead today we specified a fixed exercise price of X, then at T_1 we are awarded an option with an exercise price of X and a time to expiration of τ_{12}. Such an option is trivially equal to purchasing an option today with a time to expiration of τ_2 and an exercise price of X. If the premium is paid today and the option is received later, it is equivalent to just receiving the option today.

It is important, however, to contrast these types of options with forward contracts on options. A forward contract on an option is an agreement to purchase an option at a later date. The option will have an exercise price of X and a time to expiration of τ_{12} when granted. Being a forward contract, no money is paid today but a price that will be paid at T_1 is agreed on by the two parties. It is also a simple process to determine this option's price. If today we purchase the option with time to expiration τ_2 and exercise price X, that is, the option described in the previous paragraph, but borrow the premium, then at time T_1, we shall have replicated the payoff of the forward contract on the option.

In other words, let $c(S_t, X, \tau_2)$ be the price of a call option today struck at X with time to expiration τ_2. Let $F[c(S_t X\tau_2), T_1]$ be the forward price agreed on today to purchase the option with current price $c(S_t, X, \tau_2)$ at time T_1. The payoff at time T_1 will be

$$c(S_1, X, \tau_{12}) - F[c(S_t X \tau_{12}), T_1]. \qquad (17.53)$$

This payoff can be replicated today by purchasing the call option at $c(S_t, X, \tau_2)$ and borrowing $F[c(S_t X\tau_2), T_1] e^{-r_c \tau_1}$. At time T_1, we are holding the option worth $c(S_t, X, \tau_{12})$ and owe $F[c(S_t X\tau_2), T_1]$, which combine to replicate the payoff of the forward contract. Because the transaction is a forward contract, it requires no outlay today. Consequently, the value of the position today, $c(S_t, X, \tau_2) - F[c(S_t X\tau_2), T_1] e^{-r_c \tau_1}$, must equal zero, meaning that

$$F[c(S_t X\tau_2), T_1] = c(S_1, X, \tau_2) e^{r_c \tau_1}. \qquad (17.54)$$

Here we see the general rule that the price of a forward contract is the price of the underlying compounded to the expiration of the forward contract. This result will be derived formally in Chapter 22, when we address the pricing of forwards and futures.

If the option is written such that the exercise price is specified as αS_1, then instead of holding the call worth $c(S_t, X, \tau_2)$ today, we hold $c(1, \alpha, \tau_{12})$ units of the asset, as determined previously in our derivation of the price of a forward start option. This instrument combines elements of a forward contract on an option and a forward start option.

17.8 RECAP AND PREVIEW

In this chapter, we examined the concept of linear homogeneity and showed how it can be used in option pricing. Linear homogeneity, which occurs in certain types of functions, expresses a relationship between their differentials. Conveniently, the payoff of an option is linearly homogeneous with respect to the asset price and exercise price. As such, we can use this principle to derive the Black-Scholes-Merton model. In this chapter, we show how it can be used to also derive the value of an exchange option, as well as a forward start option.

In Chapter 18, we introduce the concept of a compound option, which is an option on an option. It will go a long way toward helping us to price American options.

APPENDIX 17A

Linear Homogeneity and the Arithmetic Brownian Motion Model

Following the approach given in this chapter and applying Euler's rule to the ABM model call price, c_t, focused on S_t, X, and σ we have

$$S_t \frac{\partial c}{\partial S} + X \frac{\partial c}{\partial X} + \sigma \frac{\partial c}{\partial \sigma} = c(S_t, X). \tag{17.55}$$

Assuming no dividends or other asset cash flows, we know

$$\Delta_c = \frac{\partial c}{\partial S} = N(d_n) \text{ and} \tag{17.56}$$

$$\frac{\partial c}{\partial X} = -e^{-r_c \tau} N(d_n). \tag{17.57}$$

$$\frac{\partial c}{\partial \sigma} = \frac{e^{-r_c \tau} \sigma_n n(d_n)}{\sigma}. \tag{17.58}$$

By substituting these derivatives, we confirm that the ABM model has the property of linear homogeneity.

$$S_t \frac{\partial c}{\partial S} + X \frac{\partial c}{\partial X} + \sigma \frac{\partial c}{\partial \sigma}$$

$$= S_t N(d_n) + X \left[-e^{-r_c \tau} N(d_n) \right] + \sigma \frac{e^{-r_c \tau} \sigma_n n(d_n)}{\sigma} = c(S_t, X). \tag{17.59}$$

Recall

$$c_t = \left(S_t - X e^{-r_c \tau} \right) N(d_n) + e^{-r_c \tau} \sigma_n n(d_n), \text{ where} \tag{17.60}$$

$$d_n = \frac{S_t - X e^{-r_c \tau}}{\sigma_n} \text{ and} \tag{17.61}$$

$$\sigma_n = \sigma \sqrt{\frac{e^{-2r_c \tau} - 1}{2r_c}}. \tag{17.62}$$

We introduce the current price of one zero-coupon bond with par value of X as $B_{1t} \equiv e^{-r_c \tau} X$ and a second zero-coupon bond with par value of X as $B_{2t} \equiv e^{-r_c \tau} \sigma$. Further, recall that a linear operation performed on a linearly homogeneous function preserves its property of linear homogeneity, hence,

$$S_t \frac{\partial c}{\partial S} + B_{1t} \frac{\partial c}{\partial B_{1t}} + B_{2t} \frac{\partial c}{\partial B_{2t}} = c(S_t, X), \tag{17.63}$$

where

$$\frac{\partial c}{\partial B_{1t}} = \frac{\partial c}{\partial X} \frac{\partial X}{\partial B_{1t}} = \frac{\partial c/\partial X}{\partial B_{1t}/\partial X} = -\frac{e^{-r_c \tau} N(d_n)}{e^{-r_c \tau}} = -N(d_n). \tag{17.64}$$

$$\frac{\partial c}{\partial B_{2t}} = \frac{\partial c}{\partial \sigma} \frac{\partial \sigma}{\partial B_{2t}} = \frac{\partial c/\partial \sigma}{\partial B_{2t}/\partial \sigma} = \frac{\frac{e^{-r_c \tau} \sigma_n n(d_n)}{\sigma}}{e^{-r_c \tau}} = \frac{\sigma_n n(d_n)}{\sigma}. \tag{17.65}$$

Alternatively, let us simply start by assuming the option price is linearly homogeneous with respect to S_t, B_{1t}, and B_{2t}. Note that we are not starting with any specified stochastic process. Using Euler's rule, we know that the current option price, which is a function of S_t and t, can be expressed as

$$c_t = \gamma S_t + \omega B_{1t} + v B_{2t}, \tag{17.66}$$

where γ is $\partial c_t / \partial S_t$, ω is $\partial c_t / \partial B_{1t}$, and v is $\partial c_t / \partial B_{2t}$. An interpretation of Equation (17.14) is that one call option can be replicated by holding γ units of the asset, each worth S_t, and $(\omega + v)$ units of a risk-free bond with current value $(X + \sigma)e^{-r_c \tau}$.

The total differential of the option price is

$$dc_t = \gamma dS_t + \omega dB_{1t} + v dB_{2t}. \tag{17.67}$$

The differential dS_t can be left in this form. Recall we obtained the differential for the risk-free bonds in Equation (17.17). Thus, we note

$$dB_{1t} = r_c B_{1t} dt. \tag{17.68}$$

$$dB_{2t} = r_c B_{2t} dt. \tag{17.69}$$

Consequently, the call price differential is

$$dc_t = \gamma dS_t + \omega r_c B_{1t} dt + v r_c B_{2t} dt. \tag{17.70}$$

At this point, it is helpful to substitute $c_t - \gamma S_t$ for $\omega B_{1t} + v B_{2t}$ from Equation (17.14), leaving us

$$dc_t = \gamma dS_t + (c_t - \gamma S_t) r_c dt. \tag{17.71}$$

Now, we assume only that the call price depends on S_t and t and that the underlying follows an Itô process expressed formally as $dS_t = \alpha(S_t, t) dt + \sigma(S_t, t) dW_t$. Allowing volatility to be nonconstant, Itô's lemma can be used to obtain an equivalent expression for the change in the price of the call,

$$dc_t = \frac{\partial c_t}{\partial S_t} dS_t + \frac{\partial c_t}{\partial t} dt + \frac{1}{2} \frac{\partial^2 c_t}{\partial S_t^2} \sigma^2(S_t, t) \, dt, \tag{17.72}$$

which we can now set Equation (17.20) equal to Equation (17.19). And by choosing γ to equal $\partial c_t / \partial S_t$, we eliminate the risky term dS_t, leaving the following non-stochastic partial differential equation,

$$c_t r_c = \frac{\partial c_t}{\partial S_t} r_c S_t + \frac{\partial c_t}{\partial t} + \frac{1}{2} \frac{\partial^2 c_t}{\partial S_t^2} \sigma^2(S_t, t), \tag{17.73}$$

which is a second order partial differential equation. This partial differential equation solution, subject to the boundary condition $c_T = \max(0, S_T - X)$, depends on the specification of the volatility term. The PDE in Equation (17.73) is not the Black-Scholes-Merton PDE unless we assume volatility is constant.

If we assume a generic arithmetic Brownian motion of the form $dS_t = \alpha(S_t, t)\, dt + \sigma dW_t$, we obtain the ABM model which is the solution to

$$c_t r_c = \frac{\partial c_t}{\partial S_t} r_c S_t + \frac{\partial c_t}{\partial t} + \frac{1}{2}\frac{\partial^2 c_t}{\partial S_t^2}\sigma^2, \tag{17.74}$$

subject to the value at expiration of

$$c_T = \max(0, S_T - X). \tag{17.75}$$

APPENDIX 17B

Multivariate Itô's Lemma

Consider n Itô processes at time t, expressed as

$$dX_t = \mu(X, t)dt + \Sigma(X, t)dW_t, \tag{17.76}$$

where

$$dX_t \equiv \begin{bmatrix} dx_1 \\ dx_2 \\ \vdots \\ dx_n \end{bmatrix}, \tag{17.77}$$

$$\mu(X, t) \equiv \begin{bmatrix} \mu_1(X, t) \\ \mu_2(X, t) \\ \vdots \\ \mu_n(X, t) \end{bmatrix}, \tag{17.78}$$

$$\Sigma(X, t) \equiv \begin{bmatrix} \sigma_{1,1}(X, t) & \sigma_{1,2}(X, t) & \cdots & \sigma_{1,m}(X, t) \\ \sigma_{2,1}(X, t) & \sigma_{2,2}(X, t) & \cdots & \sigma_{2,m}(X, t) \\ \vdots & \vdots & \ddots & \vdots \\ \sigma_{n,1}(X, t) & \sigma_{n,2}(X, t) & \cdots & \sigma_{n,m}(X, t) \end{bmatrix}, \text{and} \tag{17.79}$$

$$dW_t \equiv \begin{bmatrix} dW_{1,t} \\ dW_{2,t} \\ \vdots \\ dW_{n,t} \end{bmatrix}. \tag{17.80}$$

Note that each Itô process can depend on up to m Wiener processes or

$$dx_{k,t} = \mu_k(X, t)dt + \sum_{j=1}^{m} \sigma_{k,j}(X, t)dW_{j,t}. \tag{17.81}$$

Suppose

$$y_t = f(X, t). \tag{17.82}$$

Then

$$
\begin{aligned}
dy_t = \Bigg\{ &\frac{\partial f(X, t)}{\partial t} + \sum_{k=1}^{n} \frac{\partial f(X, t)}{\partial x_k} \mu_k(X, t) \\
&+ \frac{1}{2} \sum_{i=1}^{n} \sum_{k=1}^{n} \frac{\partial^2 f(X, t)}{\partial x_i \partial x_k} \sum_{j=1}^{m} \sum_{l=1}^{m} \sigma_{ij}(X, t) \sigma_{kl}(X, t) \rho_{jl}(X, t) \Bigg\} dt \\
&+ \sum_{k=1}^{n} \frac{\partial f(X, t)}{\partial x_k} \sum_{j=1}^{m} \sigma_{k,j}(X, t) dW_{j,t}.
\end{aligned}
\tag{17.83}
$$

APPENDIX 17C

Greeks of the Exchange Option Model

We obtain the first and second derivatives in symbolic form. The deltas with respect to the two asset prices were obtained previously but not carried out in detail. They are

$$\frac{\partial c(S_1, S_2)}{\partial S_1} = \frac{\partial S_2 c(S, 1)}{\partial S_1} = S_2 \frac{\partial c(S, 1)}{\partial S_1} = S_2 \frac{\partial c(S, 1)}{\partial S} \frac{\partial S}{\partial S_1} = S_2 N(d_1) \frac{1}{S_2} = N(d_1)$$

$$\frac{\partial c(S_1, S_2)}{\partial S_2} = \frac{\partial S_2 c(S, 1)}{\partial S_2} = S_2 \frac{\partial c(S, 1)}{\partial S} \frac{\partial S}{\partial S_2} + c(S, 1) = S_2 \frac{\partial c(S, 1)}{\partial S} \left(-\frac{S}{S_2} \right) + c(S, 1)$$

$$= c(S, 1) - S \frac{\partial c(S, 1)}{\partial S} = c(S, 1) - S N(d_1)$$

$$= S N(d_1) - N(d_2) - S N(d_1) = -N(d_2). \tag{17.84}$$

Therefore, applying what we know from Black-Scholes-Merton, the gammas are obtained as

$$\frac{\partial^2 c(S_1, S_2)}{\partial S_1^2} = \frac{\partial^2 c(S, 1)}{\partial S^2} \left(\frac{1}{S_2} \right) = \frac{n(d_1)}{S \sigma \sqrt{\tau}} \left(\frac{1}{S_2} \right) = \frac{n(d_1)}{S_1 \sigma \sqrt{\tau}}$$

$$\frac{\partial^2 c(S_1, S_2)}{\partial S_2^2} = \frac{\partial^2 c(S, 1)}{\partial S^2} \left(\frac{S^2}{S_2} \right) = \frac{n(d_2)}{S_2 \sigma \sqrt{\tau}}. \tag{17.85}$$

Hence, the theta is

$$\frac{\partial c(S_1, S_2)}{\partial t} = -\frac{S_1 \sigma e^{-d_1^2/2}}{2\sqrt{2\pi T}}.$$

(17.86)

Recall for us to arrive at theta, we first need the derivative with respect to time to expiration or

$$\frac{\partial c(S_1, S_2)}{\partial \tau} = S_2 \frac{\partial c(S, 1)}{\partial \tau} = S_2 \left[S \frac{\partial N(d_1)}{\partial d_1} \frac{\partial d_1}{\partial \tau} - \frac{\partial N(d_2)}{\partial d_2} \frac{\partial d_2}{\partial \tau} \right]$$

$$= S_2 \left[S n(d_1) \frac{\partial d_1}{\partial \tau} - n(d_2) \frac{\left(\partial d_1 - \sigma\sqrt{\tau}\right)}{\partial \tau} \right]$$

$$= S_1 n(d_1) \frac{\partial d_1}{\partial \tau} - n(d_2) \frac{\partial d_1}{\partial \tau} + n(d_2) \frac{\sigma}{2\sqrt{\tau}}.$$

(17.87)

We know the first two terms cancel and we obtain

$$\frac{\partial c}{\partial \tau} = \frac{\sigma}{2\sqrt{\tau}} n(d_2).$$

(17.88)

Recall we define theta as the derivative with respect to the point in time, that is, $\partial c/\partial t$ and we defined the time to expiration as $\tau = T - t$. Hence,

$$\Theta_c = \frac{\partial c}{\partial t} = -\frac{\sigma S}{2\sqrt{\tau}} n(d_1).$$

(17.89)

The vegas are

$$\frac{\partial c(S_1, S_2)}{\partial \sigma_1} = S_1 n(d_1) \sqrt{\tau} \left(\frac{\sigma_1 - \rho\sigma_2}{\sigma}\right)$$

$$\frac{\partial c(S_1, S_2)}{\partial \sigma_2} = S_1 n(d_1) \sqrt{\tau} \left(\frac{\sigma_2 - \rho\sigma_1}{\sigma}\right).$$

(17.90)

And the partial derivative with respect to the correlation is

$$\frac{\partial c(S_1, S_2)}{\partial \rho} = -S_1 n(d_1) \sqrt{\tau} \left(\frac{\sigma_2\sigma_1}{\sigma}\right).$$

(17.91)

Note that the risk-free rate does not appear in the price equation. Hence, there is no rho.

QUESTIONS AND PROBLEMS

1 Given the following standard ABM model, demonstrate that it does not have the property of linear homogeneity in S_t and X, alone.

$$c_t = \left(S_t - Xe^{-r_c\tau}\right)N(d_n) + e^{-r_c\tau}\sigma_n n(d_n), \text{ where}$$

$$d_n = \frac{S_t - Xe^{-r_c\tau}}{\sigma_n} \text{ and}$$

$$\sigma_n = \sigma\sqrt{\frac{e^{-2r_c\tau} - 1}{2r}}.$$

You are given the following partial derivatives:

$$\frac{\partial c}{\partial S} = N(d_n) \text{ and}$$

$$\frac{\partial c}{\partial X} = -e^{-r_c\tau}N(d_n).$$

2 Assume a stock has a \$1 price and a corresponding one year call option is at-the-money. Further, assume no dividends and the risk-free rate is zero. For this case, compare the results of the Black-Scholes-Merton model and ABM model.

3 Demonstrate that American exchange options on non-dividend-paying assets will not be exercised early. Thus, American exchange options will be worth the same as European exchange options.

4 A hedge fund recently hired a new and highly talented manager. In negotiations, they will receive 20% of any superior performance above the SPY (an exchange-traded fund that seeks to mimic an investment in the S&P 500 index) over the next year. If they underperform the SPY, then their compensation is zero. Based on this information, identify the valuation model that would provide the fair market value of this compensation scheme. Further, explain the manager's incentive based on this compensation scheme.

5 For two constants, $\alpha_1 > 0$ and $\alpha_2 > 0$, prove the following put-call parity representation:

$$c(\alpha_1 S_1, \alpha_2 S_2) - p(\alpha_1 S_1, \alpha_2 S_2) = \alpha_1 S_1 - \alpha_2 S_2.$$

6 Prove the lower bound of a call spread option is $c_t \geq \max\left(0, \alpha_1 S_{1t} - \alpha_2 S_{2t} - Xe^{-r_c\tau}\right)$.

NOTES

1. Alternatively, we could assume each asset follows arithmetic Brownian motion with geometric drift. In this case, the subtraction of two normally distributed variables is also normally distributed, hence, arriving at the pricing model is straightforward.
2. See Appendix 17B for a formal statement of the multivariate Itô's lemma.
3. For extensive discussion of spread options, see Brooks and Cline (2015).

CHAPTER 18

Compound Option Pricing

In this chapter we study compound options, which are options in which the underlying is an option. Thus, a compound option is an option on an option. Compound options are rarely seen directly in practice, but that is not an excuse to ignore them. For example, they are extremely valuable tools for understanding credit risk. Compound option models have also been deployed when addressing contingency hedging (e.g., hedging an extended bid on a project), various aspects of drug development, and captions (i.e., an option on a cap where a cap is a series of interest rate options that can be used to hedge, say, a loan portfolio for a bank). As we will show here, an option on the stock of a company that has debt is a compound option. In addition, compound option theory provides a means for pricing American options.

After introducing specific notation for compound options, we show how an option perspective on a common stock can aid in understanding the credit risk related to a firm. With this foundation, we proceed to explore options on stock as compound options. Next, we turn to a formal treatment of compound options complete with boundary conditions and parity relationships. We then turn to a careful examination of Geske's call on call model as he applied it to stock options. Next, we identify a generalized compound option model that addresses four different versions of compound options as well as incorporates cash flow yields on both the underlying instrument as well as the underlying option. We conclude this chapter with a brief look at installment options, which are variations of compound options.

The original development of the compound option model was Geske (1979b). The context in which this model was developed was that of a call option on a stock, which is itself a call option on the assets of the firm, an idea set forth by Black and Scholes and part of the foundation for the understanding of credit risk, as mentioned. Although the Black-Scholes-Merton model would appear to properly price such a call option, if all the basic assumptions are met, Geske's compound option model shows that the volatility of the stock is dynamic because it is determined by the leverage of the firm, meaning how much debt it uses. Leverage, of course, changes with changes in the value of the stock. Recall that the Black-Scholes-Merton model assumes constant volatility. And although most people use the Black-Scholes-Merton model to price options on stocks, technically the compound model is more appropriate.

The compound option model requires significantly more variables, so we briefly introduce some of the terms. Figure 18.1 illustrates three important dates with compound options: the trade date (t), the compound option expiration date (T_1), and the underlying option expiration (T_2). There are three important periods of time corresponding to the compound option: the period of time until the compound option expires (τ_1), the period of

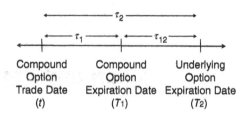

FIGURE 18.1   Compound Option Timeline

time until the underlying option expires (τ_2), and the period of time between the compound option expiration and the underlying option expiration (τ_{12}).

Further, the compound option model requires the compound option exercise price (X_C) and the underlying option exercise price (X_U). The remaining underlying parameters are as defined before with S_t denoting the underlying instrument price, r_c denoting the annualized continuously compounded risk-free rate, and σ denoting the annualized standard deviation of the continuously compounded rates of return of the underlying instrument.

Finally, we introduce an underlying option yield, $\hat{q}$. Compound options are often used in modeling complex financial valuation problems that are often not related to traditional stock options. A yield on the underlying option may be useful for many applications of compound options, including the default option embedded in bonds, drug development applications with ongoing expenses (negative option yield), and some forms of contingency hedging. Although standard stock options do not pay any dividends, there may be times when we will wish to model options that do involve cash flows. We explore one case later in this chapter.

We motivate compound options first with options on common stock.

18.1   EQUITY AS AN OPTION

Let us start by assuming the existence of a corporation that has a single issue of zero-coupon debt due at time T_2. We define the following terms:

V_t: value of the firm's assets at time t,[1]

$M = X_U$: face value of zero-coupon debt issued by the firm and maturing at time T_2,

S_t: market value of the stock of the firm at time t, and

$\sigma_V = \sigma$: volatility of the log return of the assets of the firm.

Again, we continue to use r_c as the risk-free rate and the maturity of the debt in years is τ_2. As Black and Scholes demonstrated, the equity is an option on the assets with payoff at T_2 of $\max\left(0, V_{T_2} - M\right)$. That is, at time T_2, if the stockholders pay off the creditors in the amount of their claim, M, they retain their own claim on the assets worth V_{T_2}. Thus, when the debt matures and the value of the firm is large enough to avert default, the stock is worth $V_{T_2} - M$. When the value of the assets is below the value of the debt, the stockholders will default, and their claim is worth nothing. Hence, the stock is like a call option with a payoff of max $(0, V_{T_2} - M)$. If the return on the assets can be expressed with the lognormal diffusion,

$$\frac{dV_t}{V_t} = \alpha_V dt + \sigma_V dW_t, \qquad (18.1)$$

and the usual Black-Scholes-Merton assumptions are met, the equity can be valued as

$$S_t = V_t N(d_1) - Me^{-r_c\tau_2} N(d_2)$$
$$d_1 = \frac{\ln(V_t/M) + (r_c + \sigma_V^2/2)\tau_2}{\sigma_V\sqrt{\tau_2}}$$
$$d_2 = d_1 - \sigma_V\sqrt{\tau_2}. \qquad (18.2)$$

This result is straightforward. We simply view the equity through the lens of an ordinary option in which the underlying is represented by the assets of the firm. But what if there is an option on the equity? Until now, we would value this option using the Black-Scholes-Merton model, but now we see that the option should be valued in a more meaningful way, one in which the volatility is not constant as would be the case if the firm had leverage.

Also, recall from Chapter 13 that $N(d_2)$ is the probability of the call option expiring in-the-money under the risk-neutral measure. Thus, the value of viewing equity as an option is immediately transparent as $N(d_2)$ is the probability of the firm paying off its debts when they mature under the risk-neutral measure. Alternatively, $1 - N(d_2)$ is the probability of the firm not being able to pay off its debts when they mature under the risk-neutral measure. There have been many applications of this perspective, including appraising credit risk. See, for example, the widely popular KMV model.[2]

18.2 VALUING AN OPTION ON THE EQUITY AS A COMPOUND OPTION

In valuing compound options, we must first recognize that there are calls on calls, calls on puts, puts on calls, and puts on puts. In addition, these options can be either European or American, and they can be mixed, as for example, an American call on a European put. We will use the simple notation *cc*, *cp*, *pc*, and *pp* to denote the price of a European call or put on a European call or put. We would use an uppercase *c* or *p* if either option is American, but we will not cover American options here.[3]

Now suppose there is a call option on the stock expiring at T_1 with an exercise price of X_C. Viewing stock as an option on the firm means the call option on the stock can be viewed as a compound option, specifically, a call on a call described in much more detail later. We denote this call on a call price as *cc*. Let this compound call expire before the debt matures (T_2); therefore, $T_1 < T_2$. Let us now derive the value of this compound call option in terms of the underlying asset, V_t, the value of the firm. We construct a hedge portfolio by purchasing n_1 units of the asset and n_2 compound call options. The value of this portfolio is initially[4]

$$H_t = n_1 V_t + n_2 cc_t. \qquad (18.3)$$

We shall now temporarily drop the t subscript for notational simplification. We know that the change in the value of this portfolio is given by the total differential,

$$dH = \frac{\partial H}{\partial V}dV + \frac{\partial H}{\partial cc}dcc. \tag{18.4}$$

And note that dcc is the differential of the compound call price, cc. From Equation (18.3), we know that

$$\frac{\partial H}{\partial V} = n_1$$

$$\frac{\partial H}{\partial cc} = n_2. \tag{18.5}$$

Thus, the change in the value of the hedge portfolio is

$$dH = n_1 dV + n_2 dcc. \tag{18.6}$$

Because the price of the option is a function of V and t, Itô's lemma permits us to express the change in the call price as

$$dcc = \frac{\partial cc}{\partial t}dt + \frac{\partial cc}{\partial V}dV + \frac{1}{2}\frac{\partial^2 cc}{\partial V^2}V^2\sigma_V^2 dt. \tag{18.7}$$

Now we substitute Equation (18.7) into the right-hand side of Equation (18.6). This gives

$$dH = n_1 dV + n_2\frac{\partial cc}{\partial t}dt + n_2\frac{\partial cc}{\partial V}dV + n_2\frac{1}{2}\frac{\partial^2 cc}{\partial V^2}V^2\sigma_v^2 dt. \tag{18.8}$$

Because we are free to set n_1 and n_2, then let $n_1 = -n_2\left(\partial cc/\partial V\right)$.[5] Substituting into Equation (18.8) gives

$$dH = n_2\frac{\partial cc}{\partial t}dt + n_2\frac{1}{2}\frac{\partial^2 cc}{\partial V^2}V^2\sigma_V^2 dt. \tag{18.9}$$

This expression has no stochastic terms, so it is risk free. Therefore, the value of the hedge portfolio, H, should grow at the risk-free rate. Thus, we specify that

$$dH = Hr_c dt. \tag{18.10}$$

Substituting from Equation (18.3) for H, and $-n_2\left(\partial cc/\partial V\right)$ for n_1, we obtain

$$n_2\frac{\partial cc}{\partial t}dt + n_2\frac{1}{2}\frac{\partial^2 cc}{\partial V^2}V^2\sigma_v^2 dt = -n_2\frac{\partial cc}{\partial V}Vr_c dt + n_2 ccr_c dt. \tag{18.11}$$

Dividing by n_2 and dt gives

$$ccr_c = \frac{\partial cc}{\partial V}Vr_c + \frac{\partial cc}{\partial t} + \frac{1}{2}\frac{\partial^2 cc}{\partial V^2}V^2\sigma_V^2, \tag{18.12}$$

which appears to be the same partial differential equation we obtained for the standard European call option.[6] Unfortunately, there are some complicating factors in finding the solution. The solution for the call on call price in terms of the firm's asset price, not the stock price, is more difficult because the call (modeled here as a call on call) will exercise or not at an intervening time based on the value of the stock (modeled as the underlying call in this case) relative to the underlying call exercise price, X_U. The call is *not* simply worth $\max\left(0, V_{T_1} - X_C\right)$. The call on call payoff is made at time T_1 and is $\max\left(0, c_{T_1} - X_C\right)$. Note that the value of c_{T_1} depends on S_{T_1} that is used to solve another partial differential equation because it is a standard European option expiring later at T_2.

We now turn to compound option boundaries and parities before introducing the general compound option valuation model.

18.3 COMPOUND OPTION BOUNDARY CONDITIONS AND PARITIES

We briefly review the appropriate compound option boundary conditions and parities followed by selected proofs. For completeness, we consider both the possibility of dividends on the underlying asset, denoted δ as before, as well as the possibility of dividends on the underlying option, denoted $\hat{q}$ or the option yield. In the previous section, recall equity was viewed as an option on the firm and that equity may pay a dividend. Thus, for completeness, we need a model that can address the yield paid out of the firm as a whole as well as the yield paid out specifically to equity holders. Also, recall that compound options are often used in modeling complex financial valuation problems where the capacity to capture a cash flow related to the underlying option is useful.

Going forward, we adopt the more general case where S_t is the underlying asset price. Thus, in our prior discussion, S_t is the underlying firm value. In this way, the following results are generic and can easily be applied to other assets.

We now review the detailed proof for the call on call lower and upper bounds. The remaining bounds are given without proof as they follow in a similar manner.

Call on call lower bound:

$$cc_t\left[c_t\left(S_t, X_U, T_2\right), X_C, T_1\right] \geq \max\left[0, e^{-\hat{q}\tau_1} c_t\left(S_t, X_U, T_2\right) - X_C e^{-r_c\tau_1}\right]. \tag{18.13}$$

Recall for European-style call options on the underlying instrument, we have the following lower bound:

$$c_t \geq \max\left(0, S_t e^{-\delta\tau_2} - X_U e^{-r_c\tau_2}\right). \tag{18.14}$$

Thus, in a similar fashion, for a call on a call option, we have

$$cc_t\left[c_t\left(S_t, X_U, T_2\right), X_C, T_1\right] \geq \max\left[0, e^{-\hat{q}\tau_1} c_t\left(S_t, X_U, T_2\right) - X_C e^{-r_c\tau_1}\right]. \tag{18.15}$$

Although the intuition may not be clear, the boundary condition must hold or otherwise there will be arbitrage profits available, as illustrated in the following proof.

Proof: Recall that $X_1 > 0$ and that limited liability implies observed option prices are nonnegative. Assume the opposite:

$$cc_t\left[c_t\left(S_t, X_U, T_2\right), X_C, T_1\right] < \max\left[0, e^{-\hat{q}\tau_1} c_t\left(S_t, X_U, T_2\right) - X_C e^{-r_c\tau_1}\right]. \tag{18.16}$$

TABLE 18.1 Lower Bound for Call on Call

Strategy	Today (t)	At Expiration (T_1)	
		$c_{T_1} \leq X_C$	$c_{T_1} > X_C$
Sell $e^{-\hat{q}\tau_1}$ units of underlying call	$+e^{-\hat{q}\tau_1}c_t$	$-c_{T_1}$	$-c_{T_1}$
Lend $X_C e^{-r\tau_1}$	$-X_C e^{-r\tau_1}$	$+X_C$	$+X_C$
Buy call on call	$+cc_t$	0	$+\left(c_{T_1} - X_C\right)$
Net	> 0 (By assumption)	$+X_C - c_{T_1} \geq 0$ (By column)	$= 0$

Note if $e^{-\hat{q}\tau_1}c_t\left(S_t, X_U, T_2\right) - X_C e^{-r_c\tau_1} < 0$, then $cc_t\left[c_t\left(S_t, X_U, T_2\right), X_C, T_1\right] < 0$. In this case, simply sell the compound option and you have positive cash flow today at no risk in the future. Therefore, we only need now consider the case where

$$cc_t\left[c_t\left(S_t, X_U, T_2\right), X_C, T_1\right] < e^{-\hat{q}\tau_1}c_t\left(S_t, X_U, T_2\right) - X_C e^{-r_c\tau_1}. \qquad (18.17)$$

Move terms to the greater-than side,

$$0 < e^{-\hat{q}\tau_1}c_t\left(S_t, X_U, T_2\right) - X_C e^{-r_c\tau_1} - cc_t\left[c_t\left(S_t, X_U, T_2\right), X_C, T_1\right]. \qquad (18.18)$$

Table 18.1 provides a cash flow table illustrating the arbitrage profits.

Because this trading strategy results in receiving money today and there is no chance of paying money in the future, this trading strategy is very attractive and should not appear in rational markets for very long. Therefore, the original asserted boundary condition would hold.

Call on call upper bound:

$$cc_t\left[c_t\left(S, X_U, T_2\right), X_C, T_1\right] \leq e^{-\hat{q}\tau_1}c_t\left(S, X_U, T_2\right). \qquad (18.19)$$

Recall for European call options on the underlying instrument, we have the following upper bound:

$$c_t \leq S_t e^{-\delta\tau_2}. \qquad (18.20)$$

In a similar fashion, for a compound call on call option, we have

$$cc_t\left[c_t\left(S, X_U, T_2\right), X_C, T_1\right] \leq e^{-\hat{q}\tau_1}c_t\left(S, X_U, T_2\right). \qquad (18.21)$$

The option to purchase another option should not cost more than the underlying yield-adjusted option; otherwise, one would just buy the underlying option.

Proof: Recall that $X_C > 0$ and limited liability implies observed option prices are nonnegative. Assume the opposite:

$$cc_t\left[c_t\left(S, X_U, T_2\right), X_C, T_1\right] > e^{-\hat{q}\tau_1}c_t\left(S, X_U, T_2\right). \qquad (18.22)$$

Move terms to the greater-than side:

$$cc_t\left[c_t\left(S, X_U, T_2\right), X_C, T_1\right] - e^{-\hat{q}\tau_1}c_t\left(S, X_U, T_2\right) > 0. \qquad (18.23)$$

TABLE 18.2 Upper Bound for Call on Call

Strategy	Today (t)	At Expiration (T_1)	
		$c_t \leq X_C$	$c_t > X_C$
Sell call on call	$+cc_t$	0	$-(c_t - X_C)$
Buy $e^{-\hat{q}\tau_1}$ units of underlying call	$-e^{-\hat{q}\tau_1}c_t$	$+c_{T_1}$	$+c_{T_1}$
Net	> 0	≥ 0	$X_C > 0$
	(By assumption)	(By limited liability)	(By assumption)

Table 18.2 illustrates the arbitrage opportunity with a cash flow table.

Because this trading strategy results in receiving money today and there is no chance of paying money in the future, this trading strategy is very attractive and should not appear in rational markets for very long. In other words, it is an easy arbitrage opportunity. Therefore, the original asserted boundary condition must hold.

The remaining boundary conditions are stated without proof. Proofs would follow in a similar manner to what we just covered.

Call on put lower bound:

$$cp_t \left[p_t \left(S, X_U, T_2 \right), X_C, T_1 \right] \geq \max \left[0, e^{-\hat{q}\tau_1} p_t \left(S, X_U, T_2 \right) - X_C e^{-r_c \tau_1} \right]. \qquad (18.24)$$

Call on put upper bound:

$$cp_t \left[p_t \left(S, X_U, T_2 \right), X_C, T_1 \right] \leq e^{-\hat{q}\tau_1} p_t \left(S, X_U, T_2 \right). \qquad (18.25)$$

Put on call lower bound:

$$pc_t \left[c_t \left(S, X_U, T_2 \right), X_C, T_1 \right] \geq \max \left[0, X_C e^{-r_c \tau_1} - e^{-\hat{q}\tau_1} c_t \left(S, X_U, T_2 \right) \right]. \qquad (18.26)$$

Put on call upper bound:

$$pc_t \left[c_t \left(S, X_U, T_2 \right), X_C, T_1 \right] \leq X_C e^{-r_c \tau_1}. \qquad (18.27)$$

Put on put lower bound:

$$pp_t \left[p_t \left(S, X_U, T_2 \right), X_C, T_1 \right] \geq \max \left[0, X_C e^{-r_c \tau_1} - e^{-\hat{q}\tau_1} p_t \left(S, X_U, T_2 \right) \right]. \qquad (18.28)$$

Put on put upper bound:

$$pp_t \left[p_t \left(S, X_C, T_2 \right), X_U, T_1 \right] \leq X_C e^{-r_c \tau_1}. \qquad (18.29)$$

The underlying option put-call parity is:[7]

$$e^{-\hat{q}_c \tau_2} c_t \left(S, X_U, T_2 \right) - e^{-\hat{q}_p \tau_2} p_t \left(S, X_U, T_2 \right) = S_t e^{-\delta \tau_2} - X_C e^{-r_c \tau_2}. \qquad (18.30)$$

TABLE 18.3   Call on Call–Call on Put Parity

Strategy	Today (t)	At Expiration (T_1) $c_{T_1} \leq X_C$	$c_{T_1} > X_C$
Sell call on call	$+cc_t$	0	$-\left(c_{T_1} - X_C\right)$
Buy put on call	$-pc_t$	$+\left(X_C - c_{T_1}\right)$	0
Buy $e^{-\hat{q}\tau_1}$ units of underlying call	$-e^{-\hat{q}\tau_1}c_t$	$+c_{T_1}$	$+c_{T_1}$
Borrow $e^{-r_c\tau_1}X_C$	$+e^{-r_c\tau_1}X_C$	$-X_C$	$-X_C$
Net	?	$= 0$	$= 0$

The compound option put-call parity relations include:

$$cc_t\left[c_t\left(S, X_U, T_2\right), X_C, T_1\right] - pc_t\left[c_t\left(S, X_U, T_2\right), X_C, T_1\right] = e^{-\hat{q}\tau_1}c_t - e^{-r_c\tau_1}X_C, \quad (18.31)$$

$$cp_t\left[p_t\left(S, X_U, T_2\right), X_C, T_1\right] - pp_t\left[p_t\left(S, X_U, T_2\right), X_C, T_1\right] = e^{-\hat{q}\tau_1}p_t - e^{-r_c\tau_1}X_C, \text{ and} \quad (18.32)$$

$$cc_t\left[c_t\left(S, X_U, T_2\right), X_C, T_1\right] - pc_t\left[c_t\left(S, X_U, T_2\right), X_C, T_1\right]$$

$$- \left\{cp_t\left[p_t\left(S, X_U, T_2\right), X_C, T_1\right] - pp_t\left[p_t\left(S, X_U, T_2\right), X_C, T_1\right]\right\}$$

$$= e^{-\delta\tau_2}c_t - e^{-r_c\tau_2}X_C. \quad (18.33)$$

We now sketch the proofs for the put-call parity stated in Equation (18.31).
Proof: Rearranging and simplifying notation,

$$cc_t - pc_t - e^{-\hat{q}\tau_1}c_t + e^{-r_c\tau_1}X_C = 0. \quad (18.34)$$

Consider the cash flow table shown in Table 18.3.
Because this trading strategy results in no cash flow in the future, this trading strategy would be very attractive if the net cash flow today is either positive or negative. Thus, we place a question mark for the net cash flow today. If the net cash flow today is positive, then sell the call on the call, buy the put on the call, buy $e^{-\hat{q}\tau_1}$ units of the underlying call, and borrow $e^{-r_c\tau_1}X_C$. If the net cash flow today is negative, then do the opposite set of trades, that is, buy the call on call, sell the put on call, sell call, and lend $e^{-r_c\tau_1}X_C$.

18.4   GESKE'S APPROACH TO VALUING A CALL ON A CALL

Geske uses the principle of risk-neutral valuation to price a stock option. Due to his focus solely on a call option on a stock, we follow his notation where c denotes the call option, S denotes the common stock price, and V denotes the firm value per share. Note we follow Geske's notation only in this section and will revert back to the more generic notation in the next section. Remember, our focus here is on valuing a call option on a stock as a compound option on the firm. Specifically, he evaluates the expression

$$c_t = e^{-r_c\tau_1}E\left\{\max\left[0, S_1\left(V_1\right) - X\right]\right\}. \quad (18.35)$$

This expression means that we find the expected payoff of the call at its expiration, time T_1, and discount it to the present. The expected payoff of the call is based on the stock price at T_1, which is a function of the value of the firm at T_1. More formally, the expression to be evaluated is[8]

$$c_t = e^{-r_c \tau_1} \int_{-\infty}^{\infty} \max \left[0, S_1 \left(V e^u \right) - X \right] f(u)\, du, \qquad (18.36)$$

where

$$u = \ln \left(V_1 / V_t \right)$$

$$f(u) = \frac{e^{-\frac{q^2}{2}}}{\sigma_V \sqrt{2\pi \tau_1}}$$

$$q = \frac{u - \mu_V \tau_1}{\sigma_V \sqrt{\tau_1}}$$

$$\mu_V = r_c - \sigma_V^2 / 2. \qquad (18.37)$$

These expressions arise from the fact that the value of the assets, V_1, is assumed to be lognormally distributed and the expected return is set to the risk-free rate.

Now the problem can be broken down into three parts that have logical interpretations. If the option expires in-the-money at T_1, the holder will exercise it and obtain a position in the stock. When the bonds mature at T_2, the stock value behaves like an ordinary call with a payoff equal to the expected value of the assets conditional on the bonds being paid off minus the expected payoff on the bonds. Thus, the value of the compound call can be expressed as the sum of three option parts:

(OP1) The value today of the assets of the firm at T_1

(OP2) The value today of the payment of the exercise price, M, on the compound option

(OP3) The value today of the payment of the exercise price, X, on the underlying, meaning, the payoff of the bonds

Each of these values is conditional on the compound option being exercised at T_1. These three values are written formally as

$$\text{(OP1)} \quad V e^{-r_c \tau_1} \int_{\ln(V^*/V)}^{\infty} e^u N(z) f(u)\, du$$

$$\text{(OP2)} \quad M e^{-r_c \tau_2} \int_{\ln(V^*/V)}^{\infty} N\left(z - \sigma_V \sqrt{\tau_{12}}\right) f(u)\, du$$

$$\text{(OP3)} \quad X e^{-r_c \tau_1} \int_{\ln(V^*/V)}^{\infty} f(u)\, du \qquad (18.38)$$

where

$$z = \frac{\left(\ln\left(V^*/M\right) + \left(r_c + \sigma_V^2/2\right)\tau_{12}\right)}{\sigma_V\sqrt{\tau_{12}}}.$$

The term V^* is the critical value of the assets at T_1 at which the equity would be suffi-ciently valuable to have the call option expire at-the-money. It can be found by solving the following equation for V^*:

$$V^*N(z) - Me^{-r_c\tau_{12}}N\left(z - \sigma_V\sqrt{\tau_{12}}\right) = X. \tag{18.39}$$

Observe that the left-hand solution of Equation (18.39) is the Black-Scholes-Merton value of the underlying option at the point at which the compound option is expiring. Note that the exercise price is the amount owed on the bond at time T_2, and the volatility is the volatility of the assets. The right-hand solution is the exercise price of the compound option. We are finding the value at the expiration of the compound option that forces the value of the underlying option to equal its exercise value. In other words, if the underlying is above this level, the compound option will exercise and convert to the stock, which then proceeds as an ordinary call option on the assets that will expire at T_2.

The values of the option parts, (OP1), (OP2), and (OP3), are as follows:

$$\text{(OP1)} \quad VN_2\left(x, y; \rho\right)$$

$$\text{(OP2)} \quad Me^{-r_c\tau_2}N_2\left(x - \sigma_V\sqrt{\tau_1}, y - \sigma_V\sqrt{\tau_2}; \rho\right)$$

$$\text{(OP3)} \quad Xe^{-r_c\tau_1}N_1\left(x - \sigma_V\sqrt{\tau_1}\right)$$

where

$$x = \frac{\ln\left(V/V^*\right) + \left(r_c + \sigma_V^2/2\right)\tau_1}{\sigma_v\sqrt{\tau_1}}$$

$$y = \frac{\ln\left(V/M\right) + \left(r_c + \sigma_V^2/2\right)\tau_2}{\sigma_v\sqrt{\tau_2}}$$

$$\rho = \sqrt{\tau_1/\tau_2}. \tag{18.40}$$

Note that $N_2(.,.,.)$ is the bivariate normal probability and reflects the likelihood of both events occurring. The two events are exercise of the compound option and exercise of the underlying option. The overall price of the compound option is (OP1) – (OP2) – (OP3).

18.5 CHARACTERISTICS OF GESKE'S CALL ON CALL OPTION

Geske also provides the derivatives of the compound call price with respect to the under-lying variables V, M, τ_2, r_c, σ_V, X, and τ_1.[9] These illustrate some interesting results. For example, $\partial c/\partial M < 0$, meaning that increasing the firm's leverage, which raises the variance of the equity, then increases the value of the call; however, the larger debt value combined with a fixed value of the assets lowers the value of the equity, which is the dominant effect. This lowers the call price.

Another interesting result is that the volatility of the stock is not constant. Define

$$V = B + S. \tag{18.41}$$

Then the total differential of Equation (18.41) is

$$dV = \left(\frac{\partial V}{\partial B}\right) dB + \left(\frac{\partial V}{\partial S}\right) dS. \tag{18.42}$$

Holding dB to zero, we have $dV = (\partial V/\partial S)\,dS$. Then write the equation as

$$\frac{dV}{V} = \frac{\left(\frac{\partial V}{\partial S}\right)\left(\frac{dS}{V}\right)}{S/S}$$

$$= \left(\frac{\partial V}{\partial S}\right)\left(\frac{S}{V}\right)\left(\frac{dS}{S}\right). \tag{18.43}$$

The volatility of the asset return, σ_V, is, therefore,

$$\sigma_V = \left(\frac{\partial V}{\partial S}\right)\frac{S}{V}\sigma_S, \tag{18.44}$$

where σ_S is the volatility of dS/S. Turning this around, we have

$$\sigma_S = \sigma_V \left(\frac{\partial S}{\partial V}\right)\frac{V}{S}. \tag{18.45}$$

The stock volatility is, thus, seen as a function of the asset volatility and the firm's leverage, which is picked up in the elasticity factor, $(\partial S/\partial V)(V/S)$.[10] The significance of this result is that the volatility of the stock, which is what we would usually insert into the Black-Scholes-Merton model to obtain the call price, is definitely not constant as the Black-Scholes-Merton model assumes. The volatility of the assets may well be constant, but the firm's leverage changes with any change in the market value of the equity relative to the market value of the assets.

The formulas for selected Greeks are provided in the appendix.

18.6 GESKE'S CALL ON CALL OPTION MODEL AND LINEAR HOMOGENEITY

Geske goes on to show that the compound option is linearly homogeneous with respect to the value of the firm, the face value of the debt, and the exercise price and that the compound option value is convex in the value of the firm. He also shows that if there is no debt, meaning that $M = 0$, the model will converge to the Black-Scholes-Merton formula. This will also occur if the bond is a perpetuity or if the option expiration coincides with the bond maturity. In the latter case, the two exercise prices merge to $M + X$.

18.7   GENERALIZED COMPOUND OPTION PRICING MODEL

Rubinstein (1991) generalizes the Geske result to accommodate the other possible compound options: a call on a put, a put on a call, and a put on a put, as well as the case of a continuous payout on the assets of δ. Brooks (2019) extends this generalized result to include a continuous payout on the underlying option of $\hat{q}$, which is not simply a trivial adjustment.

The generalized compound option pricing model, denoted co, is observed at time t under geometric Brownian motion based on an underlying instrument, denoted S_t. When compared to Geske's call on call model, S_t represents the firm value per share. The underlying option exercise price is denoted X_U and expires at time 2, denoted T_2. By comparison, the compound option exercise price is denoted X_C and expires at time 1 ($T_2 > T_1$). Mathematically, the generalized model can be expressed as

$$co\left(S, t, T_1, T_2, \iota_C, \iota_U\right) = \iota_C \iota_U S_t e^{-\delta \tau_2} e^{\hat{q}\tau_{12}} N_2\left(\iota_C \iota_U d_{11}, \iota_U d_{12}; \iota_C \rho\right)$$

$$-\iota_C \iota_U X_U e^{-r_c \tau_2} e^{\hat{q}\tau_{12}} N_2\left(\iota_C \iota_U d_{21}, \iota_U d_{22}; \iota_C \rho\right)$$

$$-\iota_C X_C e^{-r_c \tau_1} N\left(\iota_C \iota_U d_{21}\right), \tag{18.46}$$

where indicator functions denote

$$\iota_C = \begin{cases} +1 \text{ if compound call option} \\ -1 \text{ if compound put option} \end{cases} \tag{18.47}$$

$$\iota_U = \begin{cases} +1 \text{ if underlying call option} \\ -1 \text{ if underlying put option} \end{cases} . \tag{18.48}$$

The bivariate cumulative standard normal distribution is denoted

$$N_2\left(a, b; \rho\right) \equiv \int_{-\infty}^{a} \int_{-\infty}^{b} \frac{\exp\left\{-\frac{z_1^2 - 2\rho z_1 z_2 + z_2^2}{2(1-\rho^2)}\right\}}{2\pi\sqrt{1-\rho^2}} dz_1 dz_2. \tag{18.49}$$

The correlation coefficient used in the bivariate distribution is

$$\rho = \sqrt{\frac{\tau_1}{\tau_2}}. \tag{18.50}$$

Let $S_{T_1}^*$ be defined such that the underlying option is at-the-money or

$$\iota_U S_{T_1}^* e^{-\delta \tau_{12}} N_1\left(\iota_U d_{1,T_1,T_2}^*\right) - \iota_U X_U e^{-r_c \tau_{12}} N_1\left(\iota_U d_{2,T_1,T_2}^*\right) - X_C = 0, \tag{18.51}$$

where

$$d_{2,T_1,T_2}^* = \frac{\ln\left(\frac{S_1^*}{X_U}\right) + \left(r_c - \delta - \frac{\sigma^2}{2}\right)\tau_{12}}{\sigma\sqrt{\tau_{12}}}, \tag{18.52}$$

$$d_{1,T_1,T_2}^* = \frac{\ln\left(\frac{S_1^*}{X_U}\right) + \left(r_c - \delta + \frac{\sigma^2}{2}\right)\tau_{12}}{\sigma\sqrt{\tau_{12}}} = d_{2,T_1,T_2}^* + \sigma\sqrt{\tau_{12}}, \text{ and} \tag{18.53}$$

$$N_1(d) = \int_{-\infty}^{d} \frac{e^{-\frac{x^2}{2}}}{\sqrt{2\pi}} dx. \tag{18.54}$$

Let d_{ij} denote the upper limits used in the bivariate normal cumulative distribution function given in Equation (18.49), where $i = 1,2$ denotes whether the volatility term is added ($i = 1$) or subtracted ($i = 2$) and $j = 1,2$ denotes whether the evaluation is S^* at T_1 ($j = 1$) or X_U at T_2 ($j = 2$). We define

$$d_{21} \equiv \frac{\ln\left(\frac{S_t}{S_1^*}\right) + \left(r_c - \delta - \frac{\sigma^2}{2}\right)\tau_1}{\sigma\sqrt{\tau_1}}, \tag{18.55}$$

$$d_{11} \equiv \frac{\ln\left(\frac{S_t}{S_1^*}\right) + \left(r_c - \delta + \frac{\sigma^2}{2}\right)\tau_1}{\sigma\sqrt{\tau_1}} = d_{21} + \sigma\sqrt{\tau_1}, \tag{18.56}$$

$$d_{22} \equiv \frac{\ln\left(\frac{S_t}{X_U}\right) + \left(r_c - \delta - \frac{\sigma^2}{2}\right)\tau_2}{\sigma\sqrt{\tau_2}}, \text{ and} \tag{18.57}$$

$$d_{12} \equiv \frac{\ln\left(\frac{S_t}{X_U}\right) + \left(r_c - \delta + \frac{\sigma^2}{2}\right)\tau_2}{\sigma\sqrt{\tau_2}} = d_{22} + \sigma\sqrt{\tau_2}. \tag{18.58}$$

Extending Merton (1974), Geske (1977, 1984) has adapted his compound option pricing model to the case of coupon paying corporate bonds. Most important, however, Roll (1977) and Whaley (1981) have used the compound option model to obtain a closed-form solution for the price of an American call, and we cover this topic in Chapter 19.

18.8 INSTALLMENT OPTIONS

A variation of the compound option is the installment option. This is an option in which the premium is spread out in equal amounts over time. At each installment date, the holder of the option decides whether to exercise it, thereby paying the installment premium and continuing with the option. If the holder prefers not to continue, they simply fail to pay the installment. The option then terminates.[11] This instrument is very much like the compound option except that there are typically several installments, necessitating a more complex option pricing model requiring the evaluation of higher-order multivariate normal probability distributions. Also, the installments, which correspond to the exercise prices of the multiple underlying options, are usually all equal. Solving the pricing equation is quite difficult and usually requires a numerical iterative solution. The installment option permits the holder to change their mind later and get out of the contract by simply failing to pay later installments. All previously paid installments are, of course, forgone.

Finally, it should be noted that a bond or loan with multiple payments is somewhat like an installment option. At each payment date, the borrower decides whether to pay off, which is like deciding to make an installment payment. For more information, see Merton (1974) and Geske (1977, 1984). Once default occurs, the sequence of installment payments ends.

18.9   RECAP AND PREVIEW

In this chapter, we examined compound options, which are useful in helping us to understand how the stock of a company is actually an option itself, where the underlying is the assets of the firm.

In Chapter 19, we get into American option pricing in continuous time, where the use of a compound option will be very helpful.

APPENDIX 18A

Selected Greeks of the Compound Option

The compound option delta with respect to the underlying instrument is

$$\Delta_{co} \equiv \frac{\partial co}{\partial S} = \iota_C \iota_U e^{-\hat{q}\tau_1} e^{-\delta\tau_{12}} N_2 \left(\iota_C \iota_U d_{11}, \iota_U d_{12}; \iota_C \rho \right), \tag{18.59}$$

and the underlying option, which has a price of o, has a delta with respect to the underlying instrument, S, is

$$\Delta_o \equiv \frac{\partial o}{\partial S} = \iota_U e^{-(\delta-\hat{q})\tau_2} N_1 \left(\iota_U d_1 \right). \tag{18.60}$$

The compound option gamma with respect to the underlying instrument, S, is

$$\Gamma_{co} \equiv \frac{\partial^2 co}{\partial S^2} = \frac{e^{-\hat{q}\tau_1} e^{-\delta\tau_{12}}}{S_t} \left[N_1 \left(\iota_U \frac{d_{12} - \rho d_{11}}{\sqrt{1-\rho^2}} \right) \frac{n_1(d_{11})}{\sigma\sqrt{\tau_1}} + \iota_C N_1 \left(\iota_C \iota_U \frac{d_{11} - \rho d_{12}}{\sqrt{1-\rho^2}} \right) \frac{n_1(d_{12})}{\sigma\sqrt{\tau_2}} \right], \tag{18.61}$$

and the underlying option gamma with respect to the underlying instrument, S, is

$$\Gamma_o \equiv \frac{\partial^2 o}{\partial S^2} = \frac{e^{-(\delta-\hat{q})\tau_2} n_1 \left(d_1 \right)}{S_t \sigma \sqrt{\tau_2}}. \tag{18.62}$$

The compound option theta is

$$
\Theta_{co} \equiv \frac{\partial co}{\partial t} = -\frac{\sigma^2}{2} S_t e^{-\hat{q}\tau_1} e^{-\delta\tau_{12}} \left[\begin{array}{c} N_1 \left(\iota_U \frac{d_{12}-\rho d_{11}}{\sqrt{1-\rho^2}} \right) \frac{n_1(d_{11})}{\sigma\sqrt{\tau_1}} \\ -\iota_C N_1 \left(\iota_C \iota_U \frac{d_{11}-\rho d_{12}}{\sqrt{1-\rho^2}} \right) \frac{n_1(d_{12})}{\sigma\sqrt{\tau_2}} \end{array} \right]
$$

$$
+ \iota_C \iota_U \hat{q} S_t e^{-\hat{q}\tau_1} e^{-\delta\tau_{12}} N_2 \left(\iota_C \iota_U d_{11}, \iota_U d_{12}; \iota_C \rho \right)
$$

$$
- \iota_C \iota_U r_c e^{-r_c\tau_2} X_U N_2 \left(\iota_C \iota_U d_{21}, \iota_U d_{22}; \iota_C \rho \right) - \iota_C r_c e^{-r_c\tau_1} B_{t,T_1,r} X_C N_1 \left(\iota_C d_{21} \right), \quad (18.63)
$$

and the underlying option theta is

$$
\Theta_o \equiv \frac{\partial o}{\partial t} = \iota_U \left(\delta - \hat{q} \right) S_t e^{-(\delta-\hat{q})\tau_2} N \left(\iota_U d_1 \right)
$$

$$
- \iota_U \left(r - \hat{q} \right) X e^{-(r-\hat{q})\tau_2} N \left(\iota_U d_2 \right) - \frac{\sigma S_t e^{-(\delta-\hat{q})\tau_2}}{2} n \left(d_1 \right). \quad (18.64)
$$

QUESTIONS AND PROBLEMS

1 When equity is viewed as a call option on the firm where the par value of debt is the exercise price, prove that $N \left(-d_2 \right)$ is the probability of default on the debt under the equivalent martingale measure.

2 Within the context of compound options, explain the relationship between the volatility of the firm and the volatility of the common stock.

3 Prove the call on put option upper bound is $cp_t \left[p_t \left(S, X_U, T_2 \right), X_C, T_1 \right] \le e^{-\hat{q}\tau_1} p_t \left(S, X_U, T_2 \right)$.

4 Prove the call on put option lower bound is

$$
cp_t \left[p_t \left(S, X_U, T_2 \right), X_C, T_1 \right] \ge \max \left[0, e^{-\hat{q}\tau_1} p_t \left(S, X_U, T_2 \right) - X_C e^{-r\tau_1} \right].
$$

5 Prove the following parity:

$$
cp_t - pp_t = e^{-\hat{q}\tau_{12}} p_t - e^{-r_c\tau_1} X_C.
$$

NOTES

1. For the generic case, S_t will denote the underlying instrument price. Unfortunately, this notation would conflict with the stock price that we will see here is really an option.
2. See https://www.moodysanalytics.com/about-us/history/kmv-history.
3. Although we are not specifically covering American options in this chapter, compound options will play an important role in the pricing of American options, which we cover in Chapters 19 and 20.
4. At this point, we shall diverge slightly from the original Geske derivation.

5. Note that this is consistent with Equation (18.5).

6. Geske's alternative formulation requires that we express the stochastic process for the call as $dc/c = \alpha_c dt + \sigma_{cV} dW_c$, which will imply that the market price of risk for the call, $(\alpha_c - r_c)/\sigma_{cV}$, equals the market price of risk of the asset, $(\alpha_V - r_c)/\sigma_V$. Although this statement will be true under the assumptions of the model, introducing it without proof and relying on it can cause some confusion.

7. Note that option yields would be expected to be different between calls and puts within the context of common stocks as options, although puts in this context are difficult to interpret. Within the context of the generic compound options model, one could easily have yield paying calls and puts.

8. At this point, we begin borrowing from Rubinstein (1991), who shows more of the details of the solution.

9. A useful technique for taking these derivatives is to use the derivatives in the Black-Scholes-Merton model. Hence, $\partial c/\partial w = (\partial c/\partial S)(\partial S/\partial w)$ where w is the variable of differentiation and $\partial S/\partial w$ is obtained from the standard Black-Scholes-Merton derivatives.

10. The elasticity of S with respect to V is the percentage change in S over the percentage change in V and can be expressed as $(\partial S/\partial V)(V/S)$.

11. Because this right is built into the contract, failure to pay an installment is not viewed as a default.

CHAPTER 19

American Call Option Pricing

I t is well known, as we covered in Chapter 2, that an American call option on an asset that makes no cash payments, such as dividends on stocks, during the life of the option will not be exercised early and, hence, can be valued as a European option with the standard Black-Scholes-Merton formula. If the underlying asset makes a cash payment during the life of the option, early exercise could possibly be optimal, thereby giving the American call a premium over a European call and rendering the Black-Scholes-Merton model inappropriate.[1] As we showed when covering the binomial model, valuation of such an option can be done by numerical methods, such as the binomial model, but it is also possible to obtain a closed-form valuation model. For an asset making a single payment during the life of the option, the model is called the Roll-Geske-Whaley model after Roll (1977), Geske (1979a, 1981), and Whaley (1981), and it is based on Geske's (1979b) compound option model.

Similar to Chapter 18, the analysis in this chapter will require more precise notation as we will work our way toward more than one dividend payment. Figure 19.1 illustrates this notation for a single dividend payment. Recall with a compound option we used T_2 to denote the underlying option expiration, whereas here we will use T_A to denote the expiration of the American option. Thus, τ_A denotes the time to expiration of an American option. As we will see, interim dividend payments are creating a condition analogous to the exercise of compound options. We denote the dividend dates as T_1, T_2, and so forth to allow for generalizations. Similarly, τ_1 denotes the time until the first dividend payment and τ_{1A} denotes the time between the first dividend payment and the option time to expiration. The values of different instruments, such as assets and options, will use only the subscript to denote the observation date. For example, S_1 denotes the asset value at time T_1. Where possible, we will drop the subscript t as it will be understood, thus $S_t = S$.

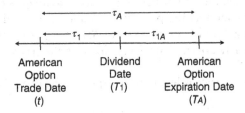

FIGURE 19.1 American Option with Single Dividend Payment Timeline

19.1   CLOSED-FORM AMERICAN CALL PRICING: ROLL-GESKE-WHALEY

The Roll-Geske-Whaley (RGW) model requires that there be only one known cash payment made over the life of the option, such as a stock that pays quarterly dividends but with only one dividend to be paid before the option expires.

The amount of the payment is assumed to be known with certainty and is the amount D, and again the time in years until the payment date is τ_1.[2] We assume that as the payment is made, the asset price falls by the amount of the payment, although this assumption can be relaxed such that it falls by a proportion of the payment. We are also given the asset volatility, σ, and the continuously compounded risk-free rate, r_c. Let us first define the following value $S - De^{-r_c\tau_1}$, which is simply the asset price minus the present value of the cash payment. We assume that this adjusted asset price follows the standard lognormal diffusion that is typically used in modeling asset price dynamics.[3] We denote the price of a European call on this asset as c and the price of an American call on this asset as C.

We can replicate the American call with three option positions, denoted OP1, OP2, and OP3. Thus, consider the following combination of option positions:

(OP1) A long position in a European call where the underlying is the stock that pays the known dividend with time to expiration τ_A and exercise price X. So this option expires at the expiration of the American option we are pricing.

(OP2) A short position in a European compound call option where the underlying option is the European call in (OP1). This compound option has a time to expiration of τ_1, meaning that it expires when the dividend payment is made, and it has an exercise price of $S_1^* + D - X$, where S_1^* is defined next.

(OP3) A long position in a European call option where the underlying is the stock that pays the known dividend with time to expiration of τ_1 and exercise price of S_1^* where S_1^* will be defined next.

The value S_1^* is the critical ex-dividend asset price above which the compound option will be exercised. The corresponding underlying call price (c^*) is also found such that

$$c^* = S_1^* + D - X. \qquad (19.1)$$

In other words, S_1^* is the ex-dividend asset price that triggers early exercise.

To understand how these options combine to replicate the American call, let us assume that we are now positioned an instant before the cash payment is to be made at time T_1. The asset price is S_1^+. As soon as the payment is made, it will fall to S_1^-. Thus, to make the exercise decision, one would compare S_1^- to S_1^*. This critical price will be found by iteration, as we will shortly cover. Now, let us look at the payoffs of the three options: (OP1), (OP2), and (OP3).

$S_1^- \leq S_1^*$ (post-payment price is less than or equal to the critical price):

(OP1)   Regardless of the post-payment price, this is a European call and cannot be exercised until its expiration T_A. It is simply worth its standard Black-Scholes-Merton value with remaining time τ_{1A}.

(OP2)   This compound option will be exercised or not depending on whether the value of (OP1) exceeds the exercise price of the compound option, $S_1^* + D - X$. The critical ex-dividend asset price is the one at which the option in (OP1) is precisely equal in value to $S_1^* + D - X$. Because the asset price S_1^- is less than S_1^*, as assumed in this case, then the value of the corresponding call (c_1^-) $c_1^- < c^*$, but by definition $c^* = S_1^* + D - X$. Thus $c_1^- < S_1^* + D - X$. Consequently, the compound option is out-of-the-money and is not exercised.

(OP3)   This is an expiring European call. Its exercise price S_1^* exceeds the ex-dividend price, so it expires worthless.

Thus if $S_1^- < S_1^*$, this combination of options produces no cash flow at this point and leaves us holding a European call with remaining expiration τ_{1A}. Note that this is similar to an American call. If it does not get exercised early at the ex-dividend date, we are left holding a European call. Now consider the other possibility. $S_1^- > S_1^*$ (ex-dividend price is greater than critical price):

(OP1)   This call will be worth c_1 when the stock goes ex-dividend. It has an exercise price of X and a remaining time of τ_{1A}.

(OP2)   The compound option is expiring. Using the presented arguments, this option is now in-the-money and is exercised. Because we are short this option, we deliver the option in (OP1) and receive the exercise price $S_1^* + D - X$.

(OP3)   This option is a European call expiring right now. Its exercise price S_1^* is less than the current asset price S_1 so it expires in-the-money. We own this option so we pay S_1^* and receive an asset worth S_1^-.

Thus if $S_1^- \geq S_1^*$ the overall cash flow is $S_1^- - S_1^* + S_1^* + D - X = S_1^- + D - X$. Thus, we paid out X and received an asset worth S_1^- and its cash payment, D. We see that this combination of two standard European options, one long and one short with different exercise prices and expirations, and one compound option produce the same results as an American option. That is, if the ex-dividend asset price is below the critical price for justification of early exercise, the American call option holder is left holding an option with exercise price X and time to expiration τ_{1A}. If the ex-dividend asset price is above the critical price for justification of early exercise, the American call option holder exercises the option, paying out X and receiving the dividend D and is left holding an asset worth S_1^-. These same results are obtained with the combination of options (OP1), (OP2), and (OP3).

The critical asset price for justification of exercise is the one such that it produces a Black-Scholes-Merton value of $S_1^* + D - X$ for a European call with exercise price of X and a time to expiration of τ_{1A}. This value must be derived iteratively by plugging in values into the Black-Scholes-Merton model with time to expiration of τ_{1A} until the option value equals $S_1^* + D - X$. A good starting point estimate of S_1^* is $X - D$ because the option would have to give a positive payoff to justify exercise, but the actual value of S_1^* will be much higher. Standard iterative equation-solving techniques such as Newton-Raphson can speed up the search. If early exercise is not justified at any price, due to too small of a cash payment relative to the other values in the model, S_1^* will be infinite. In that case, our final solution will converge to Black-Scholes-Merton.

Because our combination of the three options replicates the American call, we can easily find the value of the American call by adding the values of options (*OP1*) and (*OP3*), which are given by the Black-Scholes-Merton model, and subtracting the value of option (*OP2*), which as a compound option is given by Geske's compound option formula. The values of these three options can be written as

$$OP1 = \left(S - De^{-r_c\tau_1}\right) N_1(a) - Xe^{-r_c\tau_A} N_1(a_2)$$

$$OP2 = \left(S - De^{-r_c\tau_1}\right) N_2(b_1, a_1, \rho) - Xe^{-r_c\tau_A} N_2(b_2, a_2, \rho) - \left(S_t^* + D - X\right) e^{-r_c\tau_1} N_1(b_2)$$

$$OP3 = \left(S - De^{-r_c\tau_1}\right) N_1(b_1) - S_t^* e^{-r_c\tau_1} N_1(b_2), \tag{19.2}$$

where

$$a_1 = \frac{\ln\left[(S - De^{-r_c\tau_1})/X\right] + (r_c + \sigma^2/2)\tau_A}{\sigma\sqrt{\tau_A}}$$

$$a_2 = a_1 - \sigma\sqrt{\tau_A}$$

$$b_1 = \frac{\ln\left[(S - De^{-r_c\tau_1})/S_1^*\right] + (r_c + \sigma^2/2)\tau_{1A}}{\sigma\sqrt{\tau_{1A}}}$$

$$b_2 = b_1 - \sigma\sqrt{\tau_{1A}}$$

$$\rho = \sqrt{\tau_1/\tau_A}, \tag{19.3}$$

where $N_1(.)$ is the univariate normal probability and $N_2(.)$ is the bivariate normal probability. These formulas can be consolidated to equal

$$C = OP1 - OP2 + OP3$$

$$= (S - De^{-r_c\tau_1})N_1(a_1) - (X - D)e^{-r_c\tau_1}N_1(b_2)$$

$$+ (S - De^{-r_c\tau_1})\left[N_1(b_1) - N_2(b_1, a_1; \rho)\right] - Xe^{-r_c\tau_A}\left[N_1(a_2) - N_2(b_2, a_2; \rho)\right]. \tag{19.4}$$

The following relationships are known to exist between the univariate and the bivariate normal distributions:

$$N_1(x) - N_2(x, y; \rho) = N_2(x, -y; -\rho)$$

$$N_2(x, y; \rho) = N_2(y, x; \rho). \tag{19.5}$$

Therefore, the third expression in Equation (19.4) can be written as

$$(S - De^{-r_c\tau_1})N_2(b_1, -a_1; -\rho), \tag{19.6}$$

and the fourth expression can be written as

$$Xe^{-r_c\tau_A}\left[N_1(a_2) - N_2(a_2, b_2; \rho)\right] = Xe^{-r_c\tau_A}N_2(a_2, -b_2; -\rho). \tag{19.7}$$

Thus, our formula now becomes

$$C = (S - De^{-r_c\tau_1})N_1(a_1) + (S - De^{-r_c\tau_1})N_2(b_1, -a_1; -\rho)$$
$$- Xe^{-r_c\tau_A}N_2(a_2, -b_2; -\rho) - (X - D)e^{-r_c\tau_1}N_1(b_2). \qquad (19.8)$$

Now we can use the following relationship:

$$N_1(a_1) + N_2(b_1, -a_1; -\rho) = N_1(b_1) + N_2(a_1, -b_1; -\rho). \qquad (19.9)$$

Substituting we obtain the overall solution as

$$C = (S - De^{-r_c\tau_1})N_1(b_1) + (S - De^{-r_c\tau_1})N_2(a_1, -b_1; -\rho)$$
$$- Xe^{-r_c\tau_A}N_2(a_2, -b_2; -\rho) - (X - D)e^{-r_c\tau_1}N_1(b_2) \text{ or} \qquad (19.10)$$
$$C = (S - De^{-r_c\tau_1})N_1(b_1) - (X - D)e^{-r_c\tau_1}N_1(b_2)$$
$$+ (S - De^{-r_c\tau_1})N_2(a_1, -b_1; -\rho) - Xe^{-r_c\tau_A}N_2(a_2, -b_2; -\rho). \qquad (19.11)$$

Let us examine each of these four terms, using the interpretation based on the assumption of risk neutrality. The first term, $(S - De^{-r_c\tau_1})N_1(b_1)$, is the discounted expected value of the asset price at the payment day, conditional on the asset price exceeding the critical price. This term reflects the expected receipt of the asset on exercise of the option immediately before the dividend payment day. The term $(X - D)e^{-r_c\tau_1}N_1(b_2)$ is the discounted expected payout at the payment day from exercise of the option. The probability term is the univariate probability of early exercise. The term, $(S - De^{-r_c\tau_1})N_2(a_1, -b_1; -\rho)$, is the discounted expected value of the asset price at the expiration, given that the option was not exercised early and ends up in-the-money. The other term, $Xe^{-r_c\tau_A}N_2(a_2, -b_2; -\rho)$, is the discounted expected payout of the exercise price at expiration, conditional on the option not having been exercised early and ends up in-the-money at expiration. Each of these interpretations is based on risk neutrality, which as we know is not an assumption about investor preferences but permits correct pricing of options. Thus, when we say one of these probabilities is the probability of something occurring, we mean the probability if investors were risk neutral.

Now let us observe how the formula converges to the Black-Scholes-Merton formula for the case of zero dividends. Letting $D = 0$, we have

$$C = SN_1(b_1) + SN_2(a_1, -b_1; -\rho)$$
$$- Xe^{-r_c\tau_A}N_2(a_2, -b_2; -\rho) - Xe^{-r_c\tau_1}N_1(b_1), \qquad (19.12)$$

with $D = 0$, $S_t^* \to \infty$. Then $b_1 \to -\infty$ and $b_2 \to -\infty$. Then $N(b_1) = N(b_2) = 0$. So this leaves us with

$$C = SN_2(a_1, -b_1; -\rho) - Xe^{-r_c\tau_A}N_2(a_2, -b_2; -\rho). \qquad (19.13)$$

The first probability in Equation (19.12) is defined as

$$N_2(a_1, -b_1; -\rho) = N_1(a_1) \int_{-\infty}^{-b_1} \frac{1}{\sqrt{2\pi(1 - \rho^2)}} \exp\left(-\frac{1}{2}\frac{x^2 - 2\rho xy + y^2}{1 - \rho^2}\right) dy. \qquad (19.14)$$

With $-b_1 \rightarrow \infty$, then $N_2(a_1, -b_1; -\rho) = N_2(a_1, \infty; -\rho)$ is the joint probability that $x \leq a_1$ and $b \leq \infty$. This is simply $N_1(a_1)$. Similarly, $N_2(a_2, -b_2; -\rho) = N_2(a_2, \infty; -\rho) = N_1(a_2)$. This reduces the overall expression to

$$SN_1(a_1) - Xe^{-r_c \tau_A} N_1(a_2), \qquad (19.15)$$

which is the Black-Scholes-Merton formula, where a takes the place of d in the normal probability.

A numerical example of the one-dividend case is presented in Appendix 19A. Let us now take a look at the two-dividend case.

19.2 THE TWO-PAYMENT CASE

The single dividend model has been extended to cover the case of two known payments during the life of the option (Welch and Chen (1988) and Stephan and Whaley (1990)). Figure 19.2 illustrates the notation for the two-payment case. The notation follows closely the one-payment case and can easily be generalized.

Let the first payment be D_1 and the time to the payment date be τ_1 and the second payment be D_2 and the time to the second ex-dividend date be τ_2. The solution is[4]

$$
\begin{aligned}
C &= \left(S - D_1 e^{-r_c \tau_1}\right)\left(1 - N_3\left(-a_1, -b_1, -\hat{c}_1; \sqrt{\tau_1/\tau_A}, \sqrt{\tau_2/\tau_A}, \sqrt{\tau_1/\tau_2}\right)\right) \\
&\quad - X\left[e^{-r_c \tau_1} N_1(\hat{c}_2) + e^{-r_c \tau_2} N_2\left(b_2, -\hat{c}_2; -\sqrt{\tau_1/\tau_2}\right)\right] \\
&\quad + e^{-r_c \tau_A} N_3\left(a_2, -b_2, -\hat{c}_2; -\sqrt{\tau_1/\tau_A}, \sqrt{\tau_2/\tau_A}, \sqrt{\tau_1/\tau_2}\right) \\
&\quad + D_1 e^{-r_c \tau_1} N_1(\hat{c}_2) + D_2 e^{-r_c \tau_2}\left[N_1(\hat{c}_1) + N_2\left(\hat{b}_2, -\hat{c}_2; -\sqrt{\tau_1/\tau_2}\right)\right],
\end{aligned} \qquad (19.16)
$$

where

$$a_1 = \frac{\ln\left[\left(S - D_1 e^{-r_c \tau_1} - D_2 e^{-r_c \tau_2}\right)/X\right] + \left(r_c + \sigma^2/2\right)\tau_A}{\sigma\sqrt{\tau_A}}$$

$$a_2 = a_1 - \sigma\sqrt{\tau_A}$$

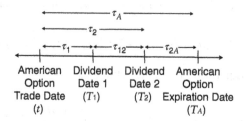

FIGURE 19.2 American Option with Two Dividend Payments Timeline

$$b_1 = \frac{\ln\left[\left(S - D_1 e^{-r_c \tau_1} - D_2 e^{-r_c \tau_2}\right)/S_{t_2}^*\right] + \left(r_c + \sigma^2/2\right)\tau_2}{\sigma\sqrt{\tau_2}}$$

$$b_2 = b_1 - \sigma\sqrt{\tau_2}$$

$$\hat{c}_1 = \frac{\ln\left[\left(S - D_1 e^{-r_c \tau_1} - D_2 e^{-r_c \tau_2}\right)/S_{t_1}^*\right] + \left(r_c + \sigma^2/2\right)\tau_1}{\sigma\sqrt{\tau_1}}$$

$$\hat{c}_2 = \hat{c}_1 - \sigma\sqrt{\tau_1}. \tag{19.17}$$

The critical stock prices are defined by the relationships,

$$c_1^* = S_{t_1}^* + D_1 + D_2 e^{-r_c(\tau_2 - \tau_1)} - X$$

$$c_2^* = S_{t_2}^* + D_2 - X. \tag{19.18}$$

Note that in the first case, the critical stock price at the first payment date must equate the exercise value, the right-hand side, to the value of an *American* call. This American call price would come from the one-payment American call formula that we derived in the previous section. At the second payment date, the critical price equates the exercise value with the Black-Scholes-Merton price of a *European* call with time $T - t_2$ remaining.

The probability $N_3(x, y, z; \rho_{xy}, \rho_{xz}, \rho_{yz})$ is the trivariate normal probability. It can be calculated by defining it in relation to the univariate normal probability as follows:

$$N_3(x, y, z; \rho_{xy}, \rho_{xz}, \rho_{yz}) = \int_{-\infty}^{y} N_1\left(\frac{y - \rho_{xy}q}{\sqrt{1 - \rho_{xy}^2}}\right) N_1\left(\frac{z - \rho_{yz}q}{\sqrt{1 - \rho_{yz}^2}}\right) N_1(q)\, dq, \tag{19.19}$$

where $n_1(q)$ is the univariate normal density function. Numerical integration can usually be used to evaluate this integral.

In the two-payment case, the problem can sometimes be simplified. Simplification can be very beneficial in coding. In particular, great savings can be amassed when not having to solve for the critical stock prices that trigger early exercise. Consider the following cases, which are collectively exhaustive:

$D_1 = D_2 = 0$: Use the Black-Scholes-Merton model.

$D_1 < X\left(1 - e^{-r_c(\tau_2 - \tau_1)}\right), D_2 = 0$: The first cash payment is too small to justify early exercise. Then use the Black-Scholes-Merton model with $S - D_1 e^{-r_c \tau_1}$ as the asset price.

$D_1 > X\left(1 - e^{-r_c(\tau_2 - \tau_1)}\right), D_2 = 0$: Early exercise is possible at the first payment date, but not at the second. Then use the one-payment American call formula.

$D_1 < X\left(1 - e^{-r_c(\tau_2 - \tau_1)}\right)$ and $D_2 < X\left(1 - e^{-r_c(\tau_A - \tau_2)}\right)$: Both payments are too small to justify early exercise. Then use $S - D_1 e^{-r_c \tau_1} - D_2 e^{-r_c \tau_2}$ as the asset price in the Black-Scholes-Merton model.

$D_1 < X\left(1 - e^{-r_c(\tau_2-\tau_1)}\right)$ and $D_2 > X\left(1 - e^{-r_c(\tau_A-\tau_2)}\right)$: The option will not be exercised at the first payment date because D_1 is too small, but it could be exercised at the second. Use the one-payment American call formula, but we will still need to subtract the present value of both payments from the asset price.

$D_1 > X\left(1 - e^{-r_c(\tau_2-\tau_1)}\right)$ and $D_2 < X\left(1 - e^{-r_c(\tau_A-\tau_2)}\right)$: The option could be exercised at the first payment date but not the second. Use the one-payment American call formula, but we will still need to subtract the present value of both payments from the asset price.

$D_1 > X\left(1 - e^{-r_c(\tau_2-\tau_1)}\right)$ and $D_2 > X\left(1 - e^{-r_c(\tau_A-\tau_2)}\right)$: The option could be exercised at either payment date. Then use the two-payment American call formula as described.

The advantage of recognizing these special cases where the two-payment model reverts to something simpler lies in programming efficiency. If any of those conditions are met, the program can execute the simpler formula.

For more than two payments, the model would extend in the same manner. For n payments, one would be required to evaluate an $(n + 1)$ variate normal probability distribution, but computation of multivariate normal integrals is quite difficult. In that case, one usually would use a numerical method such as a binomial model or finite difference approach, the latter of which we cover in Chapter 24.

In addition, Macmillan (1986) and Barone-Adesi and Whaley (1987) have developed approximation techniques for American options. They split the option price into two components: the European option value and the early exercise premium. They then obtain the partial differential equation for the early exercise premium, make some simplifying assumptions, and obtain a quadratic equation that approximates the solution. This method was very useful during the era in which computers did not have the speed they have today. In more current times, exact multivariate calculations and binomial methods are much more feasible. Research has continued to provide other formulas, but ultimately the best methods for American call option pricing are the Roll-Geske-Whaley technique and the binomial model.

19.3 RECAP AND PREVIEW

In this chapter, we looked at the pricing of American call options. We had previously learned how to price them using the binomial model, but here we see that a closed-form solution, which uses the compound option pricing model, is possible. We saw that the complicating factor is the existence of a dividend during the life of the option. The model must accommodate whether the holder exercises early to get the dividend or holds on to the option. The bivariate normal distribution is used to accommodate the existence of two exercise decisions—the one early and the one at expiration. Multiple dividends can be handled using a multivariate normal distribution.

In Chapter 20, we show that this approach can also be applied to put options.

APPENDIX 19A

Numerical Example of the One-Dividend Model

Consider a stock priced at $S = 100$ with a volatility of 20% and a time to expiration of $\tau_A = 1$ (one year). Suppose in nine months, the stock pays a dividend of 2.0. Thus, $D_1 = 2.0$, and $\tau_1 = 0.75$. The continuously compounded risk-free rate is 4%. Let us first calculate the price if this option were European. Recall that we determine the stock price minus the present value of the dividends, which is

$$S_t - De^{-r_c\tau_1} = 100 - 2e^{-0.04(0.75)} = 98.0591.$$

We then insert this value for the stock price into the Black-Scholes-Merton formula, and we obtain a value of 8.7622.

For the compound option model, we must first find the critical ex-dividend price for early exercise. Remember that we are looking for a stock price, S^*, that forces the value of a standard European option to equal, $S^* + D - X = S^* + 2 - 100 = S^* - 98$. By trial and error, we find that this price is 108.5172. To demonstrate that this is true, simply plug 108.5172 into the Black-Scholes-Merton model with time to expiration of 0.25. We would obtain 10.5197. Then find $S^* - 98 = 108.5172 - 98 = 10.5172$. These values are close enough.[5]

Recall that the American call formula is Equation (19.11), which is as follows:

$$\left(100 - 2e^{-0.04(0.75)}\right) N_1(b_1) + \left(100 - 2e^{-0.04(0.75)}\right) N_2(a_1, -b_1; -\rho)$$
$$-100e^{-0.04(1)} N_2(a_2, -b_2; -\rho) - (100 - 2)e^{-0.04(0.75)} N_1(b_2).$$

The various values that go into the normal probabilities are as follows:

$$a_1 = \frac{\ln\left[\left(100 - 2e^{-0.04(0.75)}\right)/100\right] + \left(0.04 + (0.2)^2/2\right)(1)}{0.2\sqrt{1}} = 0.202001$$

$$a_2 = 0.202001 - 0.2\sqrt{1} = 0.002001$$

$$b_1 = \frac{\ln\left[\left(100 - 2e^{-0.04(0.75)}\right)/108.5172\right] + \left(0.04 + (0.2)^2/2\right)(0.75)}{0.2\sqrt{0.75}} = -0.32527$$

$$b_2 = 0.32527 - 0.2\sqrt{1 - 0.25} = -0.49847$$

$$\rho = \sqrt{0.75/1} = 0.8660.$$

Now we calculate the probabilities using spreadsheet routines as covered in Chapter 5:

$$N_1(b_1) = N_1(-0.32527) = 0.372489$$

$$N_2(a_1, -b_1; -\rho) = N_2(0.202001, 0.32527; -0.8660) = 0.224492$$

$$N_2(a_2, -b_2; -\rho) = N_2(0.002001, 0.49847; -0.8660) = 0.209849$$

$$N_1(b_2) = N_1(-0.49847) = 0.309075.$$

The option value is, therefore,

$$98.0591\,(0.372489) + 98.0591\,(0.224492)$$

$$- 100e^{-0.04(1)}0.209849 - (100 - 2)\,e^{-0.04(0.75)}\,(0.309075)$$

$$= 8.983155.$$

QUESTIONS AND PROBLEMS

1 "An American call option on a stock that does not pay a dividend during the life of the option will not be exercised early and, hence, can be valued as a European option with the standard Black-Scholes-Merton formula." Prove this statement.

2 An American call option with one dividend payment can be valued based on the following equation. Provide an interpretation of this expression:

$$C = \left(S - De^{-r_c\tau_1}\right) N_1(b_1) - (X - D)\,e^{-r_c\tau_1}N_1(b_2)$$

$$+ \left(S - De^{-r_c\tau_1}\right) N_2(a_1, -b_1; -\rho) - Xe^{-r_c\tau_A}N_2(a_2, -b_2; -\rho).$$

3 Explain the variable, S_t^*, contained in the American call option model with one dividend presented in this chapter.

4 Explain the variables $S_{t_1}^*$ and $S_{t_2}^*$ contained in the American call option model with two dividends presented in this chapter.

5 Interpret the variable $N_3(x, y, z; \rho_{xy}, \rho_{xz}, \rho_{yz})$ contained in the American call option model with two dividends presented in this chapter.

6 [Contributed by Jason Priddle] Explain why it is necessary to use a compound option to value an American call on a dividend-paying stock.

7 [Contributed by Dennel McKenzie] Write out the equations that implicitly solve for the critical stock price for early exercise of an American call if the underlying stock pays three dividends, D_1, D_2, and D_3, with respective times with ex-dividend dates of t_1, t_2, and t_3. There will be three equations, one for each ex-dividend date.

NOTES

1. It is well known that the lower bound of an American call on an asset that makes cash payments is $S - PV(X) - PV(D)$, where S is the current asset price, $PV(X)$ is the present value of the exercise price, and $PV(D)$ is the present value of the cash payments (dividends) over the life of the option. If this option were exercised, it would give a payoff of $S - X$. Clearly if $X - PV(X) > PV(D)$, meaning, the interest forgone on paying the exercise price early exceeds the benefit of the dividends, early exercise will not be justified. This scenario will occur with a combination of small

cash payments, a high interest rate, and a high exercise price. Because such options will not be exercised early, they can, therefore, be valued using the Black-Scholes-Merton model with the asset price reduced by the present value of the dividends. Because this condition is not always met, it could be possible to justify early exercise.

2. For stocks, the relevant date is the ex-dividend date, because that is the date on which the stock price falls to reflect the fact that anyone holding the stock after that time forgoes the dividend. If the ex-dividend day is 35 days from now, $\tau_1 = 35/365$. We ignore the delay between the ex-dividend date and the actual payment date a few weeks later.

3. Roll (1977) assumed that the asset price follows the lognormal diffusion, but this cannot be true if the dividend component of the asset price is non-stochastic. This correction was made by Whaley (1981). In other words, the statistical process that follows a lognormal diffusion is the asset price minus the present value of the dividends.

4. We use $\hat{c}$ to distinguish these values from the call price.

5. How close we make them depends on what we tell the iterative technique about how close is close enough so that it can stop. Here we are within a tenth of a penny.

CHAPTER 20

American Put Option Pricing

In this chapter, we take a look at American put options. It is well known that irrespective of any dividends or cash payouts on the underlying, it could be optimal to exercise an American put option early. Dividends, in fact, reduce the likelihood of early exercise, but they do not eliminate it.

20.1   THE NATURE OF THE PROBLEM OF PRICING AN AMERICAN PUT

Based on geometric Brownian motion, the price of an American put follows the partial differential equation,

$$\frac{\partial P_t}{\partial t} = r_c P_t - r_c S_t \frac{\partial P_t}{\partial S_t} - \frac{1}{2} \frac{\partial^2 P_t}{\partial S_t^2} \sigma^2 S_t^2. \tag{20.1}$$

For a European put, in which we would use a lowercase p, the PDE would still apply with the following boundary condition:

$$p_T = \max(0, X - S_T), \tag{20.2}$$

where S_T is the asset price at expiration. For an American put we require the additional condition,

$$P_t \geq \max(0, X - S_t). \tag{20.3}$$

This condition simply means that at *any* time t prior to expiration, the American put value must be worth at least its exercise value because it can be exercised immediately. Because t and S_t change, this is a free boundary problem and had for many years been thought to be unsolvable. Geske and Johnson (1984), however, obtained a solution as an n-fold compound option, using Geske's (1979b) compound option formula and the equivalent martingale/risk neutrality assumption. We shall demonstrate this result here.

Consider the layout in continuous time: From t to $t + dt$ is time unit dt. From t to time $t + 2dt$ is time unit $2dt$, and so on. At any instant the put can be exercised. If it is not exercised at that instant, we move forward and consider whether it is optimal to exercise the put at the next instant. The exercise decision will be determined by many factors and will take into account that optimality may come from waiting until later to exercise it.

20.2   THE AMERICAN PUT AS A SERIES OF COMPOUND OPTIONS

To solve the problem, we need to know the correlation between successive cumulative Wiener processes at different times, say t_1 and t_2, where $t_1 < t_2$. For example, starting from time 0, we have

$$W_{t_1} = \int_0^{t_1} dW_j$$

$$W_{t_2} = \int_0^{t_2} dW_j = \int_0^{t_1} dW_j + \int_{t_1}^{t_2} dW_j. \tag{20.4}$$

Ultimately, we shall need the correlation between W_{t_1} and W_{t_2}, which, based on Chapter 10, can be expressed as follows:[1]

$$\rho_{12} = \frac{\text{cov}(u_1, u_2)}{\sqrt{\text{var}(u_1)}\sqrt{\text{var}(u_2)}} = \frac{\tau_1}{\sqrt{\tau_1}\sqrt{\tau_2}} = \sqrt{\frac{\tau_1}{\tau_2}}, \tag{20.5}$$

where τ_1 is the time elapsed over one increment and τ_2 is the time elapsed over two increments.[2] This correlation is important, because it is a component of a multivariate normal probability calculation, which captures the likelihood of multiple events occurring, such as exercising at one point in time, having not exercised earlier. Now we can proceed to explain how to value the option.

At any instant we exercise the option if two conditions are met:

(C1)   It has not already been exercised and
(C2)   if the payoff from exercising is greater than the value of the put if it is not exercised.

The critical asset price that triggers early exercise is defined as S_t^* in the equation

$$X - S_t^* = P_t. \tag{20.6}$$

This condition means that at t, there is an asset price, S_t^*, that produces a put price, P_t. When that put price is equal to the exercise value, any asset price below S_t^* will trigger early exercise. The time point t is clearly any time point up to the expiration, T.

At the first instant, there is no probability of prior exercise. So we integrate the exercise payoff, which is the exercise price minus the asset price, over all possible asset prices less than the critical asset price. Then we discount this value back one instant. This produces two terms, one of which is the discounted exercise price multiplied by the probability that the asset price will be below the critical asset price. At the next instant we must also discount the expected payoff, but we must consider the joint probability of exercise at that instant conditional on not having exercised at the previous instant. We follow this procedure for the remaining instants over the life of the option.

Ultimately the price of the option then becomes a discounted weighted average of all of the possible payoffs. It can be written as follows:

$$e^{-r_c dt}\left(X - S^*_{dt}\right)\text{Prob}\left(S_{dt} \le S^*_{dt}\right)$$

$$+ e^{-r_c 2dt}\left(X - S^*_{2dt}\right)\text{Prob}\left(S_{2dt} \le S^*_{2dt} \text{ and } S_{dt} > S^*_{dt}\right)$$

$$+ e^{-r_c 3dt}\left(X - S^*_{3dt}\right)\text{Prob}\left(S_{3dt} \le S^*_{3dt},\ S_{2dt} > S^*_{2dt} \text{ and } S_{dt} > S^*_{dt}\right)$$

$$+ \cdots$$

continuing until T. (20.7)

The overall solution can be written more compactly as

$$P = Xw_2 - Sw_1,$$

where

$$w_1 = N_1\left[-d_1\left(S^*_t, dt\right)\right] + N_2\left[d_1\left(S^*_{dt} dt\right), -d_1\left(S^*_{2dt} 2dt\right); -\rho_{12}\right]$$

$$+ N_3\left[d_1\left(S^*_{dt} dt\right), d_1\left(S^*_{2dt} 2dt\right), -d_1\left(S^*_{3dt} 3dt\right); \rho_{12}, -\rho_{13}, -\rho_{23}\right]$$

$$+ \cdots$$

$$w_2 = e^{-r_c dt} N_1\left[-d_2\left(S^*_{dt}, dt\right)\right] + e^{-r_c 2dt} N_2\left[d_2\left(S^*_{dt} dt\right), -d_2\left(S^*_{2dt} 2dt\right); -\rho_{12}\right]$$

$$+ e^{-r_c 3dt} N_3\left[d_2\left(S^*_{dt} dt\right), d_2\left(S^*_{2dt} 2dt\right), -d_2\left(S^*_{3dt} 3dt\right); \rho_{12}, -\rho_{13}, -\rho_{23}\right]$$

$$+ \cdots$$

and

$$d_1(q, \tau) = \frac{\ln(S/q) + \left[r_c + (\sigma^2/2)\right]\tau}{\sigma\sqrt{\tau}}$$

$$d_2(q, \tau) = d_1(q, \tau) - \sigma\sqrt{\tau} \tag{20.8}$$

for any q and τ. The correlation coefficients follow the pattern,

$$\rho_{12} = \sqrt{\frac{dt}{2dt}} = \sqrt{\frac{1}{2}} = \frac{1}{\sqrt{2}}$$

$$\rho_{13} = \sqrt{\frac{dt}{3dt}} = \sqrt{\frac{1}{3}} = \frac{1}{\sqrt{3}}$$

$$\rho_{23} = \sqrt{\frac{2dt}{3dt}} = \sqrt{\frac{2}{3}} = \sqrt{\frac{2}{3}}.$$

Geske and Johnson provide the partial derivatives with respect to each of the terms in the formula. Blomeyer and Johnson (1988) provide an extension that takes into account dividend payments, and they also conduct empirical tests of the model.

Because exercise is possible at any time instant, this formula consists of an infinite number of terms, which would seem to make the formula analytically intractable. Geske and Johnson emphasize that the formula is an exact solution, even if it does contain an infinite number of terms. They provide an analytical approximation based on a method called the *Richardson extrapolation*, which basically assumes that the put can be exercised only at three points spread over its life. This approach greatly reduces the complexity of the problem. They obtain the price of a put that can be exercised only at expiration, which is the standard Black-Scholes-Merton model price, and the price of another put that can be exercised only halfway to expiration or at expiration, which requires a bivariate normal probability computation, as well as the price of another put that can be exercised only one-third of the way through its life or two-thirds of the way through its life or at expiration, which requires a trivariate normal probability computation. The American put value is then approximated as a weighted average of the prices of these three puts. They argue that an approximation of an exact formula is better than an approximation where the solution is not known, which, other than using numerical methods, is all one can do to price an American put. Bunch and Johnson (1992) provide an improvement on this technique.

Research on American put pricing has gone well beyond the Geske-Johnson model, but it is safe to say that there is no closed-form solution with a finite number of terms and never can be. The early exercise decision itself is an infinite number of decisions. Hence, it would never be possible to reduce the complexity of the problem without making some simplifying assumptions that ultimately result in the formula being an approximation. Fortunately, advances in computing power have made the binomial model the best method of pricing American puts. Nonetheless, it is important to understand that American options are compound options. They are options in which choices are permitted that allow the option to terminate early by claiming the exercise value or to remain alive with the possibility of exercise at a later date. The Geske-Johnson model makes that clear.

20.3   RECAP AND PREVIEW

In this chapter, we showed how the American put can be viewed as a combination of an infinite number of compound options. As such, the solution has an infinite number of terms, but it does provide insights into the pricing of American puts, and it does lead to approximation formulas.

In Chapter 21, we look at min-max options, which are options with more than one underlying. The payoff will depend on either the better or worse performing of the underlyings.

QUESTIONS AND PROBLEMS

1   Derive the partial differential equation for an American put option based on geometric Brownian motion when we assume that dividends are paid in the form of a continuously compounded dividend yield.

2 The American put option model presented here is based on S_t^*. Explain this variable.

3 Based on the Geske-Johnson model, one of the bivariate terms is
$N_2\left[d_1\left(S_{dt}^*dt\right),-d_1\left(S_{2dt}^*2dt\right);-\rho_{12}\right]$. Explain why the correlation coefficient is negative.

4 Based on the Geske-Johnson model, one of the trivariate terms is
$N_3\left[d_1\left(S_{dt}^*dt\right),d_1\left(S_{2dt}^*2dt\right),-d_1\left(S_{3dt}^*3dt\right);\rho_{12},-\rho_{13},-\rho_{23}\right]$. Explain the pattern of d_1 terms as well as the pattern of correlation coefficient terms.

5 [Contributed by Dennel McKenzie] Explain why dividends on a stock reduce the likelihood that a put option on the stock will be exercised early.

6 (Contributed by Dennel McKenzie) Provide a verbal (non-quantitative) interpretation of the following representation:

$$\text{cov}\left(\int_0^{t_1}d\mathrm{W}_t,\int_0^{t_2}d\mathrm{W}_t\right).$$

NOTES

1. See Chapter 10, Section 10.6, Selected Time Series Properties.
2. The derivation of this correlation is not provided in the Geske-Johnson paper. It is added here for clarification.

CHAPTER 21

Min-Max Option Pricing

There are a number of variations of options in which there is more than one underlying asset. These options have a variety of interesting names. Nearly all of them are based on the maximum or minimum performer of two or more underlying assets or rates. The first paper to examine this type of option was Stulz (1982), who derived formulas for calls and puts that pay off based on which of two assets has the maximum or minimum value. We focus on the Stulz model initially and then show how it establishes a framework for variations of this type of option.

21.1 CHARACTERISTICS OF STULZ'S MIN-MAX OPTION

Suppose there are two assets whose current values at an arbitrary time are S_1 and S_2 and whose values at expiration T are S_{1T} and S_{2T}. Each asset may pay a cash flow yield or require a payment, such as storage costs. We generically denote these continuous cash flows as δ_1 and δ_2, consistent with dividend yields in the case where the underlying instruments are stocks. Consider a call option with exercise price of X expiring at time T that pays off based on which of the two assets has the lesser value. So what happens at expiration is two-step. We first determine which is the lesser valued of the two assets. Then we insert the value of that asset into the standard payoff formula for a call. Thus, the call payoff is

$$c_{\min,T} = \max\left[0, \min\left(S_{1T}, S_{2T}\right) - X\right].$$ (21.1)

Let us denote the current price of this option as c_{min}, where it will be understood that the call is on the minimum value of asset 1 and asset 2 and is being evaluated at time t.

Likewise, a call on the maximum has a current price of c_{max}. At expiration, we first determine the greater valued of the two assets and insert the value of that asset into the formula for the payoff of a call. The call payoff in this case is

$$c_{\max,T} = \max\left[0, \max\left(S_{1T}, S_{2T}\right) - X\right].$$ (21.2)

To establish some bounds on the prices of these options, much in the same way that we developed bounds on the prices of standard European options, let us compare the payoffs of a portfolio of long the call on the max priced at c_{max} and long the call on the min priced at c_{min} with a portfolio consisting of long the standard call on asset 1 priced at c_1 and long the standard call on asset 2 priced at c_2. We must consider two general cases, $S_{1T} < S_{2T}$ and $S_{1T} \geq S_{2T}$ and three outcomes in each case, as displayed in Table 21.1 Panels A and B.

TABLE 21.1   Min and Max Calls in Comparison to Standard Calls

Panel A. $S_{1T} < S_{2T}$

	Payoff		
Current value of instrument	$S_{1T} < S_{2T} < X$	$S_{1T} \leq X \leq S_{2T}$	$X < S_{1T} < S_{2T}$
$+c_{max}$	0	$S_{2T} - X$	$S_{2T} - X$
$+c_{min}$	0	0	$S_{1T} - X$
Total	0	$S_{2T} - X$	$S_{2T} + S_{1T} - 2X$
$+c_1$	0	0	$S_{1T} - X$
$+c_2$	0	$S_{2T} - X$	$S_{2T} - X$
Total	0	$S_{2T} - X$	$S_{2T} + S_{1T} - 2X$

Panel B. $S_{1T} \geq S_{2T}$

	Payoff		
Current value of instrument	$S_{2T} \leq S_{2T} < X$	$S_{2T} \leq X \leq S_{1T}$	$X < S_{2T} \leq S_{1T}$
$+c_{max}$	0	$S_{1T} - X$	$S_{1T} - X$
$+c_{min}$	0	0	$S_{2T} - X$
Total	0	$S_{1T} - X$	$S_{2T} + S_{1T} - 2X$
$+c_1$	0	$S_{1T} - X$	$S_{1T} - X$
$+c_2$	0	0	$S_{2T} - X$
Total	0	$S_{1T} - X$	$S_{2T} + S_{1T} - 2X$

It is apparent that the call on the max plus the call on the min has the same payoff as the combination of long calls on both assets. Therefore, the following parity must hold for their current prices:

$$c_{max} + c_{min} = c_1 + c_2. \qquad (21.3)$$

Note that the cash flow yields related to the underlying assets do not directly influence this parity. This type of *max-min* parity will continue to be useful because all we need to do is to derive a pricing model for one of the two min-max options, and the price of the other can be obtained using the previous equation.

Now let us consider a put on the minimum. Its current price evaluated at time t is expressed as p_{min}. Its payoff at expiration is

$$p_{min,T} = \max\left[0, X - \min\left(S_{1T}, S_{2T}\right)\right]. \qquad (21.4)$$

A put on the maximum has a current price of p_{max} and a payoff of

$$p_{max,T} = \max\left[0, X - \max\left(S_{1T}, S_{2T}\right)\right]. \qquad (21.5)$$

Following similar arguments as the call, we note that[1]

$$p_{max} + p_{min} = p_1 + p_2. \qquad (21.6)$$

Intuitively, the call on the max will be worth at least the call on the min as $\max(S_{1T}, S_{2T}) \geq \min(S_{1T}, S_{2T})$. Specifically, $c_{\max} \geq c_{\min}$. We demonstrate this intuition in Table 21.2. Thus, if $c_{\max} \geq c_{\min}$, then a portfolio can be created such that there is a positive probability of receiving money and no chance of losing. Clearly, this would require a nonnegative investment; otherwise, we have arbitrage resulting in disequilibrium.

Following a similar argument, the put on the min will be worth at least the put on the max. Specifically, $p_{\min} \geq p_{\max}$. We demonstrate this intuition in Table 21.3. Thus, if $p_{\min} < p_{\max}$, then again there is a portfolio that can be created such that there is a positive probability

TABLE 21.2   Max Call in Comparison to Min Call

Panel A. $S_{1T} < S_{2T}$

	Payoff		
Current value of instrument	$S_{1T} < S_{2T} < X$	$S_{1T} \leq X \leq S_{2T}$	$X < S_{1T} < S_{2T}$
$+c_{\max}$	0	$S_{2T} - X$	$S_{2T} - X$
$-c_{\min}$	0	0	$-(S_{1T} - X)$
Total	0	$S_{2T} - X$ ≥ 0	$S_{2T} - S_{1T}$ > 0

Panel B. $S_{1T} \geq S_{2T}$

	Payoff		
Current value of instrument	$S_{2T} \leq S_{1T} < X$	$S_{2T} \leq X \leq S_{1T}$	$X < S_{2T} < S_{1T}$
$+c_{\max}$	0	$S_{1T} - X$	$S_{1T} - X$
$-c_{\min}$	0	0	$-(S_{2T} - X)$
Total	0	$S_{1T} - X$ ≥ 0	$S_{1T} - S_{2T}$ > 0

TABLE 21.3   Min Put in Comparison to Max Put

Panel A. $S_{1T} < S_{2T}$

	Payoff		
Current value of instrument	$S_{1T} < S_{2T} < X$	$S_{1T} \leq X \leq S_{2T}$	$X < S_{1T} < S_{2T}$
$+p_{\min}$	$X - S_{1T}$	$X - S_{1T}$	0
$-p_{\max}$	$-(X - S_{2T})$	0	0
Total	$S_{2T} - S_{1T}$ > 0	$X - S_{1T}$ ≥ 0	0

Panel B. $S_{1T} \geq S_{2T}$

	Payoff		
Current value of instrument	$S_{2T} \leq S_{1T} < X$	$S_{2T} \leq X \leq S_{1T}$	$X < S_{2T} < S_{1T}$
$+p_{\min}$	$X - S_{2T}$	$X - S_{2T}$	0
$-p_{\max}$	$-(X - S_{1T})$	0	0
Total	$S_{1T} - S_{2T}$ ≥ 0	$X - S_{2T}$ ≥ 0	0

of receiving money and no chance of losing. Clearly, this would require a nonnegative investment; otherwise, we have arbitrage resulting in disequilibrium.

Further, the call on the max will be worth at least the maximum of the call on each asset $c_{max} \geq min(c_1, c_2)$. Without loss of generality, we let $c_1 > c_2$. We demonstrate this intuition in Table 21.4. Thus, if $c_{max} \geq c_1$, then again there is a portfolio that can be created such that there is a positive probability of receiving money and no chance of losing. Clearly, this would require a nonnegative investment; otherwise, we have arbitrage resulting in disequilibrium.

Following a similar argument, the put on the min will be worth at least the maximum of the put on each asset or $p_{max} \geq min(p_1, p_2)$.

Consider the following comparison whereby a put on the min is shown to be equivalent to a long position in risk-free bonds with face value X maturing at the option's expiration, a short position in a call on the minimum struck at zero, whose price we denote as $c_{min,X=0}$, and a long position on a call on the min with exercise price X. Note first that a call on the minimum struck at zero is an instrument that always pays off the lower-priced asset and there are effectively two exercise prices, $X = 0$ and $X > 0$. We denote the current price of a call on the min struck at zero as $c_{min,X=0}$. We illustrate these results in Table 21.5.

From Table 21.5, it is apparent that a put on the min is a perfect substitute for the combination of options and the risk-free bond. Consequently, the current value of the put on the min is the same as the current value of the combination, namely

$$p_{min} = Xe^{-r_c\tau} - c_{min,X=0} + c_{min}. \tag{21.7}$$

Thus, once we have already obtained a formula for a call on the minimum, we can simply use this formula to get a put on the minimum.

Finally, we can easily obtain the price of the put on the maximum. Let us compare it to a long position in a pure discount bond, a call on the maximum struck at zero, and a long call on the maximum with an exercise price of X illustrated in Table 21.6.

TABLE 21.4 Max Calls in Comparison to Max of Two Standard Calls

Panel A. $S_{1T} < S_{2T}$

	Payoff		
Current value of instrument	$S_{1T} < S_{2T} < X$	$S_{1T} \leq X \leq S_{2T}$	$X < S_{1T} < S_{2T}$
$+c_{max}$	0	$S_{2T} - X$	$S_{2T} - X$
$-c_1$	0	0	$-(S_{1T} - X)$
Total	0	$S_{2T} - X$ ≥ 0	$S_{2T} - S_{1T}$ > 0

Panel B. $S_{1T} \geq S_{2T}$

	Payoff		
Current value of instrument	$S_{2T} \leq S_{1T} < X$	$S_{2T} \leq X \leq S_{1T}$	$X < S_{2T} \leq S_{1T}$
$+c_{max}$	0	$S_{1T} - X$	$S_{1T} - X$
$-c_1$	0	$-(S_{1T} - X)$	$-(S_{1T} - X)$
Total	0	0	0

TABLE 21.5 Put-Call Parity for Options on the Min

Panel A. $S_{1T} < S_{2T}$

Current value of instrument	Payoff		
	$S_{1T} < S_{2T} < X$	$S_{1T} \leq X \leq S_{2T}$	$X < S_{1T} < S_{2T}$
$+p_{min}$	$X - S_{1T}$	$X - S_{1T}$	0
$+Xe^{-r_c\tau}$	X	X	X
$-c_{min,X=0}$	$-S_{1T}$	$-S_{1T}$	$-S_{1T}$
$+c_{min}$	0	0	$S_{1T} - X$
Total	$X - S_{1T}$	$X - S_{1T}$	0

Panel B. $S_{1T} \geq S_{2T}$

Current value of instrument	Payoff		
	$S_{2T} \leq S_{1T} < X$	$S_{2T} \leq X \leq S_{1T}$	$X < S_{2T} \leq S_{1T}$
$+p_{min}$	$X - S_{2T}$	$X - S_{2T}$	0
$+Xe^{-r_c\tau}$	X	X	X
$-c_{min,X=0}$	$-S_{2T}$	$-S_{2T}$	$-S_{2T}$
$+c_{min}$	0	0	$S_{2T} - X$
Total	$X - S_{2T}$	$X - S_{2T}$	0

TABLE 21.6 Put-Call Parity for Options on the Max

Panel A. $S_{1T} < S_{2T}$

Current value of instrument	Payoff		
	$S_{1T} < S_{2T} < X$	$S_{1T} \leq X \leq S_{2T}$	$X < S_{1T} < S_{2T}$
$+p_{max}$	$X - S_{2T}$	0	0
$+Xe^{-r_c\tau}$	X	X	X
$-c_{max,X=0}$	$-S_{2T}$	$-S_{2T}$	$-S_{2T}$
$+c_{max}$	0	$S_{2T} - X$	$S_{2T} - X$
Total	$X - S_{2T}$ > 0	0	0

Panel B. $S_{1T} \geq S_{2T}$

Current value of instrument	Payoff		
	$S_{2T} \leq S_{1T} < X$	$S_{2T} \leq X \leq S_{1T}$	$X < S_{2T} \leq S_{1T}$
$+p_{max}$	$X - S_{1T}$	0	0
$+Xe^{-r_c\tau}$	$+X$	$+X$	$+X$
$-c_{max,X=0}$	$-S_{1T}$	$-S_{1T}$	$-S_{1T}$
$+c_{max}$	0	$S_{1T} - X$	$S_{1T} - X$
Total	$X - S_{1T}$	0	0

Equivalence is apparent and, consequently, the price of a put on the max can be obtained by the formula,

$$p_{\max} = Xe^{-r_c\tau} - c_{\max,X=0} + c_{\max}. \tag{21.8}$$

Thus, once we obtain the price of a call on the minimum, we can then obtain the price of a call on the maximum using Equation (21.3), a put on the minimum using Equation (21.7), and a put on the maximum using Equation (21.8).

21.2 PRICING THE CALL ON THE MIN

Suppose the terminal payout on the call on the min is as given in Equation (21.1). Now suppose our two assets follow the standard lognormal diffusions,

$$\frac{dS_1}{S_1} = \alpha_1 dt + \sigma_1 dW_1$$

$$\frac{dS_2}{S_2} = \alpha_2 dt + \sigma_2 dW_2, \tag{21.9}$$

where α_i and σ_i are the drift and volatility of asset i, and dW_i is the Wiener process driving asset i, with $i = 1,2$. The correlation between assets 1 and 2 is ρ_{12}. Note that c_{min} is a function of S_1, S_2, and t. Thus, based on Itô's lemma, we know

$$dc_{\min} = \left[\begin{array}{c} \frac{\partial c_{\min}}{\partial t} + \frac{\partial c_{\min}}{\partial S_1}\alpha_1 S_1 + \frac{\partial c_{\min}}{\partial S_2}\alpha_2 S_2 \\ + \frac{1}{2}\left(\frac{\partial^2 c_{\min}}{\partial S_1^2}\sigma_1^2 S_1^2 + \frac{\partial^2 c_{\min}}{\partial S_2^2}\sigma_2^2 S_2^2 + 2\frac{\partial^2 c_{\min}}{\partial S_1 \partial S_2}\sigma_1 S_1 \sigma_2 S_2 \rho_{12} \right) \end{array} \right] dt$$
$$+ \frac{\partial c_{\min}}{\partial S_1}\sigma_1 S_1 dW_1 + \frac{\partial c_{\min}}{\partial S_2}\sigma_2 S_2 dW_2.$$

Consider a hedged portfolio where we are short one call on the minimum and long $(\partial c_{\min}/\partial S_1)$ shares of S_1 and also long $(\partial c_{\min}/\partial S_2)$ shares of S_2. Thus, the current value of the portfolio, V, is

$$V = -c_{\min} + \frac{\partial c_{\min}}{\partial S_1}S_1 + \frac{\partial c_{\min}}{\partial S_2}S_2. \tag{21.10}$$

Now let us assume constant dividend yields and denote them δ_1 and δ_2. Recall the dividend is assumed to be paid continuously based on the current asset price. Thus, the change in portfolio value is

$$dV = -dc_{\min} + \frac{\partial c_{\min}}{\partial S_1}dS_1 + \frac{\partial c_{\min}}{\partial S_2}dS_2 + \frac{\partial c_{\min}}{\partial S_1}\delta_1 S_1 dt + \frac{\partial c_{\min}}{\partial S_2}\delta_2 S_2 dt. \tag{21.11}$$

Substituting, we have

$$
dV = -\left\{
\begin{bmatrix}
\frac{\partial c_{\min}}{\partial t} + \frac{\partial c_{\min}}{\partial S_1}\alpha_1 S_1 + \frac{\partial c_{\min}}{\partial S_2}\alpha_2 S_2 \\[6pt]
+\frac{1}{2}\left(\frac{\partial^2 c_{\min}}{\partial S_1^2}\sigma_1^2 S_1^2 + \frac{\partial^2 c_{\min}}{\partial S_2^2}\sigma_2^2 S_2^2 + 2\frac{\partial^2 c_{\min}}{\partial S_1 \partial S_2}\sigma_1 S_1 \sigma_2 S_2 \rho_{12}\right) \\[6pt]
+\frac{\partial c_{\min}}{\partial S_1}\delta_1 S_1 + \frac{\partial c_{\min}}{\partial S_2}\delta_2 S_2 \\[6pt]
+\frac{\partial c_{\min}}{\partial S_1}\sigma_1 S_1 dW_1 + \frac{\partial c_{\min}}{\partial S_2}\sigma_2 S_2 dW_2
\end{bmatrix} dt
\right\}
$$
$$
+ \frac{\partial c_{\min}}{\partial S_1}\left(\alpha_1 S_1 dt + \sigma_1 S_1 dW_1\right) + \frac{\partial c_{\min}}{\partial S_2}\left(\alpha_2 S_2 dt + \sigma_2 S_2 dW_2\right). \tag{21.12}
$$

Similar to many other derivations, many terms cancel and we have

$$
dV = -\begin{bmatrix}
\frac{\partial c_{\min}}{\partial t} + \frac{1}{2}\left(\frac{\partial^2 c_{\min}}{\partial S_1^2}\sigma_1^2 S_1^2 + \frac{\partial^2 c_{\min}}{\partial S_2^2}\sigma_2^2 S_2^2 + 2\frac{\partial^2 c_{\min}}{\partial S_1 \partial S_2}\sigma_1 S_1 \sigma_2 S_2 \rho_{12}\right) \\[6pt]
+\frac{\partial c_{\min}}{\partial S_1}\delta_1 S_1 + \frac{\partial c_{\min}}{\partial S_2}\delta_2 S_2
\end{bmatrix} dt. \tag{21.13}
$$

Because the portfolio is risk free, we also know that

$$
dV = rVdt = r\left(-c_{\min} + \frac{\partial c_{\min}}{\partial S_1}S_1 + \frac{\partial c_{\min}}{\partial S_2}S_2\right) dt. \tag{21.14}
$$

Setting these two expressions equal and canceling the dt term, we have

$$
r\left(-c_{\min} + \frac{\partial c_{\min}}{\partial S_1}S_1 + \frac{\partial c_{\min}}{\partial S_2}S_2\right)
$$
$$
= -\begin{bmatrix}
\frac{\partial c_{\min}}{\partial t} + \frac{1}{2}\left(\frac{\partial^2 c_{\min}}{\partial S_1^2}\sigma_1^2 S_1^2 + \frac{\partial^2 c_{\min}}{\partial S_2^2}\sigma_2^2 S_2^2 + 2\frac{\partial^2 c_{\min}}{\partial S_1 \partial S_2}\sigma_1 S_1 \sigma_2 S_2 \rho_{12}\right) \\[6pt]
+\frac{\partial c_{\min}}{\partial S_1}\delta_1 S_1 + \frac{\partial c_{\min}}{\partial S_2}\delta_2 S_2
\end{bmatrix}. \tag{21.15}
$$

Rearranging in a more familiar form, we have the two-asset PDE,

$$
rc_{\min} = \frac{\partial c_{\min}}{\partial t} + \frac{\partial c_{\min}}{\partial S_1}(r - \delta_1)S_1 + \frac{\partial c_{\min}}{\partial S_2}(r - \delta_2)S_2
$$
$$
+ \frac{1}{2}\left(\frac{\partial^2 c_{\min}}{\partial S_1^2}\sigma_1^2 S_1^2 + \frac{\partial^2 c_{\min}}{\partial S_2^2}\sigma_2^2 S_2^2 + 2\frac{\partial^2 c_{\min}}{\partial S_1 \partial S_2}\sigma_1 S_1 \sigma_2 S_2 \rho_{12}\right). \tag{21.16}
$$

Thus, we seek a solution to this PDE that satisfies the boundary conditions.[2]

Recall from Chapter 13 that the univariate Feynman-Kac theorem provides a link between the PDE solution and the expectations solution. There is a multivariate version that provides the same link that can be applied here. Appendix 21.A provides a formal expression of the multivariate Feynman-Kac theorem. Thus, based on the multivariate Feynman-Kac theorem, we know

$$S_{iT} = S_{it}e^{\left(r_c - \delta_i - \frac{\sigma_i^2}{2}\right)\tau + \sigma_i\sqrt{\tau}\varepsilon_i} \quad ; i = 1,2 \text{ and} \tag{21.17}$$

$$c_{\min,t} = e^{-r_c\tau}E_t(c_{\min,T})$$

$$= e^{-r_c\tau}\int_0^\infty\int_0^\infty \max[0, \min(S_{1T}S_{2T}) - X]\,f(S_{1T}, S_{2T})\,dS_{1T}dS_{2T}. \tag{21.18}$$

Solving the double integral based on the bivariate normal distribution, we have the min-max option model.

The call on minimum option model can be expressed as[3]

$$c_{\min,t} = S_1e^{-\delta_1\tau}N_2\left(d_{11}, d_3, \frac{\sigma_2\rho - \sigma_1}{\sigma}\right)$$

$$+ S_2e^{-\delta_2\tau}N_2\left(d_{21}, d_4, \frac{\sigma_1\rho - \sigma_2}{\sigma}\right) - Xe^{-r_c\tau}N_2(d_{12}, d_{22}, \rho), \tag{21.19}$$

where

$$\sigma^2 \equiv \sigma_1^2 + \sigma_2^2 - 2\rho_{12}\sigma_1\sigma_2, \tag{21.20}$$

$$d_{11} \equiv \frac{\ln\left(\frac{S_1}{X}\right) + \left(r_c - \delta_1 + \frac{\sigma_1^2}{2}\right)\tau}{\sigma_1\sqrt{\tau}}, \tag{21.21}$$

$$d_{12} \equiv \frac{\ln\left(\frac{S_1}{X}\right) + \left(r_c - \delta_1 - \frac{\sigma_1^2}{2}\right)\tau}{\sigma_1\sqrt{\tau}} = d_{11} - \sigma_1\sqrt{\tau}, \tag{21.22}$$

$$d_{21} \equiv \frac{\ln\left(\frac{S_2}{X}\right) + \left(r_c - \delta_2 + \frac{\sigma_2^2}{2}\right)\tau}{\sigma_2\sqrt{\tau}}, \tag{21.23}$$

$$d_{22} \equiv \frac{\ln\left(\frac{S_2}{X}\right) + \left(r_c - \delta_2 - \frac{\sigma_2^2}{2}\right)\tau}{\sigma_2\sqrt{\tau}} = d_{21} - \sigma_2\sqrt{\tau}, \tag{21.24}$$

$$d_3 \equiv \frac{\ln\left(\frac{S_2}{S_1}\right) + \left(\delta_1 - \delta_2 - \frac{\sigma^2}{2}\right)\tau}{\sigma\sqrt{\tau}}, \text{ and} \tag{21.25}$$

$$d_4 \equiv \frac{\ln\left(\frac{S_1}{S_2}\right) + \left(\delta_1 - \delta_2 - \frac{\sigma^2}{2}\right)\tau}{\sigma\sqrt{\tau}} = -d_3 - \sigma\sqrt{\tau}. \tag{21.26}$$

Note the call on maximum can be found based on Equation (21.3). Further, the put on max is found based on Equation (21.8) and the put on min is found based on Equation (21.6).

We draw several interesting insights from the following figures generated based on the min-max option model presented here. The base parameters used in these figures are as follows: $S_1 = 100$, $S_2 = 100$, $\delta_1 = 0$, $\delta_2 = 0$, $\sigma_1 = 30\%$, $\sigma_2 = 30\%$, $\rho = 0.7$, $X = 100$, $r_c = 5\%$, and $\tau = 1$ year.

Figure 21.1 illustrates the sensitivity of min-max options with respect to changes in asset price 2 (S_2). Note that as asset price 2 declines, the call on max converges to the Black-Scholes-Merton call value for asset 1 (14.23 with these parameters), the put on the max converges to the Black-Scholes-Merton put value for asset 1 (9.35 with these parameters), the call on the min converges to zero, and the put on the min converges to the lower boundary condition for asset 2. Further, as asset price 2 increases, the call on min converges to the Black-Scholes-Merton call value for asset 1, the put on the min converges to the Black-Scholes-Merton put value, the call on the max converges to the lower boundary condition for asset 2, and the put on the max converges to zero.

Figure 21.2 illustrates the sensitivity of min-max options with respect to changes in the correlation between asset 1 and asset 2. Note that as the correlation declines, the call on max and the put on the min increase in value as the likelihood of favorable moves in one of the two assets increases due to the lower correlation. Further, the call on the min and the put on the max decrease in value as the likelihood of an unfavorable move in one of the two assets increases. Further, note that as the correlation tends to +1, the call on the min and call on the max converge to the Black-Scholes-Merton call value (14.23, same for both underlying calls) and the put on the min and put on the max converge to the Black-Scholes-Merton put value (9.35, same for both underlying puts).

Figure 21.3 illustrates the sensitivity of min-max options with respect to changes in the time to expiration. The familiar time value decay is evident with all four min-max options

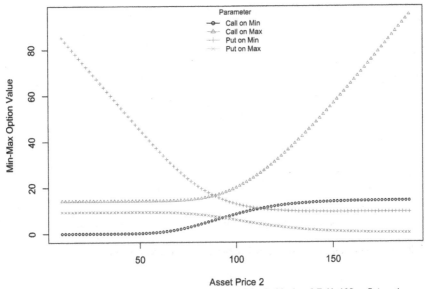

Asset Price 2

S1=100, S2=100 (Base), d1=0, d2=0, s1=30, s2=30, rho=0.7, X=100, r=5, tau=1

FIGURE 21.1 Min-Max Option Value Sensitivity to Asset Price 2

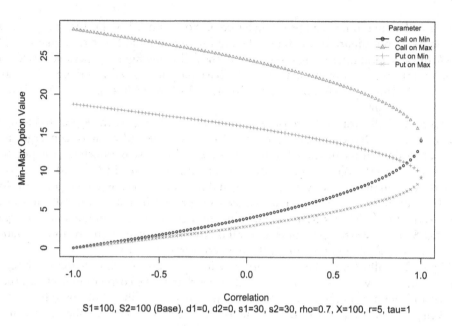

FIGURE 21.2   Min-Max Option Value Sensitivity to Correlation

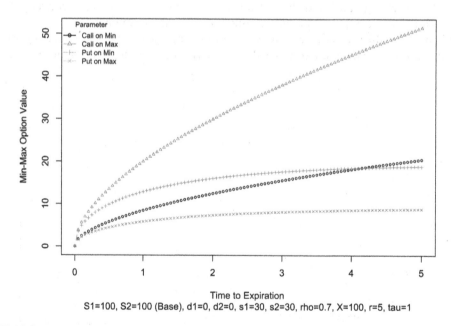

FIGURE 21.3   Min-Max Option Value Sensitivity to Time to Expiration

converging to zero as time to expiration tends toward zero. The call on the max and the put on the min have greater time value decay because they have greater value for longer maturities.

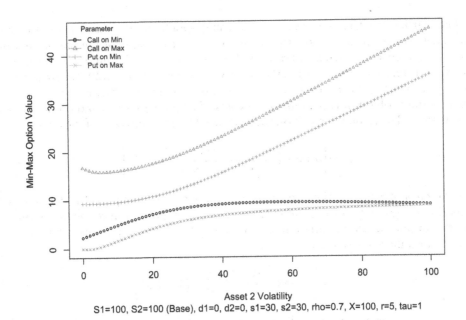

FIGURE 21.4 Min-Max Option Value Sensitivity to Asset 2 Volatility

Figure 21.4 illustrates the sensitivity of min-max options with respect to changes in asset 2 volatility. As asset 2 volatility tends toward zero, the put on the min converges to the Black-Scholes-Merton put value for asset 1 and the put on the max converges to zero. Min-max call options are less transparent. When volatility converges to zero, the Black-Scholes-Merton call value converges to the lower bound, which in this case is positive (asset price 2 less the present value of the exercise price). Based on Equation (21.3), we know the sum of min-max call options should equal the sum of the underlying calls. Thus, we observe a nonzero call on min and a call on max in excess of the Black-Scholes-Merton call value (14.23).

21.3 OTHER RELATED OPTIONS

There have been numerous extensions of the basic formula. Johnson (1987) and Rich and Chance (1993) develop the formula under the condition of more than two assets. Also, several other useful results have been obtained by Rubinstein (1991b). He first establishes a formula for an option that pays off the better of two risky assets or a fixed amount of cash. Letting X be the fixed amount of cash, we write this payoff as $\max(S_{1T}, S_{2T}, X)$. Rubinstein then derives the pricing formula for this option. Let us denote this price as c_{12X}, where the subscripts denote asset price 1, asset price 2, and some fixed amount X. Then note the following relationship, $\max(S_{1T}, S_{2T}, X) - X = \max[0, \max(S_{1T}, S_{2T}) - X]$. This equivalence implies that a long position in Rubinstein's option paying the best of two assets or X and a short position worth the present value of X is equivalent to a call on the max struck at X. Thus,

$$c_{12X} = Xe^{-r_c\tau} + c_{\max}. \tag{21.27}$$

One particular problem encountered in using options on the max or min of two or more assets is that the asset values may be far apart at the start of the option. It would hardly be interesting to own a call paying off based on the maximum of two assets if one asset were currently worth $100 and the other were currently worth $20. We already pretty much know which of the two assets will be the more highly valued asset at expiration. To overcome this problem, it is customary to express the option in terms of the assets' relative performances. For example, a call on the max would have a payoff as follows:

$$\max\left[0, \max\left(\frac{S_{1T} - S_1}{S_1}, \frac{S_{2T} - S_2}{S_2}\right) - X_R\right]. \tag{21.28}$$

Here the rates of return of the two assets are compared and the payoff is determined by comparing the greater return to an exercise rate, X_R, expressed in terms of a return. To price this option, we first express the payoff above by adding a 1 to the returns on each asset and also the exercise price:

$$\max\left[0, \max\left(\frac{S_{1T}}{S_1}, \frac{S_{2T}}{S_2}\right) - (1 + X_R)\right]. \tag{21.29}$$

This equation is the payoff of a call on the max in which the price of each asset has been normalized to a value of 1 at the start and the exercise price is expressed as 1 plus a return. This option can be valued directly with Stulz's formulas, inserting 1 as the price of each asset and using $1 + X_R$ as the exercise price. The volatilities and the correlation remain the same.[4]

Options on more than one asset are used in the over-the-counter options market. One particularly popular application is to have the assets be the returns on indexes representing different sectors of the market. Then the investor receives a return based on the better or worse performing sector. In practice, options paying off based on more than one asset are sometimes called *two-color rainbow* options. One variation is the *outperformance* option, whose payoff would be as follows:

$$\max\left[0, \left(\frac{S_{1T} - S_1}{S_1}\right) - \left(\frac{S_{2T} - S_2}{S_2}\right)\right]. \tag{21.30}$$

Note that this option pays off the difference between the return on asset 1 and the return on asset 2, if that difference is positive, and zero if the difference is negative. Of course, the option could be structured with the assets reversed. This option is an exchange option, expressed in rate of return form and can be priced by the exchange option model of Margrabe (1978). The standard version that pays off the difference in the value of the asset with the better return and the strike is commonly called an *alternative option*. Another variation is the *spread option* discussed in Chapter 17, whose payoff is

$$\max\left[0, (S_{1T} - S_{2T}) - X\right]. \tag{21.31}$$

Another variation is the *dual-strike option*, whose payoff is

$$\max\left(0, S_{1T} - X_1, S_{2T} - X_2\right), \tag{21.32}$$

which has no known closed-form solution.

21.4 RECAP AND PREVIEW

In this chapter, we looked at options in which there is more than one underlying asset. Specifically, we examined options on the greater or lesser valued or greater or lesser performing of two assets. These assets have a second feature not found in standard options in that they enable the user to bet or hedge on which asset will be worth more or less or generate a greater or lesser rate of return.

In Chapter 22, we look at two different types of derivatives: forwards and futures. We do this because they complete our knowledge base in the pricing of derivatives. Yet, they do it in a different manner. As we shall see, they do not require the dynamic hedging we are required to use with options.

APPENDIX 21A

Multivariate Feynman-Kac Theorem

Consider the multivariate partial differential equation

$$\frac{\partial f(X,t)}{\partial t} + \sum_{k=1}^{N} \frac{\partial f(X,t)}{\partial x_k} \mu_k(X,t) + \frac{1}{2} \sum_{i=1}^{N} \sum_{k=1}^{N} \frac{\partial^2 f(X,t)}{\partial x_i \partial x_k} \sum_{j=1}^{m} \sum_{l=1}^{m} \sigma_{ij}(X,t)\sigma_{kl}(X,t)\rho_{jl}(X,t)$$

$$- V(X,t)f(X,t) = g(X,t). \tag{21.33}$$

Defined for all $x \in \mathbb{R}$, $t \in (0,T)$, where $f(X,t) : \mathbb{R}^N \times (0,T) \rightarrow \mathbb{R}$ is the unknown, subject to the terminal condition,

$$f(X,t) = h(X). \tag{21.34}$$

Note that μ, σ, h, V, g, and ρ, are known functions. Then the Feynman-Kac formula provides the solution as a conditional expectation expressed as

$$f(X,t)$$

$$= E^Q \left[\int \int \cdots \int e^{-\int_t^T V(X_\tau,\tau)d\tau} g(X_\tau,\tau)\, dr_1 \ldots dr_{n-1}dr_n + e^{-\int_t^T V(X_\tau,\tau)d\tau} h(X) \Bigg| X_t = x \right]. \tag{21.35}$$

Under the probability measure Q such that X is an Itô process driven by the equation

$$dX = \mu(X,t)dt + \sigma(X,t)dW^Q. \tag{21.36}$$

With $W^Q(t)$ a Wiener process under Q and the initial condition for $X(t)$ is $X(t) = x$.

APPENDIX 21B

An Alternative Derivation of the Min-Max Option Model

Let us form a hedge portfolio currently valued at V by placing $x\%$ of our wealth in asset 1, $y\%$ of our wealth in asset 2, and $100\% - x\% - y\%$ of our wealth in the risk-free asset. For example, let x and y be expressed as percentages (e.g., $x = 0.3$ or 30%, $y = 0.6$ or 60%, $1 - x - y = 0.1$ or 10%). To express the current value of this portfolio consider that we invest xV dollars in asset 1, which will buy xV/S_1 shares. We invest yV dollars in asset 2, which will buy yV/S_2 shares. We then invest $V - xV - yV$ dollars in the risk-free asset. The current value of our portfolio can, therefore, be expressed as

$$V = \left(\frac{xV}{S_1}\right) S_1 + \left(\frac{yV}{S_2}\right) S_2 + (1 - x - y)V. \tag{21.37}$$

The change in the value of this portfolio can be expressed as

$$dV = \left(\frac{xV}{S_1}\right) dS_1 + \left(\frac{yV}{S_2}\right) dS_2 + (1 - x - y)Vrdt + \delta_1 \left(\frac{xV}{S_1}\right) S_1 dt + \delta_2 \left(\frac{yV}{S_2}\right) S_2 dt, \tag{21.38}$$

which can be written as

$$dV = x \left(\frac{dS_1}{S_1}\right) V + y \left(\frac{dS_2}{S_2}\right) V + (1 - x - y)Vrdt + \delta_1 x Vdt + \delta_2 y Vdt. \tag{21.39}$$

Because dV is driven by two stochastic processes, each of which follows a separate lognormal diffusion, we know that the change in V can also be expressed using the multivariate version of Itô's lemma. In other words,

$$dV = \frac{\partial V}{\partial S_1} dS_1 + \frac{\partial V}{\partial S_2} dS_2 + \frac{\partial V}{\partial t} dt$$

$$+ \frac{1}{2} \left(\frac{\partial^2 V}{\partial S_1{}^2} S_1{}^2 \sigma_1{}^2 + \frac{\partial^2 V}{\partial S_2{}^2} S_2{}^2 \sigma_2{}^2 + 2 \frac{\partial^2 V}{\partial S_1 \partial S_2} S_1 S_2 \sigma_1 \sigma_2 \rho_{12} \right) dt. \tag{21.40}$$

Suppose we select the value of x to be $(\partial V/\partial S_1)S_1/V$ and the value of y to be $(\partial V/\partial S_2)S_2/V$. We choose the values x and y based on current conditions, and, therefore, we can certainly set them where they should be, because we know the current values of $\partial V/\partial S_1$, S_1, $\partial V/\partial S_2$, S_2, and V.[5] Making these substitutions and setting Equations (21.39) and (21.40) equal to each other eliminates the stochastic terms, justifying use of the risk-free rate for discounting, leaving the following non-stochastic partial differential equation:

$$Vr = \frac{\partial V}{\partial S_1} S_1 (r - \delta_1) + \frac{\partial V}{\partial S_2} S_2 (r - \delta_2) + \frac{\partial V}{\partial t}$$

$$+ \frac{1}{2} \left(\frac{\partial^2 V}{\partial S_1{}^2} S_1{}^2 \sigma_1{}^2 + \frac{\partial^2 V}{\partial S_2{}^2} S_2{}^2 \sigma_2{}^2 + 2 \frac{\partial^2 V}{\partial S_1 \partial S_2} S_1 S_2 \sigma_1 \sigma_2 \rho_{12} \right). \tag{21.41}$$

This equation will apply to any min-max options. The differences in the various min-max options will come from the boundary conditions, which are Equations (21.4) and (21.5). We shall use Equation (21.41) to price a call on the minimum. Therefore its price, the solution to the above PDE, is subject to the boundary condition of Equation (21.2), which we repeat here,

$$c_{\min,T} = \max\left[0, \min\left(S_{1T}S_{2T}\right) - X\right]. \tag{21.42}$$

Stulz obtains the solution reported in the chapter by discounting the expected payoff under risk neutrality.

QUESTIONS AND PROBLEMS

1 One of the challenging aspects of creating min-max options is the need to have similarly priced underlying instruments. Explain how you can transform any two underlying instrument prices to resolve the similarly priced feature.

2 A portfolio of two call options on different stocks with the same maturity and exercise price can be replicated with options on the maximum and/or minimum. Identify the replicating min and/or max option combination and prove that this portfolio is the proper replicating portfolio.

3 Explain how a dividend yield will influence the boundary condition given in the previous problem.

4 Based on arbitrage arguments, show that $p_{\max} + p_{\min} = p_1 + p_2$.

5 Based on the previous problems as well as standard put-call parity, we have the following relationships:

$$p_{\max} + p_{\min} = p_1 + p_2,$$
$$p_{\max} = Xe^{-r_c\tau} - c_{\max,X=0} + c_{\max},$$
$$p_{\min} = Xe^{-r_c\tau} - c_{\min,X=0} + c_{\min},$$
$$p_1 = c_1 - S_1 + Xe^{-r_c\tau}, \text{and}$$
$$p_2 = c_2 - S_2 + Xe^{-r_c\tau}.$$

Demonstrate, by substitution, that $c_{\max,X=0} + c_{\min,X=0} = S_1 + S_2$. Prove with an arbitrage table that this equality holds.

NOTES

1. We leave the proof of this claim to an end-of-chapter problem.
2. An alternative derivation is provided in Appendix 21.B.
3. Haug (2007) presents a version of this solution; see pp. 211–212.
4. The properties of the lognormal diffusion hold even if the asset is normalized to a value of 1, because the diffusion describes the proportional change in the asset's value.
5. The values of the partial derivatives of V with respect to S_1 and S_2 are not known at the present. With the pricing model that solves this PDE, we can then derive these partial derivatives.

CHAPTER 22

Pricing Forwards, Futures, and Options on Forwards and Futures

To this point, we have focused exclusively on options and have paid almost no attention to other forms of derivatives. The reason for that approach is that options are somewhat more difficult to price than other derivatives. This is because option values are affected by volatility, and to model the effect of volatility, we have to incorporate characteristics of the distribution of the underlying, and we have to dynamically hedge the option. For other derivatives, the pricing process is much easier and the mathematics much simpler. In this chapter, we shall take a look at these other derivatives—forwards and futures—as well as options on forwards and futures.

22.1 FORWARD CONTRACTS

A forward contract is an agreement between two parties in which the buyer, or long, agrees that at a future date, called the *expiration*, the buyer will purchase an asset from the seller, or short, at a price they agree on today. This price they agree on is called the forward price. A forward contract is a firm commitment, which distinguishes it from an option. An option is a right, but not an obligation. The holder of the option can exercise it, but they will not do so if market conditions are not favorable. In a forward contract, the future transaction that the two parties agree on will be executed for certain.[1] No money changes hands between the two parties when the contract is initiated. Thus, the contract is neither an asset nor a liability, and in fact, it has zero value at the start. When time elapses and the value of the underlying moves, value is created for one party and destroyed for the other.

Forward contracts are completely customizable agreements that are typically created in a dealer or over-the-counter market.[2] That is, a financial institution acting as a dealer offers to take either a long or short position. As such, it quotes the price at which it would buy, called the *bid*, and the price at which it would sell, called the *ask*. Another company, called an *end user*, has a need for a forward contract, such as to lock in the purchase price of an asset it intends to buy in the future, and enters into the contract with the dealer. The end user specifies exactly what terms it wants, such as the identity of the underlying and the expiration date.

Many homes are purchased with forward contracts although typically not referenced in this way. Usually a potential buyer will sign a contract to purchase a home and also make a nonrefundable deposit. Assuming the seller meets all the contractual conditions, the buyer will purchase the home some time later at a closing meeting. The purchase terms are given in the forward contract.

We now move into the heart of the analysis: the pricing and valuation of forward contracts.

22.1.1 Pricing and Valuing Forward Contracts

Note that the title of this section uses the words *pricing* and *valuing*. When talking about most assets, including options, we use the terms somewhat interchangeably, acknowledging, however, that *value* is what something is worth and *price* is what someone pays. When markets are efficient and, therefore, process all relevant information, price equals value. In the world of forwards, futures, and swaps, however, price and value are entirely different concepts.

We start by assuming we are at time 0, at which time the underlying is at S_0. Two parties enter into a forward contract in which the buyer agrees that at the expiration, time T, it will pay an amount of money, $F_0(T)$, to the seller and receive the underlying asset. The contract expires at time T, and the time to expiration in years is τ. The parameter $F_0(T)$ is called the *forward price*. It is the price that will be paid by the buyer to the seller at the expiration. Because by now you understand options reasonably well, it might be useful to think of $F_0(T)$ as the analog of the option exercise price. Hence, the forward price will be on a similar order of magnitude to the value of the underlying, S_0. But even though $F_0(T)$ is the forward price, it is not the value of the contract. You should think of value as the amount that would be paid by one party so that this party acquires the contract from the other. Let us denote the value of a forward contract as $V_t[F_0(T)]$, which means the value of the forward contract at an arbitrary time t with the contract expiring at time T in which the forward price is $F_0(T)$. Let us first establish the principle that the value of the contract at time zero is zero:

$$V_0[F_0(T)] = 0. \qquad (22.1)$$

This result is intuitive in that neither party pays the other any money at the start. We said that value is what one party pays to acquire something from another. With forward contracts, the two parties engage in the agreement, but neither pays any money to the other. Later we shall demonstrate a nonstandard circumstance in which value is not zero at time zero.

Next, let us establish the value of the contract at expiration, which is as follows:

$$V_T[F_0(T)] = S_T - F_0(T). \qquad (22.2)$$

This result is determined by precisely what happens at expiration, at which time the buyer of the contract pays the price, $F_0(T)$, to the seller and receives the asset valued at S_T. Therefore, the value to the buyer is $S_T - F_0(T)$. Conversely, the value to the seller is $-[S_T - F_0(T)]$. Note that the payoff here is linear in the underlying, whereas with options, the payoff is nonlinear in the underlying.[3] This linearity will make a huge difference in how we construct a risk-free hedge that will provide the option value. Linearity will, in fact, make the risk-free hedge for a forward contract be static, rather than dynamic, as it is for options.

Thus, forward pricing is a relatively simple matter, especially in comparison to option pricing. Suppose we sell a forward contract at price $F_0(T)$ and simultaneously buy one unit of the asset at price S_0. At expiration, the short forward contract obligates us to deliver the asset and be paid $F_0(T)$. So what will happen at expiration is simple. We will deliver the asset and be paid $F_0(T)$. Note how the value of the asset at expiration, S_T, is irrelevant to the payoff of the overall position of long the asset and short the forward contract. In fact, there is no risk to the overall position. The value at expiration of the position is $F_0(T)$, and being risk free, this value should be discounted to the present at the risk-free rate to equal the value of the initial position. The value of the initial position is the value of the asset, S_0, minus the value of the forward contract, but the value of the forward contract is zero by Equation (22.1). Thus, the value of the asset at time 0, S_0, must equal the forward price discounted at the risk-free rate. Rearranging and solving for the forward price gives our forward pricing equation,

$$F_0(T) = S_0(1 + r)^\tau, \tag{22.3}$$

where τ is the time to expiration expressed as a fraction of a year. For forward contracts, this is the analog of the Black-Scholes-Merton equation. As you can see, it is much simpler.

Now, let us consider the special case in which we create a forward contract an instant before expiration. Then $\tau \to 0$, and the new forward price would be the spot price,

$$F_T(T) = S_T(1 + r)^0 = S_T. \tag{22.4}$$

Forward contracts are often written on assets that have costs associated with holding them, and these assets can also yield cash flows, such as dividends, or other benefits. Let $\gamma(0, T)$ be the value at time T of any accumulated costs minus any accumulated benefits. Then Equation (22.3) becomes

$$F_0(T) = S_0(1 + r)^\tau + \gamma(0, T). \tag{22.5}$$

Note that $\gamma(0, T)$ can be either positive or negative depending on whether the costs exceed the benefits or vice versa. Further, there are numerous ways to incorporate asset-related cash flows. We chose here to introduce them as simply an addition.

We have established the value of a forward contract at time 0 and the value at time T, the expiration. What we now need is the value at some arbitrary evaluation time t, sometime between times 0 and T. This result is only slightly more difficult to establish. Thus, now consider that we are at the evaluation time t, $0 < t < T$. The forward contract price was established at time 0 and set at $F_0(T)$ and is fixed. The spot price is S_t. The remaining time to expiration will be denoted τ_t. To determine the value of this forward contract, let us sell a new forward contract expiring at T. This contract will have a new price, $F_t(T)$ and a time to expiration of τ_t. Being long the original forward contract obligates us to buy the asset at time T at price $F_0(T)$, and being short a new forward contract obligates us to sell the asset at time T at price $F_t(T)$. The overall position is risk-free in that the value of the underlying at time T will have no relevance to us. We will buy the asset at $F_0(T)$ and sell it at $F_t(T)$. Thus, we are guaranteed that at time T we will have a cash flow of $F_t(T) - F_0(T)$, an amount that is known at time t. The present value from T to t equals the value of our overall position at time t. Our overall position at time t is the value of the long position in the original forward contract minus the value of the new forward contract that we are short. But the value of the new forward contract at t is zero, because it is a newly established contract and, therefore, has a value of zero. Thus, the value of the original contract at t is

$$V_t[F_0(T)] = \frac{F_t(T) - F_0(T)}{(1 + r)^{\tau_t}}. \tag{22.6}$$

TABLE 22.1 Cash Flows for Long Forward Valuation

Strategy	Cash Flow at Time 0	Value at Time t	Cash Flow at Time T
Buy forward $F_0(T)$	0	$V_t(T)$	$V_T[F_0(T)] = S_T - F_0(T)$
Sell forward $F_t(T)$	NA	0	$V_T[F_t(T)] = F_t(T) - S_T$
Net Cash Flow/Value	0	$V_t(T)$	$F_t(T) - F_0(T)$

Note: NA indicates sell forward at time t does not exist.

Table 22.1 illustrates the cash flows at the three relevant times: the initial trade date, ($\tau = 0$), the evaluation date, t, and the expiration date, T.

Adapting Equation (22.5) for the condition at time t instead of time 0, we have the price of the new forward contract as

$$F_t(T) = S_t(1+r)^{\tau_t}.$$

Using this result in Equation (22.6), we can then restate the value equation as

$$V_t[F_0(T)] = S_t - \frac{F_0(T)}{(1+r)^{\tau_t}}. \tag{22.7}$$

Or essentially, the current spot price minus the present value of the original forward price. Note that if we set $t = 0$ or $t = T$ in Equation (22.7), we obtain Equations (22.1) and (22.2):

$$t = 0 : V_0[F_0(T)] = S_0 - F_0(T)/(1+r)^{\tau} = S_0 - S_0 = 0$$
$$t = T : V_T[F_0(T)] = S_T - F_0(T)/(1+r)^0 = S_T - F_0(T). \tag{22.8}$$

For example, assume at time 0 a one-year forward contract was entered at \$25 [$F_0(T)$]. You have been tasked with determining the fair value of this contract after nine months (time t). At time t, we know the underlying instrument priced is \$26 and the interest rate is 3.902% (annual compounding). In this case we know the price of a new forward contract is

$$F_t(T) = S_t(1+r)^{\tau_t} = 26(1+0.03902)^{0.25} = 26.25.$$

Further, the value of the initial long forward contract at time t is

$$V_t[F_0(T)] = \frac{F_t(T) - F_0(T)}{(1+r)^{\tau_t}} = \frac{26.25 - 25}{(1+0.03902)^{0.25}} = \$1.238.$$

Note that the contract, which had a zero value nine months earlier, is now worth a positive value. Intuitively, its value has increased because it is the obligation to buy an asset that has since increased in value.

22.1.2 Forward Contracts and Options

Now, let us tie forward contracts back to options. First, note that because a long position in the asset, hedged with a short forward contract, can create a riskless position or essentially a risk-free bond, it follows that a position in the asset can be replicated by a long position in a forward contract and a long position in a risk-free bond. Recall that put-call parity states that

$$p_0 + S_0 = c_0 + \frac{X}{(1+r)^\tau}. \tag{22.9}$$

Given Equation (22.3), we can substitute into put-call parity for S_0 and obtain

$$p_0 + \frac{F_0(T)}{(1+r)^\tau} = c_0 + \frac{X}{(1+r)^\tau}. \tag{22.10}$$

As such, Equation (22.10) expresses the relationships among the put price, the call price, the option exercise price, and the forward price. Note that because we are free to choose the exercise price of the options, we could set the exercise price to the forward price. Then $F_0(T) = X$, and from Equation (22.10), we would have the put and call prices equal to each other.

22.1.3 Some Additional Facts on Forward Contracts

Finally, let us learn a few more important facts about forward contract pricing. We have said that a forward contract does not involve any exchange of money between buyer and seller at the start; hence, the value of the contract is zero at the start. This is the standard case, but it is not absolutely mandatory. The buyer and seller of a forward contract can set the price at whatever forward price they want. These types of contracts are called *off-market forward contracts*. First, take another look at Equation (22.3), which is repeated here:

$$F_0(T) = S_0(1+r)^\tau.$$

Suppose the parties set $F_0(T)$ to a value higher than the RHS. Now, let us see what happens. We first restate the value Equation, (22.7), with $t = 0$:

$$V_0[F_0(T)] = S_0 - \frac{F_0(T)}{(1+r)^\tau}.$$

If the forward price is set to the formula price, Equation (22.3), this value goes to zero, as we have shown. But if the forward price is set higher than the formula price, then the value becomes negative. A negative value contract means that the seller must pay the buyer the value at the start. Let us think about this. If the forward price is too high, the buyer has agreed to pay too much at time T for the asset. Hence, the seller has the advantage and must compensate the buyer for this additional value at time 0. If the forward price is set lower than the formula price, then the value in the above equation becomes positive. A positive value contract means that the buyer must pay the seller the contract value at the start. In this case, the buyer will be paying too little at time T and must compensate the seller for this additional value at time 0.

Finally, let us note several additional characteristics of forward contracts. Typical forward contracts contain provisions that allow either party to request a settlement of the other. That is, the contract can be terminated early. Suppose the long holds a position that has a value, as calculated in the manner shown in Equation (22.7), of $1.5 million. The long and the short could, thus, terminate the contract before expiration by having the short pay the long $1.5 million. If the value were negative, the long would pay the short. Either party can request an early termination, but typically one party is a dealer and dealers do not usually request early termination, because they exist to serve the needs of their clients and not to impose their own needs on their clients. The counterparty, who is usually an end user, would ordinarily be the one to request an early termination.

Forward contracts can also be written to specify that at expiration the contract calls for physical delivery of the asset or an alternative procedure called *cash settlement*. Suppose, for example, that the buyer holds a forward contract on two million barrels of crude oil with a forward price of $70 a barrel. Now, let us say we are at expiration and the price of crude oil is $75 a barrel. A physical delivery contract would result in the short delivering two million barrels of oil to the long, who would pay the short $70 a barrel. So the long benefits by acquiring two million barrels at $5 per barrel below the market price. Alternatively, a cash-settled contract would result in the short simply paying the long $10 million. By settling in cash, the two parties can avoid the high transaction costs of handling the oil.

Forward contracts are subject to default. Technically the dealer or the end user could default. In deriving these pricing formulas, we have not taken default into account, because that is a subset of the subject of credit risk. Some forward contracts are processed through a clearinghouse and are subject to margin requirements and periodic settlements, which is the standard process used in futures contracts, as we explain next.

22.2   PRICING FUTURES CONTRACTS

A futures contract is effectively a forward contract that is resettled every day. That is, at the end of the day, the clearinghouse of the exchange determines an official settlement price, which is an average of the last few trades of the day. The change in the settlement price from the previous day is then credited to the party that gained and charged to the party that lost. Thus, if the settlement price increased (decreased), the holders of long (short) positions would be credited the gain and the holders of short (long) positions would be charged the loss. This process is called *marking-to-market* and, sometimes, the *daily settlement*. It effectively results in gains and losses being collected and charged in smaller amounts over shorter periods of time than in the case of forward contracts, whereby the contract is completely settled at expiration. The process of settling daily reduces the credit risk. In fact, futures exchanges guarantee that no one will fail to be paid because of the default of the opposite party. This guarantee has been absolutely fulfilled, because no party has ever lost money on any futures exchange in the world due to default of the counterparty. The exchange, through its clearinghouse, has always either collected the money or paid up itself.

22.2.1   Daily Settlement and Futures Prices with Constant Interest Rates

The daily settlement results in a difference in the pattern of cash flows to a holder of a futures contract in comparison to that of a holder of a forward contract. As it turns out, there are some circumstances in which this difference has no economic effect, and in most cases, it is likely to have no material impact. But let us start by seeing what the effect is, however small it might turn out to be. This work draws on Jarrow and Oldfield (1981) and Cox, Ingersoll, and Ross (1981).

Let us set up the problem by denoting the futures price at time t for a contract expiring at time T as $f_t(T)$. Futures contracts trade on a futures exchange and are, therefore, updated throughout the trading session. Our first important result is the convergence of the futures price at expiration to the spot price. That is, a futures contract that is created an instant before expiration will be priced at the spot price, because it will immediately expire and result in delivery of the spot asset, thus,

$$f_T(T) = S_T. \tag{22.11}$$

Now let us take a look at how one would value a futures contract. Suppose on day t, the settlement price for the previous day was $f_{t-1}(T)$. Now, here at the end of day t the price is $f_t(T)$. The accumulated value is $f_t(T) - f_{t-1}(T)$. So, at this point, an instant before the daily settlement, the value of the futures contract is

Before the daily settlement,

$$v_t(T) = f_t(T) - f_{t-1}(T). \tag{22.12}$$

That is, the value is simply the price change from the previous settlement. As soon as the market closes, the daily settlement at day t will occur. If the previous value is positive (negative), the profit will be credited (charged) to the holder of the long position, and the loss will be charged (credited) to the holder of the short position. After the daily settlement, the value reverts to zero,

After the daily settlement,

$$v_t(T) = 0. \tag{22.13}$$

In effect, the daily settlement means that the contract is terminated and rewritten at the new settlement price. This is why we described a futures contract as a forward contract that is settled and rewritten every day.

Now, let us take a look at when forward prices would equal futures prices. Table 22.2 illustrates the pattern of cash flows from day 1 through day T, the expiration.

We want to know if the original forward price, $F_0(T)$, would equal the original futures price, $f_0(T)$. First, let us step back to one day prior to expiration, in which the futures price would be $f_{t-1}(T)$. One day later the contract will expire and effectively conduct its last daily settlement. A forward contract created on that day will also settle one day later. Hence, the two contracts will have the same cash flow one day later; thus, the forward and futures prices will be the same,

One day prior to expiration,

$$f_{T-1}(T) = F_{T-1}(T). \tag{22.14}$$

Stepping back one further day to $T - 2$, we can show that if the interest rate on day t is known on day $t - 1$, the futures price on day t would equal the forward price on

TABLE 22.2   Cash Flows from Futures and Forward Contracts

Day	Futures Cash Flow	Forward Cash Flow
0	0	0
1	$f_1(T) - f_0(T)$	0
2	$f_2(T) - f_1(T)$	0
3	$f_3(T) - f_2(T)$	0
...	...	...
$T-2$	$f_{T-2}(T) - f_{T-3}(T)$	0
$T-1$	$f_{T-1}(T) - f_{T-2}(T)$	0
T	$S_T - f_{T-1}(T)$	$S_T - F_0(T)$

day t. Let us assume that the interest rate, r, is constant from day to day, a common assumption in derivative pricing theory.[4] Let us construct the following strategy: Buy one forward contract at the price $F_{T-2}(T)$ and sell $1/(1+r)$ futures at the price $f_{T-2}(T)$. At the end of day $T-1$, we generate no cash flow on the forward contract because it is not settled daily and has one more day to expiration. Through its daily settlement, however, the futures will produce a cash flow of $-\left[f_{T-1}(T) - f_{T-2}(T)\right]$ times $1/(1+r)$ contracts. If this amount is negative, let us borrow the funds at the rate r for one day. If it is positive, we invest the funds at the rate r for one day. By borrowing or investing the daily settlement cash flow, we avoid having to put up or withdraw funds on a day-to-day basis. One day later, which is expiration, the daily settlement amount at $T-1$ plus one day's interest paid or received will be $-\left[f_{T-1}(T) - f_{T-2}(T)\right]$. In addition, the final daily settlement will be $-\left[f_T(T) - f_{T-1}(T)\right]$. Adding these two amounts gives a total cash flow at T of $-\left[f_{T-1}(T) - f_{T-2}(T)\right] - \left[f_T(T) - f_{T-1}(T)\right] = f_{T-2}(T) - f_T(T) = f_{T-2}(T) - S_T$. This is the same cash flow as on a two-day forward contract. We could successively create forwards and futures further back and similarly demonstrate that the cash flows would be the same. Hence, the price of a forward contract would equal the price of the analogous futures contract. Of course, similar to forwards, futures will also reflect carrying costs and any other potential benefits or costs of holding the asset.[5]

This result, however, was possible only because the interest rate was known in advance. In effect, we must know the entire sequence of one-day interest rates. In a world of constant interest rates, that is certainly not a problem. If interest rates are stochastic, however, that could be a problem. Let us now see why stochastic interest rates can drive the price of the futures contract above or below the price of the forward contract.

22.2.2   The Daily Settlement and Futures Prices with Stochastic Interest Rates

Consider a long position in a futures contract. If futures prices increase when interest rates increase, then gains will be realized and reinvested at higher interest rates. If futures prices increase when interest rates decrease, losses will be realized and financed at lower interest rates. Thus, when futures prices and interest rates are positively related, the daily settlement benefits parties who choose futures contracts over forward contracts, and as a result, futures prices will be above forward prices. If futures prices are negatively related to interest rates, then futures contracts will be hurt by the daily settlement, because gains will be reinvested when rates are lower and losses will be financed when rates are higher. This benefits parties who choose to take forward contracts over futures contracts. Thus, if futures prices are positively (negatively) related to interest rates, futures prices will be higher (lower) than forward prices. As you might expect, if futures prices are unrelated to interest rates, meaning essentially a zero correlation, futures and forward prices will be the same.

These points are strong statements about the relationship between futures and forwards, but they are taken without consideration for credit risk. The daily settlement that is undertaken in futures markets essentially eliminates credit risk. Forward markets do retain some element of credit risk, but that element is minimized by the careful consideration of whether to engage in a forward contract with anyone who might default and by potentially requiring margin deposits and perhaps even periodically marking to market.

If we minimize these credit considerations, futures prices will essentially equal forward prices, and therefore the futures pricing model would simply be a restatement of the forward pricing model, Equation (22.3),

$$f_t(T) = S_t(1 + r)^{\tau_t}. \tag{22.15}$$

In the next section, we briefly review some anecdotal evidence related to futures pricing.

22.2.3 Selected Futures Contracts—Gold, Wheat, and Natural Gas

Table 22.3 presents the settlement prices for gold futures contracts traded on the CME Group exchange. Note that the pattern of futures prices with respect to maturity is monotonically increasing, a property known as *contango*. The pattern of futures prices with respect to maturity that is monotonically decreasing is known as *backwardation*.

US gold futures contracts are almost always in contango. The reason is generally positive interest rates and positive storage costs as well as the ease with which to conduct arbitrage transactions. One key insight is that the upward sloping pattern does not reflect market participants' views on the future of gold prices; rather, it reflects solely carrying costs related to the underlying instrument. Interestingly, the pattern does not occur with gold futures contracts in some other countries where it is difficult to actually engage in the required arbitrage transactions.

TABLE 22.3 Selected Gold Futures Prices

Description	Price
Spot (Sep)	$1,502.20
OctY1	$1,504.50
NovY1	$1,508.10
DecY1	$1,511.10
FebY2	$1,517.40
AprY2	$1,522.90
JunY2	$1,527.60
. . .	. . .

Table 22.4 presents the settlement prices for wheat futures contracts traded on the CME Group exchange. Note that the pattern of futures prices with respect to maturity is again monotonically increasing. Due to the difficulty in conducting arbitrage-related activities with wheat, this pattern of futures prices communicates different information than gold. Wheat is perishable and consumed. The pattern of US wheat futures contracts takes many different shapes over time, unlike gold contracts. For example, global harvest expectations play a key role in wheat futures prices. Thus, one key insight is that the upward sloping pattern for wheat *does* reflect market participants' views on the future of wheat spot prices.

Table 22.5 presents the settlement prices for natural gas futures contracts traded on the CME Group exchange. Note that the pattern of futures prices with respect to maturity is neither monotonically increasing nor decreasing. US natural gas futures contracts' pattern takes many different shapes over time unlike gold contracts. As with wheat, the driving reason is that it is generally difficult to conduct the required arbitrage transactions as natural gas is volatile and difficult to store. Global natural gas production and consumption expectations also play a key role in futures prices. Natural gas is consumed in greater volumes in the summer months to generate electricity for air conditioning and in the winter months to run furnaces or generate electricity for heating. Thus, one key insight is that the dynamic pattern for natural gas *does* reflect market participants' views on the future of natural gas spot prices. In particular, winter and often summer futures prices are higher than spring and fall.

We turn now to explore options on forwards and futures contracts.

TABLE 22.4   Selected Wheat Futures Prices

Description	Price
Spot (Nov)	$4.744
DecY1	$4.744
MarY2	$4.790
MayY2	$4.814
JulY2	$4.840
SepY2	$4.914
DecY2	$5.036
. . .	. . .

TABLE 22.5   Selected Natural Gas Futures Prices

Description	Price
Spot (Oct)	$2.585
NovY1	$2.625
DecY1	$2.768
JanY2	$2.859
FebY2	$2.811
MarY2	$2.685
AprY2	$2.399
. . .	. . .

22.3   OPTIONS ON FORWARDS AND FUTURES

Options on forwards are not widely used, but options on futures are very actively traded in the US. One of the reasons for this is that futures markets are quite transparent. Prices are readily available, and there is considerable liquidity for many of these instruments. With the underlying being a futures contract and it being so widely traded in typically quite liquid markets, it is not surprising that options on futures are quite popular. A long call option that is exercised leads to a long position in the futures, whereas the counterparty is assigned a short position in the futures. A put option that is exercised leads to a short position in the underlying, whereas the counterparty is assigned a long position. If the futures expires at the same time as the option, exercise of the option converts into a position in the futures that immediately converts into a position in the underlying asset. As such, a European option on a futures that expires at the same time as the futures would have the same price as the corresponding European option on the spot.

Some options on futures, however, expire earlier than the underlying futures. In either case, however, we need an option pricing model that values the option in terms of the futures. Let us consider an option on a futures that expires in time T^* and with a time to expiration is τ^*. The underlying futures expires at T and the time to expiration of the futures is τ. To price this option, we shall construct a dynamically hedged portfolio of the option and the underlying futures. Specifically, we shall buy h call options and sell one futures. The value of this portfolio at time t is

$$w_t = hc_t. \tag{22.16}$$

This equation does not imply that the futures price is zero when this portfolio is assembled. This statement is only that the value is xero. The price is certainly nonzero, and changes in price create value. If we properly choose the value of h, we can convert this portfolio to a hedged portfolio so that the change in the value of the portfolio over the next instant can be expressed as

$$dw_t = df_t + hdc_t. \tag{22.17}$$

Note a significant difference in Equations (22.16) and (22.17) that may seem inconsistent. In the latter, the value of the overall position reflects the change in both the value of the futures and the value of the option. For the former, the value of the futures starts off at zero, in accordance with the principle that the value of a futures is zero when initiated, just as with a forward contract.[6] If the portfolio is completely hedged, its instantaneous return should be $w_t r_c dt$. Hence, we can say that the change in the futures plus the change in the value of h options must equal the initial value, w_t, plus interest,

$$df_t + hdc_t = w_t r_c dt$$
$$= hc_t r_c dt. \tag{22.18}$$

As with standard European options on the spot asset, we can assume that the call price is a function of the futures price and time. As such, we can express the call price change using Itô's lemma,

$$dc_t = \frac{\partial c_t}{\partial f_t} df_f + \frac{\partial c_t}{\partial t} dt + \frac{1}{2} \frac{\partial^2 c_t}{\partial f_r^2} df_t^2. \tag{22.19}$$

Substituting Equation (22.19) into Equation (22.18) and rearranging, we obtain

$$hc_t r_c dt = df_t + h\left(\frac{\partial c_t}{\partial f_t}df_t + \frac{\partial c_t}{\partial t}dt + \frac{1}{2}\frac{\partial^2 c_t}{\partial f_t^2}df_t^2\right). \tag{22.20}$$

We are free to set the hedge ratio at whatever we want. Thus, if we set $h = -\partial f_t/\partial c_t$, we obtain

$$-\left(\frac{\partial f_t}{\partial c_t}\right)c_t r_c dt = df_t - \left(\frac{\partial f_t}{\partial c_t}\right)\left(\frac{\partial c_t}{\partial f_t}df_t + \frac{\partial c_t}{\partial t}dt + \frac{1}{2}\frac{\partial^2 c_t}{\partial f_t^2}df_t^2\right). \tag{22.21}$$

Rearranging, we have

$$-\left(\frac{\partial f_t}{\partial c_t}\right)c_t r_c dt = -\left(\frac{\partial f_t}{\partial c_t}\right)\left(\frac{\partial c_t}{\partial t}dt + \frac{1}{2}\frac{\partial^2 c_t}{\partial f_t^2}df_t^2\right). \tag{22.22}$$

Because df_t is removed, the Equation (22.22) is non-stochastic, which means the risk is hedged away. Now, we need to simplify Equation (22.22). Recall the standard stochastic process for the asset, which we learned when studying options,

$$dS_t = S_t\alpha dt + S_t\sigma dW_t. \tag{22.23}$$

And recall the futures pricing model, Equation (22.15). Because we are now in continuous time, we should use its continuous analog,

$$f_t = S_t e^{r_c\tau}. \tag{22.24}$$

Because the futures price is known to be a function of time and the spot price, which we know follows geometric Brownian motion, we can express the futures price in terms of the spot price and time using Itô's lemma,

$$df_t = \frac{\partial f_t}{\partial S_t}dS_t + \frac{\partial f_t}{\partial t}dt + \frac{1}{2}\frac{\partial^2 f_t}{\partial S_t^2}dS_t^2. \tag{22.25}$$

Equation (22.25) contains partial derivatives that can be derived from Equation (22.24) as follows:[7]

$$\frac{\partial f_t}{\partial S_t} = e^{r_c\tau}$$

$$\frac{\partial^2 f_t}{\partial S_t^2} = 0$$

$$\frac{\partial f_t}{\partial t} = -rS_t e^{r_c\tau}, \tag{22.26}$$

and insert these into Equation (22.25),

$$df_t = e^{r_c\tau}dS_t + -rS_te^{r_c\tau} + 0. \tag{22.27}$$

Substituting the stochastic process of the asset from Equation (22.23), the expression for df_t then becomes

$$df_t = f_t(\alpha dt + \sigma dW_t) + -r_c f_t dt$$
$$= f_t(\alpha - r_c)\,dt + f_t\sigma dW_t. \tag{22.28}$$

Squaring this equation, we now know that df_t^2 will be $f_t^2\sigma^2 dt$. Then, substituting into Equation (22.22), we obtain

$$-\left(\frac{\partial f_t}{\partial c_t}\right)c_t r dt = -\left(\frac{\partial f_t}{\partial c_t}\right)\left(\frac{\partial c_t}{\partial t}dt + \frac{1}{2}\frac{\partial^2 c_t}{\partial f_t^2}f_t^2\sigma^2 dt\right). \tag{22.29}$$

Now multiply by $\partial c_t/\partial f_t$ and rearrange to obtain

$$c_t r_c = \frac{\partial c_t}{\partial t} + \frac{1}{2}\frac{\partial^2 c_t}{\partial f_t^2}f_t^2\sigma^2. \tag{22.30}$$

This equation is very similar to the partial differential equation for Black-Scholes-Merton. One difference is that in the BSM equation, there is a term $r_c S_t \partial c_t/\partial S_t$, which reflects interest on the funds tied up in the underlying. Here, however, there are no funds tied up in the underlying, because it is a futures.

The boundary condition is $c_T = \max(0, f_T - X)$. The solution is similar to the BSM model and is often referred to as the Black model, as Fischer Black provided this result in a separate paper (Black 1976):

$$c_t = e^{-r_c\tau^*}\left[f_t N(d_1) - XN(d_2)\right]$$
$$d_1 = \frac{\ln\left(\frac{f_t}{X}\right) + \left(\frac{\sigma^2}{2}\right)\tau^*}{\sigma\sqrt{\tau^*}}$$
$$d_2 = d_1 - \sigma\sqrt{\tau^*}. \tag{22.31}$$

Recall that the option expiration, τ^* is shorter than the futures expiration, τ. If the option and futures expire at the same time, we would insert τ for τ^* and using the futures pricing equation, (22.24), we would find that Equation (22.31) would be precisely the BSM equation for the value of an option on the asset. Based on put-call parity, we can easily find the put option equation or

$$p_t = c_t - e^{-r_c\tau^*}(f_t - X)$$
$$= e^{-r_c\tau^*}\{X[1 - N(d_2)] - f_t[1 - N(d_1)]\}$$
$$= e^{-r_c\tau^*}[XN(-d_2) - f_tN(-d_1)]. \tag{22.32}$$

We now illustrate this model with a numerical example. Suppose the futures price is $52, the exercise price is $52.8, the time to expiration of the option contract is 0.25, the interest rate is 2.0%, and the volatility is 35%. First, solving for the value of d_1 and d_2, we have

$$d_1 = \frac{\ln\left(\frac{f_t}{X}\right) + \left(\frac{\sigma^2}{2}\right)\tau^*}{\sigma\sqrt{\tau^*}} = \frac{\ln\left(\frac{52}{52.8}\right) + \left(\frac{0.35^2}{2}\right)0.25}{0.35\sqrt{0.25}} = 0.000$$

$$d_2 = d_1 - \sigma\sqrt{\tau^*} = 0.000 - 0.35\sqrt{0.25} = -0.175.$$

Based on Table 5.1: Standard Normal Cumulative Distribution Function Table, we have

$$c_t = e^{-r_c\tau^*}\left[f_t N(d_1) - XN(d_2)\right]$$

$$= e^{-(0.02)0.25}\left[52(0.5) - 52.8(0.430540)\right]$$

$$= 3.2512 \text{ and}$$

$$p_t = c_t - e^{-r_c\tau^*}\left(f_t - X\right)$$

$$= 3.2512 - e^{-(0.02)0.25}(52 - 52.8)$$

$$= 4.0472.$$

As the case with options on the spot, American options on futures are considerably more complex than European options on futures. What further complicates options on futures, however, is that call options could be optimally exercised early, which is not the case with options on the spot, as we have already discussed. This result occurs because of how the futures price converges to the spot price. Notice in Equation (22.3), the futures is higher than the spot. As expiration approaches, the futures must be pulled toward the spot. This pulling effect operates somewhat like a dividend, which pulls a stock price down. As we studied with options on stocks, a dividend can trigger early exercise that avoids the loss due to the downward pressure on the underlying. This downward pressure on the futures has the same effect. As such, it might be optimal to exercise a call option on a futures early. Put options can also be exercised early, just as with put options on assets, but the downward pressure that can trigger the early exercise of call options on futures will discourage the early exercise of put options on futures. If there are carrying costs or other benefits related to the futures price, the probability of early exercise will be altered, depending on the magnitude of these costs.[8] Pricing an American call on a futures will typically use a numerical method, such as the binomial model. There is also an approximation formula that is a variation of the Barone-Adesi-Whaley formula we discussed in Chapter 19.

22.4   RECAP AND PREVIEW

In this chapter, we introduced a different type of derivative, the forward contract and its variant the futures contract. We showed how these derivatives are priced, which is by the same principle as option pricing, but which does not require a dynamic hedge. As such, the math is much simpler. We also took a look at options on forwards and futures, which is our first attempt to integrate both options, which have nonlinear payoffs, and forwards and futures, which have linear payoffs.

We have now completed Part IV. In Part V, we take up two methods of obtaining numerical solutions for option prices.

QUESTIONS AND PROBLEMS

1 Assume at time 0 a five-year forward contract was entered at $75 $\left[F_0(T)\right]$. You have been tasked with determining the fair value of this contract after two years. At that time, the underlying instrument price is $70 and the interest rate is 4.89% (annual compounding). Calculate the forward price after two years as well as the value of the initial forward contract entered at time 0.

2 Assuming the underlying follows geometric Brownian motion, $dS_t = S_t \alpha dt + S_t \sigma dW_t$, derive the stochastic process of the forward price where the interest rate and dividend yield are expressed in continuous compounding, that is, $F_0(T) = S_0 e^{(r_c - \delta)\tau}$.

3 Derive the delta of a call option on a futures contract assuming

$$c_t = e^{-r_c \tau^*}\left[f_t N(d_1) - X N(d_2)\right]$$

$$d_1 = \frac{\ln\left(\frac{f_t}{X}\right) + \left(\frac{\sigma^2}{2}\right)\tau^*}{\sigma\sqrt{\tau^*}}$$

$$d_2 = d_1 - \sigma\sqrt{\tau^*}.$$

Hint: It is helpful to use the following relationship between these two probability density functions:

$$n(d_2) = \frac{e^{r_c \tau^*} f_t}{X} n(d_1).$$

4 Derive the gamma of a call option on a futures contract.

5 Derive the theta of a call option on a futures contract.

6 Demonstrate that the call option on futures model is valid when compared to the partial differential equation. Specifically, substitute the partial derivatives in the PDE and demonstrate coherence.

7 Suppose the futures price is $24.8, the exercise price is $24, the time to expiration is 4.0, the interest rate is 1.0%, continuously compounded, and the volatility is 30%. Compute the call and put futures option prices.

NOTES

1. We shall see that it is possible that the two parties can offset or negate their transaction before the expiration date, and one party can sell its obligation to another party. This is the only way in which exercise is avoided, but each party must still effectively do something for the other, and the arrangement must be fair from the perspective of both parties. In an option, however, the long can simply walk away without doing anything.
2. Options can also be created on the over-the-counter market, or on an exchange. Forwards, however, are not offered on exchanges, though they can be offered in a near-exchange manner, such as on electronic trading platforms.
3. Terminal option payoffs are piecewise linear rather than simply linear.
4. We shall discuss what happens if r is not constant.
5. Observers of futures markets note that futures prices are sometimes not equal to the spot price grossed up by the interest plus the carrying costs as a result of a rather vaguely specified benefit called the *convenience yield*. This effect is a factor in the pricing of the underlying asset that occurs sometimes in shortages, wherein the underlying asset has an abnormally high price, a result of tight market conditions. When this happens, the asset is hard to borrow because owners hold on to it. Shorting is, thus, virtually impossible and the futures could appear to violate the no-arbitrage formula, Equation (22.3), when in fact the effect is driven by an inability to short the asset. The deviation of the futures price from the no-arbitrage formula is, thus, often plugged with a concept called the *convenience yield*. For more details, see Brooks (2012).
6. As we said, the value of the futures is the accumulated price change since the last settlement. Because there has been no previous settlement, the value is zero.
7. It may be a little unclear in taking the first derivative with respect to time as to why we obtain the $-r$. We have to differentiate τ with respect to t. The parameter τ is the time to expiration in years. The parameter t is just a time indicator. We can think of it somewhat like the value of $1/365$, though it is technically a smaller unit than $1/365$. Nonetheless, the effect is the same when we differentiate τ. In other words, the derivative of τ with respect to t is simply -1.
8. These points are covered in considerable detail in Brenner, Courtadon, and Subrahmanyam (1985).

Numerical Methods

CHAPTER 23

Monte Carlo Simulation

Simulation is a procedure in which random numbers are generated according to probabilities assumed to be associated with a source of uncertainty, such as the sales of a new product or, more appropriately for our purposes, stock prices, interest rates, exchange rates, or commodity prices. Outcomes associated with these random drawings are then analyzed to determine the likely results and the associated risk. Oftentimes this technique is called *Monte Carlo simulation*, being named for the city of Monte Carlo, which is noted for its casinos.

Gambling analogy notwithstanding, Monte Carlo simulation is a powerful and widely used technique for dealing with uncertainty in many aspects of business operations. For our purposes, it has been shown to be an accurate method of pricing options and particularly useful for types of options for which no known formula exists.

To facilitate an understanding of the technique, we shall look at how Monte Carlo simulation has been used to price standard European options.[1] Of course, we know that under the appropriate assumptions, the Black-Scholes-Merton model is the correct method of pricing these options, so Monte Carlo simulation is not really needed. It is useful, however, to conduct this demonstration under the assumptions of the Black-Scholes-Merton model, because it shows the accuracy of the technique for a simple option of which the exact price is easily obtained from a known formula. That is, it gives us a benchmark from which to judge the accuracy of the simulation.

Unfortunately, Monte Carlo simulation is not very practical for American options because it works by simulating prices forward, whereas the early exercise decision is best valued by positioning oneself at expiration and working backwards, as we do in the binomial model.

23.1 STANDARD MONTE CARLO SIMULATION OF THE LOGNORMAL DIFFUSION

Recall that the Black-Scholes-Merton model is based on geometric Brownian motion with risk-neutral growth of the underlying asset,

$$dS_t = S_t r_c dt + S_t \sigma dW_t, \tag{23.1}$$

where the stochastic integral solution can be expressed as[2]

$$S_T = S_t e^{\left[\left(r_c - \frac{\sigma^2}{2}\right)\tau + \sigma\sqrt{\tau}\varepsilon\right]}, \tag{23.2}$$

where, as we have previously shown, the term ε is a normally distributed random number with mean equal to 0 and standard deviation equal to 1. Note that we do not add a time subscript to this epsilon because it is generic within the simulator and time to expiration is handled separately.

The objective of Monte Carlo simulation is to generate a sample of terminal asset prices, S_T. Following Boyle (1977), we desire a simulation that converges quickly to the annualized, continuously compounded rates of return satisfying the following two conditions:

$$r_c = \frac{E_t\left[\ln\left(\frac{S_T}{S_t}\right)\right]}{\tau} \text{ and} \tag{23.3}$$

$$\sigma^2 = \frac{\text{var}_t\left[\ln\left(\frac{S_T}{S_t}\right)\right]}{\tau}. \tag{23.4}$$

Note also that by using the risk-free rate as the expected return, we are operating under the risk-neutral measure. We shall ultimately be consistent with that assumption by discounting simulated values at the risk-free rate.

What we want to do is generate future asset prices from Equation (23.2). It is actually quite easy. We learned how to do this in Chapter 10, where we simulated Brownian motion. A standard normal random variable, denoted here as ε, can be approximated with a slight adjustment to Excel's *Rand*() function, which produces a uniform random number between 0 and 1, meaning that it generates numbers between 0 and 1 with equal probability. A good approximation for a standard normal variable is obtained by the Excel formula, = *Rand*() + *Rand*() + *Rand*() + *Rand*() + *Rand*() + *Rand*() + *Rand*() + *Rand*() + *Rand*() + *Rand*() + *Rand*() + *Rand*() − 6.0, or simply 12 uniform random numbers minus 6.0.[3]

After generating one standard normal random variable, we then simply insert it into the right-hand side of the above formula for dS. But the key question is over what period of time this price change applies. Note in Equation (23.1) the continuous increment, dt. That means the price change is instantaneous, which means an infinitesimally small unit of time. We cannot do simulation in continuous time, so we have to choose a unit of time. Let us choose one day. That is, we change Equation (23.2) to

$$S_{\Delta t} = S_t e^{\left[\left(r_c - \frac{\sigma^2}{2}\right)\Delta t + \sigma\sqrt{\Delta t}\,\varepsilon\right]} \tag{23.5}$$

and let $\Delta t = 1/365$, or approximately 0.0027. So, assuming N days in the life of the option, the expiration is $\tau = N/365$. Recall from Chapter 2, there is some debate on whether to use trading days rather than calendar days. In the US, there are approximately 252 trading days in a year. We then simulate the price change for day 1, which generates a new price, which forms the base of the price change for day 2. We continue this until we have reached the expiration, day N. Now, we have an estimate of the asset price at expiration, S_T. We then compute the price of the option at expiration according to the standard formulas, max(0, $S_T - X$) for a call or max(0, $X - S_T$) for a put, where X is the exercise price and S_T is the asset price at expiration. This operation produces one possible option value at expiration. We then repeat this procedure many thousands of times, take the average value of the call at expiration, and discount that value at the risk-free rate. Some users also compute the standard deviation of the call prices in order to obtain a feel for the possible error in estimating the price.

Alternatively, we can simply simulate the terminal asset price directly rather than the entire path. Thus, we deploy a Monte Carlo simulation and apply it within Equation (23.2). So now let us price a call option on an asset. The asset price is 55, the exercise price is 50, the continuously compounded risk-free rate is 0.05, the volatility is 0.40, and the time to expiration is 0.75; that is, the option expires in nine months, or three-quarters of a year. Inserting the previous approximation formula for a standard normal random variable in any cell in an Excel spreadsheet produces a random number. Suppose that number is 0.733449. Inserting into Equation (23.2), we obtain

$$S_{0.75} = 55e^{\left[\left(0.05 - \frac{0.4^2}{2}\right)0.75 + 0.4\sqrt{0.75}(0.733449)\right]} = 69.33.$$

Thus, the simulated value of the asset at expiration is \$69.33. At that price, the option will be worth $\max(0, 69.33 - 50) = 19.33$ at expiration. We then draw another random number. Suppose we get -0.18985. Inserting into Equation (23.2), we obtain

$$S_{0.75} = 55e^{\left[\left(0.05 - \frac{0.4^2}{2}\right)0.75 + 0.4\sqrt{0.75}(-0.18985)\right]} = 50.35.$$

This gives us an asset price of \$50.35 and an option payoff of $\max(0, 50.35 - 50) = 0.35$. Both of these cases are in-the-money, but many will be out-of-the-money, leading to an option value of zero. We repeat this procedure several thousand times, after which we take an average of the simulated option prices and then discount that average to the present using the present value formula $e^{-0.05(0.75)}$.

Recall one way to represent the call value is the present value of the expected terminal payout, where the discounting is based on the risk-free rate and the expected terminal value is based on the underlying asset growing at the risk-free rate. With Monte Carlo simulation, we replace the expected terminal value with the average terminal value or

$$c_t = PV_r[E_t(c_T)] = e^{-r_c(T-t)}\left[\frac{1}{n}\sum_{i=1}^{n}\max\left(0, S_t e^{\left[\left(r_c - \frac{\sigma^2}{2}\right)\Delta t + \sigma\sqrt{\Delta t}\varepsilon_i\right]} - X\right)\right].$$

Naturally every simulation is different because each set of random numbers is different. A Monte Carlo procedure written in the R programming language produced Table 23.1 values for this call, whose actual Black-Scholes-Merton price is 11.02, where the number of random drawings is the sample size, n.

If we reran each simulation, we would obtain different values. As the number of simulations increase, the variation declines, and we get very close to the Black-Scholes-Merton value of 11.02.

To illustrate further, we show several figures with a new set of parameters. Consider now a stock trading at \$100 with exercise price of \$100. Further, assume a 5% interest rate, 30% volatility, and one year to expiration. Figure 23.1 illustrates 200 separate sets of simulations where the sample size increases by a factor of 10. Panel A demonstrates moving from a sample size of 10 to 100 we observe a significant decline in the variation of simulated call prices when compared to the BSM model price of \$14.23. Clearly, as we

TABLE 23.1   Monte Carlo Price of Standard European Call Option

n (number of simulations)	Option Price
1,000	11.42
5,000	10.75
10,000	11.11
25,000	11.02
50,000	10.94
100,000	11.05
1,000,000	11.02

move from 10 to 10,000 illustrated in Panel A the variation consistently declines. Note that in Panel B at 100,000 (1e+05), we rescale the y-axis, enabling you to see the further decline in variation. Without rescaling, the call values would appear constant as the slight variations would be undetectable.

Given the computational time required for large sample sizes, there are several ways to reduce the errors encountered with simulation. We explore several of these methods next.

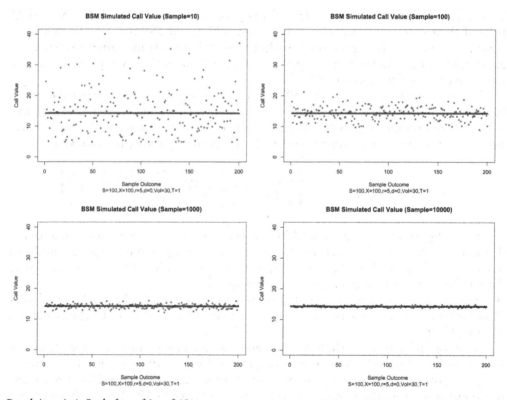

Panel A. y-Axis Scale from $0 to $40

FIGURE 23.1   Sample Error and Monte Carlo Price of Standard European Call Option

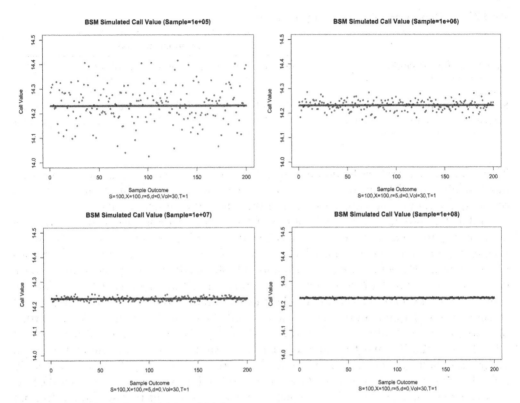

Panel B. y-Axis Scale from $14.0 to $14.5

FIGURE 23.1 *Continued*

23.2 REDUCING THE STANDARD ERROR

The option price obtained from a Monte Carlo simulation is a sample average. Thus, its standard deviation is the standard deviation of the sample divided by the square root of the sample size.[4] Consequently, the error reduces at the rate of 1 over the square root of the sample size. Notice what this means for increasing the sample accuracy by increasing the sample size. Suppose σ_S is the sample standard deviation. We first conduct a Monte Carlo simulation using n_1 random drawings. Because the option value is a sample mean, the standard deviation of our estimate of the option value is $\sigma_S/\sqrt{n_1}$, where n_1 is the sample size. Now suppose we wanted to reduce that standard deviation in half. How much larger must the sample be? Let this new sample size be n_2. Then its standard deviation of the estimate of the option price is $\sigma_S/\sqrt{n_2}$. Now note that the following specification, which reduces the standard deviation of the average in half:

$$\frac{1}{2}\frac{\sigma_S}{\sqrt{n_1}} = \frac{\sigma_S}{\sqrt{n_2}},$$

if and only if

$$n_2 = 4n_1.$$

TABLE 23.2 Monte Carlo Price of Standard European Call Option

Parameters for 20 samples	Number of Trials			
	1,000 (base case)	4,000	16,000	100,000
Average	2.9721	3.0007	2.9774	2.9795
Standard deviation	0.0921	0.0561	0.0219	0.0126
Percentage reduction from base case	NA	39.09%	76.22%	86.32%

Thus, to achieve a 50% reduction in error, that is, a 50% increase in accuracy, we must quadruple the number of random drawings. The standard error reduces only at the rate of the square root of the sample size, not at the rate of the sample size itself.

Table 23.2 provides an example of this effect. Consider a European call with an asset price of 42, an exercise price of 40, a volatility of 0.35, a risk-free rate of 5%, and a time to expiration of one month, which is $1/12 = 0.0833$. The table presents the results of simulations of 1,000, 4,000, 16,000, and 100,000 trials. For each set of these trials, the simulation is run 20 times. In other words, there are 20 simulations of 1,000 trials, 20 simulations of 4,000 trials, and so on. The mean and standard deviation of each group of 20 simulations of 1,000, 4,000, 16,000, and 100,000 trials is shown in the table. Starting off with the 1,000-trial case, the average value is 2.9721 with a standard deviation of 0.0921 across the 20 runs of 1,000 trials each. Now, suppose we want to cut that standard deviation in half. As noted, we need to quadruple the number of trials. Hence, we do 20 runs of 4,000 trials. The average is shown as 3.0007 and the standard deviation is 0.0561. This is a reduction of about 39%, which is not the 50% reduction we hoped for, but the formula is not an exact result. It is a sample result and will get more accurate with larger samples. If we want to cut the standard deviation by 75%, the formula above would say that we need 16,000 trials. In the next column, notice that the reduction in standard deviation is 76%. The final column is the case of 100,000 trials, which is the number required to cut the standard deviation by 90%. Note that it does a pretty good job, as the standard deviation is reduced by 86%.

But still, it can take a lot of trials to get a high degree of accuracy. Notice with 100,000 trials, the standard deviation in this problem is about 1.3 cents. Because the sample average is normally distributed by the central limit theorem, we can say that the mean estimate of 2.9795 is bracketed with 95% confidence in a range of plus or minus two standard deviations, which is $2(0.0126) = 0.0252$. That means the confidence interval is 2.9543 to 3.0047. For an institution with a lot of volume, a five-cent interval can be quite large.

23.2.1   The Antithetic Variate Method

With this problem in mind, it behooves the user of Monte Carlo simulation to attempt to achieve greater accuracy through other means. One method of doing so is quite simple and automatically doubles the sample size with only a minimum increase in computational time. This approach is called the *antithetic variate* method. Remember that in the standard Monte Carlo simulation, we are generating observations of a standard normal random variable. The standard normal random variable is distributed with a mean of zero, a variance of 1.0, and is symmetric. Thus, for each value we draw, there is an equally likely chance of having drawn the observed value times −1. Consequently, for each value of ε we draw, we can legitimately create an artificially observed companion observation of $-\epsilon$. This paired drawing is the antithetic variate. We simply use it the same way we use the value we drew, that is, in computing a sequence of asset prices from which we compute an option price based on the price at expiration. This procedure automatically doubles our sample size without having increased the number of random drawings. To take advantage of the antithetic variate method, we should average the option price from the standard result with the option price from the antithetic variate result. We get two observations for the price of one as well as a more accurate average.

In the case of Black-Scholes-Merton, which converges rapidly in a simulation, this may not matter that much.

23.2.2 The Control Variate Method

Another method that can be used with certain types of options is called the *control variate* method, which in this context is a somewhat similar option whose value is known. We then obtain a simulated value of that option. The difference between the value of the control variate and its simulated value is then added to the simulated value of the option we are trying to price. In this manner, the error in the control variate is added to the simulated value of the option of interest. Let us see how this method works.

Let c_s be the simulated price of the option we are trying to price. Let v_t be the value of another similar option and v_s be its simulated value. Our control variate estimate of the option price is then found as

$$c_s + (v_t - v_s).$$

What we are doing is running a simulation of $c_s - v_s$ and adding v_t to try to obtain a better estimate. Note the following result for the variance:

$$\text{var}(c_s - v_s) = \text{var}(c_s) + \text{var}(v_s) - 2\text{cov}(c_s, v_s).$$

This will be less than $\text{var}(c_s)$ if $\text{var}(v_s) < 2\,\text{cov}(c_s, v_s)$, meaning that the control variate method relies on the assumption of a large covariance between c_s and v_s. The control variate chosen should be one that is very highly correlated with the option we are pricing.

The control variate method is somewhat appropriate when pricing complex and path-dependent options. It can be challenging to use, however, in that we must know a great deal about the relationship of the price of the control variate to the price of the option of interest. Often it is difficult to know something about that relationship without knowing how to price the option of interest.

23.2.3 Other Methods for Improved Simulation

There are a number of methods for improving the accuracy of Monte Carlo simulation. Some methods improve the sampling process so that random but unusual results are balanced with more typical outcomes. These methods belong to a family called *low discrepancy sequences*. For example, if a random sample is generated, it is quite possible that numbers that ought to appear with a given frequency within a specific range might not appear with the expected frequency. Some common techniques from this family are *Halton sequences* and *Sobol numbers*, which are used to supply random outcomes that result in the sample of random numbers being much closer to a truly random sample with observed outcomes as expected.

23.3   SIMULATION WITH MORE THAN ONE RANDOM VARIABLE

In some cases, we wish to simulate more than one random variable. In certain situations, these variables are independent; in others, the two variables are related. Independent random variables are similar to rolling a pair of dice. What happens on one die is unrelated to what happens on another die. If the random variables of interest have that characteristic, it is a simple matter to run two parallel simulations, each generating random variables that are completely unrelated to each other. In some cases, however, the random variables are related, most often in a linear manner. The linear relationship between random variables is captured by the correlation.[5]

Suppose we wish to generate two normal random values that have correlation of ρ, which could be two asset prices or interest rates. First, we generate two independent, standard normal random values x_1 and x_2. We set ε_1 equal to x_1 and obtain ε_2 in the following manner:

$$\varepsilon_2 = \rho x_1 + x_2 \sqrt{1 - \rho^2}. \tag{23.6}$$

The reason this formula works is simple.

$$E(\varepsilon_1) = E(x_1) = 0$$

$$E(\varepsilon_2) = E\left(\rho x_1 + x_2 \sqrt{1 - \rho^2}\right) = pE(x_1) + \sqrt{1 - \rho^2} E(x_2) = 0$$

$$\text{var}(\varepsilon_1) = \text{var}(x_1) = 1$$

$$\text{var}(\varepsilon_2) = \text{var}\left(\rho x_1 + x_2 \sqrt{1 - \rho^2}\right)$$

$$= \rho^2 \text{var}(x_1) + (1 - \rho^2) \text{var}(x_2) + 2\rho\sqrt{1 - \rho^2}\text{cov}(x_1, x_2)$$

$$= \rho^2 + 1 - \rho^2 + 0 = 1$$

$$\text{corr}(\varepsilon_1, \varepsilon_2) = \text{corr}\left(x_1, \rho x_1 + x_2\sqrt{1 - \rho^2}\right)$$

$$= \frac{\text{cov}\left(x_1, \rho x_1 + x_2\sqrt{1 - \rho^2}\right)}{\text{var}(\varepsilon_1, \varepsilon_2)} = \text{cov}\left(x_1, \rho x_1 + x_2\sqrt{1 - \rho^2}\right)$$

$$= \text{cov}(x_1, \rho x_1) + \text{cov}\left(x_1, x_2\sqrt{1 - \rho^2}\right) = \rho\text{cov}(x_1, x_1) + \sqrt{1 - \rho^2}\text{cov}(x_1, x_2)$$

$$= \rho\text{var}(x_1) + 0 = \rho\text{var}(x_1)$$

$$= \rho.$$

These proofs show that the two independent random variables have expected values of 0, variances of 1, and a correlation of ρ. Thus, they meet our requirements. Now, armed with values of ε_1 and ε_2, we can proceed by inserting into the analog of Equation (23.5) for each of the two assets, using the respective values of the expected change and volatility of each asset we are simulating.

23.4   RECAP AND PREVIEW

In this chapter, we examined the method of Monte Carlo simulation, which can be used to price options by generating a large sample of outcomes for the underlying, calculating the corresponding option payoff, averaging that payoff, and discounting by the risk-free rate. We illustrated it for the standard Black-Scholes-Merton case, but the benefit of Monte Carlo lies in its ability to handle more complex cases.

In Chapter 24, we look at one more numerical procedure, the finite difference method, which solves the partial differential equation in a numerical manner.

QUESTIONS AND PROBLEMS

1 Let x_1 and x_2 denote two independent standard normal random variables with mean 0 and variance 1. Explain how to generate two dependent standard normal random variables. Prove your approach works.

2 Explain how one transforms correlated standard normal random variables to correlated normal random variable with nonzero means and variance not equal to one.

3 Identify and discuss two ways to reduce the standard error of the estimate when pricing options with Monte Carlo simulation.

4 The following two figures were generated with Monte Carlo simulation. One figure is based on geometric Brownian motion (GBM) and one figure is based on arithmetic Brownian motion (ABM). Identify each figure and defend your answer.

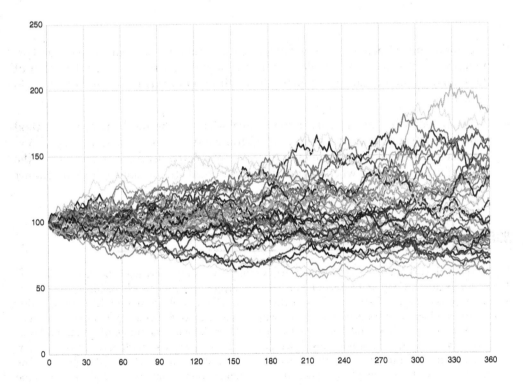

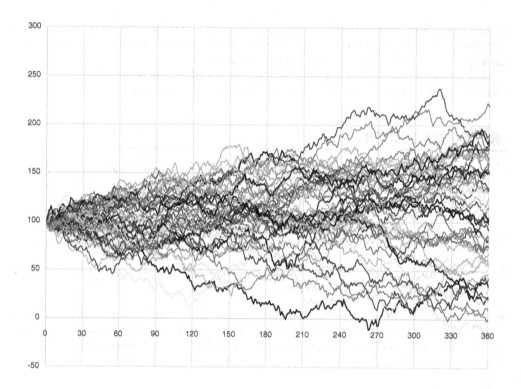

5 The option price obtained from a Monte Carlo simulation is a sample average. Distinguish between the standard deviation of a sample and the standard deviation of the sample mean estimate and explain the implication of this point for Monte Carlo simulation.

6 [Contributed by Jeremy Vasseur] Suppose you wish to use the antithetic variate method to increase the number of observations in a Monte Carlo simulation. You are not, however, using a normal distribution but rather an asymmetric distribution to generate the random values. Explain whether you can use the antithetic variate method to conduct this simulation.

NOTES

1. The seminal work on the use of Monte Carlo simulation in option pricing is Boyle (1977).
2. See, for example, Chapter 12, Section 4. Also, a dividend yield could easily be incorporated.
3. This approximation is based on the fact that the distribution of the sum of 12 uniformly distributed random numbers between 0 and 1 will have a mean of 6.0 and a standard deviation of 1. By subtracting 6.0, we shift the mean to zero without changing the standard deviation. What we obtain is technically not normally distributed, but it is symmetric, with mean zero and standard deviation of 1.0, which are three properties associated with the normal distribution. The procedure is widely accepted as a quick and reasonable approximation but might not pass the most demanding tests for normality. However, many other random number generators would also not pass the most demanding tests either. The Excel nested function normsinv(rand()) will also generate a standard normal directly from the uniform random number.

4. Although this is a point from the most elementary principles of statistics, it is important to distinguish between the standard deviation of a sample and the standard deviation of the sample mean estimate. The latter is the former divided by the square root of the sample size. Hence, the sample mean is much less volatile than the sample values themselves.

5. Although we covered this point in Chapter 4, it is worth repeating that two variables that are independent will have zero correlation, but two variables that have zero correlation can be highly dependent. As an example, the relationship $y = \sin(x)$ expresses the familiar sine function, a repeating wave in which y goes up and then down alternatively as x increases. The variable y is completely dependent on x, but the relationship is nonlinear. A graph shows a very obvious curve, not a straight line. Calculation of the correlation between y and x would result in a value of approximately zero, even though y and x are highly related, of course, in a nonlinear manner.

CHAPTER 24

Finite Difference Methods

As we have learned to this point, under the appropriate assumptions, the price of a call option is given by the solution, c, to the partial differential equation

$$\frac{\partial c}{\partial S_t} r_c S_t + \frac{\partial c}{\partial t} + \frac{1}{2} \frac{\partial^2 c}{\partial S_t^2} S_t^2 \sigma^2 = r_c c_t. \tag{24.1}$$

For standard European options and certain other options, a closed form solution is possible. In the case of European options, the boundary condition is $c_T = \max(0, S_T - X)$, where S_T is the asset price at expiration and X is the exercise price in the well-known Black-Scholes-Merton model. In some cases, however, no closed-form solution can be derived. Although the binomial model can often be used, the finite difference method is another alternative. It essentially amounts to a trinomial and in some respects is more efficient, because it allows for more possible outcomes. The approach taken here is the log transform method as suggested by Brennan and Schwartz (1978), which is known to provide a stable solution.[1]

24.1 SETTING UP THE FINITE DIFFERENCE PROBLEM

Recall that the call price depends on the underlying price and time or $c_t = c(S, t)$. Following Brennan and Schwartz, we begin by defining a new variable $y = \ln S$. Let w be defined such that $w(y, t) = c(S, t)$. The goal is to transform a multiplicative distribution (i.e., the lognormal distribution) to an additive distribution (i.e., the normal distribution). Recall that if S is lognormally distributed, then $\ln(S)$ is normally distributed. The variable $w(y, t)$ is the price of the call at time t in terms of the transformed asset price and time. In other words, we price the call in terms of the log of the asset price and time t. Based on Itô's lemma, we shall need the following:

$$\frac{\partial c}{\partial S} = \frac{\partial w}{\partial y} \frac{\partial y}{\partial S} = \frac{\partial w}{\partial y} \frac{1}{S} = \frac{\partial w}{\partial y} e^{-y}, \tag{24.2}$$

$$\frac{\partial^2 c}{\partial S^2} = \frac{\partial}{\partial S} \left(\frac{\partial w}{\partial y} e^{-y} \right) = e^{-y} \frac{\partial w / \partial y}{\partial S} + \frac{\partial w}{\partial y} \frac{\partial e^{-y}}{\partial S}$$

$$= e^{-y} \frac{\partial^2 w}{\partial y^2} \frac{\partial y}{\partial S} - \frac{\partial w}{\partial y} e^{-2y} = \left(\frac{\partial^2 w}{\partial y^2} - \frac{\partial w}{\partial y} \right) e^{-2y}, \text{ and} \tag{24.3}$$

$$\frac{\partial c}{\partial t} = \frac{\partial w}{\partial t}. \tag{24.4}$$

Substituting these results into the partial differential equation, we obtain

$$\frac{\sigma^2}{2}\frac{\partial^2 w}{\partial y^2} + \left(r_c - \frac{\sigma^2}{2}\right)\frac{\partial w}{\partial y} + \frac{\partial w}{\partial t} - r_c w = 0. \tag{24.5}$$

To get a practical solution for w in this equation, we express it as the difference equation,

$$\frac{\sigma^2}{2}\frac{\Delta^2 w}{\Delta y^2} + \left(r_c - \frac{\sigma^2}{2}\right)\frac{\Delta w}{\Delta y} + \frac{\Delta w}{\Delta t} - r_c w = 0. \tag{24.6}$$

We then partition a reasonable range of the log of the asset price into finite intervals. For example, the minimum asset price is zero and the maximum is infinity. Suppose we consider only the following prices as feasible: $0, \Delta y, 2\Delta y, \ldots, y - 2\Delta y, y - \Delta y, y, y + \Delta y, y + 2\Delta y, \ldots, \infty$. When we implement the technique, we shall need to make Δy as small as possible and we must choose a finite maximum asset price to represent infinity, so let us call it S_{max}. In addition $\ln 0$ is undefined so we let the minimum S be very close to but not equal to zero. Here we specify $\min(\ln S)$ as a very small value, η.[2]

Letting τ be the time to expiration, we partition the time remaining in the option's life into discrete intervals equaling $\tau, \tau - \Delta t, \tau - 2\Delta t, \ldots, 2\Delta t, \Delta t, 0$. These gradations of time and asset price can be arranged on a grid, as in Table 24.1.

Each dot corresponds to an option price associated with the log of the asset price in the given row and the time to expiration in the given column.

Some of the information in the grid would already be known. For example, we know the following boundary conditions,

$$c = 0 \quad \text{if} \quad S = 0 \quad \text{for all } t. \tag{24.7}$$

And because we know that $\ln S = \eta$ when S is very close to zero, we specify this condition as

$$w = 0 \quad \text{if} \quad S \text{ "close" } 0 \ (\ln S = \eta) \quad \text{for all } t. \tag{24.8}$$

This statement permits us to fill in zeroes along the bottom row.

TABLE 24.1 Finite Difference Grid

			Time to Expiration				
$\ln S$	τ	$\tau - \Delta t$	$\tau - 2\Delta t$	$\bullet$	$2\Delta t$	Δt	0
∞	$\bullet$	$\bullet$	$\bullet$	$\bullet$	$\bullet$	$\bullet$	$\bullet$
$\bullet$	$\bullet$	$\bullet$	$\bullet$	$\bullet$	$\bullet$	$\bullet$	$\bullet$
$y + 2\Delta y$	$\bullet$	$\bullet$	$\bullet$	$\bullet$	$\bullet$	$\bullet$	$\bullet$
$y + \Delta y$	$\bullet$	$\bullet$	$\bullet$	$\bullet$	$\bullet$	$\bullet$	$\bullet$
y	$\bullet$	$\bullet$	$\bullet$	$\bullet$	$\bullet$	$\bullet$	$\bullet$
$y - \Delta y$	$\bullet$	$\bullet$	$\bullet$	$\bullet$	$\bullet$	$\bullet$	$\bullet$
$y - 2\Delta y$	$\bullet$	$\bullet$	$\bullet$	$\bullet$	$\bullet$	$\bullet$	$\bullet$
$\bullet$	$\bullet$	$\bullet$	$\bullet$	$\bullet$	$\bullet$	$\bullet$	$\bullet$
$2\Delta y$	$\bullet$	$\bullet$	$\bullet$	$\bullet$	$\bullet$	$\bullet$	$\bullet$
Δy	$\bullet$	$\bullet$	$\bullet$	$\bullet$	$\bullet$	$\bullet$	$\bullet$
η	$\bullet$	$\bullet$	$\bullet$	$\bullet$	$\bullet$	$\bullet$	$\bullet$

TABLE 24.2 Partially Filled-in Finite Difference Grid

ln S				Time to Expiration			
	τ	$\tau - \Delta t$	$\tau - 2\Delta t$	$\bullet$	$2\Delta t$	Δt	0
S_{max}	$\bullet^*+$	$\bullet$	$\bullet$	$\bullet$	$\bullet$	$\bullet$	$\max(0, S_{max} - X)$
$\bullet$	$\bullet(\tau) + \Delta y e^y$	$\bullet(\tau - \Delta t) + \Delta y e^y$	$\bullet(\tau - 2\Delta t) + \Delta y e^y$	$\bullet(\bullet) + \Delta y e^y$	$\bullet(2\Delta t) + \Delta y e^y$	$\bullet(\Delta t) + \Delta y e^y$	$\max(0, e^\bullet - X)$
$y + 2\Delta y$	$\bullet$	$\bullet$	$\bullet$	$\bullet$	$\bullet$	$\bullet$	$\max(0, e^{y+2\Delta y} - X)$
$y + \Delta y$	$\bullet$	$\bullet$	$\bullet$	$\bullet$	$\bullet$	$\bullet$	$\max(0, e^{y+\Delta y} - X)$
y	$\bullet$	$\bullet$	$\bullet$	$\bullet$	$\bullet$	$\bullet$	$\max(0, e^y - X)$
$y - \Delta y$	$\bullet$	$\bullet$	$\bullet$	$\bullet$	$\bullet$	$\bullet$	$\max(0, e^{y-2\Delta y} - X)$
$y - 2\Delta y$	$\bullet$	$\bullet$	$\bullet$	$\bullet$	$\bullet$	$\bullet$	$\max(0, e^{y-2\Delta y} - X)$
$\bullet$	$\bullet$	$\bullet$	$\bullet$	$\bullet$	$\bullet$	$\bullet$	$\max(0, e^\bullet - X)$
$2\Delta y$	$\bullet$	$\bullet$	$\bullet$	$\bullet$	$\bullet$	$\bullet$	$\max(0, e^{2\Delta y} - X)$
Δy	$\bullet$	$\bullet$	$\bullet$	$\bullet$	$\bullet$	$\bullet$	$\max(0, e^{\Delta y} - X)$
η	0	0	0	0	0	0	0

When $S \to \infty$, the first derivative of the call price with respect to the asset price is 1:

$$\lim_{S \to \infty} \frac{\partial c}{\partial S} = 1 \text{ if } S \to \infty \text{ for all } t. \tag{24.9}$$

Because

$$\frac{\partial c}{\partial S} = \left(\frac{\partial w}{\partial y} \right) e^{-y},$$

we have

$$\frac{\partial w}{\partial y} = e^y = S \text{ for all } t \text{ when } \ln S \to \infty. \tag{24.10}$$

This means that once we know the second highest call value, we can obtain the highest value by adding $\Delta y e^y$ to it. This is equivalent to imposing the standard condition that the option delta is 1 at a very high stock price.

We also know the value of the option at expiration,

$$c = \max(0, S_T - X) \text{ for all } S \text{ when } t = 0. \tag{24.11}$$

In terms of y, this is simply

$$w = \max(0, e^y - X) \text{ for all } y. \tag{24.12}$$

Thus, we can fill in the entire right column. So, the grid is now partially filled in, as shown in Table 24.2.

So, let us now take a look at the explicit finite difference method.

24.2 THE EXPLICIT FINITE DIFFERENCE METHOD

It may be useful to visualize the following small section from the grid illustrated in Figure 24.3. Define any arbitrary point, t, across the columns representing time. Each column is related to the column next to it by one increment or decrement of time, Δt. Each row is the

TABLE 24.3 A Portion of the Explicit
Finite Difference Grid

	Time	
$\ln S$	t	$t + \Delta t$
$y + \Delta y$		$w_{y+\Delta y, t+\Delta t}$
y	$w_{y,t}$	$w_{y, t+\Delta t}$
$y - \Delta y$		$w_{y-\Delta y, t+\Delta t}$

natural log of the asset price. Each row above or below it increments the log of the asset price by Δy. From the point corresponding to the log of asset price, y, and time, t, look to the right one column at the option prices one asset price up, the current asset price, and one asset price down.

What we are going to do is to express the option price $w_{y,t}$ as a weighted average of these three option prices to the right, discounted one period at the risk-free rate, that is, in this general form:

$$w_{y,t} = \frac{\omega_1 w_{y+\Delta y, t+\Delta t} + \omega_2 w_{y, t+\Delta t} + \omega_3 w_{y-\Delta y, t+\Delta t}}{e^{r_c \Delta t}}. \tag{24.13}$$

The explicit finite difference method solves for $w_{y,t}$ in terms of the known values of $w_{y+\Delta y, t+\Delta t}$, $w_{y, t+\Delta t}$, and $w_{y-\Delta y, t+\Delta t}$. The three prices one time step ahead are assumed to be already known. Of course, we shall need to know the three prices to the right, but we will obviously know these prices. Recall from Table 24.2 that we know all the prices in the rightmost column. That means we can fill in the prices one column to the left. What we do not know, however, is the weights in Equation (24.13), and that is what the explicit finite difference method will provide. We shall approach this problem by examining the differentials, which represent the derivatives, which represent the differences in prices along the grid.

We can obtain finite difference estimates of the partial derivatives as follows. First, $\Delta w_{y,t} / \Delta t$ is approximately

$$\frac{\Delta w_{y,t}}{\Delta t} = \frac{w_{y, t+\Delta t} - w_{y,t}}{\Delta t}. \tag{24.14}$$

Note that we are holding the asset price constant and moving along the grid by a time step. We shall also, however, need a way to relate the price corresponding to y and t to the price one step later corresponding to $y + \Delta y$ and $y - \Delta y$. One way is called the forward difference,

$$\frac{\Delta w_{y,t}}{\Delta y} = \frac{w_{y+\Delta y, t+\Delta t} - w_{y, t+\Delta t}}{\Delta y}, \tag{24.15}$$

and another is called the backward difference,

$$\frac{\Delta w_{y,t}}{\Delta y} = \frac{w_{y, t+\Delta t} - w_{y-\Delta y, t+\Delta t}}{\Delta y}. \tag{24.16}$$

We typically average these two estimates to obtain

$$\frac{\Delta w_{y,t}}{\Delta y} = \frac{w_{y+\Delta y,t+\Delta t} - w_{y-\Delta y,t+\Delta t}}{2\Delta y}. \tag{24.17}$$

The second partial differential can be estimated as

$$\frac{\Delta^2 w_{y,t}}{\Delta y^2} = \frac{\frac{w_{y+\Delta y,t+\Delta t} - w_{y,t+\Delta t}}{\Delta y} - \frac{w_{y,t+\Delta t} - w_{y-\Delta y,t+\Delta t}}{\Delta y}}{\Delta y}.$$

$$= \frac{w_{y+\Delta y,t+\Delta t} - 2w_{y,t+\Delta t} + w_{y-\Delta y,t+\Delta t}}{\Delta y^2} \tag{24.18}$$

Substituting these estimates of the partial differentials into the difference Equation (24.6) and rearranging gives us

$$w_{y,t} = \frac{\omega_1 w_{y-\Delta y,t+\Delta t} + \omega_2 w_{y,t+\Delta t} + \omega_3 w_{y+\Delta t,t+\Delta t}}{1 + r_c \Delta t}, \tag{24.19}$$

where

$$\omega_1 = \left[\frac{1}{2}\left(\frac{\sigma}{\Delta y}\right)^2 - \frac{1}{2}\left(\frac{r_c - \frac{\sigma^2}{2}}{\Delta y}\right) \right] \Delta t$$

$$\omega_2 = 1 - \left(\frac{\sigma}{\Delta y}\right)^2 \Delta t$$

$$\omega_3 = \left[\frac{1}{2}\left(\frac{\sigma}{\Delta y}\right)^2 + \frac{1}{2}\left(\frac{r_c - \frac{\sigma^2}{2}}{\Delta y}\right) \right] \Delta t. \tag{24.20}$$

In other words, the option price at t is a discounted weighted average of the next three possible option prices at $t + \Delta t$, where the weights are given in Equation (24.20). These weights sum to unity and are the risk neutral/equivalent martingale probabilities of the three log asset prices $y + \Delta y$, y, and $y - \Delta y$ at $t + \Delta t$. This technique is, therefore, essentially a risk-neutral trinomial method.

Filling in the full grid with the option prices is now simple. As noted previously, the rightmost column is the set of option prices at expiration and is already known. Using the rightmost column to provide the three known option prices, we can then solve for the second-to-rightmost column of option prices. Other missing prices in the topmost row can be filled in by knowing the second-to-topmost row, as noted previously. The prices in the bottommost row are, as noted previously, zero. The process then continues leftward until the leftmost column is filled in. When the entire grid is filled in, the current option price can be read from the row corresponding to the log of the current asset price in the leftmost column.[3]

The log transform allows the weights to be independent of time. Thus, $\omega_i, i = 1, 2, 3$ need to be obtained only once.[4] The weights should be constrained to be nonnegative, which is important, by choosing $\Delta t \le \Delta y^2 / \sigma^2$ and $\Delta y \le \sigma^2 |r_c - \sigma^2/2|$.

24.3   THE IMPLICIT FINITE DIFFERENCE METHOD

The implicit finite difference method uses a somewhat different approach. Consider the section of the grid shown in Table 24.4.

Notice the difference with the explicit finite difference method. Here we have one price at $t + \Delta t$ and three at time t. Here we shall solve for three contemporaneous option prices by using information about the next asset price. The first-order differential with respect to the asset price is estimated as

$$\frac{\Delta w_{y,t}}{\Delta y} = \frac{w_{y+\Delta y,t} - w_{y,t}}{\Delta y}, \tag{24.21}$$

or as

$$\frac{\Delta w_{y,t}}{\Delta y} = \frac{w_{y,t} - w_{y-\Delta y,t}}{\Delta y}. \tag{24.22}$$

Averaging these two estimates we obtain

$$\frac{\Delta w_{y,t}}{\Delta y} = \frac{w_{y+\Delta y,t} - w_{y-\Delta y,t}}{2\Delta y}. \tag{24.23}$$

Note how each of these prices is at the same point in time, t. The second order differential is estimated as

$$\frac{\Delta^2 w_{y,t}}{\Delta y^2} = \frac{\frac{w_{y+\Delta y,t} - w_{y,t}}{\Delta y} - \frac{w_{y,t} - w_{y-\Delta y,t}}{\Delta y}}{\Delta y}.$$

$$= \frac{w_{y+\Delta y,t} - 2w_{y,t} + w_{y-\Delta y,t}}{\Delta y^2} \tag{24.24}$$

The time differential is estimated the same way as in the explicit finite difference method, Equation (24.14). Now we substitute these values into the difference Equation, (24.6), to obtain

$$\frac{1}{2}\sigma\left(\frac{w_{y+\Delta y,t} - 2w_{y,t} + w_{y-\Delta y,t}}{\Delta y^2}\right) + \left(r_c - \frac{\sigma^2}{2}\right)\left(\frac{w_{y+\Delta y,t} - w_{y-\Delta y,t}}{2\Delta y}\right)$$

$$+ \frac{w_{y,t+\Delta t} - w_{y,t}}{\Delta t} - r_c w_{y,t} = 0. \tag{24.25}$$

TABLE 24.4   A Portion of the Finite Difference Grid

	Time	
ln S	t	$t + \Delta t$
$y + \Delta y$	$w_{y+\Delta y,t}$	
y	$w_{y,t}$	$w_{y,t+\Delta t}$
$y - \Delta y$	$w_{y-\Delta y,t}$	

Solving for $w_{y,t+\Delta t}$ gives

$$w_{y,t+\Delta t} = \omega_1 w_{y-\Delta y,t} + \omega_2 w_{y,t} + \omega_3 w_{y+\Delta y,t}, \qquad (24.26)$$

where

$$\omega_1 = \frac{1}{2}\left(\frac{r_c - \frac{\sigma^2}{2}}{\Delta y}\right)\Delta t - \frac{1}{2}\frac{\sigma^2 \Delta t}{\Delta y}$$

$$\omega_2 = 1 + \frac{\sigma^2 \Delta t}{\Delta y^2} + r_c \Delta t$$

$$\omega_3 = -\frac{1}{2}\left(\frac{r_c - \frac{\sigma^2}{2}}{\Delta y}\right)\Delta t - \frac{1}{2}\frac{\sigma^2 \Delta t}{\Delta y}. \qquad (24.27)$$

Note how the equation above solves for the prices $w_{y+\Delta y,t}$, $w_{y,t}$, and $w_{y-\Delta y,t}$ in terms of the known value $w_{y,t+\Delta t}$. With three unknowns there must be three equations. Thus, the solution is obtained by solving simultaneously an entire column of option prices. The process starts at the next-to-rightmost column and works leftward.

The implicit finite difference method is generally considered the better approach, because it will often work when the explicit method does not. But it is slightly more difficult to implement, requiring solving simultaneous equations. The Crank-Nicholson method is also widely used. It essentially averages the explicit and implicit methods.

24.4 FINITE DIFFERENCE PUT OPTION PRICING

If the options are puts, then the procedure is exactly the same but the following three conditions replace the corresponding conditions for calls:

$$w_{y,t} = Xe^{-r_c\tau} \text{ when } \eta \to 0$$

$$w_{y,t} = 0 \text{ for all } t \text{ when } y \to \infty$$

$$w_{y,t} = \max(0, X - e^y) \text{ for all } y. \qquad (24.28)$$

24.5 DIVIDENDS AND EARLY EXERCISE

If the asset is a stock and there are dividends, the procedure is slightly more complex. If we assume a continuous, constant yield of δ_c, the first term in the partial differential equation is changed slightly with $r_c - \delta_c$ replacing r_c. Then S is redefined as $Se^{-\delta_c\tau}$. If we assume discrete dividends, the partial differential equation is the same as the no-dividend case, but the asset price is redefined as the asset price less the present value of the dividends.

If the options are American calls, then the possibility of early exercise must be considered. At each node we must determine if the asset is worth more exercised early. We simply calculate the value of the option the ordinary way and then calculate the exercise value, $\max(0, S + PV(D) - X)$, where $PV(D)$ is the present value of all remaining dividends during the life of the option. If the call is worth more exercised, then replace the calculated value of the option with the exercise value. This condition will occur with high dividends, low time to expiration, low time value, low volatility, and/or a low interest rate.

If the options are American puts, dividends are not required to justify early exercise. The same procedure as described for calls is followed, with appropriate substitution of the put boundary conditions in the topmost and bottommost rows and rightmost columns.

24.6   RECAP AND PREVIEW

In this chapter, we showed how options can be priced by finite difference methods. The finite difference approach specifies difference equations that approximate the differential equations that must hold to obtain the option price. These difference equations are solved by deriving their values over the range of possibilities, such that the option price at the start corresponds to the condition that the current underlying price and time to expiration are in effect. We showed two methods of achieving this result: the explicit and implicit finite difference methods. The former essentially creates a trinomial, which is solved in a similar manner to the binomial model. The latter solves simultaneous equations at each combination of asset price and time to expiration.

This completes Part V. In Part VI, we introduce a new type of underlying, specifically interest rates, which shall pose a special challenge in modeling the stochastic process of the underlying.

QUESTIONS AND PROBLEMS

1   Based on the standard Black-Scholes-Merton partial differential equation, derive the following transformed PDE: $\frac{\sigma^2}{2}\frac{\partial^2 w}{\partial y^2} + \left(r_c - \frac{\sigma^2}{2}\right)\frac{\partial w}{\partial y} + \frac{\partial w}{\partial t} - r_c w = 0$, when $y = \ln(S)$.

2   In general, explain the difference between the explicit finite difference method and the implicit finite difference method.

3   Explain how the finite difference methodology is changed when solving for put prices.

4   Explain the impact dividends have on finite difference methods.

5   Compare and contrast the binomial option pricing model with the explicit finite difference method.

6   [Contributed by Jeremy Vasseur] Suppose you were creating a grid for a finite difference valuation of European puts. Explain how you would fill in the rightmost column, topmost row, and bottommost row. Assume the log transform approach is used. (Base your answer on the grid structure provided in Table 24.2.)

NOTES

1. Some methods lead to an unstable solution, meaning a negative option price for certain asset prices and times to expiration.
2. If we do not let S be less than 1, then $\ln S$ would always be positive.
3. The option delta, gamma, and theta can be estimated by using approximations for the formulas for $\partial w/\partial y$, $\partial^2 w/\partial y^2$, and $\partial w/\partial t$ based on the finite differences in the grid.
4. Without the log transform, there will be a different set of weights for each time point or column.

Interest Rate Derivatives

CHAPTER 25

The Term Structure of Interest Rates

The term structure of interest rates is the relationship between the interest rate on a bond or loan and the maturity.[1] The term structure is often depicted with a graph, sometimes called a *yield curve*. A typical question raised by the term structure is whether long-term rates are more than short-term rates and, if so, why? Of course, the issue is a bit more complex than that. The maturities of loans and bonds span a range from very short term, as short perhaps as a day, to very long term, such as 30 years or more. Maturities cannot simply be dichotomized into short and long term. And it is difficult to make general statements, such as longer-term rates are higher than shorter-term rates. That might be true, but there can be more complex relationships between rates and maturities. Hence, there could be low short-term rates, high intermediate-term rates, and low long-term rates, with bumps and dips in between. And, of course, the words *low* and *high* are relative and might mean different things to different people. Figure 25.1 illustrates three potential patterns for the term structure of interest rates. The top line depicts a common upward sloping yield curve and the bottom depicts a downward sloping yield curve. The middle curve illustrates that other patterns that are not necessarily upward or downward are also possible.

In general, however, we use the term structure to think about how rates of various maturities might differ. As we discussed in Chapter 22, the rate on a loan or bond of one maturity can be related to the rate on a loan or bond of a shorter-term maturity to produce a forward rate. For example, let us say that the rate on a five-year zero-coupon bond is 7.2%, annual compounding, and the rate on a two-year zero-coupon bond is 6.9%. The forward rate is found as follows:

$$F = \sqrt[3]{\frac{(1.072)^5}{(1.069)^2}} - 1 = 0.074.$$

Much of the study of the term structure is about understanding what that forward rate might mean. We have already seen that it means that one could today enter into a contract to borrow or lend for three years with the loan starting two years from now. The rate would be agreed on today. There are some opinions that this number might tell us something about the direction of future interest rates. We explore this issue in this chapter.[2]

Almost anyone who has studied the term structure of interest rates, perhaps through a course in fixed income markets, is familiar with the unbiased expectations hypothesis (UEH),[3] which proposes that the forward rate is an unbiased predictor of the future spot rate. Thus, the forward rate is seen as the expected spot rate. The question of whether the

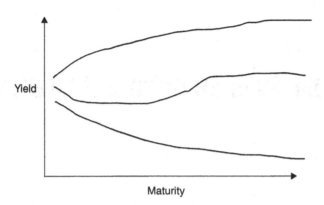

FIGURE 25.1   Term Structure of Interest Rates Illustration

UEH is true has occupied a central place in research in our body of understanding of the term structure, in particular in empirical research. There is no clear consensus empirically, but in fact it is a simple matter to prove that the unbiased expectations hypothesis cannot be true theoretically and almost surely cannot be true in reality.[4]

25.1   THE UNBIASED EXPECTATIONS HYPOTHESIS

Let us start by defining S_0 and F_0 as the spot and forward prices for an asset.[5] The forward price is for a forward contract that expires one period later and requires delivery of the underlying asset. In other words, one enters into a contract to purchase the asset one period later, paying the price F_0, which is determined when the contract is created. Unless otherwise stated, we assume annual time increments, so time 1 occurs in 1 year. Thus, we can set aside compounding issues ($\tau = 1$).

Now let us move ahead to time 1, at which time the spot price is S_1. To avoid the existence of any arbitrage opportunities, the forward price at expiration must converge to the spot price. Therefore,

$$F_1 = S_1. \qquad (25.1)$$

In a technical sense, the forward price F_1 is the price of a contract expiring instantaneously. Therefore, it has to be the spot price. That is, a forward contract calling for instantaneous payment and delivery of the asset is a spot transaction, because a spot transaction calls for instantaneous payment and delivery of the asset.

At time 0, the expected spot price at time 1 is denoted as $E_0(S_1)$. As such, with $F_1 = S_1$,

$$E_0(S_1) = E_0(F_1). \qquad (25.2)$$

The unbiased expectations hypothesis states that the forward price is the expected spot price. Formally,

$$F_0 = E_0(S_1). \qquad (25.3)$$

Let us now see why this statement cannot be true. By definition, today's spot price of any asset is the expected future spot price minus the risk premium, with this value discounted back to the present at the risk-free rate. Let us write this statement as follows:

$$S_0 = e^{-r_c}[E_0(S_1) - \lambda], \tag{25.4}$$

where r_c is the one-period continuously compounded interest rate and λ is the dollar risk premium. At this point, we do not specify what the risk premium is. For example, we do not require that it derives from the capital asset pricing model. We simply define it in general terms. It is the amount in money by which the future asset price is reduced to induce someone to buy the asset. Equation (25.4) must hold because rational investors require ex ante compensation for the opportunity cost of funds tied up in the risky asset as well as compensation for the assumption of risk. Also, note that the bracketed term is the risk-adjusted expected value at time 1, and it is, therefore, appropriate to discount this value for one period at the risk-free rate. Equation (25.4) is what we call a *certainty equivalent*, because it is the amount of money we would take for certain in lieu of facing a risk.

It is a well-known fact that to prevent arbitrage, a forward price must equal the spot price compounded at the risk-free interest rate. This is the basic cost of carry pricing model for a forward contract,[6]

$$F_0 = S_0 e^{r_c}. \tag{25.5}$$

As such, we can restate Equation (25.5) as

$$S_0 = F_0 e^{-r_c}. \tag{25.6}$$

Substituting Equation (25.6) into Equation (25.4) gives

$$F_0 = E_0(S_1) - \lambda. \tag{25.7}$$

In other words, the forward price is the expected spot price minus a risk premium. The existence of a positive risk premium makes the forward price a biased estimate of the expected spot price, thereby invalidating the unbiased expectations hypothesis. Note that the bias is on the low side, assuming a positive risk premium. That is, the forward price will understate the expected spot price by the amount of the risk premium.

For further proof, let us suppose that the forward price is unbiased. That is, assume Equation (25.3) holds. By definition, the spot price is the expected future spot price discounted at a risky rate, which we shall denote as k. That is,

$$S_0 = E_0(S_1)e^{-k}. \tag{25.8}$$

If the unbiased expectations hypothesis were correct, then substitute Equation (25.3) into Equation (25.8), and we obtain the result that

$$S_0 = F_0 e^{-k}. \tag{25.9}$$

But we know that to prevent arbitrage, Equation (25.5) must hold. This implies that $k = r_c$. That is, the rate at which a risky future value must be discounted would be the risk-free rate. Moreover, this statement would be true for any asset. So, if the unbiased expectations hypothesis were true, the risky discount rate would be the same for every asset and equal to the risk-free rate. Except in the uninteresting cases of no risk or investors being risk neutral, this statement obviously cannot be true.

As we now turn to forward interest rates, we will need slightly more precise notation because forward interest rates have a current date, the time until the interest rate period starts, and the time until the interest rate period ends. To refresh your memory, a forward contract is established at one point in time and fixes the rate on a spot transaction that begins at another point in time, accrues interest, and ends at a later point in time. Further, we distinguish between the forward price (uppercase F) and the forward rates (lowercase f). We shall standardize things by using forward rates covering only one period but up to j periods ahead. Thus, $f(t, t + j)$ denotes the continuously compounded one-period forward rate observed at time t for a spot transaction that starts at time $t + j$, $r(t + j, t + j + 1)$ denotes the continuously compounded single-period spot rate over the period $t + j$ to $t + j + 1$, and $HPR(t, t + 1, j)$ denotes the continuously compounded holding period return over the period t to $t + 1$ for a j-maturity bond.

Within the context of interest rates, the unbiased expectations hypothesis states that forward rates are unbiased predictors of future spot rates. If UEH holds, then the forward rate is the best estimate of expected future spot rates. Based on our notation, we have

$$\text{UEH: } f(t, t + j) = E_t[r(t + j, t + j + 1)]; j = 1, \ldots, n, \tag{25.10}$$

where $f(t, t + j)$ denotes the continuously compounded one-period forward rate observed at time t for a transaction that starts at time $t + j$ and $r(t + j, t + j + 1)$ denotes the continuously compounded single period spot rate over the period $t + j$ to $t + j + 1$. See, for example, Lutz (1940), Culbertson (1957), Campbell and Shiller (1987), Engsted and Tanggaard (1994, 1995), and Lund (1993).

25.2 THE LOCAL EXPECTATIONS HYPOTHESIS

A similar sounding but more valid hypothesis for our purposes is the local expectations hypothesis (LEH). In its most general form, the LEH states that all bonds have the same expected holding period return over the next period. Based on our notation, we have

$$\text{LEH: } E_t[HPR(t, t + 1, j)] = r(t, t + 1); j = 1, \ldots, n, \tag{25.11}$$

where $HPR(t, t + 1, j)$ denotes the continuously compounded holding period return over the period t to $t + 1$ for a j-maturity bond and $r(t, t + 1)$ denotes the continuously compounded single-period spot rate over the period t to $t + 1$. See Cox, Ingersoll, and Ross (1981, 1985a, 1985b) and Ingersoll (1987), where they compare the different versions of the expectations hypothesis, arguing that they are different in continuous and discrete time.

The LEH originates from a bond market in which the term structure does not permit investors to trade bonds to create arbitrage profits. In fact, the LEH is equivalent to the absence of arbitrage in the bond market. More formally, the LEH states that in the absence of arbitrage opportunities, the forward price of a zero-coupon bond will equal the expected spot price if the expected spot price is taken under the equivalent martingale probabilities, which are the artificial probabilities whose existence is guaranteed if there are no arbitrage opportunities. The equivalence of the forward price and expected spot price is, however, true only for one-period-ahead forward prices. In other words, the forward price of an

n-period bond equals the expected spot price looking only one period ahead but no further. The expected returns, taken using the martingale probabilities, of any strategies involving any bonds of any maturity, are equivalent and equal to the one-period spot rate, that is, the shortest interest rate in the market.

The reason the LEH holds only for one period ahead is that a true risk-free rate exists only for the shortest holding period. One might think that a long-term zero-coupon bond rate is risk-free, but interest rates change during the life of the bond, causing its value to fluctuate. The only true risk-free rate is over the shortest holding period. When we study interest rates in a continuous time world, we shall see that this shortest holding period is an instant.

We shall demonstrate the LEH in a simple binomial world. The extension to a more complex continuous time model is feasible but a bit more difficult. We shall use continuously compounded rates, but the proof can be easily restructured for simple rates. For our notation, let $B(a, b)$ be the price at time a of a zero-coupon bond maturing at time b. Let $r_c(a, b)$ be the rate observed on a bond at time a that matures at time b. When $a = 0$, this is a spot bond and a spot rate. For example, consider a one-period bond, whose price is given as

Price at time 0 of bond maturing at time 1

$$B(0, 1) = e^{-r_c(0,1)}. \tag{25.12}$$

The value of this bond one period later is

Value at time 1 of bond maturing at time 1

$$B(1, 1) = 1. \tag{25.13}$$

Consider a two-period bond whose price is given as

Price at time 0 of bond maturing at time 2

$$B(0, 2) = e^{-2r_c(0,2)}. \tag{25.14}$$

Its value one period later is

Price at time 1 of bond maturing at time 2

$$B(1, 2) = e^{-r_c(1,2)}. \tag{25.15}$$

Now, let us introduce some uncertainty into the market. We assume that the one-period rate, $r_c(0, 1)$ can go up to $r_c(1, 2)^+$ or down to $r_c(1, 2)^-$. Thus, one period later, the original two-period bond becomes a one-period bond with a price of

$$B(1, 2)^+ = e^{-r_c(1,2)^+}$$

or

$$B(1, 2)^- = e^{-r_c(1,2)^-}. \tag{25.16}$$

That is, we are now introducing uncertainty into the model in the form of two possible one-period future interest rates. The market moves one step forward, which produces a new interest rate, and, therefore, a new bond price, which can be either of the two previous outcomes.

Now, let us consider how we could hold a bond for one period, without investing any money of our own. That is, we want a self-financing strategy. One way is that we could buy a one-period bond, financing it by selling an equivalent dollar amount of two-period bonds. When the one-period bond matures, we would pay off the two-period bonds. Alternatively, we could buy a two-period bond, financing it by selling an equivalent dollar amount of one-period bonds, and then sell the bond after one period.

For the first strategy, an arbitrage profit would be ensured if the rate of return $r_c(0, 1)$ on the bond purchased is guaranteed to exceed the cost of financing it. Remember that the bond is financed by borrowing a two-period bond and buying it back one period later. This means that the financing strategy has an uncertain cost. That cost will be determined based on what the new one-period interest rate is. But if it were the case that the cost, in the worst-case scenario, is still less than the rate of return on the bond we buy, then we have earned an arbitrage profit. The worst-case scenario would be for the one-period rate to go down. This would drive up the price of the two-period bond that we sold to finance the position. Thus, the two-period bond, which then becomes a one-period bond, has a new price of $B(1, 2)^-$. The cost of the transaction expressed as one plus the percentage cost would be

$$\frac{B(1, 2)^-}{B(0, 2)}. \tag{25.17}$$

Hence, an arbitrage profit would be earned if

$$e^{r_c(0,1)} > \frac{B(1, 2)^-}{B(0, 2)}. \tag{25.18}$$

The left-hand side (LHS) is one plus the return from buying the one-period bond. The right-hand side (RHS) is the highest possible cost of financing the bond.

In the second strategy, we are buying a two-period bond, financing it with a one-period bond. Therefore, the financing cost is $r_c(0, 1)$ for sure, but the return on the bond purchased is unknown. The worst-case scenario is for the one-period rate to go up, driving down the price of the two-period bond to its worst value, $B(1, 2)^+$. Therefore, one plus the worst one-period return on a two-period bond would be

$$\frac{B(1, 2)^+}{B(0, 2)}. \tag{25.19}$$

Therefore, an arbitrage profit would occur if

$$e^{r_c(0,1)} < \frac{B(1, 2)^+}{B(0, 2)}. \tag{25.20}$$

Thus, to prevent arbitrage, neither of these conditions must occur. Therefore, we must have

$$\frac{B(1, 2)^+}{B(0, 2)} < e^{r_c(0,1)} < \frac{B(1, 2)^-}{B(0, 2)}. \tag{25.21}$$

This statement implies that there is a set of weights, which we shall call ϕ and $1 - \phi$, such that

$$\phi\frac{B(1,2)^+}{B(0,2)} + (1 - \phi)\frac{B(1,2)^-}{B(0,2)} = e^{r_c(0,1)}. \tag{25.22}$$

The LHS can be viewed as a type of expected return on a two-period bond held for one period, which is then set equal to the RHS, which is the one-period rate. To interpret the LHS as an expected value, however, we must recognize these weights as risk neutral or equivalent martingale weights. We call them *risk neutral* because the expression is the expected return that a risk-neutral investor would have. These are, therefore, the probabilities that would be used if current market prices held and investors were risk neutral. They are also called *equivalent martingale probabilities* because they convert the discounted bond price into a martingale, which is a stochastic process with zero expected return. You should recall that we studied this concept extensively when developing the procedures for pricing options.

Of course, a martingale is a process with an expected return of zero. Equation (25.22) does not on the surface appear to be a martingale, because it shows that the expected return is the risk-free rate. Because the risk-free rate is known and constant, it is a simple matter to convert (25.22) to a martingale by dividing by $e^{r_c(0,1)}$, which is the same thing as multiplying by $B(0,1)$. Hence, any process that has an expected return equal to the risk-free rate can be viewed as a martingale.

To generalize this result, we could let the two-period bond be a bond of any maturity, say n. Then repeat the previous exercise, and we would easily see that the same result is obtained. A more formal proof can demonstrate that there is no bond trading strategy that permits arbitrage.

The result that the expected return, under the martingale probability, equals the one-period rate does not, however, apply to a strategy that spans more than one period. As noted, the reason for this result is that arbitrage is a risk-free strategy and there is no true risk-free rate beyond one period. Although buying a zero-coupon bond of maturity $n > 1$ and holding it to maturity indeed leads to a risk-free return, arbitrage strategies require a rebalancing of the positions to properly offset the changing risk of the component instruments. This rebalancing requires trading in intermediate periods.

We have now demonstrated that the expected return, using the martingale probabilities, equals the one-period rate. Next we demonstrate the other implication of the LEH, that the expected price of the two-period bond at time 1 is the forward rate. Let us express Equation (25.22) as follows:

$$\phi\frac{B(1,2)^+}{B(0,2)} + (1 - \phi)\frac{B(1,2)^-}{B(0,2)} = \frac{1}{B(0,1)}, \tag{25.23}$$

noting that $1/B(0,1)$ is $e^{r_c(0,1)}$. Multiplying both sides by $B(0,2)$, we obtain

$$\phi B(1,2)^+ + (1 - \phi)B(1,2)^- = \frac{B(0,2)}{B(0,1)}. \tag{25.24}$$

By definition, the RHS is the forward price. The LHS is the expected spot price under the martingale probabilities. So, the forward price is the expected spot price under the martingale probabilities. In addition, this result generalizes to bonds of any maturity, but again, it applies over the next period, not beyond that horizon.

Finally, let us show that the equivalence of the forward price to the expected spot price does not imply equivalence of the forward rate to the expected spot rate. The RHS of Equation (25.24) can be written as follows:

$$\frac{B(0,2)}{B(0,1)} = \frac{e^{-r_c(0,2)}}{e^{-r_c(0,1)}} = e^{-[r_c(0,2)-r_c(0,1)]}. \tag{25.25}$$

Suppose we take the log of the RHS of Equation (25.24). So we must also take the log of the LHS, but it does not reduce to $\phi r_c(1,2)^+ + (1-\phi)r_c(1,2)^-$, the expected spot rate. It is simply $\ln[\phi B(1,2)^+ + (1-\phi)B(1,2)^-]$. Although this value will be close to the expected spot rate, it is not equivalent.

25.3 THE DIFFERENCE BETWEEN THE LOCAL AND UNBIASED EXPECTATIONS HYPOTHESES[7]

In this section, we will review some of the more technical aspects of term structure math as well as demonstrate that the LEH and UEH are equivalent in discrete time with continuously compounded interest rates. Further, we make no assumption here related to a risk-neutral framework. Our focus is on the actual probability measure. Although the LEH and UEH are not equivalent in discrete time with discretely compounded interest rates, the magnitude of the difference is indistinguishable when applied to empirical data. For our purposes, all bonds are assumed to be default free. Due to the technical nature of this discussion, we adopt more precise notation for this section.

We denote $r(t, t+j)$ as the continuously compounded spot rate observed at time t and maturing at time $t+j$, the j period spot rate. We suppress the subscript c for continuous compounding in an effort to simplify. We denote $f(t, t+j)$ as the forward rate observed at time t and for maturity $t+j$, a one-period forward rate. Thus, the spot rate spans time t to $t+j$, whereas the forward rate spans only $t+j-1$ to $t+j$. Based on this notation, the price at time t of an n-period zero-coupon bond with par value of one dollar at time $t+n$ is

$$P(t, t+n) = e^{-[r(t,t+1)+f(t,t+1)+\ldots+f(t,t+n-1)]}. \tag{25.26}$$

The continuously compounded holding period return can be expressed as

$$HPR(t, t+1, n) = \ln\left[\frac{P(t+1, t+n)}{P(t, t+n)}\right]. \tag{25.27}$$

The UEH states that

$$f(t, t+j) = E_t[r(t+j, t+j+1)]; j = 1, \ldots, n. \tag{25.28}$$

The LEH states that

$$E_t[HPR(t, t+1, j)] = r(t, t+1); j = 1, \ldots, n. \tag{25.29}$$

Substituting Equation (25.26) into Equation (25.27), we note

$$HPR(t, t+1, n) = r(t, t+1) + [f(t, t+1) - r(t+1, t+2)] + [f(t, t+2) - r(t+1, t+3)]$$

$$\ldots + [f(t, t+n-1) - r(t+1, t+n)]. \tag{25.30}$$

Based on the definition of continuously compounded holding period returns, the LEH can be expressed as

$$r(t, t+1) = E_t\{\ln[P(t+1, t+j)]\} - \ln[P(t, t+j)]; j = 2, \ldots, n. \tag{25.31}$$

Again, substituting for the price from Equation (25.26) and rearranging, we have

$$0 = \sum_{i=1}^{j} \{f(t, t+i) - E_t[r(t+1, t+i)]\}; j = 2, \ldots, n. \tag{25.32}$$

Thus, a new interpretation of LEH is related to current expectations of future spot rates over the next time period, t to $t+1$. Note that this is not a result of arbitrage forces leading one to a risk-neutral solution. UEH, however, addresses forecast accuracy over longer horizons. Note that if LEH holds at time t for two- through n-period bonds, we have

$$f(t, t+j) = E_t[r(t+j, t+j+1)]; j = 1, \ldots, n. \tag{25.33}$$

Therefore, if LEH holds for all future time periods, then UEH also holds.

This analysis breaks down with discretely compounded interest rates. Brooks and Livingston (1992) rigorously document that the difference is quite small. Based on empirical evidence they examine, the empirical difference between UEH and LEH is typically less than a basis point when using monthly rates.

In summary, with continuous compounding, if LEH holds for all future time periods, then UEH also holds. This is not true with discrete compounding. Thus, any empirical differences found when testing these two hypotheses relates solely to compounding issues. For completeness, we briefly review other hypotheses of the term structure of interest rates. Remember interest rates are at the very heart of almost all financial analysis, hence, a thorough understanding of the different explanations of interest rates of varying maturities is important.

25.4 OTHER TERM STRUCTURE OF INTEREST RATE HYPOTHESES

In this section, we review several term structure of interest rate hypotheses, excluding UEH and LEH because they have been covered already. We also provide several foundational research papers on which these hypotheses were created and evaluated.

25.4.1 Return to Maturity Expectations Hypothesis (RMEH)

The RMEH states that the holding period return from buying a long-term bond and holding it until maturity is equal to the expected return from buying one-period bonds and rolling them into the next one-period bond for the same time period as the long-term bond. Based on our notation, we have

$$\text{RMEH: } r(t, t+j) = \sum_{i=1}^{j} E_t[r(t+i-1, t+i)]; j = 1, \ldots, n. \tag{25.34}$$

See Cox, Ingersoll, and Ross (1981, 1985a, b). One variation of RMEH is the yield to maturity expectations hypothesis (YMEH). The YMEH states that the yield to maturity from buying a long-term bond and holding it until maturity is equal to the expected return from buying one-period bonds and rolling them into the next one-period bond for the same time period as the long-term bond. Note this hypothesis is just a restatement of the RMEH. See Cox, Ingersoll, and Ross (1981, 1985a, b).

25.4.2 Term Premium Hypothesis (TPH)

The TPH states that forward rates are biased predictors of future spot rates. Specifically, forward rates are biased by a known amount defined as a term premium (tp). The term premium might be positive or negative, might vary widely with maturity, and might vary across time. The word *term* is generic and might indicate a wide variety of issues depending on context. The word *term*, however, is always indicative of time to maturity. Thus, this generic TPH encapsulates all the premium hypotheses stated here. Based on our notation, we have

$$\text{TPH: } f(t, t+j) = E_t[r(t+j, t+j+1)] + tp(t, t+j); j = 1, \ldots, n. \tag{25.35}$$

See Kane (1980). The liquidity premium hypothesis (LPH) is a variation of the TPH. The LPH states that forward rates are biased predictors of future spot rates and are biased high by a known amount defined as a liquidity premium (*lp*). The idea is that investors deem shorter maturity bonds as being more liquid; hence, they demand a higher yield for longer maturity bonds. From this perspective, the liquidity premium must be positive and must be nondecreasing with long maturities. It might vary across time. Based on our notation, we have

$$\text{LPH: } f(t, t+j) = E_t[r(t+j, t+j+1)] + lp(t, t+j); j = 1, \ldots, n. \tag{25.36}$$

See Hicks (1946) and Kane (1980).

25.4.3 Money Substitute Hypothesis (MSH)

The MSH states that forward rates are biased predictors of future spot rates by a known amount defined as a money premium (*mp*). The money premium is based solely on the very short-term bonds acting as a pure substitute for money; hence, demand is high and yields and holding period returns are low. The money premium must be positive and must be nondecreasing with long maturities. It might vary across time. Based on our notation, we have

$$\text{MSH: } f(t, t+j) = E_t[r(t+j, t+j+1)] + mp(t, t+j); j = 1, \ldots, n. \tag{25.37}$$

See Thornton (1802), Mill (1923), Keynes (1935), Timberlake (1964), and Kessel (1965).

25.4.4 Segmented Markets Hypothesis (SMH)

The SMH states that forward rates are biased predictors of future spot rates based on maturity sectors representing distinct markets with their own supply and demand forces. These forces result in known allowance amounts and are defined here as a segment premium (sp). The SMH requires knowledge of where these maturity segments are located before being able to assign the appropriate segment premiums. The segment premium might be positive or negative, might vary widely across maturities, and might vary across time. Based on our notation, we have

$$\text{SMH: } f(t, t+j) = E_t[r(t+j, t+j+1)] + sp(t, t+j); j = 1, \ldots, n. \tag{25.38}$$

See Culbertson (1957) and Sundaresan (2002). Sundaresan suggests SMH driven by regulatory constraints.

The preferred habitat hypothesis (PHH) is a variation of the MSH. The core idea is that investors have preferred maturities of bonds they trade. For example, insurance companies might prefer very long-dated bonds whereas banks prefer bonds with less than a few years to maturity. These maturity ranges are their preferred habitats. For a sufficient yield differential these preferences might be overcome. The PHH states that forward rates are biased predictors of future spot rates. Specifically, forward rates are biased based on habitat displacement allowances. This known allowance amount is defined as a habitat premium (hp). The PHH requires knowledge of borrower and lender habitats before being able to assign the appropriate habitat premiums. The habitat premium might be positive or negative, might vary widely across maturities, and might vary across time. Based on our notation, we have

$$\text{PHH: } f(t, t+j) = E_t[r(t+j, t+j+1)] + hp(t, t+j); j = 1, \ldots, n. \tag{25.39}$$

See Modigliani and Sutch (1966). Cox, Ingersoll, and Ross (1981) suggest LPH is a special case of PHH where the preferred habitat is short-term bonds.

25.4.5 Risk Premium Hypothesis (RPH)

The RPH states that forward rates are biased predictors of future spot rates based on the market price of risk bearing and the systematic risk of each bond. Every bond, except the single-period bond, contains some systematic risk and its market price of risk might vary. This known compensation for risk is defined as a risk premium (rp). The RPH requires the ability to measure systematic risk as well as knowledge of the market price of risk-bearing services for various amounts of systematic risk before being able to assign the appropriate risk premiums. The risk premium must be positive, might vary widely across maturities, and might vary across time. Based on our notation, we have

$$\text{RPH: } f(t, t+j) = E_t[r(t+j, t+j) + 1] + rp(t, t+j); j = 1, \ldots, n. \tag{25.40}$$

See Modigliani and Sutch (1966).

25.5   RECAP AND PREVIEW

In this chapter, we introduced the local expectations hypothesis and distinguished it from the well-known and widely covered unbiased expectations hypothesis. The latter states that the forward rate in the term structure represents the expected future spot rate. The local expectations hypothesis says something similar, but not precisely the same. It says that when the term structure is arbitrage free, the forward rate is the expected return from any strategy over the shortest holding period, and this rate is the shortest period risk-free rate.

We further explored when the unbiased expectations hypothesis and local expectations hypothesis are different and when they are the same. Further, we briefly reviewed numerous other term structure theories that have been offered over the years.

In Chapter 26, we introduce interest rate derivatives, specifically FRAs, swaps, and interest rate options.

QUESTIONS AND PROBLEMS

1 In some countries, it is illegal to provide financing for arbitrage activities. Explain how the introduction of legal arbitrage activities would be expected to change forward prices. Assume that the hedgers and speculators are roughly balanced between buyers and sellers.

2 Explain the concept of a certainty equivalent and how it is applied in forward pricing models.

3 Compare and contrast the unbiased expectations hypothesis and the local expectations hypothesis within the context of interest rates.

4 Under what conditions are the unbiased expectations hypothesis and the local expectations hypothesis identical?

5 Other than the unbiased expectations hypothesis and the local expectations hypothesis, identify and define three other hypotheses related to the term structure of interest rates.

NOTES

1. We use rates and yields interchangeably as often occurs within the literature. Where necessary, we will distinguish between specific measures, such as yield to maturity and holding period return.
2. An excellent illustration of the US government bond term structure and how it has evolved over time is at https://stockcharts.com/freecharts/yieldcurve.php.
3. The unbiased expectations hypothesis is sometimes called the *expectations hypothesis*, but this is a tremendous misnomer that will become apparent in this chapter, whereby we cover another expectations hypothesis, the *local expectations hypothesis*. The unbiased expectations hypothesis is also sometimes called the *pure expectations hypothesis*, although it is unclear what is so pure about it.
4. For references with more details on this subject, see Longstaff (2000), Cox, Ingersoll, and Ross (1981), and Chance and Rich (2001).

5. For ease of exposition, we simplify the forward price from $F_0(T)$ to simply F_0.
6. If there were any costs of holding the asset, these costs would be added. If the asset paid out any cash, these amounts would be deducted. It is easy to prove Equation (25.5). Assume that an investor buys the asset and sells a forward contract at the price F_0, which obligates the investor to deliver the asset at time 1 and receive F_0. This transaction is risk free, because it guarantees a return per dollar invested of F_0/S_0. Being risk free, this return must equal the risk-free rate. Or, as Equation (25.5) is stated, the spot price must compound to the forward price at the risk-free rate.
7. This section is based on Brooks and Livingston (1992).

CHAPTER 26

Interest Rate Contracts: Forward Rate Agreements, Swaps, and Options

In Chapter 22, we introduced forward and futures contracts. In those cases, however, the forward and futures were on assets. One of the most popular types of derivative contracts is based on an interest rate, and an interest rate is not an asset. An asset is something that can be purchased and held in a portfolio.[1] One cannot purchase and hold an interest rate. One can, however, purchase and hold an instrument that pays an interest rate, but the difference must be accounted for when pricing a derivative off of an interest rate. In this chapter, we shall take a look at interest rate forward contracts, called *forward rate agreements*, or FRAs, interest rate swaps, and interest rate options. We shall see how we account for this important difference in which there is a payoff based on an underlying but the underlying cannot be held as an asset. For more details, there are many excellent references on this subject, among which are Buetow and Fabozzi (2001) and Corb (2012).

Before we begin looking at these instruments, however, let us set up the structure of the problem and identify the key variables. To start, we must understand how loans are done. There are a variety of ways of creating loans. Some are fixed rate, in which the interest rate is set at the start. Others are variable or floating, usually called *floating rate*, where the interest rate is set and then resets periodically. To see how this is done, let us divide the loan into interest payment periods, sometimes called *settlement periods*. For example, let us say that each period is one month. So, at the start of the loan, set the rate at the current one-month rate in the market. The interest then accrues for a month, at which time the interest is paid, and the rate resets to the then-current one-month rate. The loan continues in that manner until the final interest payment and principal are made.

These types of loans use what is called *add-on interest*. Let B_0 denote the amount borrowed. Letting L be the loan rate expressed in decimal terms, q be the number of days of the loan, and B_q denote the amount required to be repaid, the payback at the maturity of the loan is

$$B_q = B_0 \left[1 + L \left(\frac{q}{360} \right) \right]. \tag{26.1}$$

Note that the convention in this type of market is to divide the interest rate by 360. This is simply a convenience arising from an era in which calculators did not exist and loans commonly used rates such as 6% and the interest payment periods would be 30, 60, 180 days, and so on, in which case the interest could be calculated in one's head.

The rate on this type of loan is typically tied to a commonly used rate benchmark. Historically, the most popular one was LIBOR, which stood for the London interbank offered rate, the rate at which banks borrow and lend dollars and sometimes other currencies primarily in London. There was significant controversy over the accuracy of LIBOR during some historical periods due to evidence that the official published rate had been manipulated by unscrupulous traders in rate-setting banks. At the time of this writing, there is much debate over the appropriate replacement for LIBOR. Note that any alternative would still likely be used in the same manner discussed next. The lead candidate is the secured overnight financing rate and related term reference rates. LIBOR data ceased being produced on June 30, 2023.

26.1   INTEREST RATE FORWARDS

As noted, we referred to interest rate forwards as forward rate agreements, or FRAs. They are forward contracts that pay off based on what the interest rate is on the expiration date of the contract. To understand them, we shall need to add some additional specifications. There are two forms of FRAs: advanced set, settled in arrears, and advanced set, advanced settled. Advanced set, settled in arrears, assumes that the FRA cash settlement is made at the end of the embedded interest rate period (q in Figure 26.1), whereas advanced set, advanced settled, assumes that the FRA cash settlement is made at the beginning of the embedded interest rate period. When first learning the mechanics of FRAs, it is easier to focus on advanced set, settled in arrears. For credit risk reasons, FRAs are typically advanced set, advanced settled. Recall FRAs are unlike loans so there is no repayment of principal. Thus, once the floating rate is set, the settlement amount can be computed. For example, an advanced set, settled in arrears, that mimics a typical loan may require settlement in three months of say 2,000,000 (e.g., receive floating FRA and rates rose). The counterparty due the 2,000,000 would prefer to receive the present value of the settlement amount now so as not to incur the risk that the counterparty owing the money defaults in three months (e.g., due to a credit crisis).

Let us assume that the forward contract expires in m days. On day m, the payoff of the forward contract will be based on the q-day interest rate. Figure 26.1 illustrates our notation. Rather than maturity time being measured in years, with FRAs, it is typically measured in days. Note here T is simply $m + q$. In the next section, we will value FRAs at some future time t (prior to expiration).

For example, suppose the underlying is a 90-day loan rate. Then $q = 90$. If the forward contract expires in 30 days, then $m = 30$. We shall need notation for the rate on a 90-day loan on day 30 relative to the initiation date of the contract, day 0. This would be $L_{30}(90)$. In general, we shall write this rate as $L_m(q)$, the q-day rate on day m. Thus, $T = m + q = 30 + 90 = 120$. In addition, the figure shows a date t, referred to as an evaluation date. We will later value the FRA at that point in time. For now, we focus on the initiation date, 0.

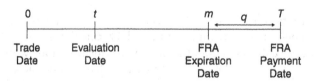

FIGURE 26.1   Interest Rate FRA Maturity Time Notation

We assume that on day 0, an FRA maturing on day m is created. We assume that the rate that the parties agree will be paid by the long to the short on day m is $R_0(m, q)$. This rate can be thought of as the forward price, in much the same manner that we think of a forward price in a standard forward contract on an asset, as described in Chapter 22.

On day m, the payoff to the long is the value of the contract at expiration, and is given as follows:

$$VFRA_m(m, q) = \frac{[L_m(q) - R_0(m, q)]\left(\frac{q}{360}\right)}{1 + L_m(q)\left(\frac{q}{360}\right)}, \tag{26.2}$$

where we will use the term $VFRA_m(m, q)$ as the value of an FRA expiring on day m where the underlying is a q-day loan rate. This amount is paid from the short to the long and represents the payment per 1.0 notional.[2] First, note the numerator, which is partly composed of the difference between the rate realized on day m, $L_m(q)$, and the rate the long agreed to pay, $R_0(m, q)$. This payment can be positive, in which case the short pays the long, or negative, in which case the long pays the short. These rates, however, are quoted as annual rates so they must be adjusted to reflect the fact that they apply to a q-day transaction in precisely the same manner that the rate on the loan is adjusted for the number of days, as in Equation (26.1). That is, we multiply by $q/360$.

Now, take a look at the denominator. It might be surprising that there even is a denominator. Notice that the denominator results in discounting the payoff over q days. And yet, the payment occurs on day m. Why do we do this discounting? The answer lies in the fact that the rate on which the payment is made, $L_m(q)$, is the q-day interest rate on day m as determined in the money market. It is the rate that a bank would expect to pay or receive if it made a loan on day m and the loan were paid back on day $m + q$. Because the FRA pays off on day m, the buyer is receiving the money before it technically should. Hence, it is appropriate to discount it. Note that if the FRA buyer were to hold on to the money and accrue interest at the rate $L_m(q)$ for q days, they would have the amount $L_m(q)$. Thus, it does not really matter if the FRA pays off on day m or day $m + q$. If it paid off on day $m + q$, it would pay $L_m(q)$. By paying off on day m, it pays $L_m(q)/(1 + L_m(q))$. These are effectively the same amounts of money.

Finally, let us again note that in Equation (26.2), the payoff amount can be positive or negative. So, if $R_0(m, q) > (<) L_m(q)$, the payoff will be negative (positive), and the buyer (seller) must pay the seller (buyer).

So, we know how the payoff of an FRA is determined. In keeping with our previous coverage of forward contracts, there are two things we need to derive: the forward price, which is the rate $R_0(m, q)$ that is established at the start, and the value of the FRA.

26.1.1 Pricing FRAs

In order to determine the fixed rate, or price, there is an important concept in valuation that we must know. First, let us establish the notation for valuation. Let $VFRA_t(m, q)$ be the value at time t of an FRA that expires at time m in which the underlying is a q-day rate. It should not be surprising that when $t = 0$, the date of initiation of the contract, the value is zero. This result is consistent with the fact that no money changes hands when the contract is initiated. What we want to know is how to set the price of the contract at time 0. In this context, price is the fixed rate we agree to pay in exchange for receiving the floating rate. Let us first establish that the payoff at m, as given in Equation (26.2), is equivalent to the following value at $m + q$:

$$[L_m(q) - R_0(m, q)]\left(\frac{q}{360}\right),$$

as we simply compounded the payoff at m to the point $m + q$ at the rate $L_m(q)$.[3] Thus, the value of an FRA is the present value of this expression. Now, let us do one simple trick that will facilitate our derivation: We add and subtract a 1 in the above equation:

$$\left[1 + L_m(q)\left(\frac{q}{360}\right)\right] - \left[1 + R_0(m, q)\left(\frac{q}{360}\right)\right].$$

That is, we add 1 to the floating payment and fixed payments and thus they offset. We can easily find the value of the fixed component, $1 + R_0(m, q)(q/360)$, by discounting it to the present at the $m + q$-day rate.

$$\frac{1 + R_0(m, q)\left(\frac{q}{360}\right)}{1 + L_0(m + q)\left(\frac{m+q}{360}\right)}.$$

Now we wish to value the floating component, $1 + L_m(q)(q/360)$. Remember that this value is positioned at time $m + q$ and we want to discount it back to time m. So we discount for q days at the rate $L_m(q)$. Its value at time m is easily found as

$$\frac{1 + L_m(q)\left(\frac{q}{360}\right)}{1 + L_m(q)\left(\frac{q}{360}\right)} = 1.$$

So the discounted value back to time m is 1. Therefore, the discounted value back to time 0 is

$$\frac{1}{1 + L_0(m)\left(\frac{m}{360}\right)}.$$

So now discounting the fixed and floating payments as indicated previously, the value of the FRA at time 0 is

$$VFRA_0(m, q) = \frac{1}{1 + L_0(m)\left(\frac{m}{360}\right)} - \frac{1 + R_0(m, q)\left(\frac{q}{360}\right)}{1 + L_0(m + q)\left(\frac{m+q}{360}\right)}.$$

Because the FRA must have a value of zero at time 0, we set this equation to zero,

$$VFRA_0(m, q) = \frac{1}{1 + L_0(m)\left(\frac{m}{360}\right)} - \frac{1 + R_0(m, q)\left(\frac{q}{360}\right)}{1 + L_0(m + q)\left(\frac{m+q}{360}\right)} = 0.$$

Solving for the fixed rate gives us our desired result,

$$R_0(m,q) = \left[\frac{1 + L_0(m+q)\left(\frac{m+q}{360}\right)}{1 + L_0(m)\left(\frac{m}{360}\right)} - 1\right]\frac{360}{q}. \tag{26.3}$$

This result is simply the forward rate for a q-day transaction to start at time m. It is the rate that the long agrees to pay the short on day m in return for receiving the floating rate. This rate guarantees that neither party can earn an arbitrage profit against the other.

For example, suppose the 270-day loan rate is 3% and the 360-day loan rate is 4%. Calculate the fixed rate on an FRA expiring in 270 days for a 90-day deposit. In this case, the FRA fixed rate is found by solving

$$\begin{aligned}R_0(m,q) &= \left[\frac{1 + L_0(m+q)\left(\frac{m+q}{360}\right)}{1 + L_0(m)\left(\frac{m}{360}\right)} - 1\right]\frac{360}{q}\\ &= \left[\frac{1 + 0.04\left(\frac{360}{360}\right)}{1 + 0.03\left(\frac{270}{360}\right)} - 1\right]\frac{360}{90} = \left(\frac{1+0.04}{1+0.0225} - 1\right)4 = 0.06846.\end{aligned}$$

Thus, the fixed rate is 6.846%.

There is a useful approximation when estimating forward rates that market participants find helpful, especially when making decisions in chaotic markets. Suppose you view the longer rate as simply a weighted arithmetic average of the shorter rate with weights based on time to maturity. In this case, we have 3% for three 90-day periods and 4% for four 90-day periods. Thus, to achieve a 4% average the missing fourth number would be 7 or $3 + 3 + 3 + \_\_\_ = 4(4)$. Note that 7% is higher than the correct answer of 6.846%. Thus, the arithmetic average is a useful starting point although slightly high. With today's computer technologies, this approximation is not as useful.

26.1.2 Valuing FRAs

From Equation (26.2) we know the value of the contract at expiration, and from Equation (26.3) we know the fixed loan rate so the value of the contract at initiation is zero. What we need to know is the value of the contract during its life. So, let us position ourselves at an arbitrary point, time t, which is between time 0 and time m. In order to value the contract, we need to create a transaction that essentially sells the contract. In other words, we need to create an offsetting contract. Keep in mind that we do not actually have to execute this new contract. We just need to be able to execute it and to know at what price it would be executed.[4]

So here at time t, we look into the market and determine that a new forward contract on the q-day day rate expiring at m has a rate of $R_t(m,q)$. If we held a long position in the original contract, we would need to go short this contract. The payoffs of the two contracts would yield an overall payoff at time m of

$$\left[\frac{L_m(q) - R_0(m,q)\left(\frac{q}{360}\right)}{1 + L_m(q)\left(\frac{q}{360}\right)}\right] - \left[\frac{L_m(q) - R_t(m,q)\left(\frac{q}{360}\right)}{1 + L_m(q)\left(\frac{q}{360}\right)}\right].$$

The first large bracketed term is the payoff from the contract we are long, the one we actually own. The second large bracketed term is the payoff from the contract we hypothetically sell. Let us rearrange this equation as follows:

$$\frac{[R_t(m,q) - R_0(m,q)]\left(\frac{q}{360}\right)}{1 + L_m(q)\left(\frac{q}{360}\right)}.$$

And this is the guaranteed payoff at time T discounted to time m. Note that the rate at time m, $L_m(q)$, is not in the numerator. We have resolved that uncertainty by hypothetically selling the new contract. Hence, the uncertainty is gone. We can thus value the FRA by discounting this amount from day m back to day t. What we see in this equation is that the amount is discounted according to the standard payoff formula for an FRA, Equation (26.2). Recall how the denominator discounts the payoff that occurs at time m but is based on a rate in which the interest is paid at time $m+q$. As we did previously, we can simply compound that payoff forward to date $m+q$ by multiplying by $1 + L_m(q)(q/360)$ to remove the denominator from the equation. We now simply discount this amount from time $m+q$ to the present, time t by the rate at time t for the period $m+q-t$ as follows:

$$VFRA_t(m,q) = \frac{[R_t(m,q) - R_0(m,q)]\left(\frac{q}{360}\right)}{1 + L_t(m+q-t)\left(\frac{m+q-t}{360}\right)}, \tag{26.4}$$

where

$$R_t(m,q) = \left[\frac{1 + L_0(m+q-t)\left(\frac{m+q-t}{360}\right)}{1 + L_0(m-t)\left(\frac{m-t}{360}\right)} - 1\right]\frac{360}{q}. \tag{26.5}$$

And this is the value of the FRA at time t. In other words, the contract we created at time zero that obligates us to pay $R_0(m,q)$ at time m is worth the above at time t. Based on the results of the previous example, suppose after 90 days, the 180-day loan rate is 2% and the 270 day rate is 3%. Calculate the value of the receive-floating FRA entered at time 0. Assume 1,000,000 notional. We first compute the new equilibrium FRA fixed rate as

$$R_{90}(270,90) = \left[\frac{1 + L_0(m+q-t)\left(\frac{m+q-t}{360}\right)}{1 + L_0(m-t)\left(\frac{m-t}{360}\right)} - 1\right]\frac{360}{q}$$

$$= \left[\frac{1 + 0.03\left(\frac{270}{360}\right)}{1 + 0.02\left(\frac{180}{360}\right)} - 1\right]\frac{360}{90} = 0.049505.$$

Now we compute the current value from the receive-floating counterparty (paying initial fixed rate and receiving the later fixed rate) as

$$VFRA_{90}(270, 90) = \$1,000,000 \left[\frac{(0.049505 - 0.06846)\left(\frac{90}{360}\right)}{1 + 0.03\left(\frac{270}{360}\right)} \right] = -\$4,634.40.$$

In this case, the fixed FRA rate fell from 6.846% to 4.9505%. In this particular case, the receive-floating counterparty is now in the hole 4,634.40.

26.1.3   Off-Market FRAs

As we discussed with forward contracts in general in Chapter 22, it is possible to create an FRA that is not set at a zero value at the start. If the fixed rate is set according to Equation (26.3), then the contract has zero value at the start and neither party pays anything to the other. If we set the fixed rate higher than that given in Equation (26.3), the buyer has agreed to pay an above-market fixed rate. Thus, the contract would have a negative value at the start, and the seller would have to pay the buyer the market value at the start. If we set the fixed rate lower than Equation (26.3), the buyer has agreed to pay a below-market fixed rate. Thus, the contract would have a positive value at the start, and the buyer would have to pay the seller the market value at the start. These versions of FRAs are called *off-market* FRAs, and although they are not common, they are implicitly embedded in interest rate swaps, as we shall see.

26.2   INTEREST RATE SWAPS

Interest rate swaps are a variation of FRAs, but they have some important distinctions. An interest rate swap is an agreement for one party to pay another a set of payments at one rate and the latter to pay the former a set of payments at another rate. Typically, one of these rates is fixed and one is variable and, thereby, floats according to a benchmark rate in the market. This standard type of swap is usually called a *vanilla swap* or sometimes *plain vanilla swap*. It is also possible to have both rates be floating, but the two rates would clearly have to be different floating rates.

Interest rate swaps are the most widely used derivative.[5] The most common application is when a company has a floating-rate loan but would prefer a fixed-rate loan. Because lenders very commonly make floating-rate loans, but borrowers prefer fixed-rate loans, lenders will usually offer a swap to enable the borrower to convert the loan from floating to fixed. That is, with the loan the borrower will be making a series of interest payments to the lender at rates that adjust periodically according to a market benchmark rate. A swap can then be attached so that the borrower receives an interest payment that is floating and makes an interest payment that is fixed. The floating interest payment the borrower makes on the loan is offset by the floating interest payment it receives on the swap, leaving a net of a fixed interest payment.

26.2.1   Pricing Interest Rate Swaps

Let us set up the problem of pricing the swap. Our objective is to determine the fixed rate the swap buyer would agree to pay in return for receiving the floating rate. We shall assume that the swap makes payments on the following dates, $j = 1, 2, \ldots, J$. Thus, at maturity, it has made the final and J^{th} payment. The number of days to each payment is designated as T_j. Thus, for example, a swap that makes payments in 180, 360, 540, and 720 days will have $J = 4$, and $j = 1, 2, 3, 4$, with the number of days to each payment being $T_1 = 180$, $T_2 = 360$, $T_3 = 540$, and $T_4 = 720$. We refer to these dates as settlement dates, reset dates, or rate-reset dates.

We shall assume that the underlying is the rate $L_t(q)$, meaning the rate on day t for a q-day loan. In the example here with four payments spaced 180 days apart, $q = 180$. This is the same type of underlying that we used to illustrate FRAs. One party agrees to pay a fixed rate of $RS_0(J, q)$, which means the fixed rate on a swap set at time 0 in which there are J payments and each payment is based on the q-day rate. The standard terms of vanilla swaps call for the rate, $L_j(q)$, to be set at the beginning of the settlement period, and the payment to be made at the end. This structure is different from the way FRAs are paid, but it is the way loan interest payments are made. With floating-rate loans, the rate is set at the beginning of the settlement period, the interest accrues through the period, and the interest is paid at the end of the period, at which time the rate is reset. The payoff determined on day j is, thus,[6]

$$[L_j(q) - RS_0(J, q)] \left(\frac{q}{360}\right), \tag{26.6}$$

and this amount is paid at date $j + 1$, meaning q days later. It should be apparent that the first floating rate on the first swap payment, $j = 1$, is known when the swap is initiated. It is known as the *stub rate* or the current q-day rate, $L_0(q)$. So, the first swap floating payment, which will be made at $j = 1$, is known for certain. All remaining payments from $j = 2$ onward are unknown, because they will be determined by the q-day rates on $j = 1$, $j = 2, \ldots, j = J - 1$. Despite this uncertainty, we shall be able to value these payments, as we show here.

First, let us make a slight alteration to the swap payments that has no net economic effect but serves a useful purpose in facilitating valuation. Usually in an interest rate swap, the parties exchange only the interest rates based on the notional. The notionals themselves are not exchanged. For example, if party A agrees to pay party B the swap fixed rate, and party B agrees to pay party A the floating rate, with both based on the same 50 million notional, there is no exchange of 50 million between the parties, because this would serve no purpose. That is, there would be no reason for A to pay 50 million to B at the start and for B to pay 50 million to A and for the two parties to again exchange 50 million at expiration. The parties simply exchange, that is, "swap" interest payments. But no harm will be done if, for valuation purposes, we simply add a notional to both the fixed and floating sides. To keep things simple, we shall set this notional at 1.0. This amount will theoretically be exchanged at the start and at the end, which is why it is called the notional, meaning to exist only in theory, not in reality. Thus, the last fixed payment will be $(1 + RS_0(J, q) q/360)$, and the last floating payment will be $(1 + L_{J-1}(q) q/360)$. We shall also refer to this notional as *hypothetical*.[7]

Let us now find the present value at time 0 of the floating payments. First, let us start with the last floating payment, at time J, which as we said is $(1 + L_{J-1}(q)\,q/360)$. Let us find its present value back to time $J - 1$. So, we are at time $J - 1$ and we are looking ahead to time J. What is the present value at $J - 1$ of the payment of $(1 + L_{J-1}(q)\,q/360)$ that will occur at J? This is a simple calculation:

$$\frac{1 + L_{J-1}(q)\left(\frac{q}{360}\right)}{1 + L_{J-1}(q)\left(\frac{q}{360}\right)} = 1.$$

In other words, discounting the upcoming payment at the rate of that payment gives a present value of 1, provided the principal is added. In what follows, we shall need to remember that 1 is the present value of the remaining payment plus the hypothetical notional.[8]

So, now we step back to $J - 2$ and value the next payment of $L_{J-2}(q)(q/360)$ plus the value of the remaining payments plus hypothetical notional beyond $J - 2$, which is 1. This valuation will equal

$$\frac{1 + L_{J-2}(q)\left(\frac{q}{360}\right)}{1 + L_{J-2}(q)\left(\frac{q}{360}\right)} = 1.$$

It should now be apparent that we can continue to step back recursively to time 0, and we shall always obtain a value of 1 at each of the settlement dates. Hence, the present value at time 0 of the floating payments, plus the hypothetical notional of 1 at expiration, equals 1.

By adding the hypothetical notional at expiration, we have effectively treated the series of floating payments like a floating-rate loan whereby the principal is repaid. We see that its value is 1 at any settlement date. This result is not a magical result. It is consistent with the purpose of a floating rate loan: the reduction of interest rate risk to the lender. Remember that when a lender makes a fixed-rate loan, it assumes interest rate risk. If rates go up, the fixed rate loan is worth less. If rates go down, the fixed rate loan is worth more. A floating-rate loan reduces this uncertainty by having the floating rate periodically reset to the current market rate. Of course, rates can adjust only periodically and not continuously, so the value of the floating-rate loan will deviate slightly from its par value, which we set here at 1. But on each settlement date, the loan rate catches up with the market rate, and the loan value returns to its par value. What we have done here is to take the floating side of the swap and make it behave like a floating-rate loan. We have also taken the fixed side of the swap and made it behave like a fixed-rate loan.

Similar to FRAs, the swap involves no net payment at the start, so it has zero value when initiated,

$$VS_0(J, q) = 0. \tag{26.7}$$

Thus, we should set the present value of the floating payments, plus hypothetical notional, to the present value of the fixed payments, plus hypothetical notional, equal to each other. The present value of the floating payments plus hypothetical notional is, as we just showed, equal to 1. Thus, we set the present value of the fixed payments, plus the hypothetical notional of 1, equal to 1,

$$\sum_{j=1}^{J}\left[\frac{RS_0\left(J,q\right)\left(\frac{q}{360}\right)}{1+L_0(t_j)\left(\frac{t_j}{360}\right)}\right]+\frac{1}{1+L_0(t_J)\left(\frac{t_J}{360}\right)}=1.$$

Solving for the fixed rate, $RS_0\left(J,q\right)$, gives

$$RS_0\left(J,q\right)=\left\{\frac{1-\dfrac{1}{1+L_0(t_J)\left(\frac{t_J}{360}\right)}}{\sum_{j=1}^{J}\left[\dfrac{1}{1+L_0(t_j)\left(\frac{t_j}{360}\right)}\right]}\right\}\frac{360}{q}.\qquad(26.8)$$

This rather complex looking equation is actually quite simple. Inside the large parentheses is a numerator and a denominator. The numerator is simply one minus the present value of 1.0 paid at the last payment date. The denominator is the sum of the present values of 1.0 at each payment date. In other words, the denominator is simply the present value of a 1.0 annuity. The ratio of numerator to denominator is then annualized by the factor $360/q$. The result is the fixed rate on the swap that equates the present value of the fixed payments to the present value of the floating payments at the start. Hence, it is the arbitrage-free fixed rate on the swap.

To facilitate finding the present value of something, as we have seen before we sometimes use the term structure of spot rates to calculate the price of a zero-coupon bond and multiply that price by the item we are discounting. Suppose the spot rates for all maturities are all 3%. The example given in Table 26.1 illustrates this process in calculating a swap rate.

With a 3% spot rate, the 1.0 par zero-coupon bond price can be found based on

$$\frac{1}{1+L_0(t_j)\left(\frac{t_j}{360}\right)}.$$

TABLE 26.1 Spot Rates, Zero-Coupon Bond Prices, and Equilibrium Swap Rate

Days	Spot Rate (%)	Zero-Coupon Bond Price
90	3.0	0.992556
180	3.0	0.985222
270	3.0	0.977995
360	3.0	0.970874
	Sum	3.926646
	Quarterly Swap Rate	0.741758%
	Annual Swap Rate	2.967032%

For example, 1.0 received on day 180 is worth

$$\frac{1}{1 + L_0(t_j)\left(\frac{t_j}{360}\right)} = \frac{1}{1 + 0.03\left(\frac{180}{360}\right)} = 0.985222.$$

The quarterly swap rate is

$$\left(\frac{1 - 0.970844}{0.992556 + 0.985222 + 0.977995 + 0.970874}\right) = 0.00741758,$$

and the annualized swap rate, based on Equation (26.8), is simply

$$RS_0(360, 90) = \left(\frac{1 - 0.970844}{0.992556 + 0.985222 + 0.977995 + 0.970874}\right)\frac{360}{90}$$
$$= (0.00741758)\,4 = 0.02967.$$

An alternative approach to pricing an interest rate swap is based on a portfolio of off-market FRAs. With this approach, we evaluate each periodic cash flow related to the swap. The risk related to each floating cash flow is offset with an off-market FRA. Recall each cash flow determined at time $j - 1$ and paid q days later based on 1.0 notional for a receive-floating swap is expressed as

$$[L_{j-1}(q) - RS_0(J, q)]\left(\frac{q}{360}\right).$$

By entering a receive-fixed FRA, the combined cash flow occurring at $t_{j-1} + q$ can be expressed as

$$[L_{j-1}(q) - RS_0(J, q)]\left(\frac{q}{360}\right) + [R_0(t_{j-1}, q) - L_{j-1}(q)]\left(\frac{q}{360}\right)$$
$$= [R_0(t_{j-1}, q) - RS_0(J, q)]\left(\frac{q}{360}\right). \tag{26.9}$$

We seek the fixed swap rate such that the current swap value is zero or

$$\sum_{j=1}^{J}\left\{\frac{[R_0(t_{j-1}, q) - RS_0(J, q)]\left(\frac{q}{360}\right)}{1 + L_0(t_{j-1} + q)\left(\frac{t_{j-1}+q}{360}\right)}\right\} = 0. \tag{26.10}$$

Solving for the equilibrium fixed swap rate, we have

$$RS_0(J, q) = \sum_{j=1}^{J} w_j R_0(t_{j-1}, q), \tag{26.11}$$

where

$$w_j = \frac{DF_j}{\sum_{j=1}^{J} DF_j} = \frac{\dfrac{1}{1 + L_0\left(t_{j-1}+q\right)\left(\frac{t_{j-1}+q}{360}\right)}}{\sum_{j=1}^{J}\left[\dfrac{1}{1+L_0\left(t_{j-1}+q\right)\left(\frac{t_{j-1}+q}{360}\right)}\right]}. \tag{26.12}$$

Note that the inverse of the accrual period, $q/360$, simply converts the periodic swap rate to an annual rate. For example, if $q = 90$, then the summation part of Equation (26.11) is the quarterly swap rate. Thus, we see that the fixed rate of an interest rate swap is simply a weighted average of discount factors, where each discount factor is expressed as

$$DF_j = \frac{1}{1 + L_0\left(t_{j-1}+q\right)\left(\frac{t_{j-1}+q}{360}\right)}. \tag{26.13}$$

Again, as before, suppose the spot rates for all maturities are all 3%. Table 26.2 illustrates both methods for finding the equilibrium swap rate, which again turns out to be about 2.967%.

The forward rates are based on Equation (26.3). For example, the 270-day forward rate is

$$R_0(m_3, q) = \left[\frac{1 + L_0\left(t_3 + q\right)\left(\frac{t_3+q}{360}\right)}{1 + L_0\left(t_3\right)\left(\frac{t_3}{360}\right)} - 1\right]\frac{360}{q}$$

$$= \left[\frac{1 + 0.03\left(\frac{270}{360}\right)}{1 + 0.03\left(\frac{180}{360}\right)} - 1\right]\frac{360}{90} = 0.029557.$$

TABLE 26.2 Two Approaches to Solving for the Equilibrium Swap Rate

Days	Spot Rate (%)	Zero-Coupon Bond Price	Forward Rate (%)	Weight (%)
90	3.0	0.992556	3.0000	25.2774
180	3.0	0.985222	2.9777	25.0907
270	3.0	0.977995	2.9557	24.9066
360	3.0	0.970874	2.9340	24.7253
	Sum	3.926646		
	Quarterly Swap Rate	0.741758%		
	Annual Swap Rate	2.967032%		2.967032

Each discount factor is simply the zero-coupon bond price (DF) divided by the sum of several zero-coupon bond prices. For example, the 270-day discount factor is

$$w_3 = \frac{DF_3}{\sum\limits_{j=1}^{4} DF_j} = \frac{\dfrac{1}{1+L_0(t_2+90)\left(\frac{t_2+90}{360}\right)}}{\sum\limits_{j=1}^{4}\left[\dfrac{1}{1+L_0(t_2+90)\left(\frac{t_2+90}{360}\right)}\right]}$$

$$= \frac{0.977995}{3.926646} = 0.249066. \tag{26.14}$$

Note that for flat spot rates, the weights are declining. Further, the weighting scheme is present value weights so near-maturity forward rates receive greater weights.

26.2.2   Valuing Interest Rate Swaps

Now we wish to determine the value of an interest rate swap at some arbitrary point in its life, denoted t. We will assume that we are between payment dates. At this point we know the next upcoming floating payment because it was set at the last payment date. Let us denote the upcoming floating payment that was set at the last date, $L_k(q)$. From the results obtained in the previous section, we know that at the next settlement date the present value of the remaining floating payments plus the hypothetical notional is 1. Thus, we can value the receive-floating swap as

$$VS_t(J,q) = \left[\frac{1+L_k(q)\left(\frac{q}{360}\right)}{1+L_t(t_{k+1}-t)\left(\frac{t_{k+1}-t}{360}\right)}\right]$$

$$-\left\{\sum_{j=k+1}^{J}\left[\frac{RS_0(q)\left(\frac{q}{360}\right)}{1+L_t(t_j-t)\left(\frac{t_{k+1}-t}{360}\right)}\right]+\frac{1}{1+L_t(t_J-t)\left(\frac{t_J-t}{360}\right)}\right\}. \tag{26.15}$$

Note that the receive-fixed swap value is simply the negation of Equation (26.15). The first term on the right-hand side (in brackets []) is the current value of a floating rate bond and the second term (in curly brackets {}) is the current value of the fixed-rate bond, where the fixed rate is based on the previously established fixed swap rate.

Let us take a closer look at each of these terms. The first term on the right-hand side is the present value of the floating payments plus hypothetical 1.0 notional discounted back from time $k+1$ to time t, that is, over $t_{k+1}-t$ days at the rate that is applicable at time t for $t_{k+1}-t$ days. The second term is the present value of the fixed payments where the discounting is from each of the upcoming payment dates at the appropriate rate for day t_j to day t for $j=k+1$ to J plus hypothetical 1.0 notional discounted from the last payment date.

Recall from Table 26.2 that the fixed swap rate was 2.967032% for a quarterly reset one-year swap. Now suppose 90 days have past and interest rates have fallen significantly. We assume the first swap payment has just been made. Table 26.3 reports the interim calculations involved in computing the current swap value. Note that all but the last row involve

TABLE 26.3 Illustrating the Swap Value for an Existing Swap

Days	Spot Rate (%)	Zero-Coupon Bond Price	Forward Rate (%)	Weight (%)
90	2.0	0.995025	2.0000	33.4986
180	2.0	0.990099	1.9900	33.3328
270	2.0	0.985222	1.9802	33.1686
Sum		2.970346		
Quarterly Swap Rate		0.497529%		
Annual Swap Rate		1.990115%		1.990115
Current Swap Value per 1.0 Notional		−0.00725445		

similar calculations as previously covered. We assume a receive floating swap. Because we are at a reset date, the present value of the floating payments equals 1.0. Substituting into Equation (26.15), we have

$$VS_t(J,q) = \left[\frac{1 + L_k(q)\left(\frac{q}{360}\right)}{1 + L_t(q)\left(\frac{t_{k+1}-t}{360}\right)}\right] - \left\{\sum_{j=k+1}^{J}\left[\frac{RS_0(J,q)\left(\frac{q}{360}\right)}{1 + L_t(t_j - t)\left(\frac{t_j-t}{360}\right)}\right] + \frac{1}{1 + L_t(t_J - t)\left(\frac{t_J-t}{360}\right)}\right\}$$

$$= \left[\frac{1 + 0.02\left(\frac{90}{360}\right)}{1 + 0.02\left(\frac{90}{360}\right)}\right] - \left\{\sum_{j=1+1}^{4}\left[\frac{0.002967392\left(\frac{90}{360}\right)}{1 + 0.02\left(\frac{t_j-t}{360}\right)}\right] + \frac{1}{1 + 0.02\left(\frac{270}{360}\right)}\right\}$$

$$= -0.00725445.$$

Note also in Table 26.3 that we provide the new swap rate based on the new term structure. This rate, 1.990115%, would be the rate on a new swap with three payments spaced 90 days apart. As it turns out, we can also value the swap by assuming that we offset the old swap with this new swap. The floating rates cancel, leaving the swap value equal to the present value of the difference in the new fixed rate and the old fixed rate:

$$(0.01990115 - 0.02967032)\left(\frac{90}{360}\right)(2.970347) = -0.00725445.$$

The use of a new hypothetical offsetting contract is how we valued FRAs, so this approach is broadly applicable.

In this problem we positioned ourselves an instant after the first swap payment, but we could have positioned ourselves completely between swap payment dates. The procedure would be the same.

Before leaving interest rate swaps, we dive into some important technical details related to understanding swaps as they are actually used in practice.

26.2.3 Technical Details Related to Actual Interest Rate Swaps

In this section, we deal with some technical aspects related to managing swaps, such as day counting and accrual periods. We elaborate further on viewing swaps as a portfolio of off-market FRAs. For example, consider a swap with payments in 90, 180, 270, and 360 days. Assume the swap is on-market, that is, it has zero value and no exchange of cash at the start, the fixed rate on this swap will come from Equation (26.8).

Now let us show how we value the swap as a series of off-market FRAs. Assume settlement dates of 90, 180, 270, and 360 days. Create an FRA expiring at each date. If each FRA is a standard FRA, it will have a different fixed rate, but the swap will have the same fixed rate. If, however, we force an FRA to have the swap fixed rate as its fixed rate, the FRA will be off-market. That is not a problem, however, and is one reason we covered off-market FRAs. Some of these FRAs will have positive values, and some will have negative values, but the values must add up to zero at the start. If they did not, it would be possible to execute a swap, hedge it with a series of FRAs, and earn an arbitrage profit.

We now examine this approach with specific notation to incorporate important practical factors. We assume 1.0 notional and the same accrual period for both legs of the swap. Each accrual period, however, can be different. For example, one accrual period may be for 90 days and another accrual period is for 92 days. Assuming a 360-day year, these two accrual periods are 90/360 and 92/360. Because the forward rate is the current market rate to lock in funds for some future period, the value of the receive-floating swap can be expressed as

$$VS_0(J, q) = \sum_{j=1}^{J} DF_{Flt,j} AP_{Flt,j} r_{FR,j-1} - r_{Fix} \sum_{j=1}^{J} DF_{Fix,j} AP_{Fix,j}, \qquad (26.16)$$

where $DF_{Flt,j}$ ($DF_{Fix,j}$) denotes the appropriate discount factor for the jth floating (fixed) rate, $AP_{Flt,j}$ ($AP_{Fix,j}$) denotes the appropriate accrual period for the jth floating (fixed) rate, and $r_{FR,j-1}$ denotes the equilibrium forward rate, which is equivalent to the fixed FRA rate. We use $j - 1$ to remind us that the forward rate is set in advance of the period and paid at the end. Solving for the fixed rate that renders Equation (26.16) equal to zero is

$$r_{Fix} = \sum_{j=1}^{J} \frac{DF_{Flt,j} AP_{Flt,j}}{\sum\limits_{j=1}^{J} DF_{Fix,j} AP_{Fix,j}} r_{FR,j-1} = \sum_{j=1}^{J} w_j r_{FR,j-1}, \qquad (26.17)$$

where

$$w_j = \frac{DF_{Flt,j} AP_{Flt,j}}{\sum\limits_{j=1}^{J} DF_{Fix,j} AP_{Fix,j}}. \qquad (26.18)$$

Note if the accrual periods are the same, then appropriate weights can be computed solely with discount factors, as we have seen before. Further, notice that the numerator is based on floating payments and the denominator is based on fixed payments. Often, swap structures have different frequencies for both legs, resulting in these weights not summing to one. For example, the plain vanilla swap so widely used has semiannual fixed payments based on a 30/360-day year, whereas the floating payments are often quarterly based on an actual/360-day year.

Table 26.4 illustrates three different sets of spot rates: flat, upward, and downward. Note that the pattern of these spot rates can be influenced by a number of factors, including expectations of future interest rates and additional premium for perceptions of riskiness. As before, we assume the same accrual periods, we observe downward sloping forward rates resulting in the equilibrium swap rate being below the 2.5% spot rate by just a few basis points. In Panel B, the upward sloping spot curve results in a steeper upward sloping forward curve. As expected, the fixed swap rate of 3.9175% is much higher than the average spot rate of 2.5%. Panel C illustrates the downward sloping spot curve with opposite results where the fixed swap rate is 1.0025%, even though the average spot rate is again 2.5%.

All three panels demonstrate that a weighted average of forward rates yields exactly the same annual fixed rate as the solution based on zero-coupon bond prices or discount factors.

TABLE 26.4　Illustration of Equilibrium Swap Rate

Panel A. Flat Spot Rate Curve

Days	Spot Rate (%)	Zero-Coupon Bond Price	Forward Rate (%)	Weight (%)
90	2.5	0.993789	2.5000	25.2317
180	2.5	0.987654	2.4845	25.0760
270	2.5	0.981595	2.4691	24.9221
360	2.5	0.975610	2.4540	24.7702
	Sum	3.938648		
	Quarterly Swap Rate	0.619254%		
	Annual Swap Rate	2.477017%		2.477017

Panel B. Upward Sloping Spot Rate Curve

Days	Spot Rate (%)	Zero-Coupon Bond Price	Forward Rate (%)	Weight (%)
90	1.0	0.997506	1.0000	25.4003
180	2.0	0.990099	2.9925	25.2117
270	3.0	0.977995	4.9505	24.9035
360	4.0	0.961538	6.8460	24.4845
	Sum	3.927139		
	Quarterly Swap Rate	0.979378%		
	Annual Swap Rate	3.917512%		3.917512

Panel C. Downward Sloping Spot Rate Curve

Days	Spot Rate (%)	Zero-Coupon Bond Price	Forward Rate (%)	Weight (%)
90	4.0	0.990099	4.0000	25.0617
180	3.0	0.985222	1.9802	24.9383
270	2.0	0.985222	0.0000	24.9383
360	1.0	0.990099	−1.9704	25.0617
	Sum	3.950641		
	Quarterly Swap Rate	0.250617%		
	Annual Swap Rate	1.002469%		1.002469

26.3   INTEREST RATE OPTIONS

As with interest rate forwards and swaps, there are also options on interest rates. An interest rate option gives the right but not the obligation to make or receive a fixed payment at an upcoming date. An interest rate call is the right to make a fixed payment and receive a floating payment. An interest rate put is the right to receive a fixed payment and make a floating payment. As with swaps, the rate is determined on the settlement date, but the payment is made on the next settlement date, at which time the rate is reset.

26.3.1   Interest Rate Option Values

Using the same notation as with interest rate swaps, on day J, the value at J of the payoff of a call with strike rate X is

$$VIRC_J(J,q) = \max\left[0, \frac{L_J(q) - X}{1 + L_J(q)\left(\frac{q}{360}\right)}\right]. \qquad (26.19)$$

Of course, this amount is paid at $j + q$, hence the discounting to obtain the value at J. The payoff of an interest rate put is

$$VIRP_J(J,q) = \max\left[0, \frac{X - L_J(q)}{1 + L_J(q)\left(\frac{q}{360}\right)}\right]. \qquad (26.20)$$

Valuing interest rate options prior to expiration, however, is much more difficult than valuing swaps or FRAs. As we have previously noted, options depend on the distribution of the underlying, in particular the volatility. Moreover, although we can typically value an option on an asset in isolation from the options on other assets, we cannot value an interest rate option in isolation from the other interest rate options on the same underlying interest rate but with different maturities and strikes. If we try to do so, we might admit arbitrage. Nonetheless, this procedure is commonly done in practice. For example, many practitioners use the Black model for pricing options on forwards and futures that we previously covered. The reasoning is that an option on an interest rate is really an option on a forward rate. So, to use the Black model one simply plugs in the forward rate as the underlying and the forward rate volatility as the volatility as required in the model. The problem with doing this, however, is that if other options on the same underlying are also offered for trading, these prices could admit arbitrage, because there is no linkage between prices from the Black model for options on the same underlying. That is, applying the Black model to two options on the same interest rate does not guarantee that one cannot arbitrage the one against the other. There is no binding force that connects the option prices, because the model is applied independently to the two options.

The best method of dealing with this concern is to develop an arbitrage-free model of the term structure. There are many such models, and we shall cover one, the Heath-Jarrow-Morton model, in Chapter 27. In Chapter 28 we shall show how to derive prices using a binomial version of that model.

26.3.2   Interest Rate Option Parity Relationships

There are some important parity relationships between interest rate options and interest rate swaps and forwards. Consider an interest rate call and an interest rate put both with an exercise rate of X. For notational simplicity, let us disregard the adjustment $(q/360)$ and state the payoffs at time j of a long call and short put as

$$\max\left[0, L_j(q) - X\right] - \max\left[0, X - L_j(q)\right].$$

Now, consider the various possible outcomes:

> If $L_j(q) > X$,
> Call pays $L_j(q) - X$.
> Put pays nothing.
> If $L_j(q) < X$,
> Call pays nothing.
> Put pays $-\left(X - L_j(q)\right) = L_j(q) - X$.

We can see that in all circumstances, the combination pays $L_j(q) - X$. This makes this combination similar to a long FRA. There is, however, a more precise specification that makes this combination identical to an FRA: that the strike X is the same as the FRA rate. In other words, if X is not the FRA rate, then the call and put would not have the same prices, so going long a call and short a put would result in either the call costing more than the put or the put costing more than the call. There would, therefore, be either a net cash outflow or inflow at the start. Hence, this synthetic FRA would be off-market. An FRA priced at the appropriate arbitrage-free rate that is based on zero value at the start would have no initial cash inflow or outflow. So, if the exercise rate of the interest rate call and put is set at the FRA rate, there would be no net cash inflow or outflow. Then the put and the call have the same price, so buying a call and selling a put would cost nothing on net.

26.3.3   Interest Rate Caps, Floors, and Collars

A combination of interest rate calls expiring at different dates is called a *cap*, because it is often used by a borrower to cap the interest rate on a floating-rate loan. The borrower takes out the loan and buys a cap, which consists of interest rate call options that expire on the dates on which the loan rates are reset. The component call options are called *caplets*. When the underlying rate is above the cap exercise rate, the caplet pays off and effectively compensates the borrower for the increase in interest rates. When the loan rate is reset below the cap exercise rate, the caplet expires unexercised, and the borrower benefits from the lower rate. The total cost of the component caplets that make up the cap does effectively raise the loan rate, but it serves as insurance that limits the overall rate paid to a maximum, hence, a cap.

A combination of interest rate puts is called a *floor* because it is often used by a lender to put a floor on the rate on a floating-rate loan. The individual put options are called *floorlets*. A lender in a floating-rate loan might buy a floor to put a minimum on the effective rate on a loan. Each floorlet would expire on the day on which the loan rate is reset.

When the rate resets below the exercise rate, the floorlet pays off and the lender is compensated for the lower rate. The overall cost of the floor lowers the rate of return on the loan but effectively sets a minimum rate.

A combination of long cap and short floor is called a *collar* because it is used by borrowers to contain the effective interest rate on a floating-rate loan within a range. The cap strike is chosen at a rate that effectively limits the loan rate to a maximum value. The borrower then sells a floor at a lower strike, which results in giving up gains when the rate falls below the lower exercise rate. The two rates are set such that the price of the cap is precisely offset by the price of the floor.[9] The effective rate on the loan will vary between the cap exercise rate and the floor exercise rate. If the call and put exercise rates are both set to the swap rate, the call premium will also equal the put premium and there will be no net cash outflow or inflow at the start. Moreover, the combination will produce the same results as an interest rate swap.

It is also possible to have forward contracts on swaps, which are called *forward swaps*, and options on swaps, which are called *swaptions*. We shall illustrate swaptions when we show how to use a binomial tree to price interest rate derivatives in Chapter 28.

26.4   RECAP AND PREVIEW

In this chapter, we looked at interest rate derivatives, which are different from the derivatives on assets that we have previously covered. An interest rate is not an asset and, therefore, cannot be owned. Nonetheless, the payoff functions of interest rate derivatives can be replicated. Hence, these derivatives can be priced by arbitrage. We showed that pricing FRAs and swaps is relatively simple, at least in comparison to pricing options in that they can be replicated in a static, rather than dynamic, manner.

In Chapter 27, we show how an arbitrage-free model of the term structure can be built. It will then form the basis for pricing derivatives on these interest rates that will prohibit arbitrage.

QUESTIONS AND PROBLEMS

1   Suppose you are quoted loan rates of 3% for both 270-day and 360-day maturities. Compute the 90-day FRA fixed rate based on this information.

2   Using the same information as in the previous problem, assume now that 90 days have passed but interest rates have not changed. Specifically, you are quoted loan rates of 3% for both the 180-day and 270-day maturities. Compute the receive-floating FRA value assuming a 1,000,000 notional.

3   Suppose you are quoted 4% for the 270-day rate and 3% for the 360-day rate. Compute the 90-day FRA fixed rate based on this information. Assuming a trader entered this receive-floating FRA with 1,000,000 notional. Now assume 90 days have passed and the 180-day rate is 3% and the 270-day rate is 4%. Compute the value of this FRA assuming all rates are quoted on an add-on interest basis on a 360-day year.

4 Suppose you are provided the following information regarding loan spot rates. Compute the equilibrium swap rate based on discount factors as well as forward rates.

Days	Spot Rate (%)
90	2.0
180	5.0
270	3.0
360	4.0

5 The following table provides three sets of forward rates each with an average forward rate of 3.5%. Compute the equilibrium swap rates by both methods (discount factors and weighted average of forward rates). Explain your results.

	Forward Rate (%)		
Year	Flat	Upward	Downward
1	3.5	2.0	5.0
2	3.5	3.0	4.0
3	3.5	4.0	3.0
4	3.5	5.0	2.0

NOTES

1. Alternatively, you can think of an asset the way accountants think of assets: as "things" that belong on the left-hand side of the balance sheet wherein the cost of acquiring these "things" is funded by either equity or debt, which comprise the right-hand side of the balance sheet.
2. All contracts are designated as having a certain underlying amount, called the *notional*. For example, if the notional were 10 million, the above amount would be multiplied by 10 million. The FRA would be a contract to pay the agreed-on fixed rate of interest and receive the underlying LIBOR interest on 10 million notional. There is no exchange of the 10 million. Only the interest payments are exchanged. In our examples, we are effectively assuming a 1.0 notional.
3. That is, we multiplied by the factor $1 + L_m(q)(q/360)$ to reflect rolling over the payoff at the rate $L_m(q)$.
4. This point is no different from the fact that if we had a liquid asset, such as stock, and wanted to know what it was worth, we would simply look up its current price in the market. We do not have to actually sell the asset. We just need to know the price at which we could sell it.
5. There are also swaps based on currencies, commodity prices, stock prices, and stock indexes, but we do not cover these other swaps in this book.
6. As with FRAs, the example here assumes 1.0 notional. In a real swap, where the notional is more than 1.0, the payments would be multiplied by whatever is the notional. Further, in practice, the accrual period and related day-count methods influence the pricing and valuation of FRAs. For our purposes here, we simply assume 30 days in a month and 360 days in a year.

7. Recall that we added a 1.0 hypothetical notional when valuing FRAs.

8. The word *notional* means in name only, but here we prepend the word *hypothetical*, which would seem to be unnecessary because these notionals are obviously not exchanged. In currency swaps however, the parties can exchange nonequivalent notionals, meaning, the notionals are not hypothetical.

9. Technically, this is not a requirement. The exercise rates could be set such that the floor generates more than enough value to result in a net cash inflow from selling the floor at a greater price than buying the cap. Or the exercise rates could be set such that the cap costs more than the floor, resulting in a net cash outflow. These cases are not common, however, and the cap and floor strikes are usually set so that the cap and floor prices are equal. Sometimes this transaction is called a zero-cost collar, but most collars are of the zero-cost type.

CHAPTER 27

Fitting an Arbitrage-Free Term Structure Model

One of the most significant discoveries in modern finance research is the Heath-Jarrow-Morton term structure model. It has revolutionized the theory and practice of interest rate derivatives as well as interest rate products in general. The Heath-Jarrow-Morton (1992) model, or HJM as it is commonly known, stands at the head of a family tree of models that take the initial term structure as given, affix the conditions required to preclude arbitrage, and produce trading strategies that lead to the valuation of all interest-dependent claims. Hence, it is in the spirit of the Black-Scholes-Merton model, which takes the price of the underlying and volatility as given and derives the arbitrage-free condition that produces the price of the option.

The HJM model is somewhat similar to a predecessor model by Ho and Lee (HL) (1986). Both models are partial equilibrium models. There are some general equilibrium models of the term structure, such as Vasicek (1977) and Cox-Ingersoll-Ross (1985). Those models derive market equilibrium conditions leading to the prices of fixed-income securities, but they do not fit the current term structure. Hence, they can admit arbitrage if used in extant markets. By accommodating the existing term structure, HJM, as well as HL, can be used in real markets to actually trade. There are a handful of other models, such as in Black, Derman, and Toy (1990), Black and Karasinski (1991), and Brace, Gatarek, and Musiela (1997), that you can, and should, study if you want to become proficient in this subject, but we shall focus on HJM, because it is probably the most well-known, and it is relatively easy to implement assuming a single risk factor for the entire term structure.

What sets HJM apart from HL is the fact that the former admits a wide range of structures for the sources driving interest rates. We should think of an interest rate as being driven by one or more "factors," which are sources of noise or uncertainty. Although HL is a one-factor model, HJM can accommodate any finite number of factors, though with each factor there is a considerable increase in complexity and practicality. In addition, HJM admits an extremely flexible structure for the volatilities of the various interest rates. On the downside, many common versions of HJM are non-Markovian, meaning that they are path dependent, which increases the computational complexity. Also, because the distribution of interest rates is assumed to be normal, HJM models can theoretically allow negative interest rates, although that occurrence would not be common.[1]

27.1 BASIC STRUCTURE OF THE HJM MODEL

In contrast to most other term structure models, which are based on movements in *spot* interest rates, the HJM model is driven by movements in *forward* interest rates. We start by specifying that the model applies over a period of time $t \in [0, T]$.[2] Let $B(t, T)$ be the price of a zero-coupon bond at time t that pays 1.0 at time T. Define $f(t, T)$ as the continuously compounded forward rate observed at time t for an instantaneous transaction to begin at time T. That is, based on the term structure in existence at time t, we observe a forward rate for a transaction to start at T and end an instant later.

In a typical situation, we find the present value of a future amount by discounting at the spot rate. We start with a simple example using annual discrete compounding. For example, suppose we want to know the present value of 1.0 two years from today. We then would discount 1.0 by the operation $1/[1 + r(0, 2)]^2$ where $r(0, 2)$ is the two-year spot rate. We can, however, discount by the one-year spot rate and the one-year ahead forward rate. That is, let $f(0, 1)$ be the forward rate observed at time 0 for a one-period transaction to start at time 1 and obviously end at time 2.[3] Thus, we can find the present value of 1.0 paid at time 2, denoted $B(0, 2)$ by the calculation,

$$B(0, 2) = \frac{1}{[1 + r(0, 1)][1 + f(0, 1)]}.$$

The substitution of the one-period spot rate and the one-period ahead forward rate is valid because

$$f(0, 1) = \frac{[1 + r(0, 2)]^2}{1 + r(0, 1)} - 1.$$

So, in a continuous time world, the price at time t of a 1.0 face value zero-coupon bond maturing at T is given as

$$B(t, T) = \exp\left(-\int_t^T f(t, s)\,ds\right). \tag{27.1}$$

That is, we can obtain the price of a zero-coupon bond by successively discounting at the forward rates. Note that we can extract a given forward rate by differentiating the previous equation with respect to T:

$$f(t, T) = -\frac{\partial \log B(t, T)}{\partial T}. \tag{27.2}$$

Although Equation (27.2) is a nice mathematical specification of a forward rate, it is not of much practical use without a formula that relates the spot price to its maturity. Such a formula can be obtained only for limited cases.[4] This formula reminds us, however, that forward rates, specifically instantaneous forward rates, are derivatives of bond prices with respect to maturity. The shortest forward rate, the one defined as $f(t, t)$, is of special significance. It is a spot rate that is sometimes called the short rate,

$$r(t) = f(t, t). \tag{27.3}$$

In the HJM model, this is the only spot rate that commands our attention.

Starting from an initial state at time 0, HJM proposes that the forward rate observed at time 0 for period T, $f(0, T)$, changes in the following manner during the time from 0 to T:

$$f(t, T) - f(0, T) = \int_0^t \alpha(v, T)dv + \sum_{i=1}^n \int_0^t \sigma_i(v, T)dW_i(v). \qquad (27.4)$$

We must now carefully examine this important equation. The term $\alpha(v, T)$ is the instantaneous drift term observed at time v for the forward rate at T. The second expression on the right-hand side begins with a summation of n terms, which means it is an n-factor model. These factors are captured by the terms, $\sigma_i(v, T)$ and $dW_i(v)$. The term $\sigma_i(v, T)$ is the volatility of factor i observed at time v for the forward rate at T.[5] The term $dW_i(v)$ is a Wiener process representing the source of uncertainty for factor i at time v. There are some formal mathematical restrictions required to uphold these assumptions, but we need not concern ourselves with them here. The expression in simple terms says that the forward rate started off at a value of $f(0, T)$ and evolved over time to a value of $f(t, T)$. These changes in the forward rate reflected the accumulation, as indicated by the integrals, of the infinitesimal changes that consist of drift and volatility that have occurred over the period 0 to T.

HJM go on to show that even though we do not really need to know it, the spot rate process can be derived and is quite similar to that of the forward rate:

$$r(t) = f(0, t) + \int_0^t \alpha(v, t)dv + \sum_{i=1}^n \int_0^t \sigma_i(v, t)dW_i(v). \qquad (27.5)$$

In addition, the process for the bond price is given as

$$\frac{dB(t, T)}{B(t, T)} = [r(t) + b(t, T)]dt + \sum_{i=1}^n a_i(t, T)dW_i(t), \qquad (27.6)$$

where

$$a_i(t, T) \equiv - \int_t^T \sigma_i(t, v)dv$$

$$b(t, T) \equiv - \int_t^T \alpha(t, v)dv + \frac{1}{2} \sum_{i=1}^n a_i(t, T)^2. \qquad (27.7)$$

This expression should look somewhat familiar because it resembles the stochastic process ordinarily used for an asset. Note that the drift, however, consists of the risk-free rate and another term, and that there are multiple volatilities representing the various factors.

HJM go on to derive their most important result, which is that the condition of no arbitrage implies that a martingale probability measure exists and implies a restriction on the drift coefficients of the forward rates. Specifically, for the n-factor model, we have

$$\alpha(t, T) = \sum_{i=1}^n \sigma_i(t, T) \int_t^T \sigma(t, v)dv. \qquad (27.8)$$

This statement means that the drift is not independent of the volatility and is in fact a specific function of the volatility.[6] At a given time point t, the drift for the forward rate to start at T is obtained by integrating, meaning to essentially add, all of the volatilities over the time periods from t to T and multiplying by the sum of the volatilities across all of the factors observed at t for the rate to start at T. Again, this restriction ensures that no arbitrage opportunities are possible. We do not explore the details of how this result is obtained in continuous time, but we will look at it more carefully in a discrete time framework later in this chapter and also in Chapter 28.

In general, the continuous time HJM model is written as

$$df(t, T) = \alpha(t, T)dt + \sum_{i=1}^{n} \sigma_i(t, T)dW_t(t), \qquad (27.9)$$

with the drift restricted as given above in Equation (27.8).

In the HJM model, the volatility structure is an input. What we mean by a volatility structure is a concept with three dimensions. Consider one volatility, $\sigma_i(a, b)$. It has three parameters. The subscript i refers to the factor that we call i. There can be many factors, each of which should be thought of as a source of risk, meaning some effect that drives volatility. Examples of factors might be Federal Reserve policy, government fiscal policy, exchange rates, the stock market, and so on. The HJM model accommodates an unlimited number of factors. Jumping to the third parameter, b identifies the maturity or expiration of the forward contract to which the rate applies. If b is, say, five, it means that we are referring to the volatility of a forward contract on the one-period rate that matures in five periods. The other parameter, a, is a time series factor and reflects how a particular volatility varies over time. For example, for factor i, maturity five, we have $\sigma_i(0, 5)$, $\sigma_i(1, 5)$, $\sigma_i(2, 5), \ldots$, and so on. These volatilities capture how the risk of that rate varies over time. There is clearly a lot of information in the volatility structure, and although this gives the model a great deal of flexibility, it also places heavy demands on the user. When we illustrate it, we shall use just one factor, a limited number of maturities, and just a few time periods.

There need not be any formal mathematical structure to these volatilities. In other words, they need not be related to each other mathematically. There are some special cases of HJM involving specific mathematical structures to the volatility, such as

$$\sigma(t, T) = \sigma e^{-\lambda(T-t)}, \qquad (27.10)$$

which is called exponentially dampened volatility. In this case a single volatility, σ, is given and all successive volatilities are related to it. The specification results in volatilities declining at an exponential rate. This particular structure is especially convenient because it permits many closed-form solutions for options and other derivatives. For details see Jarrow and Turnbull (2000, Chapters 16 and 17).

Other volatility functions include the simple case of constant volatility

$$\sigma(t, T) = \sigma, \qquad (27.11)$$

which makes this model equivalent to HL. Another case sometimes seen is the nearly proportional volatility,

$$\sigma(t, T) = \eta(t, T)\min[f(t, T), M], \qquad (27.12)$$

where $\eta(t, T)$ is a deterministic function and M is a large positive constant. This specification sets the volatility proportional to the current forward rate and caps it on the upper end so that it will not become unreasonably high. Two other structures seen are

$$\sigma(t, T) = \sigma f(t, T)^\gamma$$

$$\sigma(t, T) = \sigma r(t)^\gamma e^{-\lambda(T-t)}. \tag{27.13}$$

Note that these examples and the nearly proportional volatility case are examples of which the volatility is stochastic but completely dependent on the level of rates. Thus, although these are stochastic volatility models, they are not independent stochastic volatility models. Hence, they do not pose any additional problems not already present in the model. For more on volatility structures of the HJM model, see Jarrow (2002), Ritchken (1996), and Ritchken and Sankarasubramanian (1995).

27.2 DISCRETIZING THE HJM MODEL

With the exception of a few restrictive volatility structures, the HJM model does not produce closed-form solutions for the prices and risk measures of various instruments.[7] Hence, numerical methods are normally required. Here we will look at discretizing the HJM model for use in a binomial tree. In doing so, we shall gain a deeper understanding of the model and especially the drift restriction that prevents arbitrage. At this level, we shall focus exclusively on the one-factor version. Hence, we are given the stochastic process for the forward rate, Equation (27.9). Because we shall use only a one-factor model, Equation (27.4) becomes

$$df(t, T) = \alpha(t, T)dt + \sigma(t, T)dW(t), \tag{27.14}$$

where the arbitrage-free drift restriction, Equation (27.8), becomes

$$\alpha(t, T) = \sigma(t, T) \int_t^T \sigma(t, v)dv. \tag{27.15}$$

To generate a binomial version of the model, we must first consider the information with which we have to work. We shall need the prices of a set of bonds maturing at discrete time points, 1, 2, 3, ..., $T - 1$, T. If T is the longest maturity available, that would mean we have T forward rates available, $f(0, 0), f(0, 1), ..., f(0, T - 1)$.[8] We would then need volatilities for maturities of 1, 2, ..., $T - 1$. This set of information will be sufficient to build a binomial tree of $T - 1$ time steps.[9] First, let us write the stochastic process for the forward rates in discrete form as

$$\Delta f(t, T) = \alpha(t, T)\Delta t + \sigma(t, T)\Delta W(t). \tag{27.16}$$

To convert a Wiener process to a binomial, we simply express the random variable as a binary variable with a value of +1 or –1 at each time step and assume martingale probabilities of ½, as we did in Chapter 16. We assume each time step has a defined length of one unit. Hence, the stochastic process becomes

$$\Delta f(t, T) = \alpha(t, T) \pm \sigma(t, T). \tag{27.17}$$

Thus, at a given time, for a given forward rate $f(t, T)$, we move one step ahead to the next time in the following manner:

$$f(t + 1, T)^+ = f(t, T) + \alpha(t, T) + \sigma(t, T)$$

$$f(t + 1, T)^- = f(t, T) + \alpha(t, T) - \sigma(t, T). \tag{27.18}$$

That is, we add the drift and add or subtract the volatility. Recall that to prevent arbitrage in a term structure, the local expectations hypothesis (LEH) must hold. The LEH says that the expected return on any financial instrument over the shortest time period must be the riskless rate, where expectations are taken using a special martingale probability measure that appeals to the local expectations hypothesis. That is,

$$B(t, T) = B(t, t + h)E^Q[B(t + h, T)], \tag{27.19}$$

where the superscript Q means that expectations are taken using the martingale probabilities. The first term, $B(t, t + h)$, is the price of the bond with the shortest maturity. In multiplying by it, we are discounting at the riskless rate. The second term, $E^Q[B(t + h, T)]$, is the expectation of the bond price at time $t + h$. By discounting this expectation at the riskless rate, we obtain the current bond price.

Writing Equation (27.19) as

$$\frac{B(t, T)}{B(t, t + h)} = E^Q[B(t + h, T)], \tag{27.20}$$

and using Equation (27.1), which states that the bond price can be found by discounting at the sequence of forward rates, we substitute and obtain

$$\frac{B(t, T)}{B(t, t + h)} = e^{-\int_t^T f(t,v)dv} \, e^{\int_t^{t+h} f(t,v)dv} = e^{-\int_{t+h}^T f(t,v)dv}. \tag{27.21}$$

This result must equal the expectation in Equation (27.20), which can be found by evaluating the following expression based on Equation (27.18):

$$e^{-\int_{t+h}^T f(t,v)dv} = \left(\frac{1}{2}\right) e^{-\int_{t+h}^T [f(t,v)+\alpha(t,v)+\sigma(t,v)]dv} + \left(\frac{1}{2}\right) e^{-\int_{t+h}^T [f(t,v)+\alpha(t,v)-\sigma(t,v)]dv}. \tag{27.22}$$

This expectation reflects the binomial probabilities of ½ and the integrals represent the discounting of the sequence of forward rates. In other words, the two terms that are multiplied by ½ are the next two possible bond prices, which themselves are obtained by discounting at the sequence of forward rates over the remaining lives of the bonds. After some additional math, we obtain

$$\alpha(t, T) = \sigma(t, T) \int_t^T \sigma(t, v)dv, \tag{27.23}$$

which is the result we obtained as Equation (27.15).

To actually work with the HJM model in discrete time, however, requires that we obtain a discretized version of the drift restriction. HJM (1991) provide a version of this result, which they obtain in a clever way. Using a binomial model, they get a result that they extend to continuous time by letting the time step approach zero. They then use it to show the correct drift in a simple binomial example. Ritchken (1996: 579–580) notes this result but this is not correct for the binomial case. Starting with a binomial model, obtaining a result and taking the continuous time limit to that result is certainly correct. But the discrete time version of this continuous time limit is not correct, as shown by Grant and Vora (1999, 2006), who go on to derive the correct discrete time formula. They start with a variation of the expression we used:

$$B(t, T) = E^Q[B(t + h, T)]B(t, t + h). \tag{27.24}$$

They then make use of the fact that a Wiener process is a normal distribution, so the interaction of the volatilities has convenient properties, and the correlation between all forward rates in a one-factor model is 1.0. After considerable algebra, they find a simple expression for the drift,

$$\alpha(t, T) = \left(\frac{1}{2}\right)\left[\sigma^2(t, T) + 2\sigma(t, T)\sum_{j=t+1}^{T-1}\sigma(t, j)\right]. \tag{27.25}$$

This value can be computed very easily from the covariance matrix of forward rate volatilities, which we shall demonstrate in a numerical example below. Grant and Vora refer to this term as the *drift adjustment* term though there is really no reason to call it anything other than the *drift*. Another good treatment of how this procedure unfolds is in Jarrow and Chatterjea (2013).

27.3 FITTING A BINOMIAL TREE TO THE HJM MODEL

Let us now work a numerical example. We start with the following information for the term structure of forward rates:

$$f(0, 0) = 0.068$$

$$f(0, 1) = 0.072$$

$$f(0, 2) = 0.08$$

$$f(0, 3) = 0.082.$$

Of course, the shortest forward rate, $f(0, 0)$, is the spot rate. These rates imply the following bond prices:

$$B(0, 1) = e^{-0.068} = 0.9343$$

$$B(0, 2) = e^{-(0.068+0.072)} = 0.8694$$

$$B(0, 3) = e^{-(0.068+0.072+0.08)} = 0.8025$$

$$B(0, 4) = e^{-(0.068+0.072+0.08+0.082)} = 0.7393.$$

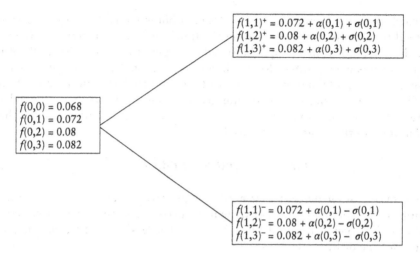

FIGURE 27.1   Structural Layout of One-Period Binomial Tree of Forward Rates

The volatilities at time 0 are given as

$$\sigma(0,1) = 0.02$$

$$\sigma(0,2) = 0.015$$

$$\sigma(0,3) = 0.01.$$

Figure 27.1 shows the structure of the problem. The volatilities are known, as just given, and could be filled in, but we use symbols here so the reader can see the distinction between the original forward rates, the drifts, and the volatilities.

We need to find the drifts, $\alpha(0,1)$, $\alpha(0,2)$, and $\alpha(0,3)$ to determine the rates at time 1 that will guarantee no arbitrage. When we have that done, we can move ahead to time 2.

The Grant-Vora formula was given as Equation (27.25). Unfortunately, this formula will be a little confusing when $t = 0$ and $T = 1$, because then we would be summing from $j = 1$ to 0. What happens is that this whole term drops out. To avoid this confusion, there is an alternative equivalent formula:

$$\alpha(t,T) = \sigma(t,T) \sum_{j=t+1}^{T} \sigma(t,j) - \left(\frac{\sigma^2(t,T)}{2} \right). \tag{27.26}$$

Because we are used to working with variances and covariances, these formulas may look familiar. Consider any two forward rates $f(t,T)$ and $f(t,T-1)$. The covariance at t between these rates is $\sigma(t,T)\sigma(t,T-1)\rho(t,T-1)$, where $\rho(t,T-1)$ is the correlation between the two forward rates. The use of a linear one-factor model, however, means that these rates are perfectly correlated. Hence, the covariance is simply the product of the volatilities, that is, $\sigma(t,T)\sigma(t,T-1)$. Thus, the covariance matrix will be of the form,

	1	2	3
1	$\sigma^2(0,1)$	$\sigma(0,1)\sigma(0,2)$	$\sigma(0,1)\sigma(0,3)$
2	$\sigma(0,2)\sigma(0,1)$	$\sigma^2(0,2)$	$\sigma(0,2)\sigma(0,3)$
3	$\sigma(0,3)\sigma(0,1)$	$\sigma(0,3)\sigma(0,2)$	$\sigma^2(0,3).$

In other words, by knowing the volatility structure we know the variances and covariances of all of the forward rates for all time points up to the final one. Grant and Vora show that the drift for any maturity T will be one-half the sum of the elements in the ith row and j^{th} column. For our example, the values in the covariance matrix are

	1	2	3
1	0.0004	0.0003	0.0002
2	0.0003	0.000225	0.00015
3	0.0002	0.00015	0.0001.

Let us begin to derive the term structure with maturity $T = 1$. We can use the formula

$$\alpha(0,1) = \sigma(0,1) \sum_{j=1}^{1} \sigma(0,1) - \frac{\sigma^2(0,1)}{2}$$

$$= \sigma(0,1)\sigma(0,1) - \frac{\sigma^2(0,1)}{2} = \frac{\sigma^2(0,1)}{2}$$

$$= \frac{0.0004}{2} = 0.0002,$$

or using the covariance matrix

$$\alpha(0,1) = \frac{0.0004}{2} = 0.0002.$$

For maturity $T = 2$, we obtain the drift by the formula

$$\alpha(0,2) = \sigma(0,2) \sum_{j=1}^{2} \sigma(0,j) - \frac{\sigma^2(0,2)}{2}$$

$$= \sigma(0,2)[\sigma(0,1) + \sigma(0,2)] - \frac{\sigma^2(0,2)}{2}$$

$$= 0.015(0.02 + 0.015) - \frac{0.000225}{2} = 0.0004125,$$

or using the covariance matrix,

$$\alpha(0,2) = \frac{0.0003 + 0.000225 + 0.0003}{2} = 0.0004125.$$

For maturity $T = 3$, we have the formula

$$\alpha(0,3) = \sigma(0,3) \sum_{j=1}^{3} \sigma(0,j) - \frac{\sigma^2(0,3)}{2}$$

$$= \sigma(0,3)[\sigma(0,1) + \sigma(0,2) + \sigma(0,3)] - \frac{\sigma^2(0,3)}{2}$$

$$= 0.01(0.02 + 0.015 + 0.01) - \frac{0.0001}{2} = 0.0004,$$

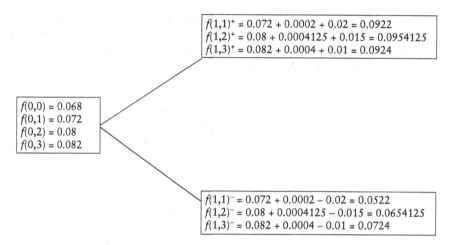

FIGURE 27.2 Numerical Illustration of One-Period Binomial Tree of Forward Rates

or from the covariance matrix,

$$\alpha(0,3) = \frac{0.0002 + 0.00015 + 0.0001 + 0.0002 + 0.00015}{2} = 0.0004.$$

So now we can fill in the first time step of the tree, which is Figure 27.2.

Now suppose we wanted to move forward another time step. If we position ourselves at the top step at time 1, where the rates are $f(1,1)^+$, $f(1,2)^+$, and $f(1,3)^+$, we can then imagine moving forward to time step 2 in the manner shown in Figure 27.3.

Note that we now require not only new drift terms but also we see volatility terms not previously seen. That is, we started with a term structure of volatility of $\sigma(0,1)$, $\sigma(0,2)$, and $\sigma(0,3)$. These were the volatilities from time point 0. Now, having moved forward to time point 1, we have a new set of volatilities: $\sigma(1,2)$ and $\sigma(1,3)$. Recall that $\sigma(0,2)$ was our value at time 0 of the volatility of the forward rate for time 2. Now, at time 1, $\sigma(1,2)$ is our value of that volatility. Should it be different? In HJM it can indeed be different. In this model, volatility can change over time, but it cannot change stochastically. That is,

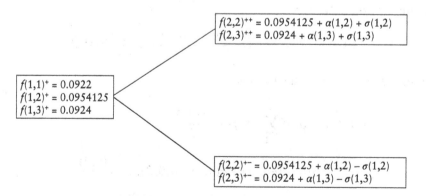

FIGURE 27.3 Binomial Tree of Forward Rates from Time 1 Up State

although $\sigma(1,2)$ can be different from $\sigma(0,2)$, we had to have known both values at time 0. In other words, volatility can change, but we have to know to what value it will change.

If we do not impose any time series changes on volatility then we simply use the value $\sigma(0,2)$ for $\sigma(1,2)$ and use $\sigma(0,3)$ for $\sigma(1,3)$. Then we would be able to calculate the drift and fit the tree. We would then step down to the lower state at time 1 and determine the rates in the next two states at time 2. As long as we make the assumption that $\sigma(0,2) = \sigma(1,2)$ and $\sigma(0,3) = \sigma(1,3)$, the tree will recombine. In other words, we require constant time series volatility. We do not, however, require equivalent cross-sectional volatility. That is, all rates do not have to have the same volatility for the tree to recombine, but a given rate must have the same volatility across time for the tree to recombine. Also, note that as we move to time step 3, we would require $\sigma(2,3)$, which might also be assumed equal to $\sigma(0,3)$ or could be treated as a different but known value.

Once we have the entire tree filled in, we have all of the information we need to price any bond or derivative. The remainder of the problem is illustrated in the next section. Recall that we never really worked with spot rates. They are related to the forward rates, but all instruments can be priced off of the forward rates as easily as off of the spot rates. In other words, there is no more information in the spot rates that we do not already have with the forward rates.

We should also note that because of the assumption of a normal distribution of interest rates, it is possible to obtain negative interest rates. This problem, as well as the problem of the tree not recombining, can be addressed in certain versions of the model. Further, negative interest rates have sometimes occurred in practice so the normal distribution may be adequate.

Here we have worked only with one-factor models. In the language of term structure movements, single-factor models seem to capture only changes in the general level of interest rates but do not reflect changes in the slope or curvature of the term structure. To capture these effects, multifactor models are required. Multifactor models are much more complex. For example, a tree version of a two-factor HJM model requires a trinomial, and the number of paths for T time steps is $3T$. Clearly some trade-offs are required before deciding to go to models of more than one factor.

So, now let us complete the tree based on all of the information we have.

27.4 FILLING IN THE REMAINDER OF THE HJM BINOMIAL TREE

Recall that we started with the rates

$$f(0,0) = 0.068$$
$$f(0,1) = 0.072$$
$$f(0,2) = 0.08$$
$$f(0,3) = 0.082.$$

From the volatilities and the no-arbitrage condition, we obtained the following rates:

$$f(1,1)^+ = f(0,1) + \alpha(0,1) + \sigma(0,1) = 0.072 + 0.0002 + 0.02 = 0.0922$$
$$f(1,2)^+ = f(0,2) + \alpha(0,2) + \sigma(0,2) = 0.08 + 0.0004125 + 0.015 = 0.0954125$$

$$f(1,3)^+ = f(0,3) + \alpha(0,3) + \sigma(0,1) = 0.082 + 0.0004 + 0.01 = 0.0924$$
$$f(1,1)^- = f(0,1) + \alpha(0,1) - \sigma(0,1) = 0.072 + 0.0002 - 0.02 = 0.0522$$
$$f(1,2)^- = f(0,2) + \alpha(0,2) - \sigma(0,2) = 0.08 + 0.0004125 - 0.015 = 0.0654125$$
$$f(1,3)^+ = f(0,3) + \alpha(0,3) - \sigma(0,3) = 0.082 + 0.0004 - 0.01 = 0.0724.$$

We will extend the tree and add the following rates:

$$f(2,2)^{++} = f(1,2)^+ + \alpha(1,2) + \sigma(1,2)$$
$$f(2,3)^{++} = f(1,3)^+ + \alpha(1,3) + \sigma(1,3)$$
$$f(2,2)^{+-} = f(1,2)^+ + \alpha(1,2) - \sigma(1,2)$$
$$f(2,3)^{+-} = f(1,3)^+ + \alpha(1,3) - \sigma(1,3)$$
$$f(2,2)^{-+} = f(1,2)^- + \alpha(1,2) + \sigma(1,2)$$
$$f(2,3)^{-+} = f(1,3)^- + \alpha(1,3) + \sigma(1,3)$$
$$f(2,2)^{--} = f(1,2)^- + \alpha(1,2) - \sigma(1,2)$$
$$f(2,3)^{--} = f(1,3)^- + \alpha(1,3) - \sigma(1,3)$$
$$f(3,3)^{+++} = f(2,3)^{++} + \alpha(2,3) + \sigma(2,3)$$
$$f(3,3)^{++-} = f(2,3)^{++} + \alpha(2,3) - \sigma(2,3)$$
$$f(3,3)^{+-+} = f(2,3)^{+-} + \alpha(2,3) + \sigma(2,3)$$
$$f(3,3)^{+--} = f(2,3)^{+-} + \alpha(2,3) - \sigma(2,3)$$
$$f(3,3)^{-++} = f(2,3)^{-+} + \alpha(2,3) + \sigma(2,3)$$
$$f(3,3)^{-+-} = f(2,3)^{-+} + \alpha(2,3) - \sigma(2,3)$$
$$f(3,3)^{--+} = f(2,3)^{--} + \alpha(2,3) + \sigma(2,3)$$
$$f(3,3)^{---} = f(2,3)^{--} + \alpha(2,3) - \sigma(2,3).$$

In order to obtain the drifts at times 1 and 2, we will need to know the volatilities $\sigma(1,2)$, $\sigma(1,3)$, and $\sigma(2,3)$. Making the assumption that the volatilities do not change, then

$$\sigma(0,2) = \sigma(1,2)$$
$$\sigma(0,3) = \sigma(1,3) = \sigma(2,3).$$

The covariance matrix at time 1 is as follows:

	2	3
2	$\sigma^2(1,2)$	$\sigma(1,2)\sigma(1,3)$
3	$\sigma(1,3)\sigma(1,2)$	$\sigma^2(1,3)$.

The term $\sigma^2(1,2)$ is the variance of the one-period-ahead forward rate at time 1, and $\sigma^2(1,3)$ is the variance of the two-period-ahead forward rate at time 1. The off-diagonal term $\sigma(1,2)\sigma(1,3)$, which equals the other off-diagonal term $\sigma(1,3)\sigma(1,2)$, is the covariance of the one- and two-period-ahead forward rates at time 1. Filling in the numbers we have

	2	3
2	$\sigma^2(1,2) = 0.00025$	$\sigma(1,2)\sigma(1,3) = (0.015)(0.01) = 0.00015$
3	$\sigma(1,3)\sigma(1,2) = (0.01)(0.015) = 0.00015$	$\sigma^2(1,3) = 0.0001.$

Recall that we explained that the drift can be obtained by taking half of the sum of the terms in the appropriate row and column. That is, if we want the drift $\alpha(i,j)$, we take half the sum of the elements in the ith row and j^{th} column, counting element ij only once. Thus, the drifts are obtained as

$$\alpha(1,2) = \frac{0.000225}{2} = 0.0001125$$

$$\alpha(1,3) = \left(\frac{1}{2}\right)(0.00015 + 0.0001 + 0.00015) = 0.0002.$$

Then the rates at time 2 will be

$$f(2,2)^{++} = f(1,2)^+ + \alpha(1,2) + \sigma(1,2) = 0.0954125 + 0.0001125 + 0.015 = 0.110525$$

$$f(2,3)^{++} = f(1,3)^+ + \alpha(1,3) + \sigma(1,3) = 0.0924 + 0.0002 + 0.01 = 0.1026$$

$$f(2,2)^{+-} = f(1,2)^+ + \alpha(1,2) - \sigma(1,2) = 0.0954125 + 0.0001125 - 0.015 = 0.080525$$

$$f(2,3)^{+-} = f(1,3)^+ + \alpha(1,3) - \sigma(1,3) = 0.0924 + 0.0002 - 0.01 = 0.0826$$

$$f(2,2)^{-+} = f(1,2)^- + \alpha(1,2) + \sigma(1,2) = 0.0654125 + 0.0001125 + 0.015 = 0.080525$$

$$f(2,3)^{-+} = f(1,3)^- + \alpha(1,3) + \sigma(1,3) = 0.0724 + 0.0002 + 0.01 = 0.0826$$

$$f(2,2)^{--} = f(1,2)^- + \alpha(1,2) - \sigma(1,2) = 0.0654125 + 0.0001125 - 0.015 = 0.050525$$

$$f(2,3)^{--} = f(1,3)^- + \alpha(1,3) - \sigma(1,3) = 0.0724 + 0.0002 - 0.01 = 0.0626.$$

Note that because of the constant volatility assumption, some of these rates are duplicates. This is a reflection of the fact that the tree recombines. For example, we have $f(2,2)^{+--} = f(2,2)^{-+}$ and $f(2,3)^{+--} = f(2,3)^{-+}$.

Moving on to extend our tree to time 3, the covariance matrix of rates at time 2 is simple. We have only the rate $\sigma^2(2,3)$, which is 0.0001. Then the drift, $\alpha(2,3) = 0.0001/2 = 0.00005$. The rates we need will be

$$f(3,3)^{+++} = f(2,3)^{++} + \alpha(2,3) + \sigma(2,3) = 0.1026 + 0.0005 + 0.01 = 0.11265$$

$$f(3,3)^{++-} = f(2,3)^{++} + \alpha(2,3) - \sigma(2,3) = 0.1026 + 0.0005 - 0.01 = 0.09265$$

$$f(3,3)^{+-+} = f(2,3)^{+-} + \alpha(2,3) + \sigma(2,3) = 0.0826 + 0.0005 + 0.01 = 0.09265$$

$$f(3,3)^{+--} = f(2,3)^{+-} + \alpha(2,3) - \sigma(2,3) = 0.0826 + 0.0005 - 0.01 = 0.07265$$

$$f(3,3)^{-++} = f(2,3)^{-+} + \alpha(2,3) + \sigma(2,3) = 0.0826 + 0.0005 + 0.01 = 0.09265$$
$$f(3,3)^{-+-} = f(2,3)^{-+} + \alpha(2,3) - \sigma(2,3) = 0.0826 + 0.0005 - 0.01 = 0.07265$$
$$f(3,3)^{--+} = f(2,3)^{--} + \alpha(2,3) + \sigma(2,3) = 0.0626 + 0.0005 + 0.01 = 0.07265$$
$$f(3,3)^{---} = f(2,3)^{--} + \alpha(2,3) - \sigma(2,3) = 0.0626 + 0.0005 - 0.01 = 0.05625.$$

Again, note that the recombining tree means that $f(3,3)^{++--} = f(3,3)^{+--+} = f(3,3)^{--++}$ and $f(3,3)^{+----} = f(3,3)^{--+--} = f(3,3)^{----+}$.

We now have the tree filled in through time 3, and it is illustrated in Figure 27.4.

Now we have fit the entire term structure through three periods. If we have done this correctly, which is guaranteed by using the proper drifts, the tree will be arbitrage free. If that is the case, then we should be able to derive a tree for the price of any instrument, such that each price is the discounted value of the expected price the next period, where expectations are taken using the martingale probability of ½. We shall illustrate this point for a four-period zero-coupon bond. First, let us establish its price at time 0. The initial information given is the set of forward rates $f(0,0)$, $f(0,1)$, $f(0,2)$, and $f(0,3)$. They imply that the price of the bond at time zero should be

$$B(0,4) = e^{-(0.068+0.072+0.08+0.082)} = 0.7393.$$

This is the price of a four-year bond at time 0, and we can verify that by going back to the initial set of bond prices. Now, we should also be able to find this price by applying the equivalent approach of rolling through the tree from maturity back to time 0. We start by pricing this bond at time 3, then working backwards. When we arrive at time zero, we should get this price, subject to perhaps a small round-off error.

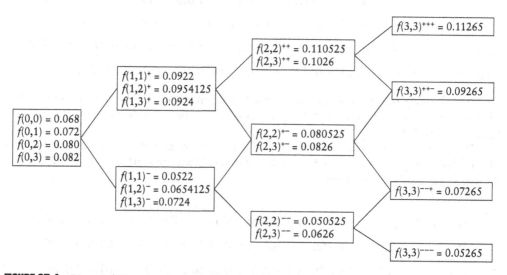

FIGURE 27.4   Binomial Tree over Four Periods

The price at time 3 is obtained by discounting the 1.0 face value at the spot rate at time 3, $f(3,3)$. There are, of course, four states at time 3. Thus, the prices in the four states are

$$B(3,4)^{+++} = e^{-f(3,3)^{+++}} = e^{-0.11265} = 0.8935$$

$$B(3,4)^{++-} = e^{-f(3,3)^{++-}} = e^{-0.09265} = 0.9115$$

$$B(3,4)^{--+} = e^{-f(3,3)^{--+}} = e^{-0.07265} = 0.9299$$

$$B(3,4)^{---} = e^{-f(3,3)^{---}} = e^{-0.05265} = 0.9487.$$

The prices of this bond at time 2 are found by weighting each of the next two possible outcomes by 0.5 and discounting by the one-period spot rate at time 2, $f(2,2)^{++}$, $f(2,2)^{+-}$, and $f(2,2)^{--}$, depending on which state we are in:

$$B(2,4)^{++} = [0.5(0.8935) + 0.5(0.9115)]e^{-0.110525} = 0.8081$$

$$B(2,4)^{+-} = [0.5(0.9115) + 0.5(0.9300)]e^{-0.080525} = 0.8495$$

$$B(2,4)^{--} = [0.5(0.9300) + 0.5(0.9487)]e^{-0.050525} = 0.8931.$$

Stepping back to time 1, we weight each of the next two values by 0.5 and discount by the one-period spot rate at time 1, $f(1,1)^+$ and $f(1,1)^-$, depending on the state:

$$B(1,4)^+ = [0.5(0.8081) + 0.5(0.8495)]e^{-0.0922} = 0.7558$$

$$B(1,4)^- = [0.5(0.8495) + 0.5(0.8931)]e^{-0.0522} = 0.8270.$$

The price at time 0 is found by weighting each of the next two values by 0.5 and discounting by the one-period spot rate at time 0, $f(0,0)$:

$$B(0,4) = [0.5(0.7558) + 0.5(0.8270)]e^{-0.068} = 0.7393.$$

This value differs from the value we originally computed by only a round-off error. Hence, the model is arbitrage free.

27.5   RECAP AND PREVIEW

In this chapter, we developed the continuous time version of the multifactor Heath-Jarrow-Morton model. We then illustrated how the model can be implemented in a binomial world with one factor. We showed that the tree is free of arbitrage.

In Chapter 28, we shall illustrate how this tree can be used to price a wide range of interest rate derivatives.

QUESTIONS AND PROBLEMS

1 In your own words, interpret the Heath-Jarrow-Morton forward rate model expressed as

$$f(t, T) - f(0, T) = \int_0^t \alpha(v, T)dv + \sum_{i=1}^n \int_0^t \sigma_i(v, T)dW(v).$$

2 Based on the HJM forward rate stochastic process, prove the spot rate process follows:

$$r(t) = f(0, t) + \int_0^t \alpha(v, t)dv + \sum_{i=1}^n \int_0^t \sigma_i(v, t)dW_i(v).$$

3 Explain how the one-factor HJM model can be discretized within a binomial model such that

$$f(t + 1, T)^+ = f(t, T) + \alpha(t, T) + \sigma(t, T)$$
$$f(t + 1, T)^- = f(t, T) + \alpha(t, T) - \sigma(t, T).$$

4 From the HJM forward rate stochastic process, the no-arbitrage result can be expressed as

$$df(t, T) = \alpha(t, T)dt + \sum_{i=1}^n \sigma_i(t, T)dW_t(t),$$

where

$$\alpha(t, T) = \sum_{i=1}^n \sigma_i(t, T) \int_t^T \sigma(t, v)dv.$$

5 Within the discretized version of the forward rate stochastic process of the HJM model, compute the appropriate forward rates at time 1 based on the following inputs.

$$f(0, 0) = 0.020$$
$$f(0, 1) = 0.021 \quad \text{and} \quad \begin{array}{l} \sigma(0, 1) = 0.04 \\ \sigma(0, 2) = 0.03. \end{array}$$
$$f(0, 0) = 0.022$$

NOTES

1. That is, given an interest rate change that is normally distributed, the original rate can eventually be driven below zero. By contrast, with proportional rate changes, the rate cannot hit or go below zero.

2. What we mean here can be expressed in simple terms. If we were just going to examine the term structure, we would pick a series of bonds whose maturities range from one to so many years. Let us use T to represent the maturity of the longest maturity bond. That might not necessarily be the longest maturity bond that exists, but it would be one likely chosen for practical reasons, such as data availability or liquidity. In some cases, we might just need to observe the shortest end of the term structure, so T would be short. Regardless, when working with the term structure, we pick a starting point, which is always time 0, and an ending point, which is the maturity of the longest maturity bond. As we shall see here, to fit the HJM model, we shall need to have bond prices and volatilities that go just past the ending point.

3. In this chapter, all forward rates are for transactions of the shortest duration. When working with the HJM model in continuous time, the forward rates are for transactions of instantaneous maturity, and for the discrete case, all forward rates are for one-period transactions. When we say *instantaneous* or *one period*, we, of course, mean the underlying. Thus, a forward rate like $f(0, 3)$ is the rate observed at time 0 for a transaction to start at time 3 and end either instantaneously or one period later.

4. In most other cases, numerical methods such as binomial models can be used to obtain the price. Partial derivatives can then be estimated by numerical approximations.

5. Note that the volatility is not necessarily constant across the term structure or across time. In other words, volatility can change, but it cannot change independent of the level of rates. That is, volatility cannot be independently stochastic. What we mean by this is that volatility is stochastic but is unrelated to the level of rates. Some volatility structures have been proposed in which the volatility is functionally related to the level of rates. In that case, the volatility will be stochastic, but all of the uncertainty is coming from the uncertainty of the interest rate. Models with stochastically dependent volatility are permitted and can oftentimes be easily accommodated because the uncertainty is not more than that already present in the model. In the case of non-stochastic volatility, all volatilities are known, but they are allowed to change as long as the change is known. Also, in the version we present here, there is no distinction between states. That is, volatility is the same at a given time regardless of what level rates are at. In the full HJM model, volatility can also differ by states, though it still must be deterministic or dependently stochastic. As we note shortly, our volatilities will be associated, not with rates, but with factors driving these rates.

6. For some reason, the literature has stressed the point that the drift under the equivalent martingale measure cannot be zero. HJM (1991) and Ritchken (1996) use a simple binomial tree example to show that if a drift of zero is assumed, there is an arbitrage opportunity. It is not clear why anyone would think that the drift should be zero. From the previous formula, it should be clear that the drift cannot be zero if interest rates are stochastic.

7. As noted, the exponentially decaying volatility structure produces a number of closed-form solutions. See Jarrow and Turnbull (2000, Chapters 16 and 17) for details. In addition, Brenner and Jarrow (1993) obtain some closed-form solutions for a special case of a two-factor model. The option pricing formulas in both of these cases bear a striking resemblance to the Black-Scholes-Merton model.

8. If we had T forward rates, this would mean we have T bond prices. Either would imply the other.

9. Technically, we can go one more step to time T but the only bond that would exist over the final time period is a one-period zero-coupon bond, which was the original T-period zero-coupon bond at time 0. With one period to go, this bond would have no uncertainty. Thus, for that last period all we would have is the riskless asset. So, all we can really build is a model with $T - 1$ maturity up to T.

CHAPTER 28

Pricing Fixed-Income Securities and Derivatives Using an Arbitrage-Free Binomial Tree

In Chapter 27, we demonstrated the Heath-Jarrow-Morton arbitrage-free model for the evolution of the term structure. Recall that the tree is specified in terms of continuously compounded forward rates. We showed how the technique of Grant and Vora (1999, 2006) could be used to discretize the model and develop a binomial tree, which we repeat here as Figure 28.1.

First, recall the notation, in which $f(x, y)$ is the continuously compounded forward rate at time x for a one-period transaction to start at time y. Thus, $f(0, 0)$ is the current spot rate (0.068), $f(0, 1)$ is the forward rate observed at time 0 for a one-period loan to start at time 1 (0.072), $f(0, 2)$ is the forward rate observed at time 0 for a one-period loan to start at time 2 (0.08), and $f(0, 3)$ is the forward rate observed at time 0 for a one-period loan to start at time 3 (0.082). The $+$ and $-$ signs in the superscripts indicate the number of up and down moves that have accumulated since time 0 to the current state. Hence, the top number at each node is the current one-period spot rate, with the forward rates below. As an example, the rate $f(2, 3)^{+-}$ rate of 0.0826 means that 8.26% is the continuously compounded forward rate when the spot rate has gone up and then down, so we are at time 2, and the rate applies to a one-period transaction to start at time 3.

In this chapter, we are going to use this model to determine the prices of a very large set of financial instruments. We will first price the simplest instruments, zero-coupon bonds. Then we will price the following instruments: coupon bonds, options on zero-coupon bonds, options on coupon bonds, callable bonds, FRAs, interest rate swaps, interest rate options, interest rate swaptions, and interest rate futures.

28.1 ZERO-COUPON BONDS

At time 0, the price of a one-period zero-coupon bond with face value 1 is easily found as

$$B(0, 1) = B(1) = e^{-0.068} = 0.9343.$$

Note that we suppress the 0 here for notational simplicity. The argument in parentheses is the maturity of the bond. Again, we do not include an argument for the time point we are at, as we shall use the $+$ and $-$ superscripts to indicate where we are. This will cut down on notational clutter.[1] There are no superscripts in this case, so we know we are at time 0.

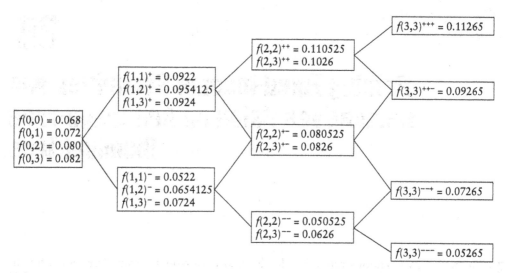

FIGURE 28.1   Heath-Jarrow-Morton Arbitrage-Free Binomial Tree Using the Method of Grant-Vora

We also need the prices at time 0 of zero-coupon bonds with maturities of two, three, and four periods. We can obtain these values by successively discounting at the forward rates as follows:

$$B(2) = e^{-0.068-0.072} = 0.8694$$

$$B(3) = e^{-0.068-0.072-0.08} = 0.8025$$

$$B(4) = e^{-0.068-0.072-0.08-0.082} = 0.7393.$$

If we want to fill in the entire tree of all zero-coupon bond prices, we need only go up through time 3, because the four-period bond will have a price of 1 at time 4. The prices of the four-period bond at time 3 are found by discounting the 1.0 payoff at time 4 at the spot rates in the forward rate tree:

$$B(4)^{+++} = e^{-0.11265} = 0.8935$$

$$B(4)^{++-} = e^{-0.09265} = 0.9115$$

$$B(4)^{--+} = e^{-0.07265} = 0.9299$$

$$B(4)^{---} = e^{-0.05265} = 0.9487.$$

For example, we read these as the prices of the four-period bond when the rate has gone up three times, up twice and down once, down twice and up once, and down three times, respectively.

Stepping back to time 2, let us obtain the prices of the three-period bond:

$$B(3)^{++} = e^{-0.110525} = 0.8954$$

$$B(3)^{+-} = e^{-0.080525} = 0.9226$$

$$B(3)^{--} = e^{-0.050525} = 0.9507.$$

Now, let us obtain the prices of the two-period bond as represented by the four-period bond at time 2. We can discount by the successive spot and forward rates. For example, for the prices of two-period bonds at time 2, we have

$$B(4)^{++} = e^{-0.110525-0.1026} = 0.8081$$

$$B(4)^{+-} = e^{-0.080525-0.0826} = 0.8495$$

$$B(4)^{--} = e^{-0.050525-0.0626} = 0.8930.$$

Of course, this is a bond that matures at time 4, and there have been two time steps so far.

And there is yet another and most important way to get the price of any such bond: We use the binomial valuation formula that weights the next two outcomes and then discounts at the one-period rate. Let us note that discounting at the one-period rate is the same as multiplying by the one-period bond price. So, again, to get the prices of two-period bonds at time 2, we do the following, using the price $B(2)$ in the appropriate state to discount:

$$B(4)^{++} = [0.5(0.8935) + 0.5(0.9115)]0.8954 = 0.8081$$

$$B(4)^{+-} = [0.5(0.9115) + 0.5(0.9299)]0.9226 = 0.8495$$

$$B(4)^{--} = [0.5(0.9299) + 0.5(0.9487)]0.9507 = 0.8930.$$

If we continue and fill in the tree, we obtain Figure 28.2.

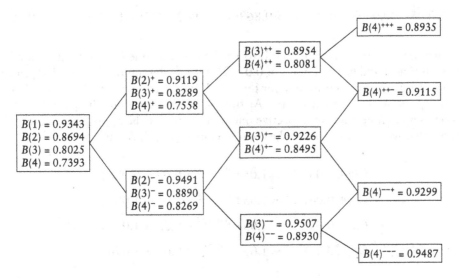

FIGURE 28.2   Binomial Tree of Zero-Coupon Bond Prices

The litmus test to determine if the tree is properly fit is to price each bond by rolling through the tree backwards using the binomial formula, but we already know the prices at time 1. So, let us see if we can obtain them from the prices at time 2:

$$B(2) = \left[0.5(0.9119) + 0.5(0.9491)\right] 0.9343 = 0.8694$$

$$B(3) = \left[0.5(0.8289) + 0.5(0.8890)\right] 0.9343 = 0.8025$$

$$B(4) = \left[0.5(0.7558) + 0.5(0.8269)\right] 0.9343 = 0.7393.$$

Because these prices match the ones we had already obtained, we know the tree is done correctly.

28.2   COUPON BONDS

Coupon bonds make a series of payments of interest and a final payment of principal. As such, they can be viewed as combinations of zero-coupon bonds. That is, each payment is a zero-coupon bond in and of itself.

Consider, for example, a four-period coupon bond with a 5% coupon per period. Its price at time 0 can be found by discounting the individual coupons and the principal at the appropriate zero-coupon bond rate. Thus, it can be viewed as zero-coupon bonds with face value 0.05, maturing at times 1, 2, and 3, and a zero-coupon bond maturing at time 4 with face value 1.05. The discounting can be done by multiplying by the time 0 zero-coupon bond prices for the various maturities in the following manner:

$$CB(4, 0.05) = 0.05(0.9343) + 0.05(0.8694) + 0.05(0.8025) + 1.05(0.7393) = 0.9066,$$

where we use the notation $CB(4, 0.05)$ to be the price at time 0 of a coupon bond that matures at time 4 and has a coupon of 0.05. We take as the standard case a 1.0 principal.

Now, let us fill in the entire tree for this bond. We shall omit time 4 because we know the bond pays off 1.05 in each state. At time 3, we can discount the payment of 1.05 that comes up in time 4 at the respective one-period rate for the appropriate state.[2] Note, however, that we also have to add the coupon paid at time 3. As such,

$$CB(4, 0.05)^{+++} = 1.05e^{-0.11265} + 0.05 = 0.9881$$

$$CB(4, 0.05)^{++-} = 1.05e^{-0.09265} + 0.05 = 1.0071$$

$$CB(4, 0.05)^{+-+} = 1.05e^{-0.07265} + 0.05 = 1.0264$$

$$CB(4, 0.05)^{---} = 1.05e^{-0.05265} + 0.05 = 1.0461.$$

We can continue to step back and obtain the prices at times 2 and 1. And we know that at time 1, the price should equal the 0.9066, the value we previously obtained. Indeed, it will. The tree of coupon bond prices is shown in Figure 28.3.

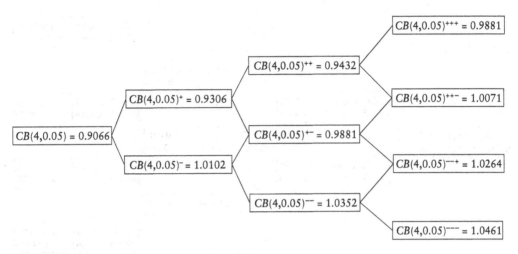

FIGURE 28.3 Binomial Tree of Coupon Bond Prices

28.3 OPTIONS ON ZERO-COUPON BONDS

Now, let us use the tree to price our first options. Specifically, we are going to price European call and put options that expire at time 3 on a four-period zero-coupon bond. Note that we do not care about four-period options on this bond, because the face value of the bond has no uncertainty at time 4: it is 1 for sure. Let us arbitrarily choose an exercise price of 0.9. The prices of the underlying zero-coupon bond are shown in Figure 28.2. They are the prices $B(4)$ in the respective nodes. To introduce options, however, we are going to have to add to the notation a bit, because with options we have two additional parameters to put in parentheses—the expiration of the option and the exercise price. In addition, we need to show information about the underlying.

Let us start at the top node of time 3, where we want the option prices, designated as $cZCB(3, 0.9, 4)$ and $pZCB(3, 0.9, 4)$ for calls and puts, respectively. This notation stands for the price of a call or put expiring at time 3 with an exercise price of 0.9 on a zero-coupon bond maturing at time 4. Let us calculate the prices in this top node:

$$cZCB(3, 0.9, 4)^{+++} = \max(0, 0.8935 - 0.9) = 0.0000$$

$$pZCB(3, 0.9, 4)^{+++} = \max(0, 0.9 - 0.8935) = 0.0065.$$

Stepping down to the three lower nodes, we follow the same pattern. To illustrate the calculation at time 2, we shall need the values:

$$cZCB(3, 0.9, 4)^{++-} = \max(0, 0.9115 - 0.9) = 0.0115$$

$$pZCB(3, 0.9, 4)^{++-} = \max(0, 0.9 - 0.9115) = 0.0000.$$

Now, we shall price the option at time 2 using the binomial discounting method:

$$cZCB(3, 0.9, 4)^{++} = \left[0.5(0.0000) + 0.5(0.0115)\right] 0.8954 = 0.0052$$

$$pZCB(3, 0.9, 4)^{++} = \left[0.5(0.0065) + 0.5(0.0000)\right] 0.8954 = 0.0029.$$

We fill in the rest of the tree and obtain Figure 28.4.

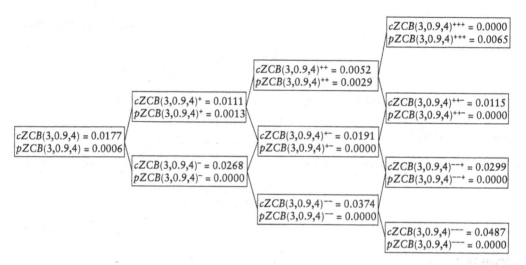

FIGURE 28.4 Binomial Tree of Options on Zero-Coupon Bond Prices

Notice that when interest rates go up, the price of the underlying goes down so puts are more highly valued in the up states, and calls are more highly valued in the down states.

If these options were American style, they could be exercised early. In each node, we would evaluate whether the exercise value would exceed the unexercised value in the node.

28.4 OPTIONS ON COUPON BONDS

Now, we shall look at options on the four-period coupon bond we analyzed in Section 28.2. Recall that its pricing tree is Figure 28.3. Let us examine an option expiring at time 3 with an exercise price of 1.0. We shall assume that at expiration, the exercise of the option would occur an instant before the coupon is paid. In other words, we would exercise in time to receive the coupon. Thus, the values in Figure 28.3 are the relevant values to compare with the exercise price. Let us illustrate the calculated values at time 3:

$$cCB(3, 1.0, 4, 0.05)^{+++} = \max(0, 0.9881 - 1) = 0.0000$$

$$pCB(3, 1.0, 4, 0.05)^{+++} = \max(0, 1 - 0.9881) = 0.0119$$

$$cCB(3, 1.0, 4, 0.05)^{++-} = \max(0, 1.0071 - 1) = 0.0071$$

$$pCB(3, 1.0, 4, 0.05)^{++-} = \max(0, 1 - 1.0071) = 0.0000$$

$$cCB(3, 1.0, 4, 0.05)^{-+-} = \max(0, 1.0264 - 1) = 0.0264$$

$$pCB(3, 1.0, 4, 0.05)^{-+-} = \max(0, 1 - 1.0264) = 0.0000$$

$$cCB(3, 1.0, 4, 0.05)^{---} = \max(0, 1.0461 - 1) = 0.0461$$

$$pCB(3, 1.0, 4, 0.05)^{---} = \max(0, 1 - 1.0461) = 0.0000.$$

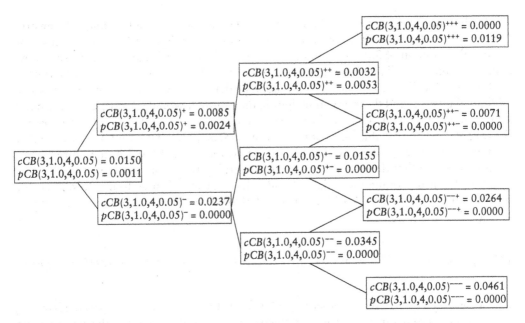

FIGURE 28.5 Binomial Tree of Options on Coupon Bond Prices

Now, let us go to the top node at time 2 and illustrate how the call and put values are calculated using the binomial formula:

$$cCB(3, 1.0, 4, 0.05)^{++} = [0.5(0.0000) + 0.5(0.0071)]0.8954 = 0.0032$$

$$pCB(3, 1.0, 4, 0.05)^{++} = [0.5(0.0119) + 0.5(0.0000)]0.8954 = 0.0053.$$

Continuing in that manner gives us Figure 28.5, the tree of European option prices for the underlying coupon bond.

If these were American options, we would compare the exercise values with the unexercised values and replace the latter with the former if early exercise were justified.

28.5 CALLABLE BONDS

A callable bond is a bond in which the issuer can choose to retire the bond early. This is sometimes done because interest rates have fallen, and a new bond can be issued at a lower rate. There are naturally some costs associated with retiring one bond and issuing another, but these costs can often be far less than the savings from issuing a bond at a lower interest rate. Firms might also retire bonds when they have unexpected available cash that can be used to pay off the remaining amount owed. A callable bond has a call price embedded in it, which is represented by the amount that is paid off to retire it. This price will often be the face value plus some additional interest, which compensates the bond holders for the inconvenience of having their investment prematurely terminated. In addition, the bondholder has to find another investment in an environment in which rates are lower. A callable bond will also have a call period, which means that it cannot be called before one date, nor after another, though the latter might simply be the maturity.

Evaluating the pricing of a callable bond is relatively simple in a binomial framework. Let us price a callable bond using the coupon bond we have previously priced, a four-year bond with a coupon of 0.05. Recall that the tree of its prices is Figure 28.3. We shall now assume the bond is callable as of time 1 at a price of 1.025. Starting off at maturity, time 4, the bond is not callable, and the issuer would not call it anyway, because it has to pay some additional interest, and there are no savings in the future. It would simply pay off the bond by paying the principal, 1.00, and the last period's interest of 0.05. So all outcomes for the callable and noncallable bond at time 4 are 1.05. Stepping back to time 3, we see that the noncallable bond values are 0.9881, 1.0071, 1.0264, and 1.0461 in the four possible outcomes. Recall that these values include the interest payment at time 3.

Now let us determine the callable bond values in each of these four states. First, for the top state, we can see that the bond is worth 0.9881, so there is no reason to call it and pay 1.025:

$$cCB(1, 1.025, 4, 0.05)^{+++} = \min[CB(4, 0.05)^{+++}, 1.025] = \min(0.9881, 1.025) = 0.9881.$$

Of course, you may wonder whether the value of the bond means anything to the issuer. It certainly means something to the holder. Remember that to the issuer it is the present value of the remaining payments. Thus, in this case, if not called, the firm is obligated to make remaining payments with a value in that state of 0.9881. If it calls the bond, it must pay 1.025 immediately. So clearly, it is not optimal to call the bond. It will also not be optimal to call it in the state in which the bond value is 1.0071. It will, however, be optimal to call it in the bottom two outcomes at time 3, as 1.025 is below 1.0264 and 1.0461:

$$cCB(1,1.025,4,0.05)^{++-} = \min[CB(4,0.05)^{++-}, 1.025] = \min(1.0071,1.025) = 1.0071$$
$$cCB(1,1.025,4,0.05)^{--+} = \min[CB(4,0.05)^{--+}, 1.025] = \min(1.0264,1.025) = 1.0250$$
$$cCB(1,1.025,4,0.05)^{---} = \min[CB(4,0.05)^{---}, 1.025] = \min(1.0461,1.025) = 1.0250.$$

Now, let us step back to time 2. At the very top state, where the bond value is 0.9432, it is clearly not optimal to call the bond and pay 1.025. Thus, $cCB(1, 1.025, 4, 0.05)^{++} = 0.9432$. Now, in the middle state, the price of the bond is 0.9881, and it is not optimal to call it and pay 1.025, but 0.9881 is not the value of the callable bond. At this point, the next two outcomes are 1.0071 and 1.0250, the latter a result of the call at time 3. Thus, we have to take a weighted average of the next two outcomes and discount by the appropriate one-period factor, $B(3)^{+-} = 0.9226$. In addition, we have to add the interest payment at time 3 of 0.05, and we have to again check and see if all of this exceeds 1.025, in which case it would be called. The remaining results for time 2 are

$$cCB(1, 1.25, 4, 0.05)^{+-} = \min\{[0.5(1.0071) + 0.5(1.025)]0.9226 + 0.05, 1.025\} = 0.9874$$
$$cCB(1, 1.25, 4, 0.05)^{--} = \min\{[0.5(1.0250) + 0.5(1.025)]0.9507 + 0.05, 1.025\} = 1.0245.$$

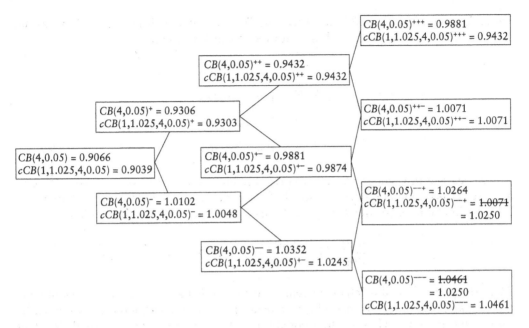

FIGURE 28.6   Binomial Tree of Noncallable and Callable Bond Prices

So clearly it would not be called at time 2. We would then step back to time 1 and apply the same procedure, and we would obtain callable bond prices of 0.9303 and 1.0048, and of course, the bond would not be called at time 1. At time 0, the bond is not callable at all and the callable bond price would be 0.9039, which is lower than the price of the bond if it were not callable, 0.9066. The lower price compensates the buyer for the risk of having it called. That risk is realized when the bond is called, the bondholder is paid back the principal, and then has to invest it elsewhere in a lower interest rate environment. The full tree is presented in Figure 28.6.

28.6   FORWARD RATE AGREEMENTS (FRAS)

Recall that an FRA, or forward rate agreement, is a forward contract on an interest rate. Let us assume that we are interested in an FRA on the one-period spot rate. Let the FRA expire at time 3. Recall that the equilibrium fixed rate that the long agrees to pay would be the forward rate. From Figure 28.1, that rate would be 8.2%. There is a slight inconsistency, however, in that 8.2% is a continuously compounded rate, inasmuch as the HJM model is based on the evolution of continuously compounded rates. FRA payoffs are always based on add-on rates. Thus, we need to respecify the term structure in terms of one-period add-on rates. Let us take the one-period continuously compounded spot rate at time 0 of 6.8%. Thus, the present value of 1.0 in one period would be $e^{-0.068} = 0.9343$, a number we have previously used as the price of a one-period zero-coupon bond at time 0. The one-period add-on rate, denoted simply as r, would, therefore, be $0.9343(1 + r) = 1$, which gives us $r = 7.0365\%$. We can get this rate directly, however, as $e^{0.068} - 1$. From Figure 28.1, we know that the one-period spot rates are the top numbers in each cell, so we can convert each to its add-on equivalent. With the FRA expiring at time 3, we would need the four possible one-period rates at time 3, which would be

$$e^{0.11265} - 1 = 0.119240$$

$$e^{0.09265} - 1 = 0.097078$$

$$e^{0.07265} - 1 = 0.075354$$

$$e^{0.05265} - 1 = 0.050461.$$

We learned in Chapter 26 that we price the FRA as the forward rate. We can actually use zero-coupon bond prices instead of forward rates. We would have the price of a three-period bond, $B(3)$, divided by the price of a four-period bond, $B(4)$, minus 1:

$$\frac{B(3)}{B(4)} - 1 = \frac{0.8025}{0.7393} - 1 = 0.085456.$$

Thus, the parties would reach an agreement at time 0 that at time 3, the long would pay this rate and receive the one-period spot rate at time 3, with the difference discounted by the current spot rate.[3] Thus, the four possible payoffs of the FRA at time 3 are their values at time 3, as follows:

$$VFRA(3,0.085456)^{+++} = (0.119240 - 0.085456)\,0.8935 = 0.0302$$

$$VFRA(3,0.085456)^{++-} = (0.097078 - 0.085456)\,0.9115 = 0.0106$$

$$VFRA(3,0.085456)^{--+} = (0.075354 - 0.085456)\,0.9299 = -0.0094$$

$$VFRA(3,0.085456)^{---} = (0.054061 - 0.085456)\,0.9487 = -0.0298.$$

Now, we step back to time 2 and calculate the possible values of the FRA based on the next two payments,

$$VFRA(3,0.085456)^{++} = \left[0.5(0.0302) + 0.5(0.0106)\right]0.8954 = 0.0183$$

$$VFRA(3,0.085456)^{+-} = \left[0.5(0.0106) + 0.5(-0.0094)\right]0.9226 = 0.0006$$

$$VFRA(3,0.085456)^{--} = \left[0.5(-0.0094) + 0.5(-0.0298)\right]0.9507 = -0.0186.$$

Usually, we would just show the remaining values in the tree, but it is important to see the following result. Let us step back to time 1 and calculate the possible values of the FRA:

$$VFRA(3,0.085456)^{+} = \left[0.5(0.0183) + 0.5(0.0006)\right]0.9119 = 0.0086$$

$$VFRA(3,0.085456)^{-} = \left[0.5(0.0006) + 0.5(-0.0186)\right]0.9491 = -0.0086.$$

Notice the symmetry of these values. This result ensures us that the value at time 1 will be zero, which is precisely what we know has to be true for an FRA priced at the equilibrium forward rate. Thus, to summarize, Figure 28.7 shows the tree of values of the FRA.

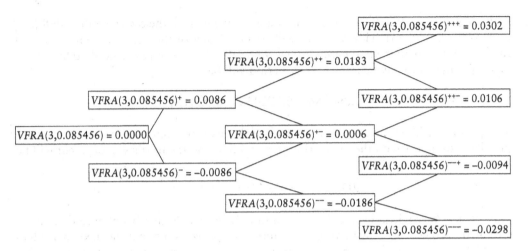

FIGURE 28.7   Binomial Tree of FRA Values

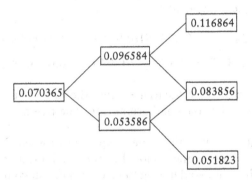

FIGURE 28.8   Tree of One-Period Add-on Spot Rates Through Time 2

28.7   INTEREST RATE SWAPS

Let us price and value a swap that expires at time 2 in which the underlying is the one-period rate. We shall now need to create the tree of one-period add-on rates. In the previous section covering FRAs, we showed how the one-period continuously compounded rate converts to the add-on rate, but FRAs have only one payment so we did not need the entire tree. Swaps have multiple payments, so we do need the full tree. Following the conversion procedure as we did in the previous section, we obtain Figure 28.8.

In Chapter 26, we covered how to obtain the equilibrium swap rate. For a swap that expires at time 3, the rate will be given from the prices of one- and two-period zero-coupon bonds.

$$\frac{1 - 0.8694}{0.9343 + 0.8694} = 0.072433.$$

We know that the first swap payment, which occurs at time 1, is based on the starting spot rate of 7.0365%. That payment will be

$$0.070365 - 0.072433 = -0.002068.$$

Note that the rate, 0.070365, is the continuous equivalent of the discrete rate 0.068 (i.e., $e^{0.068} - 1 = 0.070365$). This payment will occur in each of the next two states. Interest rates go up to where the one-period rate is 9.6584%. The swap payment of −0.002068 is made and the swap payment in the next period will be

$$0.096584 - 0.072433 = 0.024151.$$

This payment will occur regardless of which of the next two outcomes occurs. Now, stepping back to time 1 with the one-period rate at 5.3586%, the next swap payment will be

$$0.053586 - 0.072433 = -0.018847.$$

Now, let us step back and value the swap at time 1. We can do that by taking a discounted weighted average of the next two swap values. Except we know that the next two values are 0.024151 for certain, because they were determined at the last period, meaning that we do not have to take a weighted average. We simply discount that value back to the present and add the current payment:

$$VIRS(2, 0.072433)^{+} = 0.024151(0.9119) - 0.002068 = 0.019956$$

$$VIRS(2, 0.072433)^{-} = -0.018847(0.9491) - 0.002068 = -0.019956.$$

The symmetry of these two values guarantees that the value at time zero will be zero, as it should for a swap priced at the equilibrium rate. The tree of swap values is shown in Figure 28.9.

You will note that here we valued the swap a bit differently from the way we did in Chapter 26, where we added the notional to the fixed and floating sides. That was a necessary trick that helped us avoid having to value specific outcomes. But with a binomial tree, we can value specific outcomes. Just to show you that we obtain the same results, let us verify the value of 0.019956 that appears in the top state at time 1 in Figure 28.9. Remember that in Chapter 26 we showed that the present value of the remaining floating payments plus the notional is 1. The present value of the remaining fixed payment plus notional of 1 is easily found by discounting 1 plus the upcoming fixed payment at the one-period rate, which would be done by multiplying by the one-period zero-coupon bond

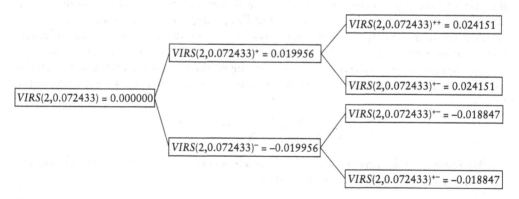

FIGURE 28.9   Binomial Tree of Swap Values

price of 0.9119. Subtracting the present value of the fixed payments plus notional from the present value of the floating payments plus notional, which is equal to 1, gives the value of the swap:

$$1.0 - 1.072433(0.9119) - 0.002068 = 0.019956.$$

And this is the same value we obtained by discounting the values as in the tree.

28.8   INTEREST RATE OPTIONS

With the tree of one-period add-on rates that we showed in Figure 28.8, we can easily value interest rate options. Remember that when interest rate options expire, the payoff is determined but not actually made until the next period.[4]

To illustrate, let us value three-period interest rate calls and puts with an exercise rate of 0.09. Following the conversion method we described previously, Figure 28.10 shows the tree of one-period add-on rates through time 3.

The values of the interest rate calls and puts at time 3 are, therefore,

$$VIRC(3, 0.09)^{+++} = \max(0, 0.119240 - 0.09)\,0.8935 = 0.0261$$
$$VIRP(3, 0.09)^{+++} = \max(0, 0.09 - 0.119249)\,0.8935 = 0.0000$$
$$VIRC(3, 0.09)^{++-} = \max(0, 0.097078 - 0.09)\,0.9115 = 0.0065$$
$$VIRP(3, 0.09)^{++-} = \max(0, 0.09 - 0.097078)\,0.9115 = 0.0000$$
$$VIRC(3, 0.09)^{--+} = \max(0, 0.075354 - 0.09)\,0.9299 = 0.0000$$
$$VIRP(3, 0.09)^{--+} = \max(0, 0.09 - 0.075354)\,0.9299 = 0.0136$$
$$VIRC(3, 0.09)^{---} = \max(0, 0.054061 - 0.09)\,0.9487 = 0.0000$$
$$VIRP(3, 0.09)^{---} = \max(0, 0.09 - 0.054061)\,0.9487 = 0.0341.$$

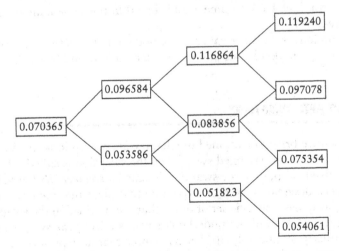

FIGURE 28.10   Binomial Tree of One-Period Add-on Spot Rates Through Time 3

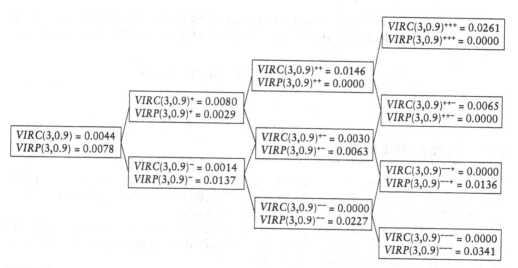

FIGURE 28.11 Binomial Tree of Interest Rate Option Values

Now, stepping back to time 2, we simply do the binomial weighted-average discounted value,

$$VIRC(3,0.09)^{++} = [0.5(0.0261) + 0.5(0.0065)]0.8954 = 0.0146$$

$$VIRP(3,0.09)^{++} = [0.5(0.0000) + 0.5(0.0000)]0.8954 = 0.0000$$

$$VIRC(3,0.09)^{+-} = [0.5(0.0065) + 0.5(0.0000)]0.9226 = 0.0030$$

$$VIRP(3,0.09)^{+-} = [0.5(0.0000) + 0.5(0.0136)]0.9226 = 0.0063$$

$$VIRC(3,0.09)^{--} = [0.5(0.0000) + 0.5(0.0000)]0.9507 = 0.0000$$

$$VIRP(3,0.09)^{--} = [0.5(0.0136) + 0.5(0.0341)]0.9507 = 0.0227.$$

We would then continue back to time 1 and time 0 in the same manner. The results are shown in Figure 28.11.

Again, if these options were American, at each node of the tree, the exercise values would be compared to each market value, and the greater value would be used.

28.9 INTEREST RATE SWAPTIONS

Swaptions, which we briefly mentioned in Chapter 26, are options to enter into a swap. If the underlying swap is a pay-fixed swap, then the swaption is called a *payer swaption*. If it is a receive-fixed swap, then the swaption is called a *receiver swaption*. In either case, the swaption has a designated fixed or strike rate that plays the role of an exercise price. The underlying swap has a specific set of terms that are specified in the swaption contract. Say the strike rate is 7%, and the underlying swap is a five-year swap with semiannual payments on January 30 and July 30. Let us assume that at expiration of the swaption, the equilibrium rate on the underlying swap is more than 7%. Then a payer swaption is in-the-money, because it allows the holder to enter into the underlying swap and pay 7% fixed to receive floating, while that same swap in the market would require paying more than 7% fixed to receive floating. The swaption holder can either maintain the 7% swap or can offset it by going into the market and entering into the opposite swap with the same counterparty, thereby paying floating and receiving more than 7% fixed. The floating sides effectively cancel and what remains is an annuity of the difference between the market rate and 7%. If that swaption were structured to settle in cash, the seller of the swap would pay the buyer the present value of that annuity. Letting the market rate at expiration of the swaption be less than 7% would result in a receiver swaption being in the money, and a similar process would apply.

Using our binomial model, let us create a swaption with a 7% strike and expiration of time 1 with the underlying swap being a three-period swap with payments made at each time step. In order to price this swap, we shall need to know the possible rates on three-year swaps at time 1. We have already covered how swap rates are determined. Now, consider the top state at time 1 where the prices of one-, two-, and three-year zero-coupon bonds were shown in Figure 28.2 as 0.9119, 0.8289, and 0.7558. Thus, the rate on a three-year swap at that point in the tree would be

$$SR(3)^+ = \frac{1 - 0.7558}{0.9119 + 0.8289 + 0.7558} = 0.097822.$$

In the bottom state at time 1, the three zero-coupon bond prices are 0.9491, 0.8890, and 0.8269. Thus, the swap rate would be

$$SR(3)^- = \frac{1 - 0.8269}{0.9491 + 0.8890 + 0.8269} = 0.064932.$$

So, assuming we are in the top state at time 1, the rate on three-year swaps would be 9.7822%, which is above the strike rate, so the payer swaption would be exercised and the receiver swaption would expire unexercised. Now, let us determine the value of the payer swaption on exercise. We are creating a swap to pay 7% and receive the floating rate, but we can offset that swap by entering into a swap in the market to pay the floating rate and receive 9.7822%. This process creates a three-year annuity of $0.097822 - 0.07 = 0.027822$. The value of this payer swap is easily derived by finding the present value of this annuity:

$$VpSW(1, 0.07, 3)^+ = (0.097822 - 0.07)(0.9119 + 0.8289 + 0.7558) = 0.0695.$$

The receiver swaption is worth nothing, because it is out-of-the-money. That is, $VrSW(1, 0.07, 3)^+ = 0.0000$. In the bottom state, it should be apparent that the payer swaption is worth nothing, $VpSW(1, 0.07, 3)^- = 0.0000$, and the receiver swaption is worth

$$VrSW(1, 0.07, 3)^- = (0.07 - 0.064932)(0.9491 + 0.8890 + 0.8269) = 0.0135.$$

Now we roll back to time 0 and discount the weighted-average of the next two possible swaption values:

$$VpSW(1, 0.07, 3) = [0.5(0.0695) + 0.5(0.0000)] 0.9343 = 0.0324$$

$$VrSW(1, 0.07, 3) = [0.5(0.0000) + 0.5(0.0135)] 0.9343 = 0.0063.$$

This information is summarized in Figure 28.12.

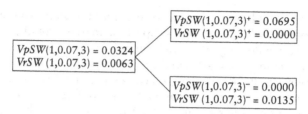

FIGURE 28.12   Binomial Tree of Swaption Values

28.10   INTEREST RATE FUTURES

We could price a futures on any underlying that we have already covered, such as add-on interest rates, continuously compounded rates, zero-coupon bonds, and coupon bonds. We shall illustrate how futures pricing works in a binomial framework by pricing a futures on the one-period add-on interest rate. This process will be very similar to how Eurodollar futures, the most widely traded futures contract, work in practice.

Let us specify that we are interested in a two-period futures on the add-on rate. At expiration, the futures price will automatically settle to the one-period add-on spot rate. The tree of one-period add-on spot rates was given in Figure 28.8. Hence, the futures will settle at either 11.6864%, 8.3856%, or 5.1823%. The final mark-to-market or settlement will be the difference between that price and the previous price.

Recall that a futures is marked-to-market, and that mark-to-market value is the value of the futures. Once the mark-to-market value is paid from one party to the other, the contract resets itself to the current futures price so that its value is pulled back to zero. At that point, the expected gain from the futures must be zero, because there is no money invested. Because a futures contract must have zero expected gain, the following conditions must hold to determine the futures price at time 1:

$$0.5\left[0.116864 - f(2)^{+}\right] + 0.5\left[0.083856 - f(2)^{+}\right] = 0.0000$$
$$0.5\left[0.083856 - f(2)^{-}\right] + 0.5\left[0.051823 - f(2)^{-}\right] = 0.0000.$$

Notice that we do no discounting of the weighted average settlement, because it would have no effect. The terms on the left-hand sides must equal zero without discounting. It is also easy to see that the futures price at time 1 in either state will be the average of the next two possible futures prices. Thus,

$$f(2)^{+} = 0.5(0.116864) + 0.5(0.083856) = 0.10036$$
$$f(2)^{-} = 0.5(0.083856) + 0.5(0.051823) = 0.06784.$$

And the futures price at time 0 will be the weighted average of these two prices:

$$0.5(0.10036) + 0.5(0.06784) = 0.0841.$$

Now let us determine the mark-to-market value of the futures as it evolves through the tree. We use an asterisk (*) in the exponent to indicate that this is the value before the settlement:

$$vf(2)^{+++} = 0.116864 - 0.10036 = 0.016504$$

$$vf(2)^{+-*} = 0.083856 - 0.10036 = -0.016504$$

$$vf(2)^{-+*} = 0.083856 - 0.06784 = 0.016016$$

$$vf(2)^{--*} = 0.051823 - 0.06784 = -0.016016.$$

Note in particular that the outcome in which rates go up then down is not the same as when rates go down first and then up. These four amounts are the payouts from the settlement at expiration. Now, let us back up to time 1 and compute the value before the settlement:

$$vf(2)^{+*} = 0.10036 - 0.0841 = 0.01626$$

$$vf(2)^{-*} = 0.067840 - 0.0841 = -0.01626.$$

Now, let us verify that all of these values are correct by determining the expectation of the next-period values through the tree:

$$vf(2)^{+} = 0.5(0.016504) + 0.5(-0.016504) = 0.0000$$

$$vf(2)^{-} = 0.5(0.016016) + 0.5(-0.016016) = 0.0000$$

$$vf(2) = 0.5(0.01626) + 0.5(-0.01626) = 0.0000.$$

Technically these values should be discounted, but clearly that is unnecessary. These results are summarized in Figure 28.13.

With the methodology learned here, we would be able to price virtually any futures following the same procedure.

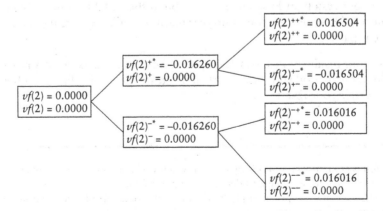

FIGURE 28.13   Binomial Tree of Futures Prices and Values

28.11   RECAP AND PREVIEW

In this chapter, we showed how to use the binomial version of the one-factor Heath-Jarrow-Morton model to price almost every type of interest rate derivative. We showed that the technique involves discounting the risk-neutral expected value at the upcoming nodes, with discounting at the risk-free rate.

This material completes Part VI on the pricing of interest rate derivatives. Part VII is a collection of miscellaneous topics. In Chapter 29, we show how option prices are related to the prices of state contingent claims.

QUESTIONS AND PROBLEMS

1   Based on material in Chapter 28 and within the discretized version of the Heath-Jarrow-Morton forward rate stochastic process, compute the appropriate forward rate tree and bond price tree at times 1, 2, and 3, based on the following inputs. The next four problems build on these results.

$$f(0,0) = 0.030 \qquad \sigma(0,1) = 0.02$$
$$f(0,1) = 0.029 \text{ and } \sigma(0,2) = 0.03$$
$$f(0,2) = 0.028 \qquad \sigma(0,3) = 0.01.$$
$$f(0,3) = 0.027$$

2   Based on your results in Problem 1, build the coupon bond price tree assuming a 1.5% stated coupon and four-year maturity. Assume 1.0 par.

3   Based on your results in Problems 1 and 2, build the call and put option on coupon bond price tree assuming a 1.0 exercise price, three-year option expiration, 1.5% stated coupon, and four-year maturity. Assume 1.0 par and expiration occurs an instant before the coupon is paid.

4   Based on your results in Problem 1 and 3, build the callable coupon bond price tree assuming a 1.5% stated coupon, four-year maturity, callable starting in year 1, and the call price is par (1.0).

5   Based on your results in Problem 1, compute the at-market fixed FRA rate expiring at time 3 for a one-period spot rate. Further, illustrate the receive floating FRA value tree and thus verify your results.

NOTES

1. The notation is going to get a bit messy later on when we have to put the arguments for both the characteristics of the underlying and derivative in parentheses.
2. Alternatively, we could simply multiply by the price of a one-period zero-coupon bond, and we shall do it this way later in this chapter.
3. Recall that with FRAs, the payoff is subject to discounting, because the payoff is made at expiration, but is based on a rate that represents interest that would be paid later if it were a loan.
4. Of course, that is also what we did with swaps.

Miscellaneous Topics

CHAPTER 29

Option Prices and the Prices of State-Contingent Claims

An option is sometimes referred to as a contingent claim. A contingent claim is a special type of asset that provides a payoff that is dependent (contingent) on something specific happening. An option is a contingent claim in that it provides a positive payoff under the condition that the option expires in-the-money. If the option does not expire in-the-money, the payoff is obviously zero. In this chapter, we look at another form of a contingent claim that happens to be an asset that pays 1.0 in a given outcome and zero otherwise. These outcomes are referred to as *states* or *states of nature*, and this particular asset is sometimes called a *state-contingent claim*. Other common names for this type of security are *pure asset*, the term we shall use, *fundamental asset*, and *Arrow-Debreu asset*, the last arising out of the work of Nobel Laureates Kenneth Arrow (1964) and Gerard Debreu (1964) and the extension to valuation by Myers (1968).[1] In this chapter, we examine some properties of pure assets and demonstrate how they relate to options in a framework called *state preference theory* and occasionally *state pricing theory*.

State pricing theory, which is also sometimes known as *time state preference theory*, provides a framework for the valuation of financial assets. It can be shown to provide a general equilibrium theory of asset pricing, consistent with a market in which assets are risky, and investors have homogeneous beliefs and aversion to risk. State preference theory was developed around the same time as the capital asset pricing model but has not received as much attention as the capital asset pricing model. This is probably because state preference theory is a more abstract theoretical framework, relying as it does on the existence of pure assets, whose prices cannot be observed in financial newspapers, from the Bloomberg, or on the internet. It is more appropriately viewed as what one would see if one took a microscopic look at the financial markets.

In order to understand this material, we shall have to tread carefully with the terminology. We shall make reference to assets that could be stocks, bonds, or commodities, and we shall call them *complex assets*, though there is nothing particularly complex about them. A share of Google stock is a complex asset, a barrel of oil is a complex asset, a bond issued by GE is a complex asset. They are complex assets in that they are made up of pure assets. In some ways, this idea is analogous to the fact that humans (and animals) are complex combinations of fundamental elements. In fact, you should think of this chapter as taking a microscopic look at a financial market and learning about the basic building blocks that enable markets to generate prices. It so happens that pure assets play a tremendous role in the functioning of a market. As such, we can gain a better appreciation for the value that options provide markets and society as a whole.

29.1 PURE ASSETS IN THE MARKET

Suppose we face a risky situation, which could be as basic as a commitment of capital to an investment in the stock market. Let us define the possible outcomes in terms of three states in which the market goes down 2% (state 1), the market is unchanged (state 2), or the market goes up 2% (state 3). Of course, in reality, the possible outcomes are infinite and cannot be reduced to such simple statements, but the framework provided by this simplification is, nonetheless, useful and generalizes to the case of a continuous spectrum of states.

Consider a complex asset that will be worth 110 in state 1, 100 in state 2, and 90 in state 3. Consider another complex asset that will be worth 105 in state 1, 101 in state 2, and 98 in state 3. Suppose the risk-free rate is 2%. Then a risk-free asset worth 100 today will have a value of 102 in each state.

Now consider a pure asset, which pays 1.0 in state 1 and zero in all other states. Another pure asset pays 1.0 in state 2 and zero in all other states. A third pure asset pays 1.0 in state 3 and zero in all other states. Our first complex asset, whose three possible future values are 110, 100, and 90, can be viewed as a portfolio of 110 units of the first pure asset, 100 units of the second pure asset, and 90 units of the third pure asset. Our second complex asset can be viewed as a portfolio of 105 units of the first pure asset, 101 units of the second pure asset, and 98 units of the third pure asset. A risk-free asset worth 100 today can be viewed as 102 units of all three pure assets. The price of a pure asset is called a *state price*. A particular pure asset pays off 1.0 if a particular state occurs and zero otherwise. The price of that pure asset, thus, is the amount of money you would need to pay to receive a fixed payoff in that state of the world.

It follows that the price of each complex asset must be the value today of the equivalent portfolio of pure assets. In other words, if we know the state prices, we can determine the prices of all complex assets. Alternatively, if we know the prices of the complex assets, we can determine the state prices. We shall see how this is done in this chapter.

These pure assets are the fundamental assets in the market. We cannot see them or trade them, but they are there, similar to financial atoms in the market. Let us now develop a formal framework for understanding these concepts.

29.2 PRICING PURE AND COMPLEX ASSETS

First, we state without proof the fact that there must always be at least as many assets as there are states. This principle is referred to as the *spanning property*, which means that the pure assets must be sufficient to reproduce any complex asset. Here we shall make the number of assets equal to the number of states. Specifically let there be n states, where each state is identified as state i, $i = 1, 2, \ldots, n$, and there are n complex assets, with each complex asset identified with the index j, $j = 1, 2, \ldots, n$. Each complex asset has price S_j. Let H_{ij} be the payoff of complex asset j in state i. A complex asset can be defined in terms of the number of units of each pure asset required to replicate the payoffs of the complex asset. We can alternatively define each pure asset in terms of the number of units of each complex asset required to replicate its outcomes. Define pure asset i as an asset that pays 1.0 in state i and zero in all other states. Then let α_{ij} be the number of units of asset j that should be held to reproduce the payoff of pure asset i. Alternatively, we can view the payoff H_{ij} as the number of units of pure asset i that are implicit in complex asset j. Let us now organize this information in a more meaningful way. We shall use both matrix and scalar notation, though the matrix notation is somewhat more useful in facilitating the solution of the simultaneous equations that will arise here.

As stated, a pure asset is a combination of complex assets. The payoffs of pure asset 1 in each of the possible states are as follows:

$$\alpha_{11}H_{11} + \alpha_{12}H_{12} + \ldots + \alpha_{1n}H_{1n} = 1 \text{ (outcome in state 1)}$$

$$\alpha_{11}H_{21} + \alpha_{12}H_{22} + \ldots + \alpha_{1n}H_{2n} = 0 \text{ (outcome in state 2)}$$

$$\ldots$$

$$\alpha_{11}H_{n1} + \alpha_{12}H_{n2} + \ldots + \alpha_{1n}H_{nn} = 0 \text{ (outcome in state } n).$$

In other words, pure asset 1 is a combination of α_{11} units of complex asset 1, α_{12} units of complex asset 2, . . ., and α_{1n} units of complex asset n.

Similarly the payoffs of pure asset 2 in each of the possible states are as follows:

$$\alpha_{21}H_{11} + \alpha_{22}H_{12} + \ldots + \alpha_{2n}H_{1n} = 0 \text{ (outcome in state 1)}$$

$$\alpha_{21}H_{21} + \alpha_{22}H_{22} + \ldots + \alpha_{2n}H_{2n} = 1 \text{ (outcome in state 2)}$$

$$\ldots$$

$$\alpha_{21}H_{n1} + \alpha_{22}H_{n2} + \ldots + \alpha_{2n}H_{nn} = 0 \text{ (outcome in state } n).$$

Pure asset 2 is, thus, a combination of α_{21} units of complex asset 1, α_{22} units of complex asset 2, and α_{2n} units of complex asset n.

The payoffs of pure asset n in each of the possible states are as follows:

$$\alpha_{n1}H_{11} + \alpha_{n2}H_{12} + \ldots + \alpha_{nn}H_{1n} = 0 \text{ (outcome in state 1)}$$

$$\alpha_{n1}H_{21} + \alpha_{n2}H_{22} + \ldots + \alpha_{nn}H_{2n} = 0 \text{ (outcome in state 2)}$$

$$\ldots$$

$$\alpha_{n1}H_{n1} + \alpha_{n2}H_{n2} + \ldots + \alpha_{nn}H_{nn} = 1 \text{ (outcome in state } n).$$

Pure asset n is, thus, a combination of α_{n1} units of complex asset 1, α_{n2} units of complex asset 2, and α_{nn} units of complex asset n.

These conditions can be easily expressed in matrix notation as follows. The H matrix represents the payoffs of each asset in each state:

$$\mathbf{H} = \begin{vmatrix} H_{11} & H_{12} & \cdots & H_{1n} \\ H_{21} & H_{22} & \cdots & H_{2n} \\ \cdot & \cdot & \cdots & \cdot \\ \cdot & \cdot & \cdots & \cdot \\ \cdot & \cdot & \cdots & \cdot \\ H_{n1} & H_{n2} & \cdots & H_{nn} \end{vmatrix}. \tag{29.1}$$

The matrix of weights is

$$
\mathbf{A} = \begin{vmatrix}
\alpha_{11} & \alpha_{12} & \cdots & \alpha_{1n} \\
\alpha_{21} & \alpha_{22} & \cdots & \alpha_{2n} \\
\cdot & \cdot & \cdots & \cdot \\
\cdot & \cdot & \cdots & \cdot \\
\cdot & \cdot & \cdots & \cdot \\
\alpha_{n1} & \alpha_{n2} & \cdots & \alpha_{nn}
\end{vmatrix}.
\tag{29.2}
$$

As is often the case in matrix algebra, we shall need the identity matrix:

$$
\mathbf{I} = \begin{vmatrix}
1 & 0 & \cdots & 0 \\
0 & 1 & \cdots & 0 \\
\cdot & \cdot & \cdots & \cdot \\
\cdot & \cdot & \cdots & \cdot \\
\cdot & \cdot & \cdots & \cdot \\
0 & 0 & \cdots & 1
\end{vmatrix}.
\tag{29.3}
$$

The key relationship is $\mathbf{HA'} = \mathbf{I}$. We can then solve for the weights using the expression[2]

$$
\mathbf{A} = (\mathbf{H}^{-1}\mathbf{I})'.
\tag{29.4}
$$

We can also obtain this result using scalar notation. Letting I_{ij} be the ij^{th} element of matrix $\mathbf{I}$, then

$$
I_{ij} = \sum_{k=1}^{n} H_{ik}\alpha_{jk}.
\tag{29.5}
$$

In other words, the rows and columns of matrix $\mathbf{I}$ are

$$
I_{11} = H_{11}\alpha_{11} + H_{12}\alpha_{12} + \ldots + H_{1n}\alpha_{1n} = 1
$$
$$
I_{12} = H_{11}\alpha_{21} + H_{12}\alpha_{22} + \ldots + H_{1n}\alpha_{2n} = 0
$$
$$
\cdot
$$
$$
\cdot
$$
$$
\cdot
$$
$$
I_{1n} = H_{11}\alpha_{n1} + H_{12}\alpha_{n2} + \ldots + H_{1n}\alpha_{nn} = 0
$$
$$
I_{21} = H_{21}\alpha_{11} + H_{22}\alpha_{12} + \ldots + H_{2n}\alpha_{1n} = 0
$$
$$
I_{22} = H_{21}\alpha_{21} + H_{22}\alpha_{22} + \ldots + H_{2n}\alpha_{2n} = 1
$$
$$
\cdot
$$
$$
\cdot
$$
$$
\cdot
$$

$$I_{2n} = H_{21}\alpha_{n1} + H_{22}\alpha_{n2} + \ldots + H_{2n}\alpha_{nn} = 0$$

$$\cdot$$

$$\cdot$$

$$\cdot$$

$$I_{n1} = H_{n1}\alpha_{11} + H_{n2}\alpha_{12} + \ldots + H_{nn}\alpha_{1n} = 0$$

$$I_{n2} = H_{n1}\alpha_{21} + H_{n2}\alpha_{22} + \ldots + H_{nn}\alpha_{2n} = 0$$

$$\cdot$$

$$\cdot$$

$$\cdot$$

$$I_{nn} = H_{n1}a_{n1} + H_{n2}a_{n2} + \ldots + H_{nn}a_{nn} = 1. \tag{29.6}$$

Now let us introduce a vector $\boldsymbol{\Psi}$, where the element ψ_i is the price today of pure asset i:

$$\boldsymbol{\Psi} = \begin{vmatrix} \psi_1 \\ \psi_2 \\ \cdot \\ \cdot \\ \cdot \\ \psi_n \end{vmatrix}. \tag{29.7}$$

The price of a complex asset can be obtained, in matrix notation, as

$$\mathbf{S} = \mathbf{H'}\boldsymbol{\Psi}. \tag{29.8}$$

or in scalar notation as

$$S = H'\Psi. \tag{29.9}$$

The equations written out are as follows:

$$S_1 = H_{11}\psi_1 + H_{21}\psi_2 + \ldots + H_{n1}\psi_n$$

$$S_2 = H_{12}\psi_1 + H_{22}\psi_2 + \ldots + H_{n2}\psi_n$$

$$\cdot$$

$$\cdot$$

$$\cdot$$

$$S_n = H_{1n}\psi_1 + H_{2n}\psi_2 + \ldots + H_{nn}\psi_n. \tag{29.10}$$

Here we see how the complex assets are combinations of the pure assets. Alternatively, we can obtain the state prices from the prices of the complex asset. This would be found as

$$\boldsymbol{\Psi} = (\mathbf{H'})^{-1}\mathbf{S}. \tag{29.11}$$

Alternatively, one could solve for the ψ_i values in the scalar equations for S_j shown in Equation (29.10).

In other words, a complex asset j can be priced by multiplying its payoff in each state by the price of the pure asset that pays off in that given state. In this way, we see that the payoffs of complex assets, or what we ordinarily just call *assets*, can be expressed in terms of the payoffs of more fundamental assets, those whose payoffs are contingent on the given states.

The risk-free asset is extremely easy to see in this context. Its payoffs are the same in all states. Denoting the risk-free asset with 1.0 current price as asset S_r and its payoff as R, we have

$$S_r = \sum_{k=1}^{n} R\psi_k = R \sum_{k=1}^{n} \psi_k. \tag{29.12}$$

One plus the risk-free rate is, by definition, R/S_r. Consequently, the risk-free rate, which we write as r, is given as[3]

$$r = \left(\frac{1}{\sum_{k=1}^{n} \psi_k} \right) - 1. \tag{29.13}$$

We see that the risk-free rate is just the inverse of the sum of the state prices minus 1. This result should make sense. A risk-free asset is one that pays 1.0 in each state. Thus, a portfolio of one unit of each pure asset will replicate the payoff of the risk-free asset. It follows that the sums of the values of one unit of each pure asset will give the present value of 1.0, which will be the price of the risk-free asset. Inverting the price gives one plus the rate.

29.3 NUMERICAL EXAMPLE

Let there be four states and four complex assets. The payoffs of these assets are shown in the columns of the matrix $\mathbf{H}$ here, where the rows are the states and the columns are the assets:

$$\mathbf{H} = \begin{vmatrix} 100 & 75 & 20 & 85 \\ 100 & 100 & 65 & 72 \\ 100 & 125 & 90 & 135 \\ 100 & 150 & 118 & 110 \end{vmatrix}.$$

It is also apparent that the first complex asset is the risk-free asset, whose payoffs are in the first column. The $\mathbf{A}$ matrix will, of course, be

$$\mathbf{A} = \begin{vmatrix} \alpha_{11} & \alpha_{12} & \alpha_{13} & \alpha_{14} \\ \alpha_{21} & \alpha_{22} & \alpha_{23} & \alpha_{24} \\ \alpha_{31} & \alpha_{32} & \alpha_{33} & \alpha_{34} \\ \alpha_{41} & \alpha_{42} & \alpha_{43} & \alpha_{44} \end{vmatrix}.$$

We can solve for **A** by obtaining $A = (H^{-1})'$.[4] The solution is

$$A = \begin{vmatrix} -0.0236 & 0.0624 & -0.0574 & -0.0020 \\ 0.0564 & -0.0822 & 0.0653 & -0.0091 \\ 0.0181 & -0.0626 & 0.0418 & 0.0242 \\ -0.0409 & 0.0825 & -0.0496 & -0.0131 \end{vmatrix}.$$

Let us assume we observe the prices of the complex assets in the market as:

$$S = \begin{vmatrix} 92.00 \\ 118.00 \\ 85.86 \\ 99.36 \end{vmatrix}.$$

Then we can find the prices of the pure assets as $\Psi = (H')^{-1}S$. The solution is

$$\Psi = \begin{vmatrix} 0.06 \\ 0.18 \\ 0.26 \\ 0.42 \end{vmatrix}. \tag{29.14}$$

In scalar notation, this would be found by solving the equations:

$$92 = 100\psi_1 + 100\psi_2 + 100\psi_3 + 100\psi_4$$
$$118 = 75\psi_1 + 100\psi_2 + 125\psi_3 + 150\psi_4$$
$$85.86 = 20\psi_1 + 65\psi_2 + 90\psi_3 + 118\psi_4$$
$$99.36 = 85\psi_1 + 72\psi_2 + 135\psi_3 + 110v_4.$$

Alternatively, if we had the prices of the pure assets, we could obtain the prices of the complex assets as $S = H'\Psi$ or in the previous scalar equations, inserting values for each ψ and leaving the left-hand sides as the unknowns.

The risk-free rate is then $(1/\iota\Psi) - 1$ or simply

$$r = \frac{1}{0.06 + 0.18 + 0.26 + 0.42} - 1 = 0.086957.$$

29.4 STATE PRICING AND OPTIONS IN A BINOMIAL FRAMEWORK

In this section, we draw on the work of Banz and Miller (1978) and Breeden and Litzenberger (1978). Consider a one-period binomial option pricing world. Let an asset worth V today be worth either Vu or Vd one period later, where u and d are one plus the return on the asset in each of the two outcomes. From what we have previously learned about state pricing, we know that V must be $Vu\psi_1 + Vd\psi_2$. Let us divide through by V and also specify the formula for the risk-free rate in terms of the state prices:

$$1 = u\psi_1 + d\psi_2$$

$$r = \left(\frac{1}{\psi_1 + \psi_2}\right) - 1. \tag{29.15}$$

Solving these equations simultaneously for ψ_1 and ψ_2, we obtain

$$\psi_1 = \frac{1 + r - d}{(1 + r)(u - d)}$$

$$\psi_2 = \frac{u - (1 + r)}{(1 + r)(u - d)}. \tag{29.16}$$

You should note the similarity of these formulas to those of the binomial probability for ϕ and $1 - \phi$, which is repeated here:

$$\phi = \frac{1 + r - d}{(u - d)}$$

$$1 - \phi = \frac{u - (1 + r)}{(u - d)}, \tag{29.17}$$

Compare the formula for ϕ in Equation (29.17), which is the risk-neutral probability of the asset going up, to the formula for ψ_1, which is the price for the state in which the asset goes up. Note that $\psi_1 = \phi/(1 + r)$. Usually when we obtain the value of ϕ, we determine $1 - \phi$ by simple subtraction, but it can be shown that $1 - \phi = (u - (1 + r))/(u - d)$. Then, it is easily seen that $\psi_2 = (1 - \phi)/(1 + r)$. Thus, in general, the state price is the risk-neutral probability discounted at the risk-free rate. Alternatively, if we are given the risk-free rate, we can easily solve for the state prices as

$$\psi_1 = \frac{p}{1 + r}$$

$$\psi_2 = \frac{1 - p}{1 + r}. \tag{29.18}$$

We know that we can obtain the price of a call option on this asset with an exercise price of X by using the standard binomial pricing formula from Chapter 7,

$$c = \frac{\phi c_u + (1 - \phi)c_d}{1 + r}, \tag{29.19}$$

where $c_u = \max(0, Su - X)$ and $c_d = \max(0, Sd - X)$. Let us demonstrate these results with an example. Consider the following scenario. We have an asset priced at 100 that can go up to either 140 or down to 75 in the next period. Thus, $u = 140/100 = 1.40$ and $d = 75/100 = 0.75$. The risk-free rate is 3%. Using the standard binomial pricing approach, we first calculate the risk-neutral probability as

$$\phi = \frac{1.03 - 0.75}{1.40 - 0.75} = 0.4308.$$

And, of course, $1 - \phi = 0.5692$. The payoffs of the call are obviously 40 and 0. The call price today is found as

$$c = \frac{0.4308(40) + 0.5692(0)}{1.03} = 16.73.$$

Now let us look at how this problem is consistent with state pricing. Given the risk-neutral probabilities of 0.4308 and 0.5692, we can find the state prices as

$$\psi_1 = \frac{0.4308}{1.03} = 0.4182$$

$$\psi_2 = \frac{0.5692}{1.03} = 0.5527.$$

The call price in terms of state prices is

$$c = c_u \psi_u + c_d \psi_d. \tag{29.20}$$

Substituting for our numerical example, we arrive at the previous value of

$$c = 40(0.4182) + 0(0.5527) = 16.73.$$

We have three financial instruments: a stock, an option, and a riskless bond. To obtain the state prices from the prices of the complex assets, we need as many instruments as there are states. Thus, we need only two of the three assets at a time. Using the formulas, we previously developed, we shall re-derive the state prices, which should be 0.4182 and 0.5527. Let us first use the stock and risk-free bond. We can set the price of the risk-free bond to anything as long as its payoff is 3% higher than its price. Let us just set it at 100, the same as the stock price. Then our **H** matrix is

$$\mathbf{H} = \begin{vmatrix} 103 & 140 \\ 103 & 75 \end{vmatrix}.$$

Our **S** vector, the prices of the assets, is

$$\mathbf{S} = \begin{vmatrix} 100 \\ 100 \end{vmatrix}.$$

Then performing the matrix operations $(\mathbf{H}')^{-1}\mathbf{S}$, we obtain

$$\mathbf{\Psi} = \begin{vmatrix} -0.0112 & 0.0154 \\ 0.0209 & -0.0154 \end{vmatrix} \begin{vmatrix} 100 \\ 100 \end{vmatrix} = \begin{vmatrix} 0.4182 \\ 0.5527 \end{vmatrix},$$

which are the correct values for the state prices.

Alternatively, we could use the option and the stock. Then our **H** matrix would be

$$\mathbf{H} = \begin{vmatrix} 40 & 140 \\ 0 & 75 \end{vmatrix}.$$

Our S vector would be

$$S = \begin{vmatrix} 16.73 \\ 100 \end{vmatrix}.$$

Then we can find the prices of the pure assets as $\Psi = (H')^{-1}S$. The solution is

$$\Psi = \begin{vmatrix} 0.0250 & -0.0467 \\ 0.0000 & 0.0133 \end{vmatrix} \begin{vmatrix} 16.73 \\ 100 \end{vmatrix} = \begin{vmatrix} 0.4182 \\ 0.5527 \end{vmatrix}.$$

Finally, using the option and the risk-free bond, our **H** matrix is

$$H = \begin{vmatrix} 40 & 103 \\ 0 & 103 \end{vmatrix}.$$

The S vector will be

$$S = \begin{vmatrix} 16.73 \\ 100 \end{vmatrix}.$$

Then we can find the prices of the pure assets as $\Psi = (H')^{-1}S$. The solution is

$$\Psi = \begin{vmatrix} 0.0250 & -0.0000 \\ -0.0467 & 0.0133 \end{vmatrix} \begin{vmatrix} 16.73 \\ 100 \end{vmatrix} = \begin{vmatrix} 0.4182 \\ 0.5527 \end{vmatrix}.$$

Thus, it does not matter which assets we use, but it is good to see that the result is internally consistent.

29.5   STATE PRICING AND OPTIONS IN CONTINUOUS TIME

In the real world, there are an infinite number of possible states. This makes it difficult, if not impossible, to identify the specific states and derive their prices. It is possible, however, to make some approximations of state prices from the prices of traded options.

As discussed in Chapter 13, a standard European call option on an asset can be decomposed into two components. One is a long position in an *asset-or-nothing option*, which pays the value of the asset if its price at expiration exceeds the exercise price and nothing otherwise. The other component is a short position in a certain number of *cash-or-nothing options*, which obligates the seller to pay a certain amount of money if the asset price at expiration exceeds the exercise price and nothing otherwise. The number of such options is the exercise price.

Letting c be the call price, S be the underlying asset price, X be the exercise price, r_c be the continuous risk-free rate, σ be the volatility of the return on the asset price, and τ be the time to expiration, the value of the European call is well known as the Black-Scholes-Merton formula,

$$c = SN(d_1) - Xe^{-r_c\tau}N(d_2)$$

$$d_1 = \frac{\ln(S/X) + (r_c + \sigma^2/2)\tau}{\sigma\sqrt{\tau}},$$

and

$$d_2 = \frac{\ln(S/X) + (r_c - \sigma^2/2)\tau}{\sigma\sqrt{\tau}} = d_1 - \sigma\sqrt{\tau}.$$

As we discussed in Chapter 13, the value of the asset-or-nothing component of a standard European option is $SN(d_1)$ and the value of the cash-or-nothing component is $Xe^{-r_c\tau}N(d_2)$. For our purposes here, we need the value of a more general cash-or-nothing option, one that pays off 1.0 if it expires with the asset value above the exercise price and zero otherwise. And as covered in Chapter 13, such an option is sometimes called a *digital* or *binary option* because it pays off zero or 1.0. Let us denote the price of the digital call option as dc, and we see that its value is

$$dc = e^{-r_c\tau}N(d_2).$$

Given the Black-Scholes-Merton put option pricing formula,

$$p = Xe^{-r_c\tau}\left[1 - N(d_2)\right] - S\left[1 - N(d_1)\right],$$

we can find the price of a digital put, which is an option that pays 1.0 if the asset value at expiration is less than the exercise price. Its formula is

$$dp = e^{-r_c\tau}\left[1 - N(d_2)\right].$$

These digital option formulas are also the partial derivatives of the Black-Scholes-Merton call and put formulas with respect to the exercise price.[5] Now let us divide the uncertain outcomes into three possibilities. Let S_T be the value of the asset at a specific future date, S_1 be one possible level of the asset, and S_2 be another possible level of the asset where $S_2 > S_1$. Thus, we know that the asset value must fall within any of a number of ranges, as specified here:

$$S_T \leq S_1$$

$$S_1 < S_T \leq S_2$$

$$S_T > S_2.$$

These ranges are collectively exhaustive. As such, they completely define the state space. Although this specification oversimplifies the real world, it does allow us to work with three easily identifiable states from which we can determine the three state prices.

We wish to find the prices of the pure assets that span the state space. Our problem, therefore, is as follows:

- What is the price of an asset that pays 1.0 if $S_T \leq S_1$?
- What is the price of an asset that pays 1.0 if $S_1 < S_T \leq S_2$?
- What is the price of an asset that pays 1.0 if $S_T > S_2$?

It turns out that we can use digital calls and puts to obtain the answer.

A digital put with an exercise price of S_1 is an asset that pays 1.0 if the first state, $S_T \leq S_1$, occurs and zero otherwise. Thus, its price is the price of the first pure asset. The second pure asset is identical to a long position in a digital call with an exercise price of S_1 and a short position in a digital call with an exercise price of S_2. To see this, note that if $S_T \leq S_1$, both options expire out-of-the-money so there is no payoff. If $S_1 < S_T \leq S_2$, the long digital call struck at S_1 pays 1.0 and the short digital call struck at S_2 pays nothing for a total payoff of 1.0. If $S_T > S_2$ the long digital call struck at S_1 pays 1.0 and the short digital call struck at S_2 will require a payment of 1.0, thereby offsetting and leaving a zero payoff. The third pure asset, which pays 1.0 if the state $S_T > S_2$ occurs, can be replicated with a long digital call with an exercise price of S_2.

Let us put some numbers on these results by deriving the prices of these pure assets from the prices of options on a particular stock index priced at 10,000, with a volatility of 20% and a yield of 2%. The continuously compounded risk-free rate is 2.5%. We consider two options that expire in 60 days, so $\tau = 60/365 = 0.1644$. The first option has an exercise price of 9,000, and the second an exercise price of 11,000. With this information, we can find the state prices for this index being below 9,000, between 9,000 and 11,000, and above 11,000.

Using the Black-Scholes-Merton option pricing model, we obtain the following values: $e^{-r_c\tau} = e^{-0.025(0.1644)} = 0.9959$, $N(d_2|X = 9,000) = 0.8978$ and $N(d_2|X = 11,000) = 0.1139$. From this, we obtain the following prices for the digital options:

digital call struck at 9,000	0.8941 (= 0.9959*0.8978)
digital put struck at 9,000	0.1018 [= 0.9959*(1 − 0.8978)]
digital call struck at 11,000	0.1135 (= 0.9959*0.1139).

Thus, our state prices are

pure asset 1	0.1018
pure asset 2	0.7806 (= 0.8941 − 0.1135)
pure asset 3	0.1135

We can then obtain the risk-free rate over that period as $1/(0.1018 + 0.7806 + 0.1135) - 1 = 0.004118$. This result is consistent with the 2.5% continuously compounded rate because the equivalent discrete compounded rate for 60 days would be $e^{0.025(60/365)} - 1 = 0.004118$.

29.8   RECAP AND PREVIEW

With the development of option pricing theory, state preference theory has stepped to the back in the family of valuation models. Although, as we have seen here, state preference theory is clearly consistent with option pricing theory, the implications of the latter are much easier to observe in the real world, and, hence, it has become more widely used in practice as well as in scholarly work. Keep in mind, however, that just as a biologist cannot simply observe a specimen with the naked eye and expect to learn much about it, so must a serious student of finance observe the internal structure of the financial pricing process. State preference theory provides the framework to accomplish that task.

In Chapter 30, we examine what option pricing theory implies for the expected returns on options and how this relates to the expected returns on the underlying assets.

QUESTIONS AND PROBLEMS

1   Compare and contrast pure assets with complex assets.

2   Suppose we have four states and four complex assets with the payoffs in the H*H* matrix. Further, assume we observe the prices of the complex assets have all been normalized to 100 (the *S* matrix). Compute the prices of the pure assets in this setting:

$$H = \begin{vmatrix} 105 & 65 & 20 & 40 \\ 105 & 95 & 150 & 130 \\ 105 & 130 & 110 & 200 \\ 105 & 150 & 180 & 110 \end{vmatrix} \quad S = \begin{vmatrix} 100.00 \\ 100.00 \\ 100.00 \\ 100.00 \end{vmatrix}.$$

3   With pure assets one must be careful with rounding. Based on the pure asset values given next, compute the implied discrete compounded interest rate when the pure asset values are rounded at the first decimal place. Repeat this exercise for rounding at the second through sixth decimal place. Discuss your findings for the six cases (rounding at first, second, third, fourth, fifth, and sixth decimal place), explaining the role of rounding in computing pure asset values:

$$\Psi = \begin{vmatrix} 0.366548 \\ 0.187780 \\ 0.143015 \\ 0.273531 \end{vmatrix}.$$

4   Suppose you have an asset priced at 50 that can go either up to 60 or down to 40 in the next year. The risk-free rate is 5% (discrete compounding). Consider a call option with exercise price of 50. First, compute the correct call option price. Identify three approaches to solving for the state prices and demonstrate that each approach results in the same state prices.

5   Using just call option prices, explain how to construct pure asset prices when given five different exercise prices for one-month options.

NOTES

1. Instead of the word *asset*, some articles and books use the term *security*, but a security is a claim issued by a company. We shall not use that term here, preferring the more general term *asset*.
2. To obtain the inverse of **H**, we require the condition that no row or column of **H** is a linear function of any other row or column. This condition will always hold if no complex asset is a linear function of any other combination of complex assets. Otherwise, that asset would be redundant.
3. To solve for r_f in matrix notation we would introduce an $n \times 1$ row vector, ι, which contains 1 as each element. Then $r_f = (1/\iota\Psi) - 1$.
4. The matrix operations of transposing, multiplying, and taking the inverse can be easily done using Excel's array formulas =transpose(), =mmult(), and =minverse().
5. The derivative of the call formula with respect to the exercise price has a minus sign, which would have to be ignored if one were using the derivative as the price of a digital option.

CHAPTER 30

Option Prices and Expected Returns

In the study of finance in general, we devote considerable time to deriving and using pricing models. Probably the two best known pricing models are the capital asset pricing model and the Black-Scholes-Merton option pricing model.[1] The former, commonly referred to as the CAPM, provides the required rate of return on the asset. If the asset is correctly priced, which the model assumes must happen in equilibrium, the return expected by investors, known as the expected rate of return, equals the required rate of return. But asset pricing models are not usually expressed in terms of the price of the asset, whereas option pricing models are almost always expressed in terms of the price of the option. In this chapter, we shall connect the expected return and price for both the asset and the option. In addition, we shall tie together the expected returns and volatilities of the option to that of the asset. Much of this work draws from Rubinstein (1984).

30.1 THE BASIC FRAMEWORK

The CAPM is as follows:

$$E(R_s) = r + [E(R_m) - r] \beta_s, \tag{30.1}$$

where

$E(R_s)$ = expected and required return on asset s

r = risk − free rate

$E(R_m)$ = expected and required return on the market portfolio of all risky assets

β_s = beta of asset s, obtained as

$$\frac{\text{cov}(R_s, R_m)}{\text{var}(R_m)}$$

and

$\text{cov}(R_s, R_m)$ = covariance betweeen return on asset s and market portfolio m

σ_m^2 = variance of the return on market portfolio m.

Note that here we use the terms *expected return* and *required return* interchangeably. As noted, the expected return is the return the investor expects, given the market price. The required return is the return that is justified given the risk. The latter assumes the market is in equilibrium, along with a number of other assumptions. Although the CAPM technically gives the required return, this rate is usually referred to as the expected return, as noted by the use of the expectations operator.

The CAPM is not typically expressed in the form of the price of an asset, though that can be done using a model specifying how the asset price is obtained from its future cash flow. For example, define the one-period return on an asset S as

$$R_s = \frac{S'}{S} - 1, \tag{30.2}$$

where S' is the asset price one period later.[2] The expected return would be

$$E(R_s) = \frac{E(S')}{S} - 1. \tag{30.3}$$

Note that Equations (30.1) and (30.3) both provide the expected return, the left-hand side, in terms of an expression on the right-hand side. Using the definition of beta as stated and equating these two specifications for the expected return, we have

$$\frac{E(R_s)}{S} - 1 = r + [E(R_m) - r]\frac{\text{cov}(R_s, R_m)}{\sigma_m^2}. \tag{30.4}$$

Given the rules for covariance, we can also say that

$$\text{cov}(R_s, R_m) = \text{cov}\left(\frac{S'}{S} - 1, R_m\right).$$

$$= \frac{1}{S}\text{cov}(S', R_m).$$

We can substitute this result into (30.4) and solve for S to obtain

$$S = \frac{E(S') - \lambda\,\text{cov}(S', R_m)}{1 + r},$$

where

$$\lambda = \frac{E(R_m) - r}{\sigma_m^2}. \tag{30.5}$$

This equation is the CAPM written in the form of a price. In the numerator, we have the expected value of the asset at the next date, $E(S')$ minus the risk premium, $\lambda\text{cov}(S', R_m)$. The term λ is the market risk premium and reflects the average level of risk in the market. The numerator is a risk-adjusted future value of the asset, which is then discounted by the risk-free rate to obtain the price. Financial economists typically refer to this form of the model as a *certainty equivalent*, a concept that we have previously encountered.[3]

Option pricing models, in contrast to asset pricing models, are nearly always written in the form of a price. In the case of Black-Scholes-Merton, which we derived in Chapter 13, the price is, of course,

$$c = SN(d_1) - Xe^{-r_c\tau}N(d_2),$$ (30.6)

where

$$d_1 = \frac{\ln(S/X) + (r_c + \sigma^2/2)\,\tau}{\sigma\sqrt{\tau}},$$

and

$$d_2 = \frac{\ln(S/X) + (r_c - \sigma^2/2)\,\tau}{\sigma\sqrt{\tau}} = d_1 - \sigma\sqrt{\tau}.$$

As you should recall, the Black-Scholes-Merton model is obtained in continuous time by forming a risk-free hedge consisting of the underlying asset and the option. The option price formula is derived as the solution given the constraint that the riskless portfolio must earn the risk-free rate to prevent arbitrage.

It is rare that the option pricing model is expressed in the form of the expected return of the option. We provide the linkage between the notion of an equilibrium expected return and an arbitrage-free price of a call option. The same result can be obtained if the option is a put with nothing but minor notational changes.

30.2 EXPECTED RETURNS ON OPTIONS

Consider a hedge portfolio consisting of h shares of the asset and one short call option. The portfolio value is $hS - c$. One period later the portfolio will be worth

$$h(S + \Delta S) - (c + \Delta c),$$ (30.7)

where the symbol Δ means the change in S or c. If this portfolio is hedged, its value should grow at the risk-free rate. Hence, the following condition must hold:

$$(hS - c)(1 + r) = h(S + \Delta S) - (c + \Delta c).$$ (30.8)

Gathering option terms on the left-hand side and asset terms on the right-hand side and dividing by c, we obtain the following result:

$$\frac{\Delta c}{c} = r + \left(\frac{\Delta S}{S} - r\right)h\left(\frac{S}{c}\right).$$ (30.9)

This expression says that the option return is the sum of the risk-free rate and an ex-post risk premium, $\Delta S/S - r$, times a risk factor $h(S/c)$. Recall that the Black-Scholes-Merton model tells us that

$$h = \frac{\partial c}{\partial S},$$ (30.10)

which should be recognized as the option's delta. Expressing the delta in discrete time, we have

$$h = \frac{\Delta c}{\Delta S}.$$ (30.11)

Then the risk factor is

$$\frac{\Delta c}{\Delta S}\frac{S}{c} = \frac{\Delta c/c}{\Delta S/c}.$$ (30.12)

Economists will recognize any term reflecting the percentage movement in one variable divided by the percentage movement in another as the concept of *elasticity*. Thus, we see that the return on the option is related to the return on the asset by the risk-free rate, a risk premium, and a term reflecting the option's elasticity.

Elasticity in this context measures the sensitivity of the option to the asset and, hence, is a reflection of the leverage of the option. Elasticity is closely related to the concept of the delta of the option, $\partial c/\partial S$, which is also the hedge ratio, but delta is an absolute measure, capturing the movement of the option price relative to the movement in the asset price. Elasticity is a relative measure and, as such, is more appropriate when dealing with rates of return.

The elasticity of a standard European call is at least equal to 1.[4] This means that the absolute value of the option return will exceed the absolute value of the asset return. The elasticity of an option is often denoted with the Greek letter omega Ω. Thus, our equation for the return on the option is

$$\frac{\Delta c}{c} = r + \left(\frac{\Delta S}{S} - r\right)\Omega.$$ (30.13)

Now with this result, we can examine the expected return on the option. Taking the expectation of this equation, we obtain

$$E\left(\frac{\Delta c}{c}\right) = r + \left[E\left(\frac{\Delta S}{S}\right) - r\right]\Omega.$$ (30.14)

The left-hand side, $E(\Delta c/c)$, is the expected return on the call, which we denote as $E(R_c)$. Within the right-hand side, the term $E(\Delta S/S)$ is the expected return on the asset, which we denote as $E(R_S)$. Now we have a simple equation for the expected return on the call:

$$E(R_c) = r + (E(R_s) - r)\Omega.$$ (30.15)

We see that the expected return on the call equals the risk-free rate plus the risk premium on the underlying asset times the elasticity of the option. This functional form is very appealing and intuitive. The option's expected return at a minimum is the risk-free rate and is increased by a risk premium, which is the product of the risk premium on the asset and the risk of the option relative to the asset. We can also write this as

$$E(R_c) = r + [E(R_s) - r]\frac{\partial c}{\partial S}\left(\frac{S}{c}\right).$$ (30.16)

Now let us try to determine the volatility of the option.

30.3 VOLATILITIES OF OPTIONS

Of course, one naturally thinks of volatility as important for an option, but the volatility we refer to in that context is the volatility of the asset. Now we are interested in the volatility of the option. Using the expression created previously for the return on the option as a function of the return on the asset, we take the variance of the option return:

$$\mathrm{var}\left(\frac{\Delta c}{c}\right) = \mathrm{var}\left\{ r + \left[\left(\frac{\Delta S}{S} \right) - r \right] \Omega \right\}$$

$$= \mathrm{var}\left\{ \left(\frac{\Delta S}{S} \right) \Omega \right\}$$

$$= \Omega^2 \, \mathrm{var}\left(\frac{\Delta S}{S} \right). \tag{30.17}$$

Expressing this result in terms of the standard deviation results in

$$\sigma_c = \Omega \sigma_S. \tag{30.18}$$

And we can also express it as

$$\sigma_c = h\left(\frac{S}{c} \right) \sigma_S. \tag{30.19}$$

We see that the volatility of the option is the volatility of the asset times the elasticity. Thus, the risk of the option is the risk of the asset times the risk of the option relative to the asset. This result should seem intuitive.

These results concerning expected returns and volatilities of options apply regardless of how expected returns are determined on the underlying asset. In the special case that the CAPM explains expected returns on assets, we can obtain some further insights.

30.4 OPTIONS AND THE CAPITAL ASSET PRICING MODEL

Recall that the CAPM is a model for the pricing of all risky assets. An option is a risky asset. Therefore, if the CAPM holds for all risky assets, the expected return on the option must also be governed by the CAPM. Hence, the following equation applies:

$$E(R_c) = r + \left[E(R_m) - r \right] \beta_c, \tag{30.20}$$

where β_c is the beta of the call and is recognized as its risk with respect to the market portfolio. Substituting our CAPM equation for the expected return on the asset into the CAPM for the expected return on the option gives

$$E(R_c) = r + \left[E(R_s) - r \right] \Omega$$

$$= r + \left(\left\{ r + \left[E(R_m) - r \right] \beta_s \right\} - r \right) \Omega$$

$$= r + \left[E(R_m) - r \right] \beta_s \Omega. \tag{30.21}$$

Hence, the option beta is given as

$$\beta_c = \beta_s \Omega, \tag{30.22}$$

indicating that the beta of the option is the beta of the asset times the elasticity. Once again, we see the role that elasticity plays. As we noted, elasticity is a relative measure, that is, the percentage change in the option return divided by the percentage change in the asset return. Because beta is a relative measure, naturally elasticity plays an important part in the relationship of the beta of the option to the beta of the asset.

30.5 OPTIONS AND THE SHARPE RATIO

A widely used measure of investment performance is the Sharpe ratio. For a portfolio with return R_p and volatility σ_p, the Sharpe ratio is

$$\text{Sharpe}_p = \frac{R_p - r}{\sigma_p}. \tag{30.23}$$

The Sharpe ratio measures the return in excess of the risk-free rate, which is the numerator, relative to the total risk, which is the denominator. Using our measures of return and volatility for an option, Equations (30.9) and (30.19), the Sharpe ratio for a call option is

$$
\begin{aligned}
\text{Sharpe}_c &= \frac{R_c - r}{\sigma_c} \\
&= \frac{r + \left(\frac{\Delta S}{S} - r \right) h \left(\frac{S}{c} \right) - r}{h \left(\frac{S}{c} \right) \sigma_s} \\
&= \frac{\left(\frac{\Delta S}{S} - r \right)}{\sigma_s}.
\end{aligned} \tag{30.24}
$$

In other words, the Sharpe ratio for an option is simply the Sharpe ratio for the asset. This result is likely to surprise some, but an explanation is simple. The Sharpe ratio measures whether an investment provided a risk premium in excess of the appropriate risk premium for its level of risk. If the option is correctly priced relative to its underlying asset, it cannot provide a risk premium beyond that already provided by the asset. The option merely leverages the risk premium of the asset. And the leverage used by the option, although an advantage in augmenting the return, is an offsetting disadvantage in augmenting the risk. That is, the Sharpe ratio is a proportional measure of performance. If the numerator and denominator change by the same proportion, the Sharpe ratio cannot change.

Of course, if the option is mispriced, then the arbitrage linkage between option and asset that enabled us to obtain the above results is broken. An option would then provide a form of excess return, and the Sharpe ratio of the option would exceed that of the asset.[5]

30.6   THE STOCHASTIC PROCESS FOLLOWED BY THE OPTION

When deriving the option pricing model, we assume that the asset follows a stochastic process of the form

$$dS = \alpha_s S dt + S \sigma_s dW, \tag{30.25}$$

where dW is a standard Brownian motion. To obtain the option pricing model, we require the stochastic process for the asset. We do not, however, specifically require a full analysis of the option's stochastic properties to obtain its price.[6] Studies of the performance of option strategies and of option pricing models, however, invariably make use of probability models and statistical rules. Hence, it may be important for some purposes to know what the stochastic process of the option would look like. In general, we would expect it to be of the form

$$dc = \alpha_c(S, t) dt + \sigma_c(S, t) dW, \tag{30.26}$$

where we note that the option's expected return and volatility are functions of S and t.

Recall that in deriving the option pricing formula, we use Itô's lemma, which expresses the change in the option price as a function of first- and second-order changes in the asset price and time,

$$dc = \frac{\partial c}{\partial S} dS + \frac{\partial c}{\partial t} dt + \frac{1}{2} \frac{\partial^2 c}{\partial S^2} dS^2. \tag{30.27}$$

Substituting the stochastic process of the asset for dS and noting that dS^2 is the well-known result $S^2 \sigma_S^2 dt$, we obtain

$$dc = \left(\frac{\partial c}{\partial S} S \alpha_s + \frac{1}{2} \frac{\partial^2 c}{\partial S^2} S^2 \sigma_s^2 + \frac{\partial c}{\partial t} \right) dt + \left(\frac{\partial c}{\partial S} S \sigma_s \right) dW. \tag{30.28}$$

Dividing by c, we now have a statement for the return on the call,

$$\frac{dc}{c} = \left(\frac{\frac{\partial c}{\partial S} S \alpha_s + \frac{1}{2} \frac{\partial^2 c}{\partial S^2} S^2 \sigma_s^2 + \frac{\partial c}{\partial t}}{c} \right) dt + \left(\frac{\partial c}{\partial S} \frac{S}{c} \sigma_s \right) dW. \tag{30.29}$$

We can write this more compactly as

$$\frac{dc}{c} = \alpha_c dt + \sigma_c dW, \tag{30.30}$$

where

$$\alpha_c = \left(\frac{1}{c} \right) \left(\frac{\partial c}{\partial S} S \alpha_s + \frac{1}{2} \frac{\partial^2 c}{\partial S^2} S^2 \sigma_s^2 + \frac{\partial c}{\partial t} \right) \text{ and} \tag{30.31}$$

$$\sigma_c = \frac{\partial c}{\partial S} \frac{S}{c} \sigma_s. \tag{30.32}$$

Now, suppose we wish to obtain the expected return, μ_c, and volatility, σ_c, of the option. We must first recognize the dimension of the parameters of the model. In the option pricing model, the return on the option, dc/c, is a measure over an infinitesimal time interval. We can take the expected return, but it will be multiplied by dt and will reflect the expectation over this very short time interval. The CAPM deals with returns over a finite interval.[7]

Taking expectations of Equation (30.29), we obtain[8]

$$E\left(\frac{dc}{c}\right) = E\left[\left(\frac{\frac{\partial c}{\partial S}S\alpha_s + \frac{1}{2}\frac{\partial^2 c}{\partial S^2}S^2\sigma_s^{\,2} + \frac{\partial c}{\partial t}}{c}\right)dt + \left(\frac{\partial c}{\partial S}\frac{S}{c}\sigma_s\right)dW\right]$$

$$= E\left[\left(\frac{\frac{\partial c}{\partial S}S\alpha_s + \frac{1}{2}\frac{\partial^2 c}{\partial S^2}S^2\sigma_s^{\,2} + \frac{\partial c}{\partial t}}{c}\right)dt\right]$$

$$= \left(\frac{\frac{\partial c}{\partial S}S\alpha_s + \frac{1}{2}\frac{\partial^2 c}{\partial S^2}S^2\sigma_s^{\,2} + \frac{\partial c}{\partial t}}{c}\right)dt$$

$$= \alpha_c dt. \tag{30.33}$$

Although this formula for the expected return does not look like the expected return for the option, it can be shown to be the same, provided we assume that Black-Scholes-Merton holds. All we are required to do is substitute the partial derivatives $\partial c/\partial S$, $\partial^2 c/\partial S^2$, and $\partial c/\partial t$ from the Black-Scholes-Merton model. We repeat these formulas from Chapter 14:

$$\frac{\partial c}{\partial S} = N(d_1). \tag{30.34}$$

$$\frac{\partial^2 c}{\partial S^2} = \frac{e^{-d_1^2/2}}{S\sigma_S\sqrt{2\pi\tau}}. \tag{30.35}$$

$$\frac{\partial c}{\partial t} = -\frac{S\sigma_S e^{-d_1^2/2}}{2\sqrt{2\pi\tau}} - r_c Xe^{-r_c\tau}N(d_2). \tag{30.36}$$

Substituting, we then obtain the expected return on the option,

$$\alpha_c = N(d_1)\,\alpha_s\left(\frac{S}{c}\right) - r_c\left(\frac{Xe^{-r_c\tau}}{c}\right)N(d_2). \tag{30.37}$$

From the Black-Scholes-Merton formula, we use the substitution, $SN(d_1) - c = Xe^{-r_c\tau}N(d_2)$,

$$\alpha_c = r_c + (\alpha_s - r_c)\,N(d_1)\left(\frac{S}{c}\right). \tag{30.38}$$

Because $N(d_1)$ is $\partial c/\partial S$ and $\mu_S = E(R_S)$, this is the same formula for the option's expected return we obtained previously, Equation (30.16).

Taking the volatility of the return on the option, we obtain[9]

$$\sigma_c^2 dt = \text{var}\left(\frac{dc}{c}\right) = \left(\frac{\partial c}{\partial S}\frac{S}{c}\sigma_s\right)^2 dt$$

$$\sigma_c = \left(\frac{\partial c}{\partial S}\right)\left(\frac{S}{c}\right)\sigma_s. \tag{30.39}$$

In this case, we have the same formula we previously obtained for the option's volatility, Equation (30.19).

Finally, we should also recognize that the risk-free rate in the finite interval CAPM and the infinitesimal interval option pricing model need to be expressed on an equivalent basis. In this book, we use r_c for continuously compounded returns and r for discrete returns. The CAPM typically uses discrete interest, and the option pricing model uses continuous interest. We would need to be sure that interest is measured in the same manner in both models to make the results equivalent.

30.7 RECAP AND PREVIEW

We learned in this chapter that option pricing is consistent with capital asset pricing. That is, the price obtained from the Black-Scholes-Merton model is consistent with the expected return from the CAPM. But even if the CAPM does not hold, the expected return on the option can be related to the expected return on the asset through the risk-free rate, the risk premium on the asset, and the risk of the option relative to the asset. Unless the option is incorrectly priced relative to the asset, the performance of the option as measured by its Sharpe ratio is no different from the performance of the asset. In short, option pricing is completely consistent with asset pricing.

In Chapter 31, we look at the concept of implied volatility, which is the volatility of the underlying that is implied by the price of an option. Implied volatility plays an extremely important role, not only in helping us to understand option pricing but also in helping us to see what opinions investors have about the level of risk.

QUESTIONS AND PROBLEMS

1 Explain why the following expression is known as the certainty equivalent version of the CAPM:

$$S = \frac{E(S') - \lambda\text{cov}(S', R_m)}{1 + r},$$

where

$$\lambda = \frac{E(R_m) - r}{\sigma_m^2}.$$

2 "The elasticity of a standard European call, $(\partial c/\partial S)(S/c)$, is at least equal to 1." Evaluate and explain whether this statement is true or false.

3 Prove that within the Black-Scholes-Merton framework, the expected return on the call can be expressed as

$$E(R_c) = r + (E(R_s) - r)\, \Omega_c,$$

where $\Omega_c \equiv \dfrac{\partial c/c}{\partial S/S}$.

4 Prove the expected return on the put can be expressed as

$$E(R_p) = r + [E(R_s) - r]\, \Omega_p,$$

where $\Omega_p \equiv \dfrac{\partial p/p}{\partial S/S}$.

5 "The Sharpe ratio of the stock will equal the Sharpe ratio of a call option on the stock." Evaluate and explain whether this statement is true or false.

6 [Contributed by Brecklyn Groce] Suppose an investment manager calculates the Sharpe ratio of an investment in a stock index as 0.35. The Sharpe ratio of a call option on the index is 0.44. In this chapter it is asserted that the Sharpe ratio of an asset and its derivative should be the same. Why are they different in this case?

NOTES

1. Other well-known models are the arbitrage pricing model and the cost of carry forward/futures pricing model, which is known as interest rate parity in the foreign currency world. The CAPM is briefly mentioned in Chapters 1 and 6.
2. We are assuming no dividends, interest, cash flows, or holding costs. These would not cause any problems, but our approach would vary depending on whether the dividends are known or random.
3. To restate what we have previously learned, the notion of a certainty equivalent is that of a value that one would accept for certain in lieu of facing a risky situation. The expected value minus the risk premium that appears in the numerator is a risk-adjusted future value, which can then be discounted at the risk-free rate.
4. It is easy to use the Black-Scholes-Merton model to see that the elasticity is not less than 1. Elasticity is defined as $(\partial c/\partial S)(S/c)$. You should recognize this as $N(d_1)\, S/c$ from the Black-Scholes-Merton model. Replacing c with the Black-Scholes-Merton formula reveals that elasticity is no less than 1 if $Xe^{-r_c\tau} N(d_2)$ is always true.
5. Using the Sharpe ratios for options is problematic, however, because option return distributions are highly nonnormal, and the Sharpe ratio characterizes performance exclusively with the expected return and volatility, ignoring higher-order moments associated with non-normal distributions. For an empirical analysis of some of the problems of evaluating the performance of covered call writing strategies, see Brooks, Chance, and Hemler (2019).
6. In other words, the Black-Scholes-Merton model is obtained without any reference to the option's expected return or volatility, nor does it directly provide the option's expected return and volatility.
7. Alternatively, we could use the continuous time version of the CAPM and adjust that equation so that it would contain a dt term and reflect expected returns over the interval dt.
8. In taking expectations of this equation, recall that the expectation of dW is zero.
9. Remember that the variance of a constant (in this case, $(\partial c/\partial S)(S/c)$), is the constant squared times the variance of the random variable. The variance of dW is dt.

CHAPTER 31

Implied Volatility and the Volatility Smile

Volatility of the underlying asset return is unquestionably the most critical parameter in option valuation. For one, it is the only parameter that is completely unobservable. Technically the risk-free rate and either the upcoming dividends or the upcoming dividend yield are unobservable, but these factors are not difficult to predict and they impart a relatively minor effect on option valuation. Volatility, however, is not easy to predict. Moreover, many option pricing models, such as the Black-Scholes-Merton and binomial models, assume that volatility is deterministic. Forcing a model with constant volatility to behave like a model with nonconstant volatility is incorrect, but it is what most researchers and practitioners do. On top of this problem, option value is highly sensitive to volatility. Thus, it is important that we take a close look at volatility.

As we previously learned, both the BSM and binomial models require five inputs: the underlying asset price, the exercise price, the risk-free rate, the volatility, and the time to expiration. If the underlying asset pays dividends or interest, or has any other cash flows, these too must be incorporated as a sixth input. But in any case, volatility is such a primary driver of option value that standard practice is to oftentimes insert the price of the option into the model and derive the volatility that makes the model price equal the market price. This volatility is called the *implied volatility* because it is implied from the market price. If there were no uncertainty about the pricing model used by investors, the implied volatility would be the volatility used by the market to price the option.

Inasmuch as the implied volatility should be the volatility of the underlying asset return, any option on the underlying asset should produce the same implied volatility. But that is not what happens. What this means is either that there is no single-option valuation model that everyone accepts and uses or that people have different opinions about volatility. Different opinions are certainly acceptable, but when the same investor uses a different volatility for two options on the same asset, there is clearly something other than a difference of opinion going on.

So, let us first contrast volatility with implied volatility illustrating with several examples. Based on this analysis, we introduce the popular volatility index (VIX) associated with the S&P 500 index options. Next, we further explore what implied volatility is, and then we shall look at the unusual behavior it exhibits in suggesting that the underlying can have more than one volatility at the same time.

31.1   HISTORICAL VOLATILITY AND THE VIX

In finance, the term *volatility* typically means some measure of dispersion, usually computed as the standard deviation. Within this context, the standard deviation is annualized just as the inputted standard deviation for the Black-Scholes-Merton model is assumed to be annualized. As previously defined, the implied volatility is derived from equating the observed option price with Black-Scholes-Merton model price by solving for the volatility implied, hence, the term *implied volatility*.

Although there are several volatility indexes, *VIX* is the term given to the Chicago Board Options Exchange reported volatility index. VIX is based on 30-day S&P 500 index options.[1] Figure 31.1 illustrates the VIX along with the 21-day rolling standard deviation based on the daily returns of the S&P 500 index ETF with ticker symbol SPY. Panel A shows the long history (12/31/2010 through 6/12/2023) of these values whereas Panel B shows the two years roughly starting with the onset of the COVID pandemic. In both panels, we observed that the two measures track closely with each other and the VIX is higher on average.

SP 500 Index ETF (SPY, Rolling SD, 21 Day) and Volatility Index (VIX)

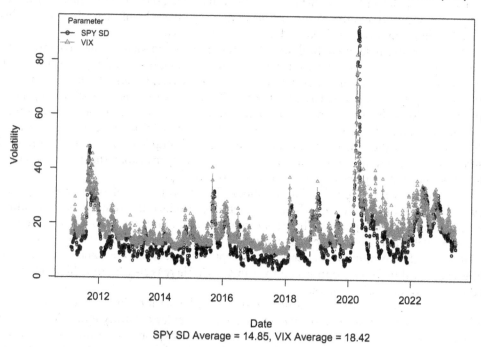

SPY SD Average = 14.85, VIX Average = 18.42

Panel A. Long History (12/31/2010–6/12/2023)

FIGURE 31.1   Rolling 21-Day Standard Deviation for SPY and the VIX

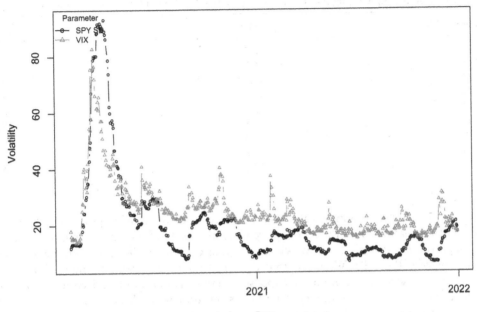

SP 500 Index ETF (SPY, Rolling SD, 21 Day) and Volatility Index (VIX)

Date
SPY SD Average = 19.77, VIX Average = 24.94

Panel B. Two Years Around COVID (12/31/2019–12/31/2021)

FIGURE 31.1 *Continued*

31.2   AN EXAMPLE OF IMPLIED VOLATILITY

Table 31.1 shows the closing ask prices on September 17 of a particular year for options on the S&P 500 expiring in approximately one month, specifically October 19. These options are among the most active of all listed options, and they are European style, so we can use the Black-Scholes-Merton model to price them. These prices were closing ask quotes, meaning that they represent the prices at the close of trading at which the dealer was offering to buy the options. Hence, these are the prices at which the public would expect to buy. At that time, the S&P 500 closed at 2,904.31, so this is the price of the underlying.

To use the Black-Scholes-Merton model, we need some additional information, which was obtained from various sources. The continuously compounded risk-free rate is estimated to be 2.02%, and the continuously compounded dividend yield is estimated at 1.73%. The option expires in 32 days, so the value inserted into the model for the time to expiration is $31/365 = 0.0849$. The implied volatilities can be derived using a number of different search routines, including the secant method, the Brent method, and the Newton-Raphson method.[2] These methods are widely available on many computer platforms. With most methods, the process is to insert the option price and all other input values except the volatility. We then make an initial guess at the volatility, whereupon the

TABLE 31.1 Option Quotes

SPX (S&P 500 Index); Closing Ask Prices for September 17, 20yy; Options Expire on October 19.

Exercise Price	Calls	Puts
2880	50.5	21.0
2885	46.7	22.3
2890	43.1	23.6
2895	39.5	25.1
2900	36.1	26.6
2905	32.8	28.4
2910	29.6	30.3
2915	26.6	32.3
2920	23.8	34.5

program computes the option value and compares it to the market price, continuing to guess at the volatility until the model price is within 0.001 of the quoted market price.[3]

Suppose we wished to know which of these options is the most expensive. First, let us consider the calls. We know that deeper-in-the-money calls, meaning those with the lowest exercise prices, are more expensive than calls that are less deep-in-the-money, at-the-money, or out-of-the-money. That fact is certainly natural. But after taking moneyness into account, we wish to know if some of the calls are more expensive than others. Likewise, for puts where the most expensive ones are those with high exercise prices. After taking moneyness into account, are some puts still more expensive than others?

Now, it should be the case that because all of the options are on the same underlying, the same volatility would be used by the market to price all of the options. Hence, we cannot use volatility to determine which option is more expensive. Or can we? It turns out that we can. Table 31.2 shows that indeed some options are more expensive than others, even after taking into account the logical reasons why certain ones would be more expensive. Those with higher volatilities are more expensive after accounting for other factors.

From Table 31.2, we see that the call implied volatilities range from 9.08% to 10.89%, and the put implied volatilities range from 7.79% to 9.52%. The results are graphed in Figure 31.2.

TABLE 31.2 Implied Volatilities of SPX Options

Exercise Price	Calls	Puts
2880	0.1089	0.0952
2885	0.1066	0.0931
2890	0.1044	0.0907
2895	0.1018	0.0886
2900	0.0996	0.0862
2905	0.0973	0.0843
2910	0.0949	0.0822
2915	0.0928	0.0800
2920	0.0908	0.0779

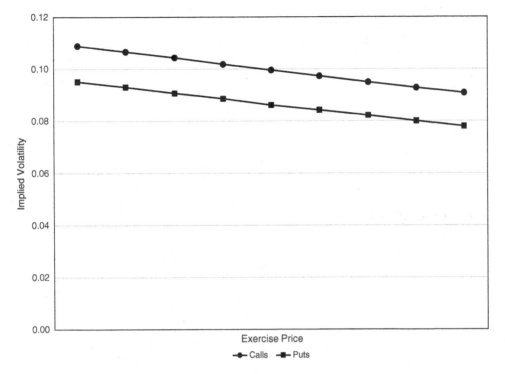

FIGURE 31.2   Graph of Implied Volatilities by Exercise Price

Note that we observe a pattern in which the implied volatilities monotonically decrease with a higher exercise price. The relationship between the option exercise price and the implied volatility has been documented at the least since the time of the massive stock market crash of October 19, 1987. When first observed, the implied volatilities were *u*-shaped, giving the appearance of a smile. Hence this relationship was named the *volatility smile*.[4] In more recent years, the smile has mostly disappeared, and the relationship has sometimes been referred to as a skew or even a smirk, reflecting the appearance of a sort of snarky half smile.[5] Note also that we obtain different implied volatilities depending on whether we are looking at calls or puts. There is no a priori reason why puts and calls should have different implied volatilities but obviously they do.

We now turn to options on the ETF SPY, a very popular index option. Recall from Chapter 1, SPY is the exchange-traded fund of the S&P 500 Index.[6] Options on SPY are American style; hence, early exercise must be incorporated into the option pricing model. The results reported here are based on a dividend-adjusted binomial model that captures potential early exercise.

Figure 31.3 illustrates actual SPY option prices and exercise values on the left and reported implied volatilities on the right for selected dates before and during the great financial crisis starting in 2008.[7] As is common, we select the nearest option that expires at least more than one month from the observation date. The days to maturity range from 47 to 53. Recall that the SPY is an exchange-traded fund that seeks to mimic the total return of the S&P 500 index. We normalize the exercise price by dividing by the stock price to ease comparison across time. Further, we fix the axis range, improving the ability to see changes over time. Thus, the call (put) option is out-of-the-money (in-the-money) when

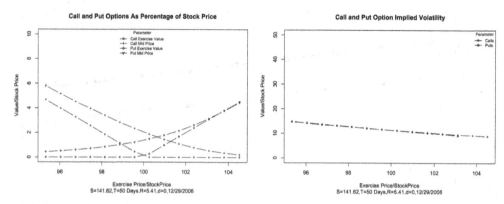

Panel A. December 29, 2006

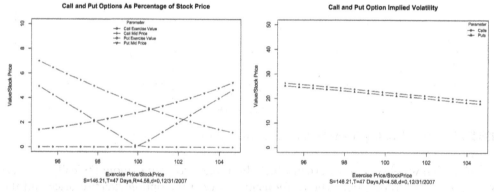

Panel B. December 31, 2007

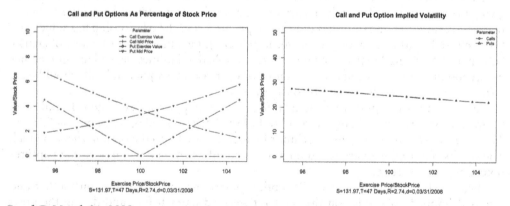

Panel C. March 31, 2008

FIGURE 31.3 Selected Graphs of SPY Option Prices and Implied Volatilities by Normalized Exercise Price, 2006–2008

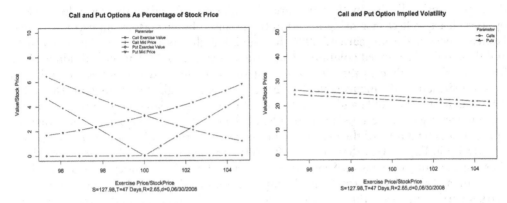

Panel D. June 30, 2008

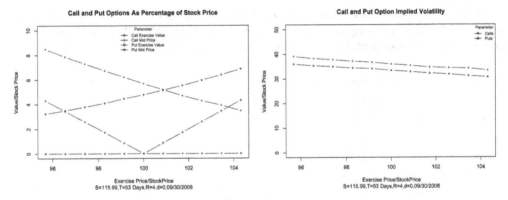

Panel E. September 30, 2008

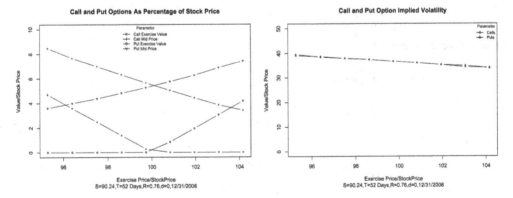

Panel F. December 31, 2008

FIGURE 31.3 *Continued*

the exercise price divided by the stock price is high (low). Based on these graphs, the SPY options clearly reflect increasing uncertainty measured by increasing implied volatilities. Notice as we move from year-end 2006 through 2008, the option prices increase (left side) and thus so do the implied volatilities (right side). It seems likely that the unfolding of the financial crisis resulted in these increases. At times the implied volatilities for calls and puts are virtually indistinguishable. See, for example, the right-side figures on December 29, 2006, March 31, 2008, and December 31, 2008. On these dates, both puts and calls for various exercise prices resulted in nearly identical implied volatilities. Finally, note that the option prices tend toward the exercise value as we move further in- or out-of-the-money.

Notice in Figure 31.3 that on June 30, 2008, the put implied volatilities were above those of the calls, whereas on September 30, 2008, the call volatilities were above those of the puts. This behavior results in the intersection of calls and puts being lower. Some market participants view this intersection point as "lean" in the options market. Thus, on September 30, 2008, the market lean was bullish, whereas on June 30, 2008, the market lean was bearish.

Figure 31.4 illustrates the same type of input data as Figure 31.3 but in this case for quarter-end observations in 2009. We see that as the financial crisis ended, the options market prices dropped and, hence, so did implied volatilities.

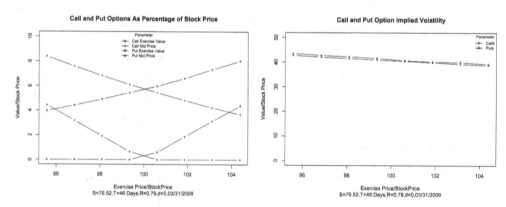

Panel A. March 31, 2009

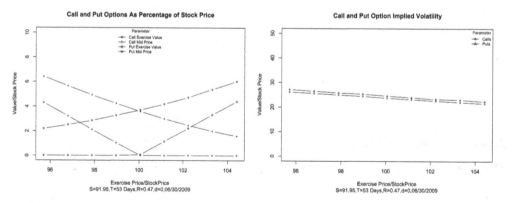

Panel B. June 30, 2009

FIGURE 31.4 Selected Graphs of SPY Option Prices and Implied Volatilities by Normalized Exercise Price, 2009

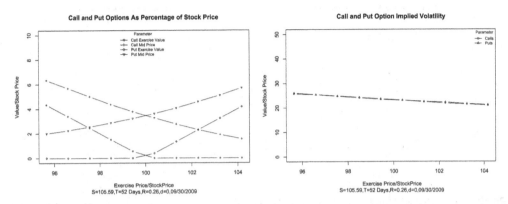

Panel C. September 30, 2009

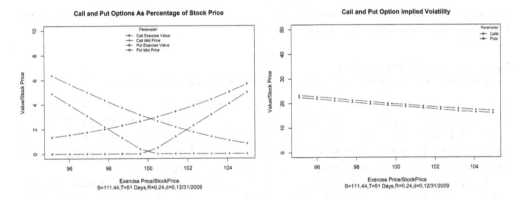

Panel D. December 31, 2009

FIGURE 31.4   *Continued*

Figure 31.5 illustrates the same data as Figures 31.3 and 31.4 for year-end 2017, a relatively normal time period. We see that well after the financial crisis was over, the options market prices dropped and, hence, so did implied volatilities. Also, notice that the put option implied volatilities are nonexistent when the put is deep-in-the-money because the put prices are essentially at their exercise values.

The existence of multiple implied volatilities, regardless of whether they arrange themselves in a smile-like pattern, should be somewhat disconcerting. How can the market be telling us that there is more than one volatility for the S&P 500? Clearly there is something wrong with the Black-Scholes-Merton model, which is that it fails to consider all of the factors that enter into the pricing of an option. It accounts for the asset price, the exercise price, the time to expiration, the dividends, and the risk-free rate. The *implied volatility* is more or less a catchall term, capturing whatever variables are missing, as well as the possibility that the model is improperly specified or blatantly wrong.

What we learn, however, is what we wanted to know: Which options are the most expensive? We see that the calls and puts with the lowest exercise price are the most expensive options. But what is so puzzling is that in the Black-Scholes-Merton world, no option should be more expensive than any other option after accounting for the exercise price and

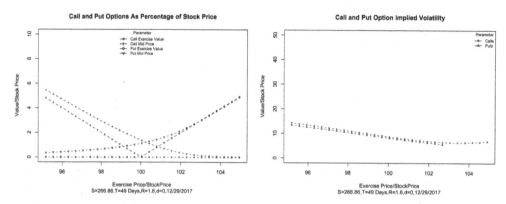

FIGURE 31.5 Selected Graphs of SPY Option Prices and Implied Volatilities by Normalized Exercise Price, December 29, 2017

time to expiration. Any option should be a perfect substitute for any other option, given the ability to replicate using the concept of delta.

31.3 THE VOLATILITY SURFACE

In the prior section, we focus on options with approximately one month to expiration and documented various volatility smiles and skews. Volatility smiles and skews are based on various measures relating to only the exercise price. We now illustrate the three-dimensional perspective of the implied volatilities, known as the volatility surface, where we address the exercise price as well as the time to expiration.

Figure 31.6 illustrates two volatility surfaces for McDonalds: Panel A when markets were relatively normal and Panel B when markets are stressed during the COVID pandemic. Note the vertical axis is fixed between 15% and 85% implied volatility to ease

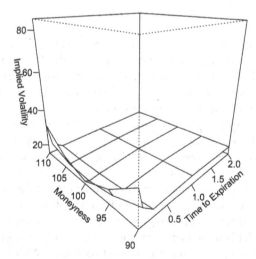

Panel A. November 11, 2019, Normal Market

FIGURE 31.6 Illustrates the Volatility Surface for McDonalds

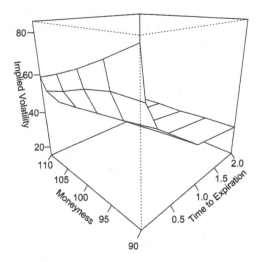

Panel B. April 1, 2020, Stressed Market

FIGURE 31.6 *Continued*

comparison between the two surfaces. The time to expiration is expressed in years and moneyness is expressed as a percentage of the stock price. During normal markets, the volatility surface is relatively flat with a pronounced smile for short maturity options. During stressed markets, the volatility surface is dramatically higher overall and dramatically skewed for short maturity options.

Thus, analyzing volatility surface changes over time often provides a perspective on option participants' market views. Further, analyzing volatility surfaces across assets provides a perspective on the relative riskiness of different assets.

31.4 THE PERFECT SUBSTITUTABILITY OF OPTIONS

Within the Black-Scholes-Merton framework, any option on the same underlying should be a perfect substitute for any other option. To demonstrate this claim, consider an option priced at w_1 and a second option on the same asset priced at w_2. These options could have different exercise prices and/or times to expiration. They could differ in that one is a call and one is a put. It does not matter. They are simply two options on the same asset, which itself is priced at S.

Suppose we wished to buy the first option. Under the assumptions of the Black-Scholes-Merton model, we could create the same result by combining the second option with risk-free bonds. The math to demonstrate this result requires essentially a derivation of the Black-Scholes-Merton model. We can avoid these technical details by using some simple results from the Black-Scholes-Merton formula itself.

We can write a general expression for the values of these options as follows:

$$w_1 = h_{s1}S + h_{b1}X_1 e^{-r_c \tau_1}$$
$$w_2 = h_{s2}S + h_{b2}X_2 e^{-r_c \tau_2}.$$

(31.1)

The quantities h_{s1} and h_{s2} represent the holdings of the underlying asset necessary to replicate options 1 and 2, and the quantities h_{b1} and h_{b2} represent the holdings of risk-free bonds necessary to replicate options 1 and 2.

If option 1 is a call, the Black-Scholes-Merton formula is, of course,

$$w_1 = SN\left(d_1^{(1)}\right) - X_1 e^{-r_c \tau_1} N\left(d_2^{(1)}\right), \tag{31.2}$$

where $N(d_1^{(1)})$ and $N(d_2^{(1)})$ are the well-known standard cumulative normal probabilities, typically identified as $N(d_1)$ and $N(d_2)$ when working with a single option. These values are computed using the appropriate exercise price X_1 and time to expiration τ_1 in the standard formula for d_1 and d_2.[8] Then h_{s1} is $N(d_1^{(1)})$ and h_{b1} is $-N(d_2^{(1)})$.

If option 1 is a put, the formula is

$$w_1 = X_1 e^{-r_c \tau_1} N\left(-d_2^{(1)}\right) - SN\left(-d_1^{(1)}\right). \tag{31.3}$$

Then h_{s1} would be $-N(-d_1^{(1)})$, and h_{b1} would be $N(-d_2^{(1)})$. If option 2 is a call, the Black-Scholes-Merton formula is

$$w_2 = SN\left(d_1^{(2)}\right) - X_2 e^{-r_c \tau_2} N\left(d_2^{(2)}\right), \tag{31.4}$$

with the $N(d_1^{(2)})$ and $N(d_2^{(2)})$ terms defined in the manner previously described. Then h_{s2} would be $N(d_1^{(2)})$, and h_{b2} would be $-N(d_2^{(2)})$. If option 2 is a put, the formula is

$$w_2 = X_2 e^{-r_c \tau_2} N\left(-d_2^{(2)}\right) - SN\left(-d_1^{(2)}\right), \tag{31.5}$$

and h_{s2} would be $-N\left(-d_1^{(2)}\right)$ and h_{b2} would be $N\left(-d_2^{(2)}\right)$.

From these general expressions, we can see how to replicate any option with any other option on the same underlying. Take the general expression for w_2 and solve for S to obtain

$$S = \frac{w_2 - h_{b2} X_2 e^{-r_c \tau_2}}{h_{s2}}. \tag{31.6}$$

Then substitute this result for S into the expression for w_1:

$$w_1 = \left(\frac{h_{s1}}{h_{s2}}\right) w_2 - \left(\frac{h_{s1}}{h_{s2}}\right) h_{b2} X_2 e^{-r_c \tau_2} + h_{b1} X_1 e^{-r_c \tau_1}. \tag{31.7}$$

We see that the first option can be replicated by holding (h_{s1}/h_{s2}) units of option 2, $-(h_{s1}/h_{s2})h_{b2}$ units of a risk-free bond paying X_2 at the expiration of option 2, and h_{b1} units of a risk-free bond paying X_1 at the expiration of option 1.[9] Of course, this position must be dynamically adjusted through time. Nonetheless, any option can be used to replicate any other option on the same underlying, regardless of exercise price, time to expiration, or whether it is a call or a put. Hence, any option is a perfect substitute for any other option on the same underlying, at least in the Black-Scholes-Merton world.

That being the case, there is no rationale within the Black-Scholes-Merton framework for why any one option should be more expensive than any other option, after accounting for time to expiration, exercise price, and whether it is a put or a call. But clearly the existence of multiple implied volatilities tells us that some options are more expensive than others, and the smile tells us that there is a pattern to the relative costs of these options. Typically the most expensive options are the deep-in-the-money calls and deep out-of-the-money puts as illustrated in the previous figures.

Unfortunately, no one really knows why certain options are more expensive than others. Deep out-of-the-money puts have oftentimes been viewed as a form of insurance against large drops in the market. Fear of a crash, it is said, leads some investors to pay relatively more for the protection these options afford than for other options. Yet, the Black-Scholes-Merton theory says that any option should substitute perfectly for any other option. Yet, theory and reality diverge. Perhaps the dynamic replicating strategy is too complex or too costly to implement. For whatever reason, perfect substitution does not hold in reality.

The Black-Scholes-Merton model is a wonderful theory, but it tells us nothing about why anyone would hold an option. It cannot motivate the holding of options because any option serves as well as any other option. In reality, some options are more desirable than others. Whatever factors motivate the holding of options are simply not captured in the Black-Scholes-Merton model. Hence, these factors show up subsumed within the implied volatility.

One must then wonder why we do not simply throw out the Black-Scholes-Merton model. There are three reasons why we do not. First, the model is attractively simple. Although the mathematical details that support the derivation of the model are complex, there are a number of simple conceptual approaches to understanding the model.[10] Practitioners have shown not only that they can understand the model but also that they can accept it. Second, the computational demands of the model are also quite modest. Perhaps its greatest virtue is that it requires so little information. But therein lies its vice: by requiring so little information, the model almost surely misses factors that explain why options are held and why some investors would pay more for certain options.

But the model is admired so much for its simplicity that practitioners continue to use it. The volatility smile and related surface is the price paid for oversimplification.

31.5 OTHER ATTEMPTS TO EXPLAIN THE IMPLIED VOLATILITY SMILE

When the volatility smile was first observed, some researchers believed that the explanation was liquidity. The true "smile" appearance meant that deep out-of-the-money put options and deep in-the-money call options had the highest implied volatilities. These options have low liquidity; hence, it was argued that the prices observed for these options of low liquidity reflected the thinness of their markets. But this explanation would suggest that highly liquid options—typically those trading nearly at-the-money—would have the same implied volatilities. In fact, they do not and never did. Moreover, when the smile turned into a skew, the moneyness argument fell by the wayside. In addition, liquidity would not explain the implied volatilities of deep in-the-money put options and deep out-of-the-money call options, which also have low liquidity, and yet have high implied volatilities.

Other researchers believe that the smile reflects stochastic volatility.[11] Volatility is surely not constant as assumed in the Black-Scholes-Merton model. If volatility is stochastic, researchers argue that the smile reflects the failure of the Black-Scholes-Merton model to capture the random nature of volatility. Others argue that the Black-Scholes-Merton model, which assumes that asset prices fluctuate in a smooth and continuous manner, fails to capture the true nature of asset price movements, which are observed to have discrete jumps.[12]

These arguments about stochastic volatility and jumps have a great deal of appeal because they in some sense preserve many of the essential features of the Black-Scholes-Merton model. These arguments do not require that the model motivate the holding of options and the preference for some investors for certain options over others. And if these models were used, the smile would presumably go away, which it does not. Unfortunately, once these looser assumptions are introduced, the mathematical tractability of the model is lost, and the process of pricing an option becomes one of making other strong assumptions or imposing severe computational demands.[13] It is fair to say that mathematicians have devoted many hours of human and machine time to researching the smile with little if any regard to the reasons why the smile exists.[14]

Finally, it should be noted that there is a portion of the volatility smile that is completely artificial and is driven by the algorithmic and computational choices made by researchers and practitioners. That is, choosing the algorithm for convergence and the criterion that define whether convergence has occurred can lead to a smile-like effect.[15]

31.6   HOW PRACTITIONERS USE THE IMPLIED VOLATILITY SURFACE

Practitioners seem capable of operating in a world of complex volatility smiles and volatility surfaces. They even use the surface to simplify how they trade. That is, they oftentimes quote option prices not in terms of the actual price but in terms of the implied volatility. For example, a dealer might indicate an intention to sell a January 50 call by quoting a volatility of 45. This statement is interpreted to mean that the actual price of this option that expires in January and has an exercise price of 50 is derived from the Black-Scholes-Merton model using a volatility of 45, technically 0.45. By quoting prices this way, traders can immediately see which options are truly more expensive, that is, after accounting for moneyness, time to expiration, and whether the option is a call or a put.

Table 31.3 Panel A illustrates mid-market call option quotes for McDonalds during stressed markets (see Figure 31.6 Panel B and related discussion). Panel B illustrates the implied volatilities related to the call prices. Due to the variation in exercise prices (expressed as a percentage of the $158.77 stock price), it is difficult to appraise the relative valuations.[16] With implied volatilities, however, one can opine that short-dated options (1 week) are more expensive than longer time to expiration options.

Traders also will often express the smile in terms of the implied volatility associated with the option's delta. For example, an at-the-money call has a delta of about 0.50. Traders would often then refer to this as a 50 delta call. The volatilities are then graphed with the deltas on the horizontal axis. The resulting smile would be like the images we have seen.

But in general, traders use the smile as a way of determining the values of options relative to each other, after taking into account obvious differences related to moneyness,

TABLE 31.3 McDonalds Option Quotation, April 1, 2021, Stressed Market

Panel A. Mid-Market Call Option Prices

Tenor/Moneyness	90%	95%	100%	105%	110%
1 Week	17.48	11.14	6.03	2.59	0.80
1 Month	20.55	15.51	10.83	6.69	3.73
3 Months	24.26	19.03	14.50	10.58	7.44
1 Year	26.97	22.21	18.01	14.56	11.16
2 Years	28.49	24.77	20.75	18.26	15.29

Panel B. Implied Volatilities

Tenor/Moneyness	90%	95%	100%	105%	110%
1 Week	82.90	75.71	69.18	63.72	58.87
1 Month	64.40	63.81	60.49	55.24	51.15
3 Months	53.10	50.31	47.96	45.56	43.57
1 Year	35.58	34.02	32.70	31.86	30.35
2 Years	30.72	30.35	29.18	29.42	28.70

time to expiration, and the type of option, call or put. They give traders insights into which options are the most and least expensive, after accounting for the obvious distinguishing factors.

31.7 RECAP AND PREVIEW

The entire notion of implied volatility and the existence of the volatility smile are results of using a model that does not capture every factor that affects the prices of options. Practitioners and academics largely accept the limitations of the model and consider the smile a means of forcing the model to reveal information it is not designed to reveal. As flawed as the model may be, the advantages of the Black-Scholes-Merton model and its attendant defects may outweigh the disadvantages of other more complex models.

In the final chapter, we look at foreign currency options. There are, of course, options on a wide range of underlyings, of which foreign currency is just one. Foreign currency, however, is a very different type of underlying in that it can be viewed from the perspective of either one currency or the other. Thus, there are two mirror-image underlyings.

QUESTIONS AND PROBLEMS

1   As the financial crisis unfolded, the implied volatilities rose as illustrated in the following two figures. Note that on June 30, 2008, the put implied volatilities were always above the call implied volatilities. On September 30, 2008, the put implied volatilities were always below the call implied volatilities. Based solely on these patterns, what does this imply regarding the exercise price such that the call and put prices are the same?

Call and Put Option Implied Volatility

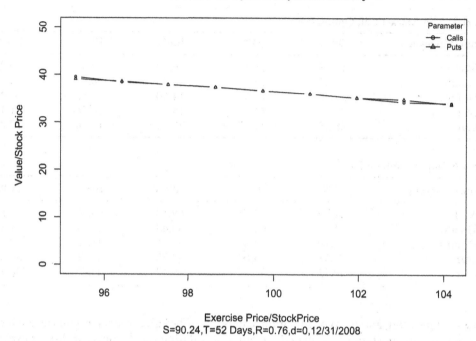

Exercise Price/StockPrice
S=90.24,T=52 Days,R=0.76,d=0,12/31/2008

Call and Put Option Implied Volatility

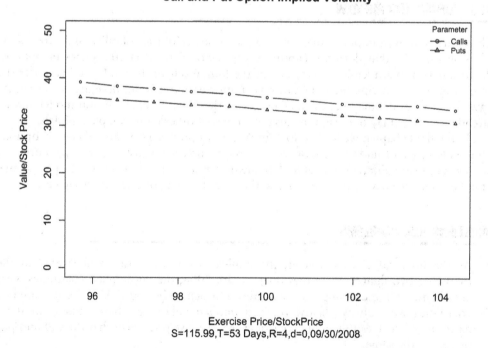

Exercise Price/StockPrice
S=115.99,T=53 Days,R=4,d=0,09/30/2008

2 "Within the Black-Scholes-Merton framework, any option on the same underlying should be a perfect substitute for any other option." Demonstrate this result by designing a perfect substitute for one call option with another call option.

3 Identify and explain four reasons that have been offered for the observed implied volatility smile.

4 Explain how practitioners use the implied volatility smile.

5 "One measure of the internal consistency of an option pricing model is deviations from the average implied volatility for options with the same maturity." Evaluate and explain whether this statement is true or false.

NOTES

1. There are several technical issues in the VIX calculations that do not materially change VIX's interpretation.
2. A great source for various algorithms, such as solving for embedded parameters, can be found at www.numerical.recipes.
3. The choice of 0.001 is arbitrary.
4. The classic papers on the smile are Derman and Kani (1994), Dupire (1994), and Rubinstein (1994).
5. In some cases, the smile is illustrated in a three-dimensional graph in which the third dimension is the time to expiration. The volatilities could also differ by time to expiration, giving rise to the notion of a term structure of volatilities. When the implied volatilities are illustrated with exercise price along one axis and time to expiration along the other, the relationship is known as the *volatility surface*.
6. In Chapter 10, we documented the important role of dividends for SPY holding period returns.
7. The software used to generate these graphs linearly interpolates between observations. Thus the exercise value does not break exactly at 100 as it should.
8. Given an exercise price of X, time to expiration of τ, and volatility of σ, the formulas for d_1 and d_2 are $d_1 = \frac{\ln(S/X)+(r_c+\sigma^2/2)\tau}{\sigma\sqrt{\tau}}$
$$d_2 = d_1 - \sigma\sqrt{\tau}.$$
9. If the two options had either the same time to expiration or the same exercise price, the formula for the holdings would simplify a little.
10. As, for example, the binomial analog and the Black-Scholes-Merton model expressed as a discounted expected future value under risk neutrality.
11. See, for example, Renault and Touzi (1996).
12. See, for example, Andersen and Andreasen (2000).
13. For example, one strong assumption is that the risk arising from stochastic volatility is non-priced risk. That is, the risk associated with uncertain volatility is a risk that does not concern investors. They are either neutral toward that risk or the risk is uncorrelated with their other holdings, meaning that the risk is diversifiable. Once these assumptions are invoked, the financial economics is lost and what remains is simply an exercise in computational finance.
14. Perhaps if these reasons were found, the mathematicians would be out of work.
15. See Chance, Hanson, Li, and Muthuswamy (2017). They estimate that a minimum of 16% of the smile can be explained by these purely artificial effects.
16. Other parameters required include the risk-free rate of 9 basis points and dividend yield of 3.12%.

CHAPTER 32

Pricing Foreign Currency Options

One of the most popular option underlyings is foreign currency. The right to buy or sell a foreign currency is a widely used instrument by global corporations, institutional investors, and speculators. Yet, options on foreign currencies can be a bit more difficult to understand than options on stocks, bonds, and commodities. The reason primarily lies in the fact that every currency contract involves not one but two currencies. Options on stocks, bonds, and commodities, by contrast, do not involve more than one currency. They do, however, implicitly involve more than one asset. For example, an option to buy a stock is an option to exchange a currency for the stock. From the opposite perspective, however, it is clearly an option to exchange stock for the currency. We may speak of a share of stock as costing 50 but we could just as easily say that the cost of 1.0 is 0.02 shares of stock. That is, one could theoretically exchange 0.02 shares of stock and receive 1.0. Although the notion of exchanging stock for money is a valid economic concept and explains precisely the same idea as if we described the transaction as exchanging money for stock, it is not the form we typically use in commerce. But in the currency world, things are just a bit different.

Hence, exchanging dollars for euros is the mirror image of exchanging euros for dollars. Likewise, exchanging dollars for stock is the mirror image of exchanging stock for dollars. Because we never describe the stock transaction as the latter, we may have trouble understanding the currency transaction from that point of view.

In the world of currency trading, however, it is necessary to understand the transaction from either point of view. In particular, foreign currency option transactions have an important characteristic that sheds light on some interesting issues in option valuation. That characteristic is that a foreign currency call can be viewed as a foreign currency put from the mirror image perspective. We describe this idea as the concept of *isomorphism*, the notion that two things have essentially the same form. This point would also hold for calls on stocks because they can be viewed as puts on cash, but because we virtually never view a stock sale as the purchase of money paid for by tendering stock, we almost never use this concept elsewhere in financial markets.

In foreign currency transactions, we shall refer to the two currencies as the base currency and the quoted currency. The base currency is just the underlying currency. For example, a call option quoted in dollars in which the underlying is euros has the dollar as the quoted currency and the euro as the base currency.[1] In addition, the instruments we examine are sometimes called *foreign currency options*, *currency options*, and *foreign exchange options*, so be prepared for alternative nomenclature.

32.1 DEFINITION OF TERMS

Let us start by giving a clear definition of the terms we shall use:

S_0, S_T = spot exchange rate at times 0 or T, quoted in the domestic currency, that is, units of the domestic currency or quote currency per unit of the foreign currency, or base currency

$R_0 = 1/S_0$ and $R_T = 1/S_T$, the spot exchange rate at times 0 or T, quoted in the foreign currency, that is, units of the foreign currency per unit of domestic currency, or the reciprocal of the spot exchange rate

X = exercise price, sometimes called the *exercise rate*, expressed in the quote currency of a call option to buy one unit of the foreign currency

T = expiration date

σ = volatility of the log return on the quote currency

r_D = domestic risk-free interest rate; note that we have often used the symbol r_f, but we do not wish to suggest that the f represents the foreign currency

r_F = foreign risk-free interest rate; note that the subscript is capitalized

32.2 OPTION PAYOFFS

At this point in the book, you will be familiar with option payoff functions. Nonetheless, it will be useful to repeat them here. The payoff of the call at expiration is

$$
\begin{aligned}
S_T - X &\quad \text{if } S_T \geq X \\
0 &\quad \text{if } S_T < X.
\end{aligned}
\tag{32.1}
$$

Of course, we frequently wrote this compactly as $\max(0, S_T - X)$. We will now show that this option is isomorphic to a put option denominated in the foreign currency to sell X units of the domestic currency at an exercise price of $1/X$. Let $X_R = 1/X$, the reciprocal strike. We shall call this put the reciprocal put. Because the reciprocal put has an exercise price of $1/X$ and the underlying value at expiration is $1/S_T$, which we designate as R_T, the put will exercise when $1/S_T < 1/X$. Rearranging, we have

$$
1/S_T \leq 1/X
$$
$$
\Rightarrow S_T \geq X.
\tag{32.2}
$$

Thus, when the reciprocal put is in-the-money, the spot call is in-the-money. The out-of-the-money put condition is obviously

$$
1/S_T > 1/X
$$
$$
\Rightarrow S_T < X.
\tag{32.3}
$$

Thus, the reciprocal put out-of-the-money condition is the same as the call out-of-the-money condition. Thus, to summarize, we have

The condition	$1/S_T \leq 1/X$	$1/S_T > 1/X$
is mathematically equivalent to	$S_T \geq X$	$S_T < X$
And, therefore, implies that		
Reciprocal put	In-the-money	Out-of-the-money
Call	In-the-money	Out-of-the-money

Incidentally, it should also be apparent that a standard put is equivalent to a reciprocal call.

Having established that the moneyness conditions of a standard call and reciprocal put are the same, we now proceed to show that the payoffs are the same, provided we adjust the quantity of the reciprocal put and translate the payoff value into the domestic currency. The payoff of the reciprocal put option is

$$X_R - R_T = (1/X) - (1/S_T) \quad \text{if } S_T \geq X$$
$$0 \qquad\qquad\qquad\qquad \text{if } S_T < X. \tag{32.4}$$

If we assume there are X put options, this payoff in aggregate is

$$X(1/X) - X(1/S_T) = 1 - X/S_T \quad \text{if } S_T \geq X$$
$$0 \qquad\qquad\qquad\qquad\qquad \text{if } S_T < X. \tag{32.5}$$

This payoff is in units of the foreign currency. To convert to domestic currency, we multiply by S_T:

$$S_T(1 - X/S_T) = S_T - X \quad \text{if } S_T \geq X$$
$$0 \qquad\qquad\qquad\qquad \text{if } S_T < X. \tag{32.6}$$

Thus, we see that the payoff of the X reciprocal puts is the same as that of the spot call.

So, to repeat, a spot call on a foreign currency with an exercise price of X quoted in terms of the domestic currency is equivalent to X reciprocal puts on the domestic currency with an exercise price of $1/X$ in terms of the foreign currency. As such, in the world of foreign currency options, every call is also a put. Now, just to remind you, this would be true for any option if we were comfortable talking in terms of calls say on a stock as being puts on money in terms of stock. But we typically do not characterize other types of options in this manner.

32.3 VALUATION OF THE OPTIONS

Now we demonstrate how the premiums of these two options are equivalent in the Black-Scholes-Merton model. The BSM value of the call option is

$$c = S_0 e^{-r_F \tau} N(d_1^D) - X e^{-r_D \tau} N(d_2^D), \tag{32.7}$$

where

$$d_1^D = \frac{\ln\left(\frac{S_0 e^{-r_F \tau}}{X e^{-r_D \tau}}\right) + \left(\frac{\sigma^2}{2}\right)\tau}{\sigma\sqrt{\tau}}$$

$$d_2^D = d_1^D - \sigma\sqrt{\tau}. \tag{32.8}$$

Note that we use the D superscript on the normal distribution function to indicate that this is the normal probability for the call written on the foreign currency and denominated in the domestic currency. That is, it is being viewed from the domestic standpoint.

We now provide a numerical example: Suppose the euro exchange rate is \$1.15 per euro, the exercise price is \$1.14 per euro, the time to maturity is 0.25, the domestic US interest rate is 0.8815%, the foreign euro interest rate is 0.4%, and the volatility is 15%. First, solving for the values of d_1^D and d_2^D, we have

$$d_1^D = \frac{\ln\left(\frac{S_0 e^{-r_F \tau}}{X e^{-r_D \tau}}\right) + \left(\frac{\sigma^2}{2}\right)\tau}{\sigma\sqrt{\tau}} = \frac{\ln\left(\frac{1.15 e^{-(0.004)0.25}}{1.14 e^{-(0.008815)0.25}}\right) + \left(\frac{0.15^2}{2}\right)0.25}{0.15\sqrt{0.25}} = 0.17$$

$$d_2^D = \frac{\ln\left(\frac{S_0 e^{-r_F \tau}}{X e^{-r_D \tau}}\right) - \left(\frac{\sigma^2}{2}\right)\tau}{\sigma\sqrt{\tau}} = \frac{\ln\left(\frac{1.15 e^{-(0.004)0.25}}{1.14 e^{-(0.008815)0.25}}\right) - \left(\frac{0.15^2}{2}\right)0.25}{0.15\sqrt{0.25}} = 0.095.$$

Based on Table 5.1: Standard Normal Cumulative Distribution Function Table, we have

$$c = S_0 e^{-r_F \tau} N(d_1^D) - X e^{-r_D \tau} N(d_2^D)$$

$$= 1.15 e^{-(0.004)0.25}(0.567495) - 1.14 e^{-(0.008815)0.25}(0.537843) = 0.0402.$$

As we stated, this call is equivalent to X puts written on the domestic currency denominated in the foreign currency with a strike of $X_R = 1/X$. We now validate that this result is upheld in the BSM model.

From the foreign investor's perspective, the value of a single unit of the reciprocal put is

$$p_R = X_R e^{-r_F \tau}\left[1 - N(d_2^F)\right] - R_0 e^{-r_D \tau}\left[1 - N(d_1^F)\right], \tag{32.9}$$

where p_R is denominated in the foreign currency and[2]

$$d_1^F = \frac{\ln\left(\frac{R_0 e^{-r_D \tau}}{X_R e^{-r_F \tau}}\right) + \left(\frac{\sigma^2}{2}\right)\tau}{\sigma\sqrt{\tau}}$$

$$d_2^F = \frac{\ln\left(\frac{R_0 e^{-r_D \tau}}{X_R e^{-r_F \tau}}\right) - \left(\frac{\sigma^2}{2}\right)\tau}{\sigma\sqrt{\tau}}. \tag{32.10}$$

Before reconciling this put formula to the prior call formula, we circle back to our example. Recall the euro exchange rate is \$1.15 per euro, or €0.869565 per dollar, the exercise price is \$1.14 per euro, or €0.877193 per dollar, the time to maturity is 0.25, the domestic euro interest rate is 0.4%, the foreign US interest rate is 0.8815%, and the volatility is 15%. Again, solving for the value of d_1^F and d_2^F, we have

$$d_1^F = \frac{\ln\left(\dfrac{R_0 e^{-r_D \tau}}{X_R e^{-r_F \tau}}\right) + \left(\dfrac{\sigma^2}{2}\right)\tau}{\sigma\sqrt{\tau}} = \frac{\ln\left(\dfrac{0.869565 e^{-(0.004)0.25}}{0.877193 e^{-(0.008815)0.25}}\right) + \left(\dfrac{0.15^2}{2}\right)0.25}{0.15\sqrt{0.25}} = -0.095$$

$$d_2^F = \frac{\ln\left(\dfrac{R_0 e^{-r_D \tau}}{X_R e^{-r_F \tau}}\right) - \left(\dfrac{\sigma^2}{2}\right)\tau}{\sigma\sqrt{\tau}} = \frac{\ln\left(\dfrac{0.869565 e^{-(0.004)0.25}}{0.877193 e^{-(0.008815)0.25}}\right) - \left(\dfrac{0.15^2}{2}\right)0.25}{0.15\sqrt{0.25}} = -0.17.$$

Note how these are related to their corresponding call terms, a point we return to later. Again based on Table 5.1 we have[3]

$$p_R = X_R e^{-r_F \tau}\left[1 - N(d_2^F)\right] - R_0 e^{-r_D \tau}\left[1 - N(d_1^F)\right]$$

$$= 0.877193 e^{-(0.004)0.25}(1 - 0.432505) - 0.869565 e^{-(0/008815)0.25}(1 - 0.462157)$$

$$= 0.0306.$$

Clearly, this put value of €0.0306 is not equal to the call value of \$0.0402, even after adjusting for the exchange rate.

Now let us examine why this is still the correct put formula at the core. In words, the BSM formula for the value of a put is the strike times the present value factor for the investor's currency times $1 - N(d_2^F)$ minus the underlying price in the investor's currency times the present value factor for the yield on the currency, times $1 - N(d_2^F)$. Thus, in the previous formula for p, the strike is properly designated as $X_R = 1/X$. The strike is then multiplied by the present value factor for the investor's currency. Because this put is being valued from the foreign investor's perspective, it is appropriate to discount the strike at the foreign rate. Then we use the normal probability based on $-d_2^F$, with the F superscript to distinguish it from d_2^D in the call. From the product of these three terms, we subtract the exchange rate in foreign currency, $R_0 = 1/S_0$, times the present value factor for the opposite currency. From the foreign investor's perspective, this rate is r_D. Then this product is multiplied by one minus the normal probability for d_1^F.

Now let us work with d_1^F and d_2^F. First, note the following:

$$\ln\left(\frac{R_0 e^{-r_D \tau}}{X_R e^{-r_F \tau}}\right) = \ln\left(\frac{\dfrac{1}{S_0} e^{-r_D \tau}}{\dfrac{1}{X} e^{-r_F \tau}}\right)$$

$$= \ln 1 - \ln S_0 - r_D \tau - \ln 1 + \ln X + r_F \tau$$

$$= -\ln S_0 - r_D \tau + \ln X + r_F \tau$$

$$= -\ln\left(\frac{S e^{-r_F \tau}}{X e^{-r_D \tau}}\right). \tag{32.11}$$

Thus,

$$d_1^F = \frac{-\ln\left(\frac{S_0 e^{-r_F\tau}}{Xe^{-r_D\tau}}\right)+\left(\frac{\sigma^2}{2}\right)\tau}{\sigma\sqrt{\tau}} = -\left[\frac{\ln\left(\frac{S_0 e^{-r_F\tau}}{Xe^{-r_D\tau}}\right)-\left(\frac{\sigma^2}{2}\right)\tau}{\sigma\sqrt{\tau}}\right] = -d_2^D$$

$$d_2^F = \frac{-\ln\left(\frac{S_0 e^{-r_F\tau}}{Xe^{-r_D\tau}}\right)-\left(\frac{\sigma^2}{2}\right)\tau}{\sigma\sqrt{\tau}} = -\left[\frac{\ln\left(\frac{S_0 e^{-r_F\tau}}{Xe^{-r_D\tau}}\right)+\left(\frac{\sigma^2}{2}\right)\tau}{\sigma\sqrt{\tau}}\right] = -d_1^D. \tag{32.12}$$

We saw this in the numerical calculations already. Therefore,

$$N(-d_1^F) = N(d_2^D)$$
$$N(-d_2^F) = N(d_1^D). \tag{32.13}$$

Our put formula will, thus, be

$$p_R = X_R e^{-r_F\tau}N(-d_2^F) - R_0 e^{-r_D\tau}N(-d_1^F)$$
$$= X_R e^{-r_F\tau}N(d_1^D) - R_0 e^{-r_D\tau}N(d_2^D)$$
$$= \left(\frac{1}{X}\right)e^{-r_F\tau}N(d_1^D) - \left(\frac{1}{S_0}\right)e^{-r_D\tau}N(d_2^D). \tag{32.14}$$

Recall that the isomorphic put is actually X puts, so the aggregate put premium is

$$Xp_R = X\left(\frac{1}{X}\right)e^{-r_F\tau}N(d_1^D) - X\left(\frac{1}{S_0}\right)e^{-r_D\tau}N(d_2^D)$$
$$Xp_R = e^{-r_F\tau}N(d_1^D) - \left(\frac{X}{S_0}\right)e^{-r_D\tau}N(d_2^D) \text{ (in units of the foreign currency).} \tag{32.15}$$

This value is in units of the foreign currency. To convert to the domestic currency, we simply multiply by S_0.

$$Xp_D = S_0 e^{-r_F\tau}N(d_1^D) - Xe^{-r_D\tau}N(d_2^D) \text{ (in units of the domestic currency).} \tag{32.16}$$

This is the same as the right-hand side of the call formula.

Once again, we illustrate the correspondence between models with our numerical example. Recall

$$c = S_0 e^{-r_F T}N(d_1^D) - Xe^{-r_D T}N(d_2^D)$$
$$= 1.15e^{-(0.004)0.25}(0.567495) - 1.14e^{-(0.008815)0.25}(0.537843) = \$0.0402$$

and

$$p_R = X_R e^{-r_F T} \left[1 - N(d_2^F) \right] - R_0 e^{-r_D T} \left[1 - N(d_1^F) \right]$$

$$= 0.877193 e^{-(0.004)0.25}(1 - 0.432505) - 0.869565 e^{-(0.008815)0.25}(1 - 0.462157)$$

$$= 0.0306 €.$$

Assuming we purchase $X = 1.14$ puts and then convert to USD, we note that $S_0 X p = (\$1.15/€)(1.14) €0.0306 = \$0.0402$. This numerical example highlights the two different ways to express currency options.

32.4 PROBABILITY OF EXERCISE

It is well known that the risk-neutral probability of exercise of the spot call option is $N(d_2^D)$ and correspondingly the risk-neutral probability of exercise of the spot put option is $N(-d_2^D)$. For spot calls, this is the risk-neutral probability that $S_T > X$. The corresponding probability for exercise of the reciprocal put is the probability that $1/S_T < 1/X$, which must be the same as the probability that $S_T > X$. Thus, $N(d_2^D)$ is the risk-neutral probability of exercise of the call, and this equals $N(-d_1^F)$, which is the probability of exercising the reciprocal put. Note that the probability of exercising the reciprocal put is not the term in the BSM put formula that is multiplied by the present value of $1/X$, or $N(-d_2^F)$. There is an important reason for that, which will become clearer when we examine the actual, as opposed to risk-neutral, probability of exercise.

Let the true stochastic process for S be given as

$$\frac{dS_t}{S_t} = \alpha dt + \sigma dW_t. \tag{32.17}$$

Here α and σ are the expected return and volatility of dS_t/S_t. It is important to remember that the stochastic process for the log return is

$$d \ln S_t = \mu dt + \sigma dW_t, \tag{32.18}$$

with the log expected return given as

$$\mu = \alpha - \sigma^2/2, \tag{32.19}$$

which holds by definition for normal distributions.

Now we need the stochastic process of the reciprocal exchange rate. We will apply Itô's lemma for the reciprocal exchange rate, $R_t = 1/S_t$. Thus, we will need certain partial derivatives:

$$\frac{\partial R_t}{\partial S_t} = -\frac{1}{S_t^2}$$

$$\frac{\partial^2 R_t}{\partial S_t^2} = \frac{2}{S_t^3}$$

$$\frac{\partial R_t}{\partial t} = 0. \tag{32.20}$$

Now, applying Itô's lemma to R_t,

$$dR_t = \frac{\partial R_t}{\partial S_t} dS_t + \frac{\partial R_t}{\partial t} dt + \frac{1}{2} \frac{\partial^2 R_t}{\partial S_t^2} dS_t^2. \tag{32.21}$$

Substituting Equations (32.17) and (32.20), and using the fact that $dS_t^2 = S_t^2 \sigma^2 dt$, we obtain

$$dR_t = \left(\frac{1}{S_t}\right)\left(-\alpha dt - \sigma dW + \sigma^2 dt\right)$$

$$= \left(\frac{1}{S_t}\right)\left[\left(-\alpha + \sigma^2\right) dt - \sigma dW\right]. \tag{32.22}$$

Dividing by R_t and multiplying by S_t, we obtain

$$\frac{dR_t}{R_t} = \left(-\alpha + \sigma^2\right) dt - \sigma dW_t. \tag{32.23}$$

Thus, the expected return on the reciprocal exchange rate is $-\alpha + \sigma^2$. Note, however, the minus term on the volatility, which contributes to the reason why the probability of exercise of the put is not given by the $N(-d_2)$ term in its formula, as it usually is in the BSM model.[4]

The log return on R_t is

$$-\alpha + \sigma^2 - \sigma^2/2 = -\alpha + \sigma^2/2. \tag{32.24}$$

Substituting for α, this can then be expressed as

$$-\alpha + \sigma^2/2 = -\left(\mu - \sigma^2/2\right) + \sigma^2/2 = -\mu. \tag{32.25}$$

This is the term that must be inserted into the formula for the normal probability.[5] It is the expected log return on R_t expressed in terms of the log return on S_t. Making the substitution into $-d_1^F$, and using a* to denote that we are dealing with the true probability distribution, not the risk-neutral distribution, we have

$$N\left(-d_1^{F*}\right) = N\left[-\frac{\ln(R_0/X_R) + \left(-\mu + \sigma^2/2\right)\tau}{\sigma\sqrt{\tau}}\right] = N\left[-\frac{\ln(X/S_0) + \left(-\mu + \sigma^2/2\right)\tau}{\sigma\sqrt{\tau}}\right]$$

$$= N\left[\frac{-\ln(X/S_0) + \left(\mu - \sigma^2/2\right)\tau}{\sigma\sqrt{\tau}}\right] = N\left[\frac{\ln(S_0/X) + \left(\mu - \sigma^2/2\right)\tau}{\sigma\sqrt{\tau}}\right]. \tag{32.26}$$

This is, of course, $N\left(d_2^{D*}\right)$ because

$$d_2^{D*} = \frac{\ln(S_0/X) + \left(\mu - \sigma^2/2\right)\tau}{\sigma\sqrt{\tau}}. \tag{32.27}$$

Thus, we see $N\left(d_2^{D*}\right)$ is the risk-neutral probability of exercise of the call, and this equals $N\left(-d_1^{F*}\right)$, which is the probability of exercising the reciprocal put.

32.5 SOME TERMINOLOGY CONFUSION

Some foreign currency option traders do not like the terminology *put* and *call*. They see every option as being both a put and a call. Hence, it is common in foreign currency option transactions to use slightly different terminology than in other types of options. Foreign currency options, thus, are often written with two values referred to as the *call currency amount* and the *put currency amount*. As with any option, the holder of the option makes the exercise decision. If they choose to exercise, they tender the put currency amount and receive the call currency amount.[6] As an example, suppose an option written on the pound designates the call currency amount as £10,000,000 and the put currency amount as $15,000,000. At expiration, if the exchange rate is higher than $1.50, the holder of the option tenders (*puts*) $15,000,000 to the writer and claims (*calls*) £10,000,000. With the exchange rate higher than $1.50, the option holder can exchange the £10,000,000 and receive more than $15,000,000.

In addition, there is a further matter in the world of currencies that causes some confusion. It involves the use of ratios in expressing exchange rates. It can be quite confusing.

Consider an ordinary ratio, say 3/8. In elementary math, this ratio is easily interpreted. We mentally consider an object divided into eighths in which we reference three of those eighths. For example, we might bake a cake and cut it into eight pieces. Suppose we have three guests and we give each a piece. We have just used up three-eighths of the cake. Ratios can be easily multiplied. For example, $(2/3)(1/5) = 2/15$. When we express the price of milk, we frequently say something like $3.00 a gallon. This can be written as $3/gal. where the / is interpreted as *per*. We say that $3 is the price per gallon of milk. These certainly seem like simple concepts.

In the currency world, we can do a similar operation. We can indicate the ratio USD/ EUR. In conformance with the notion that this is a ratio, / is interpreted as *per* and we read that ratio as "the number of dollars per euro." Thus, if USD/EUR = $1.30, we say that one euro costs $1.30.

As it turns out, however, the currency world does just the opposite. The notation USD/ EUR means "the number of euros per US dollar." The first term (before the slash) is the base currency and the second term (after the slash) is the quote currency. The multiplication works just the same and is just as valid. For example, using the dollar (USD), euro (EUR), and Japanese yen (JPY), we have $(USD/EUR) * (EUR/JPY) = USD/JPY$. This is read as euros per dollar times yen per euro equals yen per dollar, which is USD/JPY in the industry's notation. This notational form is standard practice but does appear reversed. If it were not, we could write the price of milk as gal./$3, which we would never do. Thus, you should be aware of this format.

So, in short, if the euro costs $1.30, it would be written as $EUR/USD = 1.30$.

32.6 RECAP

In this final chapter of the book, we look at foreign currency puts and calls. They are viewed somewhat differently from standard underlyings in that a call on one currency can be viewed from the perspective of a put on the other currency. We showed that pricing options on foreign currency is, however, fairly easy to implement using the Black-Scholes-Merton model.

This chapter completes the book. We have laid the foundations of the theory of derivative pricing within the framework of the much broader discipline of quantitative finance. With the caveat that no theory ever works perfectly in practice, you can now proceed to study further and begin applying what you know to gain a better understanding of how options are used directly and how they apply indirectly to a variety of problems in finance. In addition, you are well positioned for further study of other derivatives, such as forwards, futures, and swaps.

QUESTIONS AND PROBLEMS

1 In your own words, define what is meant by isomorphism within the context of foreign exchange options.

2 Prove that a call option on a foreign currency with an exercise price of X quoted in terms of the domestic currency is equivalent to X puts on the local currency with an exercise price of $1/X$ in terms of the foreign currency.

3 Demonstrate that $d_1^F = -d_2^D$ and $d_2^F = -d_1^D$, where

$$d_1^D = \frac{\ln\left(\frac{S_0 e^{-r_F \tau}}{X e^{-r_D \tau}}\right) + \left(\frac{\sigma^2}{2}\right)\tau}{\sigma\sqrt{\tau}}$$

$$d_2^D = \frac{\ln\left(\frac{S_0 e^{-r_F \tau}}{X e^{-r_D \tau}}\right) - \left(\frac{\sigma^2}{2}\right)\tau}{\sigma\sqrt{\tau}}$$

and

$$d_1^F = \frac{\ln\left(\frac{R_0 e^{-r_D \tau}}{X_R e^{-r_F \tau}}\right) + \left(\frac{\sigma^2}{2}\right)\tau}{\sigma\sqrt{\tau}}$$

$$d_2^F = \frac{\ln\left(\frac{R_0 e^{-r_D \tau}}{X_R e^{-r_F \tau}}\right) - \left(\frac{\sigma^2}{2}\right)\tau}{\sigma\sqrt{\tau}}.$$

4 The risk-neutral probability of exercise of the call option is $N(d_2^D)$. Demonstrate that $N(d_2^D) = N(-d_1^F)$.

5 The actual probability of exercising a foreign exchange put option is not $N(-d_2^F)$. Explain.

6 Suppose the Australian dollar exchange rate is $0.72 per AUD, the exercise price is $0.75 per AUD, the time to maturity is 1.0, the domestic U.S. interest rate is 1.05%, the foreign AUD interest rate is 2.97%, and the volatility is 20%.
   a. Calculate the value of a foreign exchange call option (US $ per AUD) based on the Black-Scholes-Merton model from a US-based investor perspective.
   b. Calculate the value of a foreign exchange put option (AUD per US $) based on the Black-Scholes-Merton model from an Australian-based investor perspective.
   c. Demonstrate how to reconcile these two results.

NOTES

1. The primary works that uncovered the unique characteristics of currency options and the application of many of the standard principles of option pricing to currency options are Grabbe (1983) and Garman and Kohlhagen (1983).
2. Keep in mind that the reference to an interest rate being D for domestic or F for foreign must maintain consistency. Even though we are viewing this transaction from the foreign investor's point of view, the rate denoted F is their rate and D is the foreign rate from their point of view.
3. Remember that the final answer is sensitive to the degree of rounding. The answer reported here assumes no rounding.
4. The probability in the BSM formula is based on a stochastic process in which the noise term, σdW_t, is added, not subtracted.
5. This result is easy to see in the risk-neutral context. The risk-free rate term in the numerator of d_1^D is $r_D - r_F$ where under risk neutrality, r_D is the risk-neutralized expected return on the domestic currency and r_F is the risk-neutralized expected return on the foreign currency. From the foreign investor's point of view, this corresponding term would be $r_F - r_D$, which is $-(r_D - r_F)$. That is, the expected return is converted to its negative value.
6. In this context, it is easy to see why currency options are referred to as both puts and calls. The holder of the option *puts* one currency to the writer of the option and *calls* the other currency. In the context of assets other than currencies, the holder of the option *puts* the asset and *calls* the agreed-on cash, the exercise price, or the other way around.

References

Chapter numbers where the references can be found are shown in brackets at the end of each source.

Abramowitz, M., and I. Stegun. *Handbook of Mathematical Functions with Formulas, Graphs, and Mathematical Tables*. Washington, DC: United States Department of Commerce, National Bureau of Standards, 1972. [5]

Alexander, S. "Price Movements in Speculative Markets: Trends or Random Walks." *Industrial Management Review* 2 (1961): 7–26. [12]

Andersen, L., and J. Andreasen. "Jump-Diffusion Processes: Volatility Smile Fitting and Numerical Methods for Option Pricing." *Review of Derivatives Research* 4 (2000): 231–262. [31]

Arrow, K. "The Role of Securities in the Optimal Allocation of Risk-Bearing." *Review of Economic Studies* 31 (1964): 91–96. [29]

Bachelier, L. "Théorie de la Spéculation." *Annales Scientifiques de l'École Normale Supérieure* 3, no. 17 (1900): 21–86. [10, 12, 13]

Banz, R. W., and M. H. Miller. "Prices for State-Contingent Claims: Some Extensions and Applications." *The Journal of Business* 51 (1978): 653–72. [29]

Barone-Adesi, G., and R. W. Whaley. "Efficient Analytic Approximation of American Option Values." *The Journal of Finance* 42 (1987): 301–20. [19]

Baxter, M., and A. Rennie. *Financial Calculus: An Introduction to Derivative Pricing*. Cambridge, UK: Cambridge University Press, 1996. [10, 11, 16]

Bernstein, P. *Against the Gods: The Remarkable Study of Risk*. New York: John Wiley & Sons, 1996. [1, 13]

Bernstein, P. *Capital Ideas Evolving*. New York: John Wiley & Sons, 2007. [1]

Bernstein, P. *Capital Ideas: The Improbable Origins of Modern Wall Street*. New York: The Free Press, 1992. [1]

Billingsley, R. S. *Understanding Arbitrage: An Intuitive Approach to Financial Analysis*. Upper Saddle River, NJ: Wharton Publishing, 2006. [6]

Björk, T. *Arbitrage Theory in Continuous Time*. Oxford, UK: Oxford University Press, 1998. [6]

Black, F. "How to Use the Holes in Black-Scholes." *Journal of Applied Corporate Finance* 1 (Winter 1989a): 67–73. [13]

Black, F. "How We Came Up with the Option Formula." *Journal of Portfolio Management* 15 (Winter 1989b): 4–8. [13]

Black, F. "The Pricing of Commodity Contracts." *Journal of Financial Economics* 3 (1976): 167–79. [22]

Black, F., E. Derman, and W. Toy. "A One-Factor Model of Interest Rates and Its Application to Treasury Bond Options." *Financial Analysts Journal* 46 (January–February 1990): 33–39. [27]

Black, F., and P. Karasinski. "Bond and Option Pricing When Short Rates Are Lognormal." *Financial Analysts Journal* 47 (July–August 1991): 52–59. [27]

Black, F., and M. Scholes. "The Pricing of Options and Corporate Liabilities." *Journal of Political Economy* 81 (1973): 637–59. [12, 13]

Black, F., and M. Scholes. "Valuation of Options and a Test of Market Efficiency." *The Journal of Finance* 27 (1972): 399–418. [13]

Blomeyer, E. C., and H. Johnson. "An Empirical Examination of the Pricing of American Put Options." *Journal of Financial and Quantitative Analysis* 23 (1988): 13–22. [20]

Boness, A. J. "Elements of a Theory of Stock-Option Value." *Journal of Political Economy* 72 (1964): 163–83. [13]

Boyle, P. B. "Options: A Monte Carlo Approach." *Journal of Financial Economics* 4 (1977): 323–38. [23]

Brace, A., D. Gatarek, and M. Musiela. "The Market Model of Interest Rate Dynamics." *Mathematical Finance* 7 (1997): 127–54. [27]

Breeden, D. T., and R. H. Litzenberger. "Prices of State-Contingent Claims Implicit in Option Prices." *The Journal of Business* 51 (1978): 621–51. [29]

Brennan. M. J., and E. S. Schwartz. "Finite Difference Methods and Jump Processes Arising in the Pricing of Contingent Claims." *Journal of Financial and Quantitative Analysis* 13 (1978): 461–74. [24]

Brenner, M., G. Courtadon, and M. G. Subrahmanyam. "Options on the Spot and Options on Futures." *The Journal of Finance* 40 (1985): 1303–17. [22]

Brenner, R. J., and R. A. Jarrow. "A Simple Formula for Options on Discount Bonds." *Advances in Futures and Options Research* 6 (1993): 45–51. [27]

Brooks, R. "Compound Option Valuation with Maturity Varying Volatility, Maturity Varying Yields, and Maturity Varying Interest Rates." (September 23, 2019). https://ssrn.com/abstract=3458918. [18]

Brooks, R. "Samuelson Hypothesis and Carry Arbitrage." *Journal of Derivatives* 20 (2012): 37–65. [22]

Brooks, R., and J. Brooks. "An Option Valuation Framework Based on Arithmetic Brownian Motion: Justification and Implementation Issues." *Journal of Financial Research* 40 (2017): 401–27. [12]

Brooks, R., D. Chance, and M. Hemler. "The 'Superior Performance' of Covered Calls on the S&P 500: Rethinking an Anomaly." *Journal of Derivatives* (2019). [30]

Brooks, R., and B. N. Cline. "Spread Options and Risk Management: Lognormal Versus Normal Distribution Assumption." *Financial Services Review* 24 (2015): 15–35. [17]

Brooks, R., and M. Livingston. "The Difference Between the Local and Unbiased Expectations Hypotheses." *Review of Quantitative Finance and Accounting* 2 (1992): 377–89. [25]

Buetow, G. W., Jr., and F. J. Fabozzi. *Valuation of Interest Rate Swaps and Swaptions.* New York: Frank J. Fabozzi & Associates, 2001. [26]

Bunch, D. S., and H. Johnson. "A Simple and Numerically Efficient Valuation Method for American Puts Using a Modified Geske-Johnson Approach." *The Journal of Finance* 47 (1992): 809–16. [20]

Campbell, J., and R. Shiller. "Cointegration and Tests of Present Value Models." *Journal of Political Economy* 95 (1987): 1062–88. [25]

Carmona, R., and V. Durrleman. "Generalizing the Black-Scholes Formula to Multivariate Contingent Claims." *Journal of Computational Finance* 9 (2006): 43–67. [17]

Carmona, R., and V. Durrleman. "Pricing and Hedging Spread Options." *SIAM Review* 45 (2003): 627–85. [17]

Chance, D. M. "A Synthesis of Binomial Option Pricing Models for Lognormally Distributed Assets." *Journal of Applied Finance* 18 (2008): 38–56. [9]

Chance, D. M. "Translating the Greek: The Real Meaning of Call Option Derivatives." *Financial Analysts Journal* 50 (July–August 1994): 43–49. [14]

Chance, D. M., and Ş. Ağca. "Speed and Accuracy Comparisons of Bivariate Normal Approximations for Option Pricing." *The Journal of Computational Finance* 5 (2003): 67–90. [5]

Chance, D. M., and R. Brooks. *An Introduction to Derivatives and Risk Management.* 10th ed. Mason, OH: Cengage, 2016. [2]

Chance, D. M., T. A. Hanson, W. Li, and J. Muthuswamy. "A Bias in the Volatility Smile." *Review of Derivatives Research* 20 (2017): 47–90. [31]

Chance, D. M., and D. Rich. "The False Teachings of the Unbiased Expectations Hypothesis." *Journal of Portfolio Management* 27 (Summer 2001): 83–95. [25]

Chiang, A., and K. Wainwright. *Fundamental Methods of Mathematical Economics*. 4th ed. New York: McGraw-Hill, 2005. [3]

Corb, H. *Interest Rate Swaps and Other Derivatives*. New York: Columbia University Press, 2012. [26]

Cox, J. C., J. E. Ingersoll, Jr., and S. A. Ross. "A Re-Examination of Traditional Hypotheses About the Term Structure of Interest Rates." *The Journal of Finance* 36 (1981): 769–99. [22, 25]

Cox, J. C., J. E. Ingersoll, Jr., and S. A. Ross. "The Relation Between Forward Prices and Futures Prices." *Journal of Financial Economics* 9 (1981): 321–46. [22]

Cox, J. C., J. E. Ingersoll, Jr., and S. A. Ross. "An Intertemporal General Equilibrium Model of Asset Prices." *Econometrica* 53 (1985a): 363–84. [25]

Cox, J. C., J. E. Ingersoll, Jr., and S. A. Ross. "A Theory of the Term Structure of Interest Rates." *Econometrica* 53 (1985b), 385–407. [25, 27]

Cox, J. C., S. A. Ross, and M. Rubinstein. "Option Pricing: A Simplified Approach." *Journal of Financial Economics* 7 (1979): 229–63. [7, 9]

Cox, J. C., and M. Rubinstein. *Options Markets*. Englewood Cliffs, NJ: Prentice-Hall, 1985. [2]

Culbertson, J. M. "The Term Structure of Interest Rates." *Quarterly Journal of Economics* 71 (1957): 485–517. [25]

Debreu, G. *The Theory of Value*. New York: John Wiley & Sons, 1964. [29]

Derman, E. *My Life as a Quant: Reflections on Physics and Finance*. New York: John Wiley & Sons, 2004. [1]

Derman, E., and I. Kani. "Riding on a Smile." *Risk* 7 (February 1994): 32–39. [31]

Drezner, Z. "Computation of the Bivariate Normal Integral." *Mathematics of Computation* 32 (1978): 277–79. [5]

Dupire, B. "Pricing with a Smile." *Risk* 7 (January 1994): 18–20. [31]

Engsted, T., and C. Tanggaard. "Cointegration and the US Term Structure." *Journal of Banking and Finance* 18 (1994): 167–81. [25]

Engsted, T., and C. Tanggaard. "The Predictive Power of Yield Spreads for Future Interest Rates: Evidence from the Danish Term Structure." *The Scandinavian Journal of Economics* 97 (1995): 145–59. [25]

Fama, E. F. *Foundations of Finance*. New York: Basic Books, 1976. [6]

Fama, E. F., and M. H. Miller. *The Theory of Finance*. Hinsdale, IL: The Dryden Press, 1972. [6]

Fischer, S. "Call Option Pricing When the Exercise Price Is Uncertain, and the Valuation of Indexed Bonds." *The Journal of Finance* 33 (1978): 169–76. [17]

Garman, M. B., and S. W. Kohlhagen. "Foreign Currency Option Values." *Journal of International Money and Finance* 2 (1983): 231–37. [32]

Geske, R. "A Note on an Analytical Valuation Formula for Unprotected American Call Options on Stocks with Known Dividends." *Journal of Financial Economics* 7 (1979a): 375–80. [19]

Geske, R. "On the Valuation of American Call Options on Stocks with Known Dividends: A Comment." *Journal of Financial Economics* 9 (1981): 213–15. [19]

Geske, R. "The Valuation of Compound Options." *Journal of Financial Economics* 7 (1979b): 63–81. [18, 19, 20]

Geske, R. "The Valuation of Corporate Liabilities as Compound Options." *Journal of Financial and Quantitative Analysis* 12 (1977): 541–52. [18, 19]

Geske, R. "The Valuation of Corporate Liabilities as Compound Options: A Correction." *Journal of Financial and Quantitative Analysis* 19 (1984): 231–32. [18]

Geske, R., and H. E. Johnson. "The American Put Option Valued Analytically." *The Journal of Finance* 39 (1984): 1511–24. [20]

Grabbe, O. "The Pricing of Call and Put Options on Foreign Exchange." *Journal of International Money and Finance* 2 (1983): 249–53. [32]

Grant, D., and G. Vora. "Extending the Universality of the Heath-Jarrow-Morton Model." *Review of Financial Economics* 15 (2006): 129–57. [27, 28]

Grant, D., and G. Vora. "Implementing No-Arbitrage Terms Structure of Interest Rate Models in Discrete Time When Interest Rates Are Normally Distributed." *The Journal of Fixed Income* 8 (May 1999): 85–98. [27, 28]

Harrison, J. M., and D. Kreps. "Martingales and Arbitrage in Multiperiod Securities Markets." *Journal of Economic Theory* 20 (1979): 381–408. [15]

Harrison, J. M., and S. R. Pliska. "Martingales and Stochastic Integrals in the Theory of Continuous Trading." *Stochastic Process and Their Applications* 11 (1981): 215–60. [15]

Haug, E. G. *The Complete Guide to Option Pricing Formulas.* 2nd ed. New York: McGraw-Hill, 2007. [21]

Heath, D., R. Jarrow, and A. Morton. "Bond Pricing and the Term Structure of Interest Rates: A New Methodology." *Econometrica* 60 (1992): 77–105. [27]

Heath, D., R. A. Jarrow, and A. Morton. "Contingent Claims Valuation with a Random Evolution of Interest Rates." *Review of Futures Markets* 9 (1991): 54–76. [27]

Hicks, J. R. *Value and Capital.* 2nd ed. London: Oxford University Press, 1946. [25]

Ho, T.S.Y., and S. B. Lee. "Term Structure Movements and Pricing Interest Rate Contingent Claims." *The Journal of Finance* 41 (1986): 1011–19. [27]

Hsia, C.-C. "On Binomial Option Pricing." *The Journal of Financial Research* 6 (1983): 41–46. [9]

Huang, C.-F., and R. H. Litzenberger. *Foundations for Financial Economics.* New York: North-Holland, 1988. [6]

Ingersoll, J. E., Jr. *Theory of Financial Decision Making.* Totowa, NJ: Rowman & Littlefield, 1987. [6, 25]

Jarrow, R. *Finance Theory.* Englewood Cliffs, NJ: Prentice-Hall, 1988. [6]

Jarrow, R. *Modelling Fixed Income Securities and Interest Rate Options.* 2nd ed. Stanford, CA: Stanford University Press, 2002. [27]

Jarrow, R., and A. Chatterjea. *An Introduction to Derivative Securities, Financial Markets, and Risk Management.* New York: W. W. Norton & Company, 2013. [27]

Jarrow, R. A., and G. Oldfield. "Forward Contracts and Futures Contracts." *Journal of Financial Economics* 9 (1981): 373–82. [22]

Jarrow, R., and A. Rudd. *Option Pricing.* Homewood, IL: Irwin, 1983. [9]

Jarrow, R., and S. Turnbull. *Derivative Securities.* Cincinnati: South-Western College Publishing, 2000. [27]

Johnson, H. "Options on the Maximum or the Minimum of Several Risky Assets." *Journal of Financial and Quantitative Analysis* 22 (1987): 277–83. [21]

Karatzas, I., and S. E. Shreve. *Brownian Motion and Stochastic Calculus.* New York: Springer-Verlag, 1991. [11, 15]

Karlin, S., and H. M. Taylor. *A Second Course in Stochastic Processes.* Boston: Academic Press, 1981. [11]

Kessel, R. A. "The Cyclical Behavior of the Term Structure of Interest Rates." *NBER*, Occasional Paper No. 91 (1965). [25]

Kutner, G. W. "Black-Scholes Revisited: Some Important Details." *The Financial Review* 23 (1988): 95–104. [13]

Li, M., S. Deng, and J. Zhou. (2008). "Closed-Form Approximations for Spread Option Prices and Greeks." *Journal of Derivatives* 15: 58–80. [17]

Longstaff, F. A. "Arbitrage and the Expectations Hypothesis." *The Journal of Finance* 60 (2000): 989–94. [25]

Lutz, F. A. "The Structure of Interest Rates." *Quarterly Journal of Economics* 55 (1940): 36–63. [25]

Macmillan, L. W. "Analytic Approximations for the American Put Option." *Advances in Futures and Options Research* 1 (1986): 119–39. [19]

Malliaris, A. G., and W. A. Brock. *Stochastic Methods in Economics and Finance.* Amsterdam: North-Holland, 1982. [10, 11, 12]

Margrabe, W. "The Value of an Option to Exchange One Asset for Another." *The Journal of Finance* 33 (1978): 177–86. [17, 21]

Mehrling, P. *Fischer Black and the Revolutionary Idea of Finance.* New York: John Wiley & Sons, 2005. [1, 13]

Merton, R. C. "On the Pricing of Corporate Debt: The Risk Structure of Interest Rates." *Journal of Finance* 29 (1974): 449–70. [18]

Merton, R. C. "The Relationship Between Put and Call Option Prices: Comment." *The Journal of Finance* 28 (1973a): 183–84. [2]

Merton, R. C. "Theory of Rational Option Pricing." *Bell Journal of Economics and Management Science* 4 (1973b): 141–83. [2, 13, 17]

Modigliani, F., and R. Sutch. "Innovations in Interest Rate Policy." *American Economic Review* 56 (1966):178–97. [25]

Murphy, A. E. "Corporate Ownership in France: The Importance of History," *NBER*, Working Paper 10716 (2004). http://www.nber.org/papers/w10716. [12]

Myers, S. C. "A Time-State Preference Model of Security Valuation." *Journal of Financial and Quantitative Analysis* 3 (1968): 1–33. [29]

Neftci, S. N. *An Introduction to the Mathematics of Financial Derivatives.* 2nd ed. San Diego: Academic Press, 2006. [10, 11, 12, 15]

Osborne, M. "Brownian Motion in the Stock Market." *Operations Research* 7 (1959): 145–73. [12]

Pearson, N. D. "An Efficient Approach for Pricing Spread Options." *The Journal of Derivatives* 3 (1995): 76–91. [17]

Renault, E., and N. Touzi. "Option Hedging and Implied Volatilities in a Stochastic Volatility Model." *Mathematical Finance* 6 (1996): 279–302. [31]

Rendleman, R. J., Jr., and B. J. Bartter. "Two-State Option Pricing." *The Journal of Finance* 34 (1979): 1093–110. [7, 9]

Rich, D. R., and D. M. Chance. "An Alternative Approach to the Pricing of Options on Multiple Assets." *The Journal of Financial Engineering* 2 (1993): 271–85. [21]

Ritchken, P. *Derivative Markets.* New York: HarperCollins, 1996. [27]

Ritchken, P., and L. Sankarasubramanian. "Volatility Structures of Forward Rates and the Dynamics of the Term Structure." *Mathematical Finance* 5 (1995): 55–72. [27]

Roll, R. "An Analytic Valuation Formula for Unprotected American Call Options on Stocks with Known Dividends." *Journal of Financial Economics* 5 (1977): 251–58. [18, 19]

Ross, S. M. *An Introduction to Mathematical Finance: Options and Other Topics.* Cambridge, UK: Cambridge University Press, 1999. [12]

Rubinstein, M. "Compound Options." The Review of Financial Studies 4, no. 1 (1991): 87–120. [18]

Rubinstein, M. "Implied Binomial Trees." *The Journal of Finance* 69 (1994): 771–818. [31]

Rubinstein, M. "Pay Now, Choose Later." *Risk* 4 (1991a): 13. [17]

Rubinstein, M. "A Simple Formula for the Expected Rate of Return of an Option over a Finite Holding Period." *The Journal of Finance* 39 (1984): 1503–09. [30]

Rubinstein, M. "Somewhere over the Rainbow." *Risk* 4 (1991b): 63–66. [13, 21]

Rubinstein, M. "The Valuation of Uncertain Income Streams and the Pricing of Options." *Bell Journal of Economics* 7 (1976): 407–25. [13]

Samuelson, P. A. "Rational Theory of Warrant Pricing." *Industrial Management Review* 6 (1965): 13–39. [12, 13]

Shafer, G., and V. Vovk. *Probability and Finance: It's Only a Game.* New York: John Wiley & Sons, 2001. [4]

Shimko, D. *Finance in Continuous Time: A Primer.* Miami: Kolb Publishing, 1992. [12]

Smith, C. W., Jr. "Option Pricing: A Review." *Journal of Financial Economics* 3 (January–March 1976): 3–51. [2]

Sprenkle, C. M. "Warrant Prices as Indicators of Expectations and Preferences." *Yale Economic Essays* 1 (1961): 178–231. [12, 13]

Stephan, J., and R. E. Whaley. "Intraday Price Changes and Trading Volume Relations in the Stock and Stock Options Market." *The Journal of Finance* 45 (1990): 191–220. [19]

Stoll, H. R. "The Relationship Between Put and Call Option Prices." *The Journal of Finance* 24 (December 1969): 802–24. [2]

Stulz, R. M. "Options on the Minimum or the Maximum of Two Risky Assets." *Journal of Financial Economics* 10 (1982): 161–81. [21]

Szpiro, G. G. *Pricing the Future: Finance, Physics, and the 300-Year Journey to the Black-Scholes Equation.* New York: Basic Books, 2011. [13]

Taleb, N. N. *The Black Swan: The Impact of the Highly Improbable.* New York: Random House, 2007. [1]

Taleb, N. N. *Fooled by Randomness: The Hidden Role of Chance in the Markets and in Life.* New York: Texere, 2001. [1]

Thorp, E. O., and S. Kassouf. *Beat the Market: A Scientific Stock Market System.* New York: Random House, 1967. [13]

van der Hoek, J., and R. J. Elliott. *Binomial Models in Finance.* New York: Springer, 2006. [7]

Vasicek, O. "An Equilibrium Characterization of the Term Structure." *Journal of Financial Economics* 5 (1977): 177–88. [27]

Welch, R. L., and D. M. Chen. "On the Properties of the Valuation Formula for an Unprotected American Call Option with Known Dividends and the Computation of Its Implied Standard Deviation." *Advances in Futures and Options Research* 3 (1988): 237–56. [19]

Whaley, R. E. "On the Valuation of American Call Options on Stocks with Known Dividends." *Journal of Financial Economics* 9 (1981): 207–11. [18, 19]

Wilmott, P., S. Howison, and J. DeWynne. *The Mathematics of Financial Derivatives: A Student Introduction.* Cambridge, UK: Cambridge University Press, 1995. [13]

Symbols Used

Notation	Definition
$\forall$	For all, e.g., $x > 0 \forall x$. Read: x is greater than zero for all x.
Σ	Summation, e.g., $\sum_{j=1}^{N} p_d(x_j) = 1$. Read: Sum from $j = 1$ to N of the discrete probabilities of outcomes x_j equals to one.
!	Factorial, e.g., $n!$. Read: n factorial or $n(n-1)(n-2) \ldots (2)1$.
π	Mathematical constant pi, approximately 3.14159.
$\prod$	Product, e.g., $\prod_{i=1}^{n} \left(\frac{S_i}{S_{i-1}} \right)$. Read: Product from $i = 1$ to n of the asset price relative over time.
∞	Mathematical infinity or something larger than the largest real number.
$\varnothing$	The null set within probability theory.
Ψ	Vector where the element ψ_i is the price today of pure asset i.
H	Matrix where the element h_{ij} is the payoffs of each asset in each state.
A	Matrix where the element α_{ij} is the weights of each asset in each state.
I	Identity matrix.
S	Matrix of complex asset prices.
ι	Row vector which contains 1 as each element.

TIME-RELATED NOTATION

Notation	Definition
t	Calendar time, e.g., 5/22/2020, represented at a fraction of the year, such as $t = 0$. Depending on context, $t = 7$ months or $t = 4.25$ years. Often, t is generically "today."
T	Another calendar time, e.g., 5/22/2021, represented as a fraction of the year, such as $T = 1$. For example, the expiration date of the derivative.
$\hat{T}$	A finite time horizon used in Girsanov's theorem.
τ	Maturity time, e.g., time to expiration of an option. Note $\tau = T - t$. Thus, if $t = 0.25$ and $T = 1.0$, then $T = 0.75$.
τ_1, τ_2	Maturity times, e.g., time to expiration of an option, where $\tau_2 > \tau_1$.

Notation	Definition
t'	Calendar time, e.g., 6/15/2020, represented at a fraction of the year, such as $t = 0.1$, where the option holder chooses to exercise an American option early.
tD	Calendar time, for example, 6/15/2020, represented at a fraction of the year, such as $tD = 0.1$, when a dividend is paid.

INSTRUMENT-RELATED NOTATION

Notation	Definition
$\alpha(v,T)$	The instantaneous drift term observed at time v for the forward rate at T within the Heath–Jarrow–Morton model.
α_{ij}	The number of units of asset j that should be held to reproduce the payoff of pure asset i.
A	Denotes a scalar within the economic production function.
$AP_{Flt,j}$, $AP_{Fix,j}$	The appropriate accrual period for the jth floating and fixed rate related to swaps.
α	Generic component of the drift term with different forms of Brownian motion, expressed in percentage.
$\alpha_\$$	Generic component of the drift term with different forms of arithmetic Brownian motion, expressed in units of the asset price (typically $).
B_0	The amount initially borrowed at time 0 on a loan.
B_q	The amount required to be repaid, the payback at the maturity of the loan, q days later.
B_1	Present value adjusted binomial probability used in binomial model.
B_2	Binomial probability used in binomial model.
B_c	Specified borrowing amount from the binomial model for call options.
B_p	Specified borrowing amount from the binomial model for put options.
$B(j;n,p)$	Binomial probability of observing j up moves given n time steps with probability p.
$B(t,T)$	The price of a zero-coupon bond at time t that pays $1 at time T.
β_c	The beta of the call and is recognized as its risk with respect to the market portfolio within a CAPM framework.
β_s	The beta of an underlying asset and is recognized as its risk with respect to the market portfolio within a CAPM framework.
CB	Coupon-bearing bond.
cc_t	Call on call compound option value or price at time t.
co	Compound option value or price.
cp_t	Call on put compound option value or price at time t.

Notation	Definition
cCB, pCB	Call and put option prices on coupon-bearing bonds, including callable bonds.
$cZCB$, $pZCB$	Call and put option prices on zero-coupon bonds.
c_q, c, c_t	European call price (quoted price or model price) observed at calendar time t.
c_{tH}, c_{tM}, c_{tL}	European call prices for high, medium, and low exercise price call options observed at calendar time t.
c_{t1}, c_{t2}	European call prices for two options with differing maturity dates, $\tau_2 > \tau_1$, observed at calendar time t.
c_d	European call price in binomial framework after one down move.
c_u	European call price in binomial framework after one up move.
c_{ud}	European call price in binomial framework after one up move and one down move.
c_{u^2}	European call price in binomial framework after two up moves.
c_{d^2}	European call price in binomial framework after two down moves.
c_{min}, c_{max}	European call options on the minimum or maximum of two assets, respectively.
C_t	American call price observed at calendar time t.
C_{tH}, C_{tM}, C_{tL}	American call prices for high, medium, and low exercise price call options observed at calendar time t.
C_{t1}, C_{t2}	American call prices for two options with differing maturity dates, $\tau_2 > \tau_1$, observed at calendar time t.
cov(,)	Covariance of two random variables.
d	The down factor that represents one plus the percentage change in the asset price assuming the down event occurred within the binomial framework.
dt	Increment of time that tends to zero.
d, d_1, d_2	Parameters of the Black-Scholes-Merton option pricing model, the upper limit of the cumulative distribution function of the standard normal distribution.
$dW_i(v)$	A Wiener process representing the source of uncertainty for factor i at time v within the Heath-Jarrow-Morton model.
δ	Dividend yield, expressed in decimal terms.
D_1	Value of dividend payment at time 1 within binomial framework.
D_t	Present value of all cash payments paid on an asset over the life of an option, valued at calendar time t.
D_{tD}	Dividend payment amount valued at calendar time tD, when dividend is paid.
$DF_{Flt,j}$, $DF_{Fix,j}$	The appropriate discount factor for the jth floating and fixed rate related to swaps.
Δ	Delta or change in some variable such as asset price.
Δ_c, Δ_p	Delta of call and put, the partial derivative of the option price with respect to the underlying asset price.
ΔS	Change in underlying asset value.

Notation	Definition
e^x	Exponential of x.
ε_t	Standard normal random variable measured at time t.
$E(x)$	Expected value of x.
$E(y\|x)$	The conditional expected value of y given the value of x.
$f(x)$	Generic mathematical function of variable x.
$f(x, y, \rho)$	The probability density function for the bivariate normal.
$f_t(T)$	Futures price at time t for a contract expiring at time T.
$f, f(t, t+j)$	Forward interest rate.
$f(t, t)$	The shortest forward rate that is the spot rate and is sometimes called the short rate.
$f(t, T)$	The continuously compounded forward rate observed at time t for an instantaneous transaction to begin at time T.
$F_0(T), F_0$	Forward price that matures at time T, observed at time 0.
FV	Future value of some cash flow.
Γ_c, Γ_p	Gamma of call and put, the second partial derivative of the option price with respect to the underlying asset price.
γ	Location shift parameter within a distribution.
γ_t	An adapted process with respect to probability measure.
Γ_t^s	Ratio of probabilities at time t and state s.
$\gamma(0, T)$	Related to forward and futures contracts, the value at time T of any accumulated costs minus any accumulated benefits from owning the underlying instrument.
G_t	Natural log of the asset value or price.
H_{ij}	The payoff of complex asset j in state i.
h, h_c	Call hedge ratio and call hedge ratio in the binomial model.
h_p	Put hedge ratio in binomial model.
$HPR(t, t+1, j)$	The continuously compounded holding period return over the period t to $t+1$ for a j maturity bond
I	The sigma field of distinguishable events within a complete probability space and keeps track of the information along the complete past sample path.
I_{ij}	The ij^{th} element of the identity matrix I.
ι	Filler variable (Greek lowercase iota) used in demonstrating convergence of the binomial option model to the Black-Scholes-Merton option model.
ι_C	Indicator variable for whether the compound option is a call (+1) or a put (−1).
ι_U	Indicator variable for whether the underlying option within a compound option is a call (+1) or a put (−1).
I	The sigma field of distinguishable events representing information at time $\widehat{T}$.
j	The start date for a forward start option.
K	Denotes capital in the economic production function.

Notation	Definition
k_o	The discount rate of an option.
k_R	The discount rate that reflects an adjustment for risk.
$Kurtosis(\Delta S_t) =$ $Kurtosis(S_t)$	The fourth central moment of the change in asset price.
L	Denotes labor in the economic production function.
L	The interest rate on a loan expressed in decimal terms.
$L_m(q)$	Loan rate observed on day m that matures in q days.
λ	The Poisson distribution only requires one parameter, the average number of observed outcomes during an interval of time.
λ	The market price of risk of the stock expressed as $\lambda = [E(R) - r]/\sigma$.
λ	Factor related to homogeneous functions.
$\lambda(x)$	Probability density function of a lognormal distribution.
μ	Mean, often annualized.
μ_p	Periodic mean.
$\mu_n(x)$	The nth central moment.
$\mu_n'(x)$	The nth noncentral moment.
m	The number of compounding periods per year.
m	FRA expiration date expressed in number of days.
$\max(\ )$	Maximum of a set of variables.
$\min(\ )$	Minimum of a set of variables.
$n(x)$	Probability density function of a standard normal distribution.
N	Time to maturity expressed in compounding periods.
$N(d_1), N(d_2)$	The value of the cumulative distribution function of the standard normal distribution evaluated at d_1 or d_2.
$N_2(x,y;\rho)$	The bivariate normal probability for the variables x and y, which can be normalized or not, with correlation ρ.
$N_3(x,y,z;\rho_{xy},\rho_{xz},\rho_{yz})$	The trivariate normal probability for the variables x, y, and z, which can be normalized or not, with correlations ρ.
Ω	Within a state space, the set of all possible realizations of the stochastic economy between time 0 and time $\hat{T}$.
Ω	The elasticity of an option is often denoted with the Greek letter omega Ω.
Φ	A vector where the element ϕ_i is the price today of pure asset i.
p_i	Probability of observing outcome i.
p_q, p, p_t	European put price (quoted price or model price) observed at calendar time t.
p_d	Discrete probability.
p_d	European put price in binomial framework after one down move.
p_u	European put price in binomial framework after one up move.
p_{ud}	European put price in binomial framework after one up move and one down move.
p_{u^2}	European put price in binomial framework after two up moves.
p_{d^2}	European put price in binomial framework after two down moves.
p_{tH}, p_{tM}, p_{tL}	European put prices for high, medium, and low exercise price call options observed at calendar time t.

Notation	Definition
p_{t1}, p_{t2}	European put prices for two options with differing maturity dates, $\tau_2 > \tau_1$, observed at calendar time t.
p_{min}, p_{max}	European put options on the minimum or maximum of two assets, respectively.
P	A probability measure defined on the elements of I within a state space, short for $\Pr(x)$.
P_t	American put price observed at calendar time t.
P_{tH}, P_{tM}, P_{tL}	American put prices for high, medium, and low exercise price call options observed at calendar time t.
P_{t1}, P_{t2}	American put prices for two options with differing maturity dates, $\tau_2 > \tau_1$, observed at calendar time t.
pc_t	Put on call compound option value or price at time t.
pp_t	Put on put compound option value or price at time t.
ϕ	Risk-neutral probability of an up move in the binomial model and sometimes martingale or equivalent martingale probability or the null set within a probability space.
$\%MCTR$	Percentage marginal contribution to risk.
$\Pr(x)$	Probability of some event or outcome x occurring.
PV	Present value of some cash flow.
q	The number of days within a loan.
$\hat{q}$	Yield paid on an option, e.g., equity viewed as an option where this option pays a dividend yield.
Q	A probability measure defined on the elements of I within a state space, short for $\Pr(x)$.
q	Actual probability of an event or price or an alternative probability of an event or price.
r	Periodic, discretely compounded rate of return.
$r(t+j-1, t+j)$	The continuously compounded single-period spot rate over the period $t + j - 1$ to $t + j$.
r_c	Annualized, continuously compounded risk-free interest rate.
$r_{FR,j-1}$	The equilibrium forward rate, which is equivalent to the fixed FRA rate.
$r(t)$	The spot rate or the short rate, also the shortest forward rate or $f(t, t) = r(t)$.
$R_0(m, q)$	Fixed rate on FRA entered on day 0 with m days to FRA expiration date and q days to FRA payment date from the FRA expiration date.
$R_t(m, q)$	Fixed rate on FRA entered on day t with m days to FRA expiration date and q days to FRA payment date from the FRA expiration date.
$RS_0(J, q)$	The fixed rate on a swap set at time 0 in which there are J payments and each payment is based on the q-day rate.
ρ_c, ρ_p	Rho of call and put, the partial derivative of the option price with respect to the interest rate.

Notation	Definition
$\rho(x,y)$	Correlation between two variables, say x and y.
R_t^c	Periodic, continuously compounded rate of return observed at time t.
R_t^d	Periodic, discretely compounded rate of return observed at time t.
$Sharpe_c$	Sharpe ratio for long call option.
$Sharpe_p$	Sharpe ratio for a generic investment portfolio.
$\sigma, \sigma(\ )$	Annualized standard deviation of continuously compounded percentage changes in the underlying instrument's price. Also known as volatility.
$\sigma_\$$	Annualized standard deviation of price changes of the underlying.
$\sigma_{y\|x}^2$	The conditional variance of y given x.
$\sigma_i(v,T)$	The volatility of factor i observed at time v for the forward rate at T in the Heath-JarrowMorton model.
$Skew(\Delta S_t)$	Third central moment of the changes in asset price.
$SD(\widetilde{S}_T)$	Lognormal standard deviation.
S_t	Asset price or instrument value observed at time t. For example, Apple stock price on 5/22/2020 is \$318.89, thus $S_t = S_0 = 318.89$.
$\widehat{S}$	An adjusted asset price, that is, the future S value is the future price already discounted.
$S_{T_1}^*, S_t^{\,*}$	The underlying instrument value where the underlying option is at-the-money within the compound option valuation model.
SR	Equilibrium swap rate within an interest rate binomial lattice.
t	Calendar time slider, generally measured in fractions of a year, e.g., $T - t$ would be time to say option expiration in years. Also, the evaluation date of an FRA.
θ_c, θ_p	Theta of call and put, the partial derivative of the option price with respect to calendar time.
T	Expiration of some financial instrument expressed in the fraction of a year, e.g., $T - t$ would be time to say option expiration in years.
T	Number of days until the payment date of a FRA.
T_1	Expiration of compound option expressed in the fraction of a year.
T_2	Expiration of the underlying option within a compound option expressed in the fraction of a year.
$\widehat{T}$	$\widehat{T} < \infty$, finite measure of time related to state spaces.
TR_t	Periodic, total return observed at time t.
u	The up factor that represent one plus the percentage change in the asset price assuming the up event occurred within the binomial framework.
ω	Percentage or weight allocated to a particular instrument.
x, x_1, x_2	Generic variables.
X	Exercise price (also known as strike price) of an option. Also used to represent the strike rate for interest rate options.

Notation	Definition
X_H, X_L	High and low exercise prices of an option, where $X_H > X_L$.
X_C	Compound option exercise price.
X_U	Underlying option exercise price within a compound option.
V_B	Bond value at time 0.
$VFRA(m,q)$	The value of an FRA expiring on day m where the underlying is q-day-rate.
$VIRC, VIRP$	The value of interest rate call and put options.
$VIRS$	Value of an interest rate swap.
$VpSW, VrSW$	Value of a payer and receiver interest rate swap.
V_d	Value of some portfolio when down occurs in the binomial model.
V_u	Value of some portfolio when up occurs in the binomial model.
V, V_t	Instrument, asset, or portfolio value at time t.
vf	The mark-to-market value of an interest rate futures contract as it evolves through an interest rate binomial tree.
σ_p^2	Periodic variance.
$var(\ ), \sigma^2$	Variance of some random variable.
$VS_0(J,q)$	Value of swap initiated at time 0 in which there are J payments and each payment is based on the q-day rate.
v_c, v_p	Vega of call and put, the partial derivative of the option price with respect to the volatility.
ϕ_i	The risk neutral/equivalent martingale probabilities of the log asset prices.
dW_t	Standard Wiener process at time t.
W_t	Wiener variable observed at time t.
$w(y,t)$	The price of the call at time t in terms of the transformed asset price ($y = \ln S$) and time.
ψ_i	The price today of pure asset i.
y	Transformed asset price ($y = \ln S$).
Δy	Change in transformed asset price.
Z	Assuming equivalent probability measures for P and Q, the Radon-Nikodym derivative of Q with respect to P.

About the Website

Thank you for purchasing this book. You may access the Solutions Manual by visiting www.wiley.com\go\brooks\financialderivatives.

The Solutions Manual provides detailed solutions to the end-of-chapter problems provided in the book. This document will be helpful to adopting professors when seeking to appraise student understanding of the content within each chapter.

Note: Page numbers with *f* and *t* refer to figures and tables, respectively